ADVANCED MATHEMATICAL AND COMPUTATIONAL TOOLS IN METROLOGY AND TESTING XII

Series on Advances in Mathematics for Applied Sciences – Vol. 90

ADVANCED MATHEMATICAL AND COMPUTATIONAL TOOLS IN METROLOGY AND TESTING XII

Editors

F Pavese
IMEKO TC21, Italy

A B Forbes
National Physical Laboratory, UK

N F Zhang
National Institute of Standards and Technology, USA

A G Chunovkina
Institute for Metrology "D I Mendeleyev", Russia

World Scientific

NEW JERSEY • LONDON • SINGAPORE • BEIJING • SHANGHAI • HONG KONG • TAIPEI • CHENNAI • TOKYO

Published by

World Scientific Publishing Co. Pte. Ltd.
5 Toh Tuck Link, Singapore 596224
USA office: 27 Warren Street, Suite 401-402, Hackensack, NJ 07601
UK office: 57 Shelton Street, Covent Garden, London WC2H 9HE

Library of Congress Control Number: 2021060087

British Library Cataloguing-in-Publication Data
A catalogue record for this book is available from the British Library.

Series on Advances in Mathematics for Applied Sciences — Vol. 90
ADVANCED MATHEMATICAL AND COMPUTATIONAL TOOLS IN METROLOGY AND TESTING XII

ISBN 978-981-124-237-3 (hardcover)
ISBN 978-981-124-238-0 (ebook for institutions)
ISBN 978-981-124-239-7 (ebook for individuals)

For any available supplementary material, please visit
https://www.worldscientific.com/worldscibooks/10.1142/12422#t=suppl

F Pavese, A B Forbes, N F Zhang, A G Chunovkina (Eds):
Advanced Mathematical and Computational Tools in Metrology and Testing XII
Series on Advances in Mathematics for Applied Sciences, Vol. 90

https://doi.org/10.1142/9789811242380_fmatter

The AMCTM XII Book is dedicated to
Prof. Maurice G. Cox, NPL
to mark his 80th birthday in 2020

The AMCTM conference series can trace its origins to conferences on Algorithms for Approximation, started in the 1980s and co-organised by Maurice Cox and John Mason. At those conferences, there were a number of sessions dedicated to metrology applications. From these sessions, Franco Pavese and others recognised the need for a conference along similar lines dedicated to metrology and related fields, leading to the first Advanced Mathematical Tools for Metrology (AMTM) conference chaired by Franco and held in Torino in 1993. Maurice was strongly involved in this first conference and organised and chaired the AMTM II conference in Oxford in 1999. His involvement has continued in all later editions, including this current volume, as they have evolved into the AMCTM conferences and books, now organised by IMEKO TC21. We are very pleased to honour his outstanding contribution to the AMCTM community and to mark the occasion of his 80th birthday with this dedication.

The Editors

F Pavese, A B Forbes, N F Zhang, A G Chunovkina (Eds):
Advanced Mathematical and Computational Tools in Metrology and Testing XII
Series on Advances in Mathematics for Applied Sciences, Vol. 90

https://doi.org/10.1142/9789811242380_fmatter

Foreword

Due to the Covid 19 pandemic that started in early 2020, the AMCTM XII Conference, organised by IMEKO TC21 and due to be held in Sarajevo in September 2020, had to be cancelled. However, TC21 proposed to not interrupt the tradition of the book that follows each Conference. This volume contains the refereed contributions that researchers from institutions and laboratories across the world involved in metrology and testing have sent to the Editors, in response to their proposal to issue an AMCTM XII Book.

The aims of the IMEKO Technical Committee 21 "Mathematical Tools for Measurements" (http://www.imeko.org/index.php/tc21-homepage) are:

- To present and promote reliable and effective mathematical and computational tools in metrology and testing.
- To understand better the modelling, statistical and computational requirements in metrology and testing.
- To provide a forum for metrologists, mathematicians, software and IT engineers that will encourage a more effective synthesis of skills, capabilities and resources.
- To promote collaboration in the context of EU and International Programmes, Projects of EURAMET, EMPIR, EA and of other world Regions, and address requirements relating to the CIPM Mutual Recognition Arrangement.
- To support young researchers in metrology, testing and related fields.
- To address industrial and societal requirements.

The themes in this volume fully reflect the aims of TC21 and the importance of mathematical, statistical and numerical tools and techniques in metrology and testing and their role in addressing the challenges associated with the Metre Convention relating to the demand for measurement standards of ever increasing accuracy, range and diversity, and the need to demonstrate equivalence between national measurement standards.

Torino, January 2021 *The Editors*

Book Editors

F Pavese (Chair, IMEKO TC21, Torino, Italy),
frpavese@gmail.com
A B Forbes (National Physical Laboratory, Teddington, UK),
alistair.forbes@npl.co.uk
N F Zhang (National Institute of Standards and Technology, Gaithersburg, MD, USA),
nien-fan.zhang@nist.gov
A G Chunovkina (Institute for Metrology "D I Mendeleyev", St. Petersburg, Russia),
a.g.chunovkina@vniim.ru

For the first eleven Volumes see this Series vol. 16 (1994), vol. 40 (1996), vol. 45 (1997), vol. 53 (2000), vol. 57 (2001), vol. 66 (2004), vol. 72 (2006), vol. 78 (2009), vol. 84 (2012), vol. 86 (2015) and vol. 89 (2018)

ℵ

IMEKO TECHNICAL COMMITTEE 21 "MATHEMATICAL TOOLS FOR MEASUREMENTS".
Chairperson: F Pavese, Chair, IMEKO TC21, Torino, Italy,
http://www.imeko.org/index.php/tc21-homepage, http://www.imeko.org

ℵ

THIS VOLUME WAS PROMPTED TO REPLACE THE AMCTM 2020 CONFERENCE

ORGANISED BY THE OFFICERS OF THE IMEKO TC21
http://www.imeko.org/index.php/tc21-homepage.

THANKS TO THE REVIEWERS:

Seda Aytekin (UME, Turkey),
Emil Bashkanski (ORT Braude College, Israel),
Eric Benoit (Université de Savoie, France),
Dolores del Campo (CEM, Spain),
Sascha Eichstädt (PTB, Germany),
Eduarda Filipe (IPQ, Portugal),
Rainer Göb (University of Würzburg, Germany),
Valery A Granovski (Electropribor, Russia),
Kavya Jagan (NPL, UK),
Gregory Kyriazis (INMETRO, Brazil),
Valerie Livina (NPL, UK),
Katsuhiro Shirono (AIST-NMIJ, Japan),
Ian Smith (NPL, UK),
Rod White (Callagan, New Zealand).

F Pavese, A B Forbes, N F Zhang, A G Chunovkina (Eds):
Advanced Mathematical and Computational Tools in Metrology and Testing XII
Series on Advances in Mathematics for Applied Sciences, Vol. 90

https://doi.org/10.1142/9789811242380_fmatter

Contents

F Pavese, A B Forbes, N F Zhang, A G Chunovkina (Eds.):
Advanced Mathematical and Computational Tools in Metrology and Testing XII
Series on Advances in Mathematics for Applied Sciences, Vol. 90

https://doi.org/10.1142/9789811242380_0001

Explicit unconditionally numerically stable solution of a class of cubic equations

Maurice G. Cox

National Physical Laboratory, Teddington, United Kingdom
E-mail: maurice.cox@npl.co.uk

Explicit algebraic expressions are derived for the roots of a cubic equation having one real and two complex roots. As opposed to traditional methods for direct (that is, non-iterative) solution, evaluation of the expressions is unconditionally numerically stable. A floating-point error analysis of the computations is given, which shows that the computed roots have extremely small relative errors. The computations are proved to be backward stable, that is, the computed roots are exact for a closely neighbouring cubic equation. Applications and uncertainty propagation are also considered.

Keywords: cubic equations, direct solution, floating point, error bounds, numerical analysis, numerical stability, uncertainties

1. Introduction

Direct (non-iterative) methods of solution of a full cubic equation

$$t^3 + a_2t^2 + a_1t + a_0 = 0, \tag{1}$$

with real coefficients have been provided over centuries since Italian mathematicians in the 16th century first worked on the problem[6].

A simpler form of cubic equation is the so-called depressed or reduced form

$$y^3 + py = q, \tag{2}$$

with p and q real, which is obtained from equation (1) by a simple change of variable, and for which much analysis has been carried out[6].

In application areas such as thermodynamics there have been complaints that explicit solutions of cubic equations provide inadequate numerical accuracy[24]. Such a loss of accuracy often results from massive numerical cancellation when the formulæ typically used for the roots are implemented.

When the discriminant

$$D = \frac{p^3}{27} + \frac{q^2}{4} \tag{3}$$

is positive, equation (2) has one real root and two complex roots[16]. We give a treatment of that case in this paper. The case $D < 0$, when the cubic equation has three real roots, is not considered.

We dismiss two trivial cases from further consideration. We take $q > 0$ and $p \neq 0$ in equation (2). The condition $q > 0$ is not restrictive, since setting $q = -\widetilde{q}$ and $y = -\widetilde{y}$ gives the same equation in $\widetilde{y}$. Moreover, when $p = 0$, the three roots are $q^{1/3}$ (the real cube root of q) multiplied by the cube roots of unity,

$$\cos\frac{2k\pi}{3} + \mathrm{i}\sin\frac{2k\pi}{3}, \quad k = 0, 1, 2,$$

where $\mathrm{i}^2 = -1$ (De Moivre's theorem), that is

$$1, \qquad \omega = -\frac{1}{2} + \frac{\sqrt{3}}{2}\mathrm{i}, \qquad \omega^2 = -\frac{1}{2} - \frac{\sqrt{3}}{2}\mathrm{i}. \tag{4}$$

By making a further elementary change of variable, equation (2) can be reduced to an even more fundamental form, which we use for analysis and computational purposes, and which we term the *primitive form*:

$$x^3 + 3gx = 2h, \qquad g = \pm 1, \qquad \text{for all } h > 0. \tag{5}$$

We distinguish between the two cases $x^3 - 3x = 2h$ and $x^3 + 3x = 2h$, corresponding to $g = \mp 1$, respectively, since they require different treatments regarding the provision of expressions that can be computed stably.

Unconditionally numerically stable explicit algebraic expressions are derived for the real root and the two complex roots of the primitive form. These expressions do not suffer from the mentioned catastrophic subtractive numerical cancellation often associated with the traditional approach. Although a cubic equation can of course be solved iteratively, and many software libraries offer facilities for its solution, a new direct method of solution has interest in its own right for comparison with other direct methods.

Furthermore, such a solution, particularly if it has excellent numerical properties – fast computation and extremely small error bounds – is important when used in the context of numerical optimization. Suppose a computed root is to be used as a variable in an optimization routine, that variable depending on further variables that define the coefficients in the cubic equation. In the neighbourhood of the solution to the optimization problem, errors in the values of the computed root will give rise to errors in the values of the function that is being optimized. Non-systematic errors will give rise to 'non-smooth' function values with a detrimental effect on the routine producing an optimum since many such routines rely on continuity and differentiability of the function[12]. In an instance of optimization and advanced control of bioprocesses, in the context of thermodynamic modelling for the vapourliquid equilibrium of binary systems of super-critical fluids and ionic liquids[18], the coefficients in the cubic equation of state are functions of the concentration of the different components in the mixture.

A floating-point error analysis is given of the calculation of the roots, which shows the roots obtained are correct to within a small multiple of the machine round-off. The computed roots satisfy exactly (in a relative sense) the cubic equation (5) with h perturbed slightly. Thus the computation is backward stable[21], implying the floating-point errors introduced during the computation have the same effect as a small perturbation in the data[13], here h. If this perturbation is comparable to any uncertainty in h, the algorithm produces as good a result as can be expected.

The formulæ in this paper involve terms such as $\phi^{1/3}$, where ϕ is real, in which case $\phi^{1/3}$ denotes the real root known as the principal cube root. Similarly, some formulæ involve $\phi^{1/2}$, ϕ real and positive, in which case the positive root is taken.

Section 2 of this paper introduces the fundamental concepts used. It also covers the nature of the solution and the manner in which the depressed form is obtained from the full form of the cubic equation and in turn treats the transformation from the depressed form to the primitive form. Section 3 summarizes the traditional method of solution as applied to the primitive form of the cubic equation, variants of which have existed for over 450 years. Section 4 presents our numerically stable method. In section 5, we give simple algorithms for the implementation of the method. Section 6 uses the standard model of floating point arithmetic to analyse these algorithms, yielding error bounds for the relative error in the computed roots. Section 7 gives numerical examples, comparing the results with those obtained by the traditional method. Section 8 briefly discusses uncertainty propagation, from data used to specify the coefficients in the cubic equation, to the roots. Some applications of cubic equations are mentioned in section 9. Conclusions and scope for further work are given in section 10.

2. Fundamentals

2.1. *Primitive form*

The primitive form (5) exposes the essential nature of the computation, especially since it involves a single parameter, and is obtained from the depressed form (2) by the change of variable (also see section 2.3):

$$y = kx, \qquad k = \left(\frac{|p|}{3}\right)^{1/2}, \qquad h = \frac{3\sqrt{3}q}{2|p|^{3/2}}, \tag{6}$$

giving for all $h > 0$,

$$x^3 - 3x = 2h, \qquad \text{(Case 1)}, \tag{7}$$

$$x^3 + 3x = 2h, \qquad \text{(Case 2)}. \tag{8}$$

The real root of the primitive form (5) is straightforwardly transformed to that of the depressed equation (2) or that of the full form (1) by reversing the transformations carried out.

For the primitive form, the discriminant condition (3) becomes

$$D = h^2 + g > 0. \tag{9}$$

2.2. *Nature of the solution*

We examine the nature of a real root in the two cases.

Case 1: $x^3 - 3x = 2h$. Because

$$\frac{\mathrm{d}x}{\mathrm{d}h} = \frac{2}{3(x^2 - 1)} > 0, \quad x > 1,$$

x is a monotonically increasing function of h for $x > 1$. Since the discriminant condition (9) for $g = -1$ implies $h > 1$, x is such that $x^3 - 3x > 2$ and so, by monotonicity, $x > 2$.

Case 2: $x^3 + 3x = 2h$. A real positive solution exists for all $h > 0$ since the factor $x^2 + 3$ in the form $x(x^2 + 3) = 2h$ is positive and so the other factor x must be positive. Further, since $g = 1$ in this case, discriminant condition (9) is automatically satisfied.

2.3. *Transformations to the depressed and primitive forms*

The depressed form (2) is obtained from equation (1) by the change of variable

$$t = y - \frac{a_2}{3}$$

giving

$$p = a_1 - \frac{a_2^2}{3}, \qquad q = a_0 - \frac{a_1 a_2}{3} + \frac{2a_2^3}{27}.$$

We dispense with the trivial case $q = 0$, since then $y = 0$ or $y = \pm\sqrt{p}$.

The reduction to primitive form is then achieved by making the further elementary change of variable $y = kx$ as in expressions (6), giving for all $h > 0$,

$$\frac{|p|^{3/2}}{3\sqrt{3}}x^3 + \frac{|p|^{3/2}\,\mathrm{sgn}(p)}{\sqrt{3}}x = q.$$

Thus,

$$x^3 + 3gx = 2h, \qquad g = \mathrm{sgn}(p), \qquad h = \frac{3\sqrt{3}q}{2|p|^{3/2}},$$

yielding the primitive form in expressions (7) and (8).

The solution to this form leads through inverse transformation to the solution of the reduced or full form of the cubic equation.

The use of a transformation such as (6) to reduce the depressed cubic equation to a form involving only one parameter is not new. Birkhoff and Mac Lane[4] used such a transformation to reduce the depressed form to

$$4x^3 \pm 3x = C, \tag{10}$$

for some constant C, which can be solved by applying either of the trigonometric identities

$$\sinh 3\theta = 4\sinh^3\theta + 3\sinh\theta, \qquad \cos 3\theta = 4\cos 3\theta - 3\cos\theta$$

depending on the sign in equation (10). They did not, however, consider a numerical analysis of the approach.

3. Traditional solution

We consider the solution of the primitive form (5) of the cubic equation, where $g = \pm 1$ and $h > 0$, for single real root. Applying the approach given in Dickson[9] to the primitive form (5), the complex roots can be determined as follows. Substituting

$$x = z - \frac{g}{z} \tag{11}$$

in equation (5) gives

$$z^3 - \frac{g}{z^3} = 2h,$$

yielding

$$z^6 - 2hz^3 - g = 0$$

with solution (for z^3)

$$z^3 = h \pm D^{1/2}, \qquad D = h^2 + g.$$

Choose the plus sign to yield the real root:

$$z_1 = (h + D^{1/2})^{1/3}.$$

The complex roots are given by the product of z_1 and the other two, complex roots of unity:

$$z_2 = \omega z_1, \qquad z_3 = \omega^2 z_1,$$

with ω as in expressions (4).

Now,

$$\frac{1}{z_1} = \frac{1}{(h + D^{1/2})^{1/3}} = \frac{(h - D^{1/2})^{1/3}}{(h^2 - D)^{1/3}} = -g(h - D^{1/2})^{1/3} \tag{12}$$

using $D = h^2 + g$ from inequality (9).

Hence, using substitution (11), the three roots are

$$x_1 = A + B, \qquad x_2 = \omega A + \omega^2 B, \qquad x_3 = \omega^2 A + \omega B, \tag{13}$$

where

$$A = (h + D^{1/2})^{1/3}, \qquad B = (h - D^{1/2})^{1/3}, \tag{14}$$

which can be verified by substitution.

Separating real and imaginary parts by using (4), expressions (13) become

$$x_1 = A + B, \qquad x_2,\ x_3 = -\frac{A+B}{2} \pm \frac{\sqrt{3}}{2}(A - B)\,\mathrm{i}. \tag{15}$$

4. Numerically stable solution

The consideration of the stability of the calculation of the roots is reduced to that of the terms $A + B$ and $A - B$ in expressions (15).

There are non-pathological situations in practice, particularly in thermodynamics, when obtaining the real root as $A + B$, where A and B are computed according to the traditional formulæ (14) or one of its variants causes a disastrous loss of numerical accuracy[7]. This loss is manifested as follows. In the use of formula (15), since the discriminant $D = h^2 \pm 1$, there is the possibility of subtractive cancellation in forming A and B and then in forming $A + B$ and $A - B$. Such cancellation can be severe when $h \gg 1$.

There is also a problem due to subtraction in forming h^2-1 when h is close to unity, which can simply be overcome (see section 6 and appendix A).

We give expressions for the roots that are mathematically equivalent to formulæ (15) and free from any subtractive cancellation. Then, we give informal statements regarding the stability of calculations based on these expressions, repeatedly using the result given in appendix B concerning the sum of positive terms each of which has been computed with small relative error. Finally, we provide floating-point error analyses to yield bounds on the relative error in the computed roots.

4.1. *The case* $x^3 - 3x = 2h$

The calculation of the discriminant D when $g = -1$ can be carried out stably using $D = h^2 + g = h^2 - 1 = (h-1)(h+1)$ (see appendix A). Hence, $A = (h + D^{1/2})^{1/3}$ in formulæ (14), which involves the sum of two positive terms, with $D^{1/2}$ computed stably, is itself computed stably (see appendix B).

The evaluation of $B - (h - D^{1/2})^{1/3}$ in formulæ (14) suffers from subtractive cancellation when $h \gg 1$ since $h - D^{1/2} = h - (h^2 - 1)^{1/2}$. To avoid this cancellation, we replace $h - D^{1/2}$ by $1/(h + D^{1/2})$ [compare with expression (12)]. Thus $B = 1/A$ and its calculation is hence stable.

Regarding $A+B$, its calculation is stable being the positive combination of two stably-computed positive terms (again see appendix B).

The evaluation of $A - B$, which involves the difference between two positive terms, can be accomplished in a stable fashion as follows. Consider the identity

$$(\alpha+\beta)^{1/3} - (\alpha-\beta)^{1/3} = \frac{2\beta}{(\alpha+\beta)^{2/3} + (\alpha-\beta)^{2/3} + (\alpha^2-\beta^2)^{1/3}}, \quad (16)$$

the validity of which can readily be confirmed. Associating α with h and β with $D^{1/2}$, and using $\alpha^2 - \beta^2 = h^2 - D = 1$,

$$A - B = \frac{2D^{1/2}}{(h+D^{1/2})^{2/3} + (h-D^{1/2})^{2/3} + 1} = \frac{2D^{1/2}}{A^2+B^2+1},$$

which can be computed stably, no subtractive cancellation being involved.

4.2. *The case* $x^3 + 3x = 2h$

For this case, $A = (h + D^{1/2})^{1/3}$ can be computed stably as the sum of two terms, one of which is computed stably. Further, again using expression (12), $B = -1/A$ and so its computation is stable.

Now,

$$A + B = (D^{1/2} + h)^{1/3} - (D^{1/2} - h)^{1/3}, \tag{17}$$

involving possibly damaging subtraction. However, formula (17) can be re-expressed using identity (16) with $\alpha = D^{1/2}$ and $\beta = h$:

$$A + B = \frac{2h}{(D^{1/2} + h)^{2/3} + (D^{1/2} - h)^{2/3} + 1} = \frac{2h}{A^2 + B^2 + 1},$$

which can be computed stably.

Finally,

$$A - B = (D^{1/2} + h)^{1/3} + (D^{1/2} - h)^{1/3},$$

which is already in stable form.

5. Simple algorithms for the traditional and stable approaches

We state an algorithm for the traditional method for calculating the roots of the primitive form (5) of the cubic equation based on the use of formulæ (15), where A and B are defined by expressions (14).

We give simple algorithms based on the considerations of sections 4.1 and 4.2 for calculating the roots. In any implementation, checks on the suitability of the 'input' data for the calculation should of course initially be carried out. Relevant checks are that $g = \pm 1$, $D = h^2 + g > 0$, $h > 1$ when $g = -1$ and $h > 0$ when $g = 1$ (see section 2.2).

In section 6, we give floating-point error analyses of these algorithms.

5.1. *Algorithm T: Traditional approach for the real root*

(1) $D = h^2 + g$
(2) $E = D^{1/2}$
(3) $A = (h + E)^{1/3}$
(4) $B = |h - E|^{1/3} \operatorname{sgn}(h - E)$ (to ensure real cube root)
(5) $x_1 = A + B$
(6) $x_2,\ x_3 = -\dfrac{x_1}{2} \pm \dfrac{\sqrt{3}}{2}(A - B)\,\mathrm{i}$

The total operation count (that is, for the real parts and the imaginary parts) for this algorithm is $5A + 5M + 2S + 2C$, where A denotes addition or subtraction, M multiplication or division, S square root and C cube root.

5.2. *Algorithm S−: Stable approach for all roots of $x^3 - 3x = 2h$*

(1) $D = (h-1)(h+1)$
(2) $E = D^{1/2}$
(3) $A = (h+E)^{1/3}$
(4) $B = \dfrac{1}{A}$
(5) $x_1 = A + B$
(6) $G = \dfrac{E}{A^2 + B^2 + 1}$
(7) $x_2,\ x_3 = -\dfrac{x_1}{2} \pm \sqrt{3}G\,\mathrm{i}$

The total operation count is $6A + 6M + 2S + C$.

5.3. *Algorithm S+: Stable approach for all roots of $x^3 + 3x = 2h$*

(1) $D = h^2 + 1$
(2) $E = D^{1/2}$
(3) $A = (h+E)^{1/3}$
(4) $B = -\dfrac{1}{A}$
(5) $x_1 = \dfrac{2h}{A^2 + B^2 + 1}$
(6) $G = \dfrac{A-B}{2}$
(7) $x_2,\ x_3 = -\dfrac{x_1}{2} \pm \sqrt{3}G\,\mathrm{i}$

The total operation count is $5A + 8M + 2S + C$.

For the stable approaches, the total operation count is competitive with that for the traditional method.

5.4. *Properties of the algorithms*

Qualitatively, taking account of the informal remarks in section 4, it can be seen that the algorithms in sections 5.2 and 5.3 are numerically stable. They involve only basic arithmetic operations (together with square roots and cube roots) on (stably computed) positive numbers carried out in such a way that no cancellation occurs. We provide in section 6 a quantitative assessment of the maximum error committed by the approaches, which enables them to be used as a benchmark for other approaches for the class of cubic equations considered.

A general conclusion (also see section 6) is that the computed roots are in error by at most a few units in the last place (ulps). We also show that the calculations are unconditionally stable in a forward error sense.

6. Floating-point error analysis

Our error analysis follows a model of floating-point arithmetic used by Wilkinson[23] and later standardized by the IEEE[15]. For floating-point numbers x and y in base 2 arithmetic,

$$\mathrm{fl}(x \text{ op } y) = (x \text{ op } y)(1+\varepsilon)^{\pm 1}, \qquad |\varepsilon| \leq \eta, \qquad \text{op} = +, -, *, /,$$

where η is the unit roundoff, $2^{-52} \approx 2 \times 10^{-16}$ on many computers. Generally, the symbol ε, often subscripted, denotes a floating-point error satisfying $|\varepsilon| \leq \eta$.

Further, it is assumed that x^n for some integer $n > 2$ is calculated by repeated multiplication and that for a single floating point number x,

$$\mathrm{fl}(\sqrt{x}) = \sqrt{x}(1+\varepsilon)^{\pm 1}, \qquad \mathrm{fl}(\sqrt[3]{x}) = \sqrt[3]{x}(1+\varepsilon)^{\pm 1}, \qquad |\varepsilon| \leq \eta.$$

Multiplication or division by an integral power of 2 is taken as exact.

$\bar{y}$ is used to denote the computed value of y. Thus, for example, for $y = x^3$,

$$\bar{y} = \mathrm{fl}(x^3) = \mathrm{fl}(x \times x \times x) = \mathrm{fl}(x \times x^2(1+\varepsilon_1)) = y(1+\varepsilon_1)(1+\varepsilon_2).$$

The computed result can also be expressed as

$$\bar{y} = y(1+\varepsilon)^2.$$

In the same vein, consider the calculation $D = h^2 + 1$. The floating-point counterpart is

$$\bar{D} = [h^2(1+\varepsilon_1) + 1](1+\varepsilon_2), \qquad |\varepsilon_i| \leq \eta, \quad i = 1, 2.$$

$\bar{D}$ can be replaced by the weaker expression (also see appendix B)

$$\bar{D} = [h^2(1+\varepsilon_1) + (1+\varepsilon_1)](1+\varepsilon_2) = (h^2+1)(1+\varepsilon)^2 = D(1+\varepsilon)^2, \qquad |\varepsilon| \leq \eta.$$

Despite employing such often conservative bounds, we establish that the computed roots have very small relative errors bounded by a few ulps.

In our analyses, we sometimes replace $(1+\varepsilon)^n$ by $1 + n\varepsilon$. In fact, n must be larger than $\approx 6 \times 10^7$ before *computationally* this replacement is invalid.

We also use the approximation $(1+\varepsilon_1)^{-1} = 1 - \varepsilon_1 = 1 + \varepsilon_2$, accurate to $\mathcal{O}(\varepsilon_1^2)$.

In step 1 of Algorithm S−, D is evaluated from $(h-1)(h+1)$ rather than the 'more natural' h^2-1. Evaluation of the former expression is unconditionally stable whereas the latter can suffer catastrophic cancellation. The subtraction involved in computing $(h-1)(h+1)$ is benign. The relative error in this computation is bounded in magnitude for any h by 3η. Conversely, the relative error in the evaluation of h^2-1 is bounded in magnitude by $[h^2/(h^2-1)+1]\eta$, which for h close to unity could be extremely large. See appendix A.

The bounds given below are upper bounds and are generally pessimistic. Probabilistic error bounding would make an improvement[14] but the more important aspect is the quantitative proof of numerical stability.

A modern variant of the 'Wilkinson' approach to floating-point error analysis is due to Higham[13], which yields comparable results to those here.

6.1. *Calculation of the root using the stable method*

6.1.1. $x^3 - 3x = 2h$

A detailed floating-point error analysis of Algorithm S− (section 5.2) is given in appendix C.1. It is concluded that the computed value of the real root x_1 has a relative error bound of 5 ulps (units in the last place). For the complex roots x_2 and x_3, the bounds are 5 ulps for the real part and 15 ulps for the imaginary part.

6.1.2. $x^3 + 3x = 2h$

The corresponding floating-point error analysis of Algorithm S+ (section 5.3) in appendix C.2 yields a computed value of x_1 having a relative error bound of 10 ulps. For x_2 and x_3, the bounds are 10 ulps for the real part and 6 ulps for the imaginary part.

6.2. *Backward error interpretation*

Suppose $\widetilde{x}$ is an approximation to a root of the primitive form of the cubic equation. Substitution of $\widetilde{x}$ into equation (5) gives

$$\widetilde{x}^3 + 3g\widetilde{x} = 2h + 2\delta h,$$

say. Thus, $\widetilde{x}$ satisfies *exactly* the neighbouring or perturbed cubic equation

$$x^3 + 3gx = 2\widetilde{h},$$

where

$$\widetilde{h} = h + \delta h,$$

which has the interpretation that if $|\delta h| = |\widetilde{h} - h|$ is small compared with any error in h, the computed solution is consistent with the data upon which the cubic equation is based.

Using our knowledge of a relative error bound for a computed root, we can determine a bound for $|\delta h|$ or the relative perturbation $|\delta h|/h$. Differentiation of the primitive form (5) of the cubic equation with respect to x gives

$$\frac{\mathrm{d}h}{\mathrm{d}x} = \frac{3}{2}(x^2 + g). \tag{18}$$

So, denoting a calculated root by $\widetilde{x} = x + \delta x$, to a close approximation, since $|\delta x| \ll x$,

$$\delta h = \frac{3}{2}(x^2 + g)\delta x$$

and hence

$$\frac{\delta h}{h} = 3\frac{x^2 + g}{x^3 + 3gx}\delta x = 3\frac{x(x^2 + g)}{x^3 + 3gx}\frac{\delta x}{x} = 3\frac{x^2 + g}{x^2 + 3g}\frac{\delta x}{x}.$$

This, using $|\delta x| \leq Cx\eta$, C being a constant $\mathcal{O}(1)$ (see section 6.1), and recalling that $x > 2$ for $g = -1$ and $x > 0$ for $g = 1$,

$$\frac{1}{3C\eta}\frac{|\delta h|}{h} \leq f(x),$$

where $1 < f(x) < 3$ for $g = -1$ and $1/3 < f(x) < 1$ for $g = 1$. Thus, somewhat weakly,

$$\frac{|\delta h|}{h} \leq \begin{cases} 9C\eta, & g = -1, \\ 3C\eta, & g = 1. \end{cases} \tag{19}$$

The computed x satisfies the cubic equation with h replaced by a value that departs from it by a very small relative error. Using the bounds for $|\delta x|/x$ in section 6.1, the (relative) backward error bound (BEB) for the stable computation of the real root is 45η for the equation $x^3 - 3x = 2h$ and 30η for $x^3 + 3x = 2h$. Comparable statements apply to the complex roots.

7. Examples

We give examples to illustrate aspects of the preceding discussion for an increasing set of values of h, for which it is expected that the traditional method will give increasing loss of numerical accuracy. Table 1 relates to Case 1 and table 2 to Case 2. In the tables, x_{stab} denotes the real root computed using the stable method and x_{trad} the real root computed using the traditional method, e denotes $x_{\text{trad}} - x_{\text{stab}}$. BEB($x_{\text{trad}}$) and BEB($x_{\text{stab}}$) denote the relative backward error bounds $(x_{\text{trad}}^3 - 3x_{\text{trad}} - 2h)/(2h)$ and $(x_{\text{stab}}^3 - 3x_{\text{stab}} - 2h)/(2h)$, respectively. The improvement in the numerical accuracy of the stably computed roots over the traditionally computed roots is evident.

Table 1. Examples of the calculation of the real root of $x^3 - 3x = 2h$

$2h$	x_{trad}	e	e/x_{stab}	BEB(x_{trad})	BEB(x_{stab})
10^2	10.10000	6.87e–13	6.81e–14	2.08e–13	–1.14e–16
10^3	21.59076	1.73e–11	8.01e–13	2.41e–12	–5.46e–16
10^4	46.43743	2.43e–09	5.23e–11	1.57e–10	–4.37e–16
10^5	100.01000	2.54e–08	2.54e–10	7.62e–10	–9.31e–16
10^6	215.44811	–5.40e–06	–2.51e–08	–7.52e–08	–9.31e–16
10^7	464.16104	–2.01e–04	–4.34e–07	–1.30e–06	–7.45e–16

For the computations in table 1, the relative backward error bound (BEB) (19) with $C = 5$ gives a BEB of about 1×10^{-14} and a probabilistic BEB of approximately 1×10^{-15}, numbers that compare favourably with those in the last column. In terms of the relative error in the computed root, some 10 significant digits are lost with the traditional computation for $2h = 10^7$, whereas the method introduced has proven stability.

Table 2. Examples of the calculation of the real root of $x^3 + 3x = 2h$

$2h$	x_{trad}	e	e/x_{stab}	BEB(x_{trad})	BEB(x_{stab})
10^2	9.90000	4.99e–13	5.04e–14	1.49e–13	5.68e–16
10^3	21.49793	–4.53e–11	–2.11e–12	–6.29e–12	1.09e–15
10^4	46.39434	–2.43e–09	–5.24e–11	–1.57e–10	1.31e–15
10^5	99.99000	–2.54e–08	–2.54e–10	–7.61e–10	2.10e–15
10^6	215.43883	5.40e–06	2.51e–08	7.52e–08	1.86e–15
10^7	464.15673	2.01e–04	4.34e–07	1.30e–06	2.38e–15

For the computations in table 2, expression (19) with $C = 10$ gives a BEB of approximately 7×10^{-15} and a probabilistic BEB of about 2×10^{-15}, numbers that again compare favourably with those in the last column.

Like the previous computations, in terms of the relative error in the computed root, some 10 significant digits are lost with the traditional computation for $2h = 10^7$.

Similarities in the results in the two tables are due to the fact that as h becomes larger, the x^3 term in the primitive form of the equation becomes more dominant and $x \to (2h)^{1/3}$.

8. Uncertainties

In practice there will be standard measurement uncertainties[1] associated with the values of the coefficients $\boldsymbol{a} = [a_2, a_1, a_0]^\top$ in the full form (1) of the cubic equation. Rarely will the coefficients equate to fundamental physical quantities but rather to terms based on *combinations* of such quantities and so be correlated even if the original physical quantities were independent.

An important application is to cubic equations of state in thermodynamics[17] in which the case of a single real root corresponds to conditions above the critical temperature. There will be covariances associated with pairs of values of these coefficients since the elements of $\boldsymbol{a}$ will have been obtained from application-specific data such as P, R, T, b_1 and b_2 in the thermodynamic equation of state (section 9). These standard uncertainties and covariances are represented by a covariance matrix $\boldsymbol{V_a}$ of dimension 3×3 containing the squared standard uncertainties as its diagonal elements and covariances as its off-diagonal elements[3].

Using established rules for uncertainty propagation[1], assuming here that the appropriate equations can adequately be linearized, the covariance matrix $\boldsymbol{V}_{[p,q]}$ associated with p and q in the depressed form can be evaluated using $\boldsymbol{V}_{[p,q]} = \boldsymbol{S}_1 \boldsymbol{V_a} \boldsymbol{S}_1^\top$,[2] where S_1 is the sensitivity matrix of dimension 2×3 of partial derivatives of p and q with respect to the elements of $\boldsymbol{a}$. In turn, the standard uncertainty $u(h)$ associated with h in the primitive form of the cubic equation can be evaluated from $u^2(h) = \boldsymbol{S}_2 \boldsymbol{V}_{[p,q]} \boldsymbol{S}_2^\top$, where $\boldsymbol{S}_2$ is the sensitivity matrix of dimension 1×2 of partial derivatives of h with respect to p and q. Finally, the standard uncertainty associated with a computed root x of the primitive form of the cubic equation can be evaluated from $u(x) = su(h)$, where $s = \mathrm{d}x/\mathrm{d}h$ [see expression (18)].

A clear advantage in having only one derived parameter, h, in the primitive form (5) is that there is uncertainty but no associated correlation. The above process for uncertainty propagation delivers as stated the covariance matrix $\boldsymbol{V}_{[p,q]}$ associated with p and q. This matrix might be of interest on its own account if some analysis is required using the depressed form. Otherwise, the sensitivity matrices can be combined, that is, $\boldsymbol{S} = \boldsymbol{S}_2\boldsymbol{S}_1$, and $u(x)$ evaluated directly using $u^2(x) = s^2\boldsymbol{S}\boldsymbol{V}_{\boldsymbol{a}}\boldsymbol{S}^{\top}$.

9. Applications

The determination of the real root of a cubic equation having only one such root arises in many applications including the dynamic behaviour of linear systems[5], tunnel-diode design[22], computer-aided design[10] and computer graphics[11]. One of the most important applications is in thermodynamics where an equation of state relating variables describing the state of matter under given physical conditions is used to describe properties of fluids.

Such equations describe the behaviour of fluids as a function of pressure P, volume V and temperature T. The simplest equation of state describing the equilibrium behaviour of a pure fluid originated in the work of Van der Waals dating from 1873 (see[20]):

$$P = \frac{RT}{V - b_1} - \frac{b_2}{V^2}, \tag{20}$$

where b_1 and b_2 are parameters that depend on the specific fluid. Many empirical improvements to the equation have been made subsequently. As a family, the improved equations can still be described by equation (20), but with V^2 replaced by a general quadratic in V. A cubic equation with one real root and two complex roots, as treated here, corresponds in thermodynamics to temperatures and pressures higher than a critical point where a substance is considered neither as a gas nor a liquid.

Rationalization enables equation (20) or its generalization to be expressed as a full cubic equation (1).

In practice the variable (the root) can cover a large range of numerical values. Calculating real roots of these equations is a key issue when constructing engineering diagrams or running flow processes. Numerical errors in the obtained values of course propagate in subsequent calculations. Any enhancement of the stability and efficiency of the root determination for cubic equations of state is valuable in huge simulations such as compositional reservoir simulations (used for forecasting oil production, decision making, and reservoir management). In such simulations, the Peng-Robinson variant[19,20] of the equation of state is used to determine thermodynamic properties thousands of times where the polynomial coefficients varies with tens of variables: temperature, pressure, composition (dozens of components), space and time within dynamic simulations. In these cases an important aspect is the time taken to obtain the root to a prescribed numerical accuracy[8].

10. Concluding remarks and scope for further work

Traditional methods are often used in a number of fields for computing directly (non-iteratively) the three roots of cubic equations possessing one real root and two complex roots. The methods comprise formulæ that when evaluated 'naturally' can give rise to loss of numerical accuracy that can be considerable in important cases arising in practice. Here, we have re-expressed these formulæ in a manner that ensures that the roots are computed in a numerically stable manner. The relative errors in the computed roots have the property that they are bounded in magnitude by a small multiple of the computer's round-off (of the order of 2×10^{-16} on many computers). It is also shown that the process is backward stable, that is, the computed roots satisfy exactly a neighbouring cubic equation. Examples illustrate the points made.

Our formulation is in terms of a primitive form of the cubic equation that possesses just one defining parameter h. The primitive form is obtained from the full form of the cubic equation under a simple transformation and has several advantages. One advantage is that analysis of the primitive form is simpler than that of the full form or even the so-called depressed form that is also obtained by a change of variable from the full form and contains two parameters. A second advantage is that the formulæ for the roots of the primitive form are simpler than those for the full and depressed forms.

The statement made about the stability of the calculation of the roots of the primitive form of the cubic equation and the floating-point bounds on the relative error apply to the data defining that form as held on the computer. The source of that data should also be considered since an error in h is propagated through the calculation to give (further) errors in the roots. When cubic equations are used in practice, their coefficients are derived from 'raw data' possessing uncertainties. Since they are functions of these data, the coefficients in turn will be uncertain and also be correlated since in general each coefficient depends on common data. Although the full and depressed forms will have correlation associated with them because they contain more than one coefficient, the primitive form has the advantage that its single coefficient implies no correlation in that representation. The analysis of uncertainty propagation through the computational process is hence simplified.

A further paper is planned concentrating on the detailed application of the results here to equations of state arising in thermodynamic applications.

There is another advantage in having a single parameter in the form of cubic equation under analysis. There is a one-to-one relationship between h and the real root x_1. Thus, in principal it would be possible to construct a functional form $x_1 = F(h)$ using polynomial or rational functions, for example. Should it be possible to construct F in a sufficiently compact form enabling its stable and accurate evaluation to within a modest number of ulps, the use of such a representation would be a viable alternative to the approach presented here.

A considerable challenge is to extend the ideas presented here to cubic equations possessing three real roots. Although such an extension may not be viable without imposing some restrictions on the coefficients of the equation, the author intends investigating possibilities.

Appendix A. Floating-point error analysis of the discriminant calculation

We first establish a floating-point error bound for the evaluation of the discriminant D from the formula h^2-1. Applying the principles in section 6,

$$\bar{D} = [h^2(1+\varepsilon_1) - 1](1+\varepsilon_2) \approx h^2 - 1 + h^2\varepsilon_1 + (h^2-1)\varepsilon_2.$$

So,

$$\frac{\delta D}{D} = \frac{\bar{D} - D}{D} = \frac{h^2\varepsilon_1 + (h^2-1)\varepsilon_2}{h^2-1} = \frac{h^2}{h^2-1}\varepsilon_1 + \varepsilon_2,$$

implying the relative error bound

$$\frac{|\delta D|}{D} \leq \left(\frac{h^2}{h^2-1} + 1\right)\eta. \tag{A.1}$$

For h close to one this bound on the relative error in the evaluation of h^2-1 could be extremely large. For instance, when $h-1$ is 1×10^{-8}, the relative error in the computed D is -5×10^{-9} (using the stably computed value as a reference), thus losing about half the number of available significant digits. The bound given by inequality (A.1) in this case is 1×10^{-8}.

In contrast, a floating-point analysis of the use of $D = (h-1)(h+1)$ yields

$$\bar{D} = (h-1)(1+\varepsilon_4)(h+1)(1+\varepsilon_5)(1+\varepsilon_6) = D(1+\varepsilon_7)^3 \approx D(1+3\varepsilon_7),$$

giving

$$\frac{\delta D}{D} = \frac{D(1+3\varepsilon_7) - D}{D} = 3\varepsilon_7$$

and so

$$\frac{|\delta D|}{D} \le 3\eta,$$

an extremely small relative error bound, which means that this computation is unconditionally numerically stable.

Appendix B. Sum of two stably computed positive terms

Let ε_r, $r = 1, 2, \ldots$, denote floating-point errors satisfying $|\varepsilon_r| \le \eta$. Suppose

$$Q = S + T,$$

where $S > 0$ and $T > 0$, and computed values $\bar{S}$ and $\bar{T}$ are available having small relative errors, that is,

$$\bar{S} = S(1+\varepsilon_1)^m, \qquad \bar{T} = T(1+\varepsilon_2)^n,$$

with m and n (not necessarily integral) of modest size, less than 100, say. Now,

$$\begin{aligned}\bar{Q} &= (\bar{S} + \bar{T})(1+\varepsilon_3)\\ &= S(1+\varepsilon_1)^m(1+\varepsilon_3) + T(1+\varepsilon_2)^n(1+\varepsilon_3)\\ &= S(1+\varepsilon_4)^p + T(1+\varepsilon_5)^p,\end{aligned}$$

where $p = \max(m, n) + 1$. Since S and T are positive,

$$\bar{Q} = (S+T)(1+\varepsilon_6)^p = Q(1+\varepsilon_6)^p,$$

showing that $\bar{Q}$ has a small relative error and thus its evaluation is stable.

Appendix C. Floating-point error analysis

A detailed step-by-step floating-point error analysis of the method is given. Appendix C.1 treats the equation $x^3 - 3x = 2h$ as solved by the algorithm in section 5.2. Appendix C.2 is the counterpart for the equation $x^3 + 3x = 2h$ (algorithm in section 5.3). Relative error bounds on the computed roots are given in terms of 'ulps', units in the last place. The T_r denote values intermediate to the computation.

Appendix C.1. *Primitive cubic equation* $x^3 - 3x = 2h$

Step 1: $D = (h-1)(h+1)$ [using $T_1 = h - 1$, $T_2 = h + 1$, $D = T_1 T_2$]

(a) $\bar{T}_1 = (h-1)(1+\varepsilon_1)$
(b) $\bar{T}_2 = (h+1)(1+\varepsilon_2)$
(c) $\bar{D} = \bar{T}_1\bar{T}_2(1+\varepsilon_3) = (h-1)(1+\varepsilon_1)(h+1)(1+\varepsilon_2)(1+\varepsilon_3)$
$= D(1+\varepsilon_4)^3$

Step 2: $E = D^{1/2}$

$$\bar{E} = \bar{D}^{1/2}(1+\varepsilon_5) = D^{1/2}(1+\varepsilon_4)^{3/2}(1+\varepsilon_5) = E(1+\varepsilon_6)^{5/2}$$

Step 3: $A = (h+E)^{1/3}$ $[T_3 = h+E,\ A = T_3^{1/3}]$

(a) $\bar{T}_3 = (h+\bar{E})(1+\varepsilon_7) = [h+E(1+\varepsilon_6)^{5/2}](1+\varepsilon_7)$
$= (h+E)(1+\varepsilon_8)^{5/2}(1+\varepsilon_7) = T_3(1+\varepsilon_9)^{7/2}$
(b) $\bar{A} = \bar{T}_3^{1/3}(1+\varepsilon_{10}) = T_3^{1/3}(1+\varepsilon_9)^{7/6}(1+\varepsilon_{10})$
$= A(1+\varepsilon_{11})^{13/6}$

Step 4: $B = 1/A$

$$\bar{B} = \frac{1}{\bar{A}}(1+\varepsilon_{12}) = \frac{1}{A}(1+\varepsilon_{11})^{-13/6}(1+\varepsilon_{12})$$
$$= B(1+\varepsilon_{13})^{19/6}$$

Step 5: $x_1 = A+B$

$$\bar{x}_1 = (\bar{A}+\bar{B})(1+\varepsilon_{14})$$
$$= [A(1+\varepsilon_{11})^{13/6} + B(1+\varepsilon_{13})^{19/6}](1+\varepsilon_{14})$$
$$= (A+B)(1+\varepsilon_{15})^{25/6} = x_1(1+\varepsilon_{15})^{25/6}$$

Step 6: $G = \dfrac{E}{A^2+B^2+1}$ $[T_4 = A^2,\ T_5 = B^2,\ T_6 = T_4+T_5,$
$T_7 = T_6+1,\ G = E/T_7]$

(a) $\bar{T}_4 = \bar{A}^2(1+\varepsilon_{16}) = A^2(1+\varepsilon_{11})^{13/3}(1+\varepsilon_{16})$
$= A^2(1+\varepsilon_{17})^{16/3}$
(b) $\bar{T}_5 = \bar{B}^2(1+\varepsilon_{18}) = B^2(1+\varepsilon_{13})^{19/3}(1+\varepsilon_{18})$
$= B^2(1+\varepsilon_{19})^{22/3}$
(c) $\bar{T}_6 = (\bar{T}_4+\bar{T}_5)(1+\varepsilon_{20})$
$= [A^2(1+\varepsilon_{17})^{16/3} + B^2(1+\varepsilon_{19})^{22/3}](1+\varepsilon_{20})$
$= (A^2+B^2)(1+\varepsilon_{21})^{25/3}$
(d) $\bar{T}_7 = (\bar{T}_6+1)(1+\varepsilon_{22}) = [(A^2+B^2)(1+\varepsilon_{21})^{25/3}+1](1+\varepsilon_{22})$
$= (A^2+B^2+1)(1+\varepsilon_{23})^{28/3} = T_7(1+\varepsilon_{23})^{28/3}$
(e) $\bar{G} = \dfrac{\bar{E}}{\bar{T}_7}(1+\varepsilon_{24})$
$= \dfrac{E(1+\varepsilon_6)^{5/2}}{T_7(1+\varepsilon_{23})^{28/3}}(1+\varepsilon_{24}) = G(1+\varepsilon_{25})^{77/6}$

Step 7: $x_2,\ x_3 = -x_1/2 + \sqrt{3}G\,\mathrm{i}$ $[T_8 = \sqrt{3},\ T_9 = T_8G,$
$x_2,\ x_3 = -x_1/2 \pm T_9\,\mathrm{i}]$

(a) $\bar{T}_8 = \sqrt{3}(1+\varepsilon_{26})$
(b) $\bar{T}_9 = \bar{T}_8\bar{G}(1+\varepsilon_{27}) = \sqrt{3}(1+\varepsilon_{26})G(1+\varepsilon_{25})^{77/6}(1+\varepsilon_{27})$
$= \sqrt{3}G(1+\varepsilon_{28})^{89/6}$
(c) $\bar{x}_2,\ \bar{x}_3 = -\dfrac{x_1}{2}(1+\varepsilon_{15})^{25/6} \pm \sqrt{3}G(1+\varepsilon_{28})^{89/6}\,\mathrm{i}$

Appendix C.2. *Primitive cubic equation $x^3 + 3x = 2h$*

The analysis uses different T_r and different ε_r from those in appendix C.1.

Step 1: $D = h^2 + 1$ $[T_1 = h^2,\ D = T_1 + 1]$

(a) $\bar{T}_1 = h^2(1+\varepsilon_1)$
(b) $\bar{D} = (\bar{T}_1 + 1)(1+\varepsilon_2) = [h^2(1+\varepsilon_1)+1](1+\varepsilon_2)$
$= (h^2+1)(1+\varepsilon_4)^2 = D(1+\varepsilon_4)^2$

Step 2: $E = D^{1/2}$

$$\bar{E} = \bar{D}^{1/2}(1+\varepsilon_5) = D^{1/2}(1+\varepsilon_4)(1+\varepsilon_5) = E(1+\varepsilon_6)^2$$

Step 3: $A = (h+E)^{1/3}$ $[T_2 = h + E,\ A = T_2^{1/3}]$

(a) $\bar{T}_2 = (h+\bar{E})(1+\varepsilon_7) = [h + E(1+\varepsilon_6)^2](1+\varepsilon_7)$
$= (h+E)(1+\varepsilon_8)^3 = T_2(1+\varepsilon_9)^3$
(b) $\bar{A} = \bar{T}_2^{1/3}(1+\varepsilon_9) = T_2^{1/3}(1+\varepsilon_8)(1+\varepsilon_9) = A(1+\varepsilon_{10})^2$

Step 4: $B = -\dfrac{1}{A}$

$$\bar{B} = -\frac{1}{\bar{A}}(1+\varepsilon_{11}) = -\frac{1}{A}(1+\varepsilon_{10})^{-2}(1+\varepsilon_{11}) = B(1+\varepsilon_{12})^3$$

Step 5: $x_1 = \dfrac{2h}{A^2+B^2+1}$ $[T_3 = A^2,\ T_4 = B^2,\ T_5 = T_3 + T_4,\ T_6 = T_5 + 1,$
$x_1 = 2h/T_6]$

(a) $\bar{T}_3 = \bar{A}^2(1+\varepsilon_{13}) = A^2(1+\varepsilon_{14})^5$
(b) $\bar{T}_4 = \bar{B}^2(1+\varepsilon_{15}) = B^2(1+\varepsilon_{16})^7$
(c) $\bar{T}_5 = (\bar{T}_3+\bar{T}_4)(1+\varepsilon_{17}) = [A^2(1+\varepsilon_{14})^5 + B^2(1+\varepsilon_{16})^7](1+\varepsilon_{17})$
$= (A^2+B^2)(1+\varepsilon_{18})^8$
(d) $\bar{T}_6 = (\bar{T}_5+1)(1+\varepsilon_{19}) = [(A^2+B^2)(1+\varepsilon_{18})^8 + 1](1+\varepsilon_{19})$
$= (A^2+B^2+1)(1+\varepsilon_{20})^9 = T_6(1+\varepsilon_{20})^9$
(e) $\bar{x}_1 = \dfrac{2h}{\bar{T}_6}(1+\varepsilon_{21}) = x_1(1+\varepsilon_{22})^{10}$

Step 6: $G = \dfrac{A-B}{2}$ (noting $B < 0$)

$$\bar{G} = \frac{1}{2}(\bar{A} - \bar{B})(1+\varepsilon_{23}) = \frac{1}{2}[A(1+\varepsilon_{10})^2 - B(1+\varepsilon_{12})^3](1+\varepsilon_{23})$$
$$= \frac{1}{2}(A - B)(1+\varepsilon_{24})^4 = G(1+\varepsilon_{24})^4$$

Step 7: $x_2,\ x_3 = -\dfrac{x_1}{2} \pm \sqrt{3}G\,\mathrm{i}$ $[T_7 = \sqrt{3},\ T_8 = T_7 G,\ x_2, x_3 = -x_1/2 \pm T_8\,\mathrm{i}]$

(a) $\bar{T}_7 = \sqrt{3}(1+\varepsilon_{25})$
(b) $\bar{T}_8 = \bar{T}_7\bar{G}(1+\varepsilon_{26}) = \sqrt{3}G(1+\varepsilon_{24})^4(1+\varepsilon_{25})(1+\varepsilon_{26})$
$= \sqrt{3}G(1+\varepsilon_{27})^6$
(c) $\bar{x}_2,\ \bar{x}_3 = -\dfrac{x_1}{2}(1+\varepsilon_{22})^{10} \pm \sqrt{3}G(1+\varepsilon_{27})^6\,\mathrm{i}$

Acknowledgments

The National Measurement Office of the UKs Department for Department for Business, Energy and Industrial Strategy supported the work as part of NPLs Data Science programme. Christian Bouchot of the Escuela Superior de Ingeniería Química e Industrias Extractivas, Mexico provided information on applications arising in thermodynamic equations of state.

References

1. BIPM, IEC, IFCC, ILAC, ISO, IUPAC, IUPAP, and OIML. Evaluation of measurement data — Guide to the expression of uncertainty in measurement. Joint Committee for Guides in Metrology, JCGM 100:2008.
2. BIPM, IEC, IFCC, ILAC, ISO, IUPAC, IUPAP, and OIML. Evaluation of measurement data — Supplement 2 to the "Guide to the expression of uncertainty in measurement" — Models with any number of output quantities. Joint Committee for Guides in Metrology, JCGM 102:2011.
3. BIPM, IEC, IFCC, ILAC, ISO, IUPAC, IUPAP, and OIML. Evaluation of measurement data — Supplement 1 to the "Guide to the expression of uncertainty in measurement" — Propagation of distributions using a Monte Carlo method. Joint Committee for Guides in Metrology, JCGM 101:2008, 2008.
4. Birkhoff, G., and Lane, S. M. *A survey of modern algebra.* A.K. Peters, Wellesley, Mass, 1997.
5. Boris Hasselblatt, A. K. *Introduction to the Modern Theory of Dynamical Systems.* Cambridge University Press, 2011.
6. Burnside, W. S., and Panton, A. W. *The theory of equations: with an introduction to the theory of binary algebraic forms, 2nd edition.* Longmans, Green and Co., London, 1886.
7. Carrero-Mantilla, J. I. The numerical problems within analytical methods of solution for cubic equations of state. *Ingeniera y Competitividad 14*, 1 (2013).

8. Deiters, U. K., and Macías-Salinas, R. Calculation of densities from cubic equations of state: Revisited. *Industrial & Engineering Chemistry Research 53*, 6 (2014), 2529–2536.
9. Dickson, L. *First course in the theory of equations.* J Wiley & Sons, New York, 1922.
10. Farin, G., Hoschek, J., and Kim, M. S. *Handbook of Computer Aided Geometric Design.* Elsevier Science, 2002.
11. Foley, J. *Computer graphics : principles and practice.* Addison-Wesley, Reading, Mass, 1995.
12. Gill, P. E., Murray, W., and Wright, M. H. *Practical Optimization.* Academic Press, London, 1981.
13. Higham, N. J. *Accuracy and Stability of Numerical Algorithms, 2nd edition.* Society for Industrial and Applied Mathematics, Philadelphia, PA, 2002.
14. Higham, N. J., and Mary, T. A new approach to probabilistic rounding error analysis. *SIAM Journal on Scientific Computing 41*, 5 (2019), A2815–A2835.
15. IEEE 754-2008. Standard for Floating-Point Arithmetic. Standard, Institute of Electrical and Electronics Engineers, New York, USA, 2008.
16. Irving, R. S. *Integers, Polynomials, and Rings: A Course in Algebra (Undergraduate Texts in Mathematics).* Springer, 2004.
17. Kingsolver, J. G., and Umbanhowar, J. The analysis and interpretation of critical temperatures. *The Journal of Experimental Biology 221*, 12 (2018), jeb167858.
18. Lazzús, J. Optimization of a cubic equation of state and van der Waals mixing rules for modeling the phase behavior of complex mixtures. *Revista Mexicana de Física* (2012), 510–514.
19. Peng, D.-Y., and Robinson, D. B. A new two-constant equation of state. *Industrial & Engineering Chemistry Fundamentals 15*, 1 (1976), 59–64.
20. Reid, R., Poling, B. E., and Prausnitz, J. M. *The Properties of Gases and Liquids*, fourth ed. McGraw-Hill Companies, 1987.
21. Robert, M., and Corless, N. F. *A Graduate Introduction to Numerical Methods.* Springer-Verlag GmbH, 2013.
22. Turner, L. *Electronics engineer's reference book.* Newnes-Butterworths, London, 1976.
23. Wilkinson, J. H. *Notes in Applied Science No. 32. Rounding Errors in Algebraic Processes.* Her Majesty's Stationery Office, London, 1963.
24. Zhi, Y., and Lee, H. Fallibility of analytic roots of cubic equations of state in low temperature region. *Fluid Phase Equilibria 201*, 2 (2002), 287–294.

F Pavese, A B Forbes, N F Zhang, A G Chunovkina (Eds.):
Advanced Mathematical and Computational Tools in Metrology and Testing XII
Series on Advances in Mathematics for Applied Sciences, Vol. 90

https://doi.org/10.1142/9789811242380_0002

Effective number of degrees of freedom associated with regression models

Alistair B. Forbes

National Physical Laboratory, Teddington, Middlesex, UK
E-mail: alistair.forbes@npl.co.uk

This paper is concerned with the approximation of data using linear models for which there is additional information to be incorporated into the analysis. This situation arises, for example, in Tikhonov regularisation, model averaging, Bayesian inference and approximation with Gaussian process models. In a least-squares approximation process, the m-vector of data $\boldsymbol{y}$ is approximated by a linear function $\hat{\boldsymbol{y}} = H\boldsymbol{y}$ of the data, where H is an $m \times m$ matrix related to the observation matrix. The effective number of degrees of freedom associated with the model is given by the sum of the eigenvalues of H. For standard linear least-squares regression, the matrix H is a projection and has n eigenvalues equal to 1 and all others zero, where n is the number of parameters in the model. Incorporating prior information about the parameters reduces the effective number of degrees of freedom since the ability of the model to approximate the data vector $\boldsymbol{y}$ is partially constrained by the prior information. We give a general approach for providing bounds on the effective number of degrees of freedom for models with prior information on the parameters and illustrate the approach on common problems. In particular, we show how the effective number of degrees of freedom depends on spatial or temporal correlation lengths associated with Gaussian processes. The correlation lengths are seen to be tuning parameters used to match the model degrees of freedom to the (generally unknown) number of degrees of freedom associated with the system giving rise to the data. We also show how Gaussian process models can be related to Tikhonov regularisation for an appropriate set of basis functions.

Keywords: Degrees of Freedom, Least Squares Regression, Regularisation

1. Introduction

This paper is concerned with the approximation of data using linear models for which there is prior information associated with some or all of the model parameters. This situation arises, for example, in Tikhonov regularisation, model averaging, Bayesian inference and approximation with Gaussian process models.[10] The latter case is particularly important in model calibration[8] in which the measured response of a system is modelled as

$$y = Ba + e + \epsilon, \quad e \in \mathrm{N}(\mathbf{0}, V), \quad \epsilon \in \mathrm{N}(\mathbf{0}, \sigma_M^2 I),$$

where Ba is the response predicted by a hypothesised model, ϵ are random effects associated with measurement process, and e are systematic effects that account for the mismatch between the modelled response Ba and the actual response of the system. Prior information about e is encoded by the variance matrix V. In a least-squares approximation process, the m-vector of data y is approximated by a linear function $\hat{y} = Hy$ of the data, where H is an $m \times m$ matrix related to the observation matrix B. The effective number of degrees of freedom[7] associated with the model is given by the sum of the eigenvalues of H. For standard linear least-squares regression, the matrix H is a projection and has n eigenvalues equal to 1 and all others zero, where n is the number of parameters in the model. Incorporating prior information about the parameters reduces the effective number of degrees of freedom since the ability of the model to approximate the data vector y is partially constrained by the prior information.[3] We give a general approach for providing bounds on the effective number of degrees of freedom for models with prior information on the parameters and illustrate the approach on common problems. In particular, we show how the effective number of degrees of freedom depends on spatial or temporal correlation lengths associated with Gaussian processes. The correlation lengths are seen to be tuning parameters used to match the model degrees of freedom to the (generally unknown) number of degrees of freedom associated with the system giving rise to the data. We also show how Gaussian process models can be related to Tikhonov regularisation for an appropriate set of basis functions.

The remainder of this paper is organised as follows. In section 2 we overview linear regression while in section 3 we discuss Tikhonov regularisation and the effective number of degrees of freedom associated with such problems. Section 4 looks at incorporation prior information into a model while section 5 discusses model averaging. In section 6 we generalise from standard linear regression to Gauss-Markov regression involving correlated effects and apply the results to Gaussian process models in section 7. An numerical example involving a reference function in thermometry is given in section 8 and our concluding remarks in section 9.

2. Linear regression

Consider the standard model

$$y_i = f(x_i, \boldsymbol{a}) + \epsilon_i, \quad f(x_i, \boldsymbol{a}) = \sum_{j=1}^{n} a_j f_j(x_i), \quad \epsilon_i \in \mathrm{N}(0, \sigma_M^2), \quad i = 1, \ldots, m,$$

or in matrix terms, $\boldsymbol{y} \in \mathrm{N}(B\boldsymbol{a}, \sigma_M^2 I)$, where $B_{ij} = f_j(x_i)$. Given data (x_i, y_i), $i = 1, \ldots, m$, the linear least-squares estimate $\hat{\boldsymbol{a}}$ of the parameters $\boldsymbol{a}$ is found by solving

$$\min_{\boldsymbol{a}} \; (\boldsymbol{y} - B\boldsymbol{a})^\top (\boldsymbol{y} - B\boldsymbol{a}).$$

If B has QR factorisation[5] $B = Q_1 R_1$ where Q_1 is an $m \times n$ orthogonal matrix and R_1 is an $n \times n$ upper-triangular matrix, then

$$\hat{\boldsymbol{a}} = (B^\top B)^{-1} B^\top \boldsymbol{y} = R_1^{-1} Q_1^\top \boldsymbol{y},$$

and $\hat{\boldsymbol{a}}$ can be calculated by solving the upper-triangular system of linear equations $R_1 \hat{\boldsymbol{a}} = Q_1^\top \boldsymbol{y}$. The model predictions $\hat{\boldsymbol{y}}$ are given by

$$\hat{\boldsymbol{y}} = B\hat{\boldsymbol{a}} = B(B^\top B)^{-1} B^\top \boldsymbol{y} = Q_1 Q_1^\top \boldsymbol{y} = H\boldsymbol{y}, \quad H = Q_1 Q_1^\top.$$

The matrix H is a projection, so that $H = H^\top = H^2$, and projects the data vector $\boldsymbol{y}$ on to the n-dimensional subspace spanned by the columns of B; the columns of Q_1 define an orthogonal axes system for this subspace. In Hastie et al.,[7] the sum of the eigenvalues of the matrix H is taken to be the *effective number of degrees of freedom* associated with the model. In this case, H is a projection and has n eigenvalues equal to 1 and all other 0 so that the effective number of degrees of freedom is n, the number of parameters in the model. We will denote by $n_d = n_d(H)$ the sum of the eigenvalues of H.

3. Tikhonov regularisation

Tikhonov regularisation is used to convert an ill-posed or ill-conditioned problem to a near-by, well-posed problem. For linear regression, it has the form

$$\min_{\boldsymbol{a}} \; (\boldsymbol{y} - B\boldsymbol{a})^\top (\boldsymbol{y} - B\boldsymbol{a}) \; + \; \kappa^2 \boldsymbol{a}^\top \boldsymbol{a}, \tag{1}$$

where $\kappa \geq 0$ controls the degree of regularisation.

Let C and $\boldsymbol{z}$ be defined by

$$C = \begin{bmatrix} B \\ \kappa I \end{bmatrix}, \quad \boldsymbol{z} = \begin{bmatrix} \boldsymbol{y} \\ \boldsymbol{0} \end{bmatrix},$$

then the solution $\hat{\boldsymbol{a}}$ of (1) is given by

$$\hat{\boldsymbol{a}} = (C^\top C)^{-1} C^\top \boldsymbol{z} = (C^\top C)^{-1} B^\top \boldsymbol{y}$$

and the vector model predictions $\hat{\boldsymbol{y}}$ is given by

$$\hat{\boldsymbol{y}} = B\hat{\boldsymbol{a}} = H\boldsymbol{y}, \quad H = B(C^\top C)^{-1} B^\top. \tag{2}$$

Suppose B has (lean) singular value decomposition[5] (SVD) $B = Q_1 S V^\top$, where V is an $n \times n$ orthogonal matrix, Q_1 is an $m \times n$ orthogonal matrix and S is a diagonal matrix with diagonal elements $s_1 \geq s_2 \geq \cdots \geq s_n \geq 0$. Then

$$\begin{aligned}(C^\top C)^{-1} &= (B^\top B + \kappa^2 I)^{-1}, \\ &= (V S^2 V^\top + \kappa^2 I)^{-1}, \\ &= V(S^2 + \kappa^2 I)^{-1} V^T,\end{aligned}$$

and

$$\begin{aligned}H &= Q_1 S V^\top V (S^2 + \lambda^2 I)^{-1} V^T V S^\top Q_1^\top, \\ &= Q_1 S_H^2 Q_1^\top\end{aligned}$$

where $S_H = S_H(\kappa)$ is the $n \times n$ diagonal matrix with diagonal elements

$$s_{H,i} = \frac{s_i}{\sqrt{s_i^2 + \kappa^2}}.$$

If the singular values are bounded away from zero so that $s_i \geq s_0 > 0$, then for $\kappa \ll s_0$, $s_{H,i} \approx 1$ and the effective degrees of freedom is close to the limiting value of n. If, on the other hand, there are a number of very small singular values, then the presence of a regularisation term can reduce the effective degrees of freedom associated with the model.

4. Prior information associated with linear regression

In Bayesian inference, a sequence of experiments relating to the same physical system allows the knowledge from a previous experiment to be updated by knowledge available from a new experiment. Suppose the current experiment is modelled as $\boldsymbol{y}|\boldsymbol{a} \in \mathrm{N}(B\boldsymbol{a}, I)$, and that from previous experiment or otherwise we have prior information encoded as $\boldsymbol{a} \sim \mathrm{N}(\boldsymbol{a}_0, V_0)$. The posterior estimate of $\hat{\boldsymbol{a}}$ is found from solving

$$\min_{\boldsymbol{a}} \ (\boldsymbol{y} - B\boldsymbol{a})^\top (\boldsymbol{y} - B\boldsymbol{a}) \ + \ (\boldsymbol{a} - \boldsymbol{a}_0)^\top V_0^{-1} (\boldsymbol{a} - \boldsymbol{a}_0). \tag{3}$$

Set $V_0^{-1} = B_0^\top B_0$, $\quad \boldsymbol{b}_0 = B_0 \boldsymbol{a}_0$, and let C and $\boldsymbol{z}$ be defined by

$$C = \begin{bmatrix} B \\ B_0 \end{bmatrix}, \quad \boldsymbol{z} = \begin{bmatrix} \boldsymbol{y} \\ \boldsymbol{b}_0 \end{bmatrix},$$

then the solution $\hat{\boldsymbol{a}}$ of (3) is given by

$$\hat{\boldsymbol{a}} = (C^\top C)^{-1} B^\top \boldsymbol{y}, \hat{\boldsymbol{y}} = H\boldsymbol{y}, \quad H = B(C^\top C)^{-1} B^\top.$$

If C has QR decomposition $C = Q_1 R_1$ then the eigenvalues of H are the squares of the singular values of $R_1^{-\top} B^\top$.

The matrix B_0 can be thought of as representing, or in fact may be, the observation matrix from a previous experiment. If this previous experiment involved a similar design to that represented by B and involved m_0 observations, then

$$B_0^\top B_0 \approx \frac{m_0}{m} B^\top B.$$

Based on this approximation

$$n_d(H) \approx \frac{mn}{m + m_0}.$$

If B represents a repeat of the experiment represented by m, then $n_d(H) = n/2$; half of the available degrees of freedom are used to match the prior information and half to match the new data.

5. Model averaging

Suppose we have data (x_i, y_i), $i = 1, \ldots, m$, where it as assumed that the data can be expressed as linear combinations of basis functions evaluated at x_i. We may have a number of basis functions $f_k(x)$, $k = 1, \ldots, n$, that we can use to approximate $\boldsymbol{y}$ but that we are unsure about which ones to include. Let $\mathcal{M}_j$ be the model that specifies a particular subset $I_j \subset 1, \ldots, n$ and models the measured response as

$$y_i = \sum_{k \in I_j} a_k f_k(x_i) + \epsilon_i, \quad \epsilon_i \in \mathrm{N}(0, \sigma_M^2),$$

where σ_M is known. Given the data $\boldsymbol{y}$, we wish to be able to estimate the response of the system at x. If we know that the data arose according to model $\mathcal{M}_j$, maximum likelihood estimates $\boldsymbol{a}_j$ of the model parameters are given by

$$\hat{\boldsymbol{a}}_j = \left(B_j^\top, B_j\right)^{-1} B_j^\top \boldsymbol{y},$$

and the model predictions $\hat{\boldsymbol{y}}_j$ at $\boldsymbol{x}$ are

$$\hat{\boldsymbol{y}}_j = B_j \hat{\boldsymbol{a}}_j = B_j \left(B_j^\top, B_j\right)^{-1} B_j^\top \boldsymbol{y}.$$

If B_j has QR factorisation

$$B_j = Q_j R_j = [Q_{j,1}\ Q_{j,2}] \begin{bmatrix} R_{j,1} \\ \mathbf{0} \end{bmatrix} = Q_{j,1} R_{j,1},$$

then $\hat{\boldsymbol{y}}_j = H_j \boldsymbol{y}_j$ where

$$H_j = B_j \left(B_j^\top, B_j\right)^{-1} B_j^\top = Q_{j,1} Q_{j,1}^\top.$$

It may be that after fitting the various models to the data, there a number of models that are plausible, and instead of selecting one, we choose a weighted average of the models[9] where the weights p_j reflect some measure of the plausibility of each model:

$$\hat{\boldsymbol{y}} = \sum_j p_j \hat{\boldsymbol{y}}_j, \quad 0 \le p_j \le 1, \quad \sum p_j = 1.$$

For such an average,

$$\hat{\boldsymbol{y}} = H\boldsymbol{y}, \quad H = \sum_j p_j H_j,$$

and the effective number of degrees of freedom $n_d(H) = \sum_j p_j n_j$, where n_j is the number of parameters associated with the jth linear model. Ideally we would want the weights p_j to represent the posterior probability for each model $p(\mathcal{M}_j|\boldsymbol{y})$ given the observed data $\boldsymbol{y}$.

5.1. *Averaging polynomial models*

As an example of model averaging, we consider the case of model averaging where the basis function $f_k(x)$ is an order k, degree $k-1$, polynomial, $k = 1, \ldots, n$, and the jth model $\mathcal{M}_j$ involves the basis functions f_k, $k = 1, \ldots, j$, $j = 1, \ldots, n$. In practice, is it important to use a set of basis functions with good numerical stability such as Chebyshev polynomials.

The sequence $\mathcal{M}_1 \subset \mathcal{M}_2 \subset \ldots \mathcal{M}_j \ldots \mathcal{M}_n$ represents a nested sequence of model spaces in the sense that any function represented by a linear combination of basis functions associated with $\mathcal{M}_j$ can also be represented using model $\mathcal{M}_{j+1}$. Algebraically, this means that all of the model fits associated with each model can be determined from the fit associated with model $\mathcal{M}_n$. Let B be the $m \times n$ observation matrix associated with model $\mathcal{M}_n$ and let $B = Q_1 R_1$ its QR decomposition, $\boldsymbol{z} = Q_1^\top \boldsymbol{y}$, and $\boldsymbol{q}_j$ be the

jth column of Q_1. Then the model prediction $\hat{\boldsymbol{y}}_1$ associated with model $\mathcal{M}_1$ is given by $\hat{\boldsymbol{y}}_1 = z_1\boldsymbol{q}_1$ and, for $j = 2, \ldots, n$, $\hat{\boldsymbol{y}}_j = \hat{\boldsymbol{y}}_{j-1} + z_j\boldsymbol{q}_j$ are the model predictions associated with $\mathcal{M}_j$. Given a weighting vector $\boldsymbol{p} = (p_1, \ldots, p_n)^\top$, the averaged model prediction is given by

$$\hat{\boldsymbol{y}} = \sum_{j=1}^{n} p_j \hat{\boldsymbol{y}}_j.$$

If $\boldsymbol{c}$ is the n-vector defined by

$$c_j = \sum_{k=j}^{n} p_j, \tag{4}$$

then $\tilde{\boldsymbol{z}}$ is the n-vector defined by $\tilde{z}_j = c_j z_j$, and $\tilde{Q}_1$ is the $m \times n$ matrix with columns $\tilde{\boldsymbol{q}}_j = c_j\boldsymbol{q}_j$. The model prediction is given by $\hat{\boldsymbol{y}} = Q_1\tilde{\boldsymbol{z}} = \tilde{Q}_1\boldsymbol{z}$. The standard uncertainty $u(\hat{y}_i)$ associated with the ith model prediction $\hat{y}_i$ given by

$$u^2(\hat{y}_i) = \sigma_M^2 \sum_{j=1}^{n} \tilde{Q}_1(i,j)^2.$$

We note that $\hat{\boldsymbol{y}} = B\hat{\boldsymbol{a}}$ and $R_1\hat{\boldsymbol{a}} = \tilde{\boldsymbol{z}}$, so that $\hat{\boldsymbol{y}}$, being an average of polynomials of order n, is itself a polynomial of order n with coefficients $\hat{\boldsymbol{a}}$.

5.2. *Model average as a regularised solution*

In general, $\hat{\boldsymbol{a}}$ defining the averaged model prediction $B\hat{\boldsymbol{y}}$ will be different from $\hat{\boldsymbol{a}}_n$ associated with the model $\mathcal{M}_n$. Some insight as to why this should be can be provided as follows. Suppose the basis functions $f_k(x)$ are chosen to be orthogonal on the support vector $\boldsymbol{x}$. Such orthogonal polynomials can be generated using a Forsythe scheme.[4] In this case the observation matrix B is orthogonal,

$$\boldsymbol{z} = B^\top\boldsymbol{y}, \quad \hat{\boldsymbol{a}} = \tilde{\boldsymbol{z}}, \quad \tilde{z}_j = c_j z_j,$$

where c_j is as in (4). Assuming that $c_j > 0$ for all j, then we can form weights $\kappa_j \geq 0$ according to

$$\kappa_j^2 = \frac{c_j - 1}{c_j}.$$

With this definition it is straight forward to show that $\hat{\boldsymbol{a}}$ is the solution to the regularised problem

$$\min_{\boldsymbol{a}} \; (\boldsymbol{y} - B\boldsymbol{a})^\top(\boldsymbol{y} - B\boldsymbol{a}) + \sum_j \kappa_j^2 a_j^2. \tag{5}$$

The weight κ_j will be small if c_j is close to one. This will be the case if the jth basis function appears in the majority of the models that are plausible according to the weight p_j. Conversely, the weight κ_j will be large if few of the plausible models contain the jth basis function. The effect of a large κ_j is to force the coefficient $\hat{a}_j$ to be close to zero, reflecting that the behaviour represented by $f_j(x)$ is not a good predictor of the behaviour of the response $\boldsymbol{y}$.

For c_j near zero the problem formulated as (5) is likely to be numerically unstable. Setting

$$C = \begin{bmatrix} BD \\ K \end{bmatrix}, \quad \boldsymbol{z} = \begin{bmatrix} \boldsymbol{y} \\ \boldsymbol{0} \end{bmatrix}$$

where D is the $n \times n$ diagonal matrix with diagonal elements $d_j = \sqrt{c_j}$ and K is the $n \times n$ diagonal matrix with diagonal elements $k_j = (1 - c_j)^{1/2}$, then C is orthogonal, representing a stable re-parametrization of the problem in terms of $\tilde{\boldsymbol{a}}$ with $\boldsymbol{a} = D\tilde{\boldsymbol{a}}$, and the solution $\hat{\boldsymbol{a}}$ is simply $D^2 B^\top \boldsymbol{y}$.

5.3. *Weighting according to the Bayesian Information Criterion*

In order to provide an averaged model prediction, it is required to set the weights p_j that reflect how well each model is supported by the data. A number of approaches rely on a measure of the goodness of fit of each model to the data to assign these weights.[2,11,13] Let

$$\mathrm{RSS}_n = \frac{1}{\sigma_M^2}(\boldsymbol{y} - \hat{\boldsymbol{y}}_n)^\top(\boldsymbol{y} - \hat{\boldsymbol{y}}_n) = \frac{1}{\sigma_M^2}\left(\boldsymbol{y}^\top\boldsymbol{y} - \boldsymbol{q}_n^\top\boldsymbol{q}_n\right),$$

the (weighted) residual sum of squares associated with the nth model fit. Then, for $j = n-1, \dots, 1$,

$$\mathrm{RSS}_j = \mathrm{RSS}_{j+1} + \left(q_{j+1}/\sigma_M\right)^2,$$

is the residual sum of squares associated with the jth model fit. The quantities RSS_j can be used to determine a posterior probability distribution $p(\mathcal{M}_j|\boldsymbol{y})$ using the Bayesian Information Criterion,[6] for example. Defining BIC_j by

$$\mathrm{BIC}_j = \mathrm{RSS}_j + j \log m \tag{6}$$

the posterior probability for $p(\mathcal{M}_j|\boldsymbol{y})$ is estimated by

$$p_j = p(\mathcal{M}_j|\boldsymbol{y}) \propto p(\mathcal{M}_j)\exp(-\mathrm{BIC}_j/2),$$

where $p(\mathcal{M}_j)$ is the prior probability of model $\mathcal{M}_j$. The first term in the definition of BIC in (6) means that models that provide a good fit to the data achieve a greater posterior weight while the second term penalises, to some extent, models with larger numbers of the parameters.

6. Gauss-Markov regression

In this section we consider the case where the data $\boldsymbol{y}$ arises according to the model

$$\boldsymbol{y} \in \mathrm{N}(B\boldsymbol{a}, V), \tag{7}$$

where the variance matrix V reflects correlation due, for example, to common systematic effects associated with the measurement system. The Gauss-Markov estimate of $\boldsymbol{a}$ is found by solving

$$\min_{\boldsymbol{a}} \; (\boldsymbol{y} - B\boldsymbol{a})^\top V^{-1}(\boldsymbol{y} - B\boldsymbol{a}). \tag{8}$$

If V can be factored as $V = LL^\top$ where L is square, e.g., using a Cholesky factorisation,[5] then we can solve the ordinary linear least-squares regression problem

$$\min_{\boldsymbol{a}} (\boldsymbol{z} - C\boldsymbol{a})^\top (\boldsymbol{z} - C\boldsymbol{a}),$$

involving the transformed observation matrix C and data vector $\boldsymbol{z}$ where $LC = B$ and $L\boldsymbol{z} = \boldsymbol{y}$. If $C = Q_1 R_1$, then the transformed model predictions $\hat{\boldsymbol{z}}$ are given in terms of the transformed data vector $\boldsymbol{z}$ by $\hat{\boldsymbol{z}} = G\boldsymbol{z}$ where $G = Q_1 Q_1^\top$. The matrix G is a projection matrix, as discussed above, and has n eigenvalues equal to 1 and all others are zero.

The unweighted model predictions $\hat{\boldsymbol{y}}$ are given in terms of the original data vector $\boldsymbol{y}$ by $\hat{\boldsymbol{y}} = H\boldsymbol{y}$, where $H = LGL^{-1}$. The matrix H is not in general a projection but it is conjugate to the projection matrix G and therefore has the same eigenvalues: if $G\boldsymbol{v} = \lambda\boldsymbol{v}$, then $H(L\boldsymbol{v}) = LG\boldsymbol{v} = \lambda(L\boldsymbol{v})$. Thus, the effective number of degrees of freedom associated with the model is n, the number of parameters in the model, as for the case of ordinary least squares regression.

6.1. *Calculation of the effective degrees of freedom for Gauss-Markov problems*

We now investigate in more detail the calculation of the effective degrees of freedom associated with Gauss-Markov problems. Suppose V is factored as

$V = UD^2U^\top$ where U is and orthogonal matrix of eigenvectors and D^2 is the diagonal matrix of eigenvalues d_i^2 of V. We assume the eigenvalues are ordered so that $d_i \geq d_{i+1} \geq 0$. We can write the model (7) equivalently as

$$\boldsymbol{y} = UD\boldsymbol{e} + B\boldsymbol{a} + \boldsymbol{\epsilon}, \quad \boldsymbol{e} \in \mathrm{N}(\boldsymbol{0}, I), \quad \boldsymbol{\epsilon} \in \mathrm{N}(\boldsymbol{0}, \sigma_M^2 I), \tag{9}$$

or

$$\tilde{\boldsymbol{y}} = D\boldsymbol{e} + \tilde{B}\boldsymbol{a} + \tilde{\boldsymbol{\epsilon}}, \tag{10}$$

where $\tilde{\boldsymbol{y}} = U^\top \boldsymbol{y}$, $\tilde{B} = U^\top B$, and $\tilde{\boldsymbol{\epsilon}} = U^\top \boldsymbol{\epsilon}$. We note that, since U is orthogonal, $\tilde{\boldsymbol{\epsilon}} \in \mathrm{N}(\boldsymbol{0}, \sigma_M^2 I)$.

For the moment, let us rename $\boldsymbol{y} = \tilde{\boldsymbol{y}}$, $B = \tilde{B}$ and $\boldsymbol{\epsilon} = \tilde{\boldsymbol{\epsilon}}$ and set $\sigma_M = 1$, and consider the least squares system $C\boldsymbol{b} \approx \boldsymbol{z}$ where

$$C = \begin{bmatrix} D & B \\ I & \boldsymbol{0} \end{bmatrix}, \quad \boldsymbol{b} = \begin{bmatrix} \boldsymbol{e} \\ \boldsymbol{a} \end{bmatrix}, \quad \boldsymbol{z} = \begin{bmatrix} \boldsymbol{y} \\ \boldsymbol{0} \end{bmatrix}. \tag{11}$$

Define the orthogonal matrix M by

$$M^\top = \begin{bmatrix} C_M & S_M \\ -S_M & C_M \end{bmatrix}$$

constructed from plane rotations where C_M and S_M are the diagonal matrices with diagonal elements $c_i = d_i/r_i$ and $s_i = 1/r_i$, respectively, where $r_i^2 = 1 + d_i^2$. We note that $c_i \geq c_{i+1} \geq 0$ and $0 \leq s_i \leq s_{i+1}$. The orthogonal matrix M is defined so that

$$M^\top C = \begin{bmatrix} R_{11} & C_M B \\ \boldsymbol{0} & -S_M B \end{bmatrix}$$

where R_{11} is the diagonal matrix with diagonal elements r_i. Thus, the matrix M can be used to perform a partial upper-triangularisation of the matrix C. Let N be the $m \times m$ orthogonal matrix such that

$$N^\top(-S_M B) = \begin{bmatrix} T \\ \boldsymbol{0} \end{bmatrix} \tag{12}$$

is upper triangular and set $N = [N_1 \; N_2]$ where N_1 is made up of the first n columns of N and N_2 the remaining $m - n$ columns. Then

$$Q = \begin{bmatrix} C_M & -S_M \\ S_M & C_M \end{bmatrix} \begin{bmatrix} I & \\ & N \end{bmatrix}$$

is such that

$$C = Q \begin{bmatrix} R_{11} & C_M B \\ \boldsymbol{0} & T \\ \boldsymbol{0} & \boldsymbol{0} \end{bmatrix}$$

is a QR decomposition of C. Setting,

$$Q_1 = \begin{bmatrix} C_M & -S_M N_1 \\ -S_M & C_M N_1 \end{bmatrix}, \quad G = Q_1 Q_1^\top,$$

then $\hat{\boldsymbol{z}} = G\boldsymbol{z}$ and setting

$$Q_{11} = \begin{bmatrix} C_M & -S_M N_1 \end{bmatrix}, \quad H = Q_{11} Q_{11}^\top,$$

we have $\hat{\boldsymbol{y}} = H\boldsymbol{y}$.

The sum of the eigenvalues of H is the trace of the matrix

$$C_M^2 + S_M N_1 N_1^\top S_M.$$

We wish to show that

$$n + \sum_{i=n+1}^{m} c_i^2 \leq n_d(H) \leq n + \sum_{i=1}^{m-n} c_i^2. \tag{13}$$

These bounds are attained if the first or last n rows of N_1 is the identity matrix and all other rows are zero. The trace of $S_M N_1 N_1^\top S_M$ is given by

$$S = \sum_{i=1}^{m} \left(s_i^2 \sum_{j=1}^{n} n_{ij}^2 \right). \tag{14}$$

The matrix N_1 is formed from n columns of an orthogonal matrix so that

$$\sum_{i=1}^{m} n_{ij}^2 = 1, \quad j = 1, \ldots, n. \tag{15}$$

We also know that

$$\sum_{j=1}^{n} n_{ij}^2 \leq n, \quad i = 1, \ldots, m. \tag{16}$$

Suppose we wish to minimise S in (14) subject to the constraints (15). The solution n_{ij} define a critical point of the Lagrangian

$$\sum_{i=1}^{m} \left(s_i^2 \sum_{j=1}^{n} n_{ij}^2 \right) - \sum_{j=1}^{n} \lambda_j \left(\sum_{i=1}^{m} n_{ij}^2 - 1 \right)$$

leading to the equations

$$s_i^2 n_{ij} - \lambda_j n_{ij} = 0, \quad i = 1, \ldots, m, \quad j = 1, \ldots, n,$$

so that either $n_{ij} = 0$ or $\lambda_j = s_i^2$. If all the s_i are distinct, $0 < s_i < s_{i+1}$, $i = 1, \ldots, m-1$, for $n_{ij} \neq 0$, we must have $\lambda_j = s_i^2$ and, since λ_j can only be equal to exactly one s_i^2, the jth column of N_1 must be zero apart from one element $n_{ij} = 1$ for some i. It follows from the constraints (16), that ith row is zero apart from $n_{ij} = 1$. For this case, the contribution of the ith row to S is s_i^2. To minimise S, then, the columns of N_1 can be reordered so that it has a copy of the $n \times n$ identity matrix in the first n rows and be zero otherwise. Similarly to maximise S, the first $m - n$ rows of N_1 must be zero and the remaining n rows contain a copy of the identity matrix, up to column order. If $s_{p-1} < s_p = s_{p+1} = \cdots = s_{p+k-1} < s_{p+k}$, $p + k - 1 \leq n$, then N_1 that minimises S can have non-zero elements in rows p to $p + k - 1$. A minimising N_1 can be column ordered so that the first n rows of N_1 is block diagonal with each diagonal block an orthogonal matrix, the size of which is the multiplicity associated the corresponding s_i; in the case above, $k \times k$. The exception to this can be the final block. If $s_{p-1} < s_p = s_{p+1} = \cdots = s_{p+k-1}$, $p \leq n$, $p + k - 1 > n$, the final block can be a $k \times (n - p + 1)$ orthogonal matrix with nonzero elements in rows p to $p + k - 1$. In all cases a minimising N_1 leads to a value of

$$S = \sum_{i=1}^{n} s_i^2.$$

Similarly, a maximising N_1 corresponds to

$$S = \sum_{i=m-n+1}^{m} s_i^2.$$

Recalling that for all i, $c_i^2 + s_i^2 = 1$, the bounds on the trace of H in (13) follow.

We now go back to the case for general σ_M and we suppose also that the eigenvalues are written as $\sigma_E^2 d_i^2$. We can think of σ_E as the standard deviation associated with the prior Gaussian distribution for the parameters $\mathbf{e}$. In this case

$$c_i^2 = \frac{\sigma_E^2 d_i^2}{\sigma_M^2 + \sigma_E^2 d_i^2}, \quad s_i^2 = \frac{\sigma_M^2}{\sigma_M^2 + \sigma_E^2 d_i^2}$$

and the bounds on the trace of H in (13) become

$$n + \sum_{i=n+1}^{m} \frac{\sigma_E^2 d_i^2}{\sigma_M^2 + \sigma_E^2 d_i^2} \leq n_d(H) \leq n + \sum_{i=1}^{m-n} \frac{\sigma_E^2 d_i^2}{\sigma_M^2 + \sigma_E^2 d_i^2}. \tag{17}$$

Where the effective degrees of freedom are represented by the trace of H lies within these constraints is determined by the matrix N_1, the submatrix of N in (12). The matrix N_1 is determined in part by the basis functions represented in the matrix $\tilde{B} = U^\top B$ in (10). Suppose

$$N_1 = \begin{bmatrix} I \\ \mathbf{0} \end{bmatrix}$$

so that S in (14) is minimised. Since N is an orthogonal matrix, N must have the form

$$N = \begin{bmatrix} I & \mathbf{0} \\ \mathbf{0} & N_{22} \end{bmatrix}$$

where N_{22} is an $(m-n) \times (m-n)$ orthogonal matrix. Partitioning $S_M B$ in (12) as

$$S_M B = \begin{bmatrix} S_1 B_1 \\ S_2 B_2 \end{bmatrix},$$

it follows from (12) that $S_2 B_2 = \mathbf{0}$. Since $s_i^2 > 0$ it follows that $B_2 = \mathbf{0}$. Recalling that B_2 is in fact the bottom $m-n$ rows of $U^\top B$, $B_2 = \mathbf{0}$ says that the basis function in B can be constructed from the first n eigenvectors of V associated with the n largest eigenvalues d_i^2. Writing the model predictions $\hat{\boldsymbol{y}}$ as $\hat{\boldsymbol{y}} = UD\hat{\mathbf{e}} + B\hat{\boldsymbol{a}}$, the diagonal elements d_i of D represent the prior likelihood that the measurement response $\boldsymbol{y}$ can be explained by the ith eigenvector $\boldsymbol{u}_i$. If B is a linear combination of the first n eigenvectors in U, the extra approximating power as represented by

$$\sum_{i=1}^{n} s_i^2 = \sum_{i=1}^{n} \frac{\sigma_M^2}{\sigma_M^2 + \sigma_E^2 d_i^2}$$

is small if $\sigma_E d_i$ is large relative to σ_M: any response $\boldsymbol{y}$ that can be explained by $B\hat{\boldsymbol{a}}$ can also be plausibly explained through $UD\hat{\mathbf{e}}$. If on the other hand,

$$N_1 = \begin{bmatrix} \mathbf{0} \\ I \end{bmatrix}$$

the basis vectors B explain behaviour that can be least plausibly explained by the effects $\boldsymbol{e}$ and contribution to the effective degrees of freedom is maximised.

6.2. *Gauss-Markov models as regularised models*

Suppose the eigenvalues $d_i^2 > 0$ so that we can compute D^{-1}. In this case, we can write the model (9) equivalently as

$$\boldsymbol{y} = U\boldsymbol{e} + B\boldsymbol{a} + \boldsymbol{\epsilon}, \quad \boldsymbol{e} \in \mathrm{N}(\mathbf{0}, \sigma_E^2 D^2), \quad \boldsymbol{\epsilon} \in \mathrm{N}(\mathbf{0}, \sigma_M^2 I), \qquad (18)$$

and best estimates of the model parameters are found by solving

$$\min_{\boldsymbol{a},\boldsymbol{e}} \frac{1}{\sigma_M^2}(\boldsymbol{y} - U\boldsymbol{e} - B\boldsymbol{a})^\top(\boldsymbol{y} - U\boldsymbol{e} - B\boldsymbol{a}) + \frac{1}{\sigma_E^2}\sum_{i=1}^{m}\frac{e_i^2}{d_i^2}, \tag{19}$$

a form of ridge regression[7] regularisation. The extent that the response $\boldsymbol{y}$ can be explained in terms of the ith eigenvector $\boldsymbol{u}_i$ depends of the size of d_i, relative to σ_M. The smaller d_i is, the more strongly e_i is driven towards zero. A more numerically stable formulation of (19) is given by

$$\min_{\boldsymbol{a},\boldsymbol{e}} \frac{1}{\sigma_M^2}(\boldsymbol{y} - UD\boldsymbol{e} - B\boldsymbol{a})^\top(\boldsymbol{y} - UD\boldsymbol{e} - B\boldsymbol{a}) + \sum_{i=1}^{m} e_i^2. \tag{20}$$

7. Gaussian processes

Gaussian process models are typically used to estimate the value of a response at a particular spatio-temporal location on the basis of measurements of the response at nearby locations under the assumption that the response behaviour exhibits spatio-temporal correlation. Here we consider the simple case of spatial correlation in one dimension involving data (x_i, y_i), $i = 1, \ldots, m$, generated according to

$$\boldsymbol{y} = \sum_{j=1}^{n} a_j f_j(x_i) + e_i + \epsilon_i, \quad \epsilon_i \in \mathrm{N}(0, \sigma_M^2)$$

where effects e_i are correlated according to

$$\mathrm{cov}(e, e') = \sigma_E^2 k(x, x')$$

for some correlation kernel, for example, specified by

$$k(x, x') = \exp\left\{-\frac{1}{\lambda^2}(x - x')^2\right\}. \tag{21}$$

The parameter σ_E governs the likely size of the effect while the parameter λ determines the spatial length scale over which the correlation applies. Given $\boldsymbol{x} = (x_1, \ldots, x_m)^\top$, let V be the variance matrix given by $V(i, q) = k(x_i, x_q)$. If V has eigenvalue decomposition $V = UD^2U^\top$ then we can write the model equivalently as

$$\boldsymbol{y} = UD\boldsymbol{e} + B\boldsymbol{a} + \boldsymbol{\epsilon}, \quad \boldsymbol{e} \in \mathrm{N}(\boldsymbol{0}, I), \quad \boldsymbol{\epsilon} \in \mathrm{N}(\boldsymbol{0}, \sigma_M^2 I), \tag{22}$$

where $B(i, j) = f_j(x_i)$, the matrix of basis functions evaluated at $\boldsymbol{x}$.

The theory above applies directly and the effective degrees of freedom for the model is bounded according to (17). To evaluate these bounds it is necessary to calculate the eigenvalues associated the variance matrix V given by the correlation kernel. For large m, the eigenvalue decomposition will be computationally expensive to carry out and we are interested in determining estimates of the eigenvalues more cheaply. We are also interested in how the eigenvalues and hence the effective number of degrees of freedom depends on the length scale parameter λ.

7.1. *Approximating the eigenvalues of a spatial correlation variance matrix*

The correlation kernel $k(x, x')$ in (21) can be written in terms of a stationary correlation function $\rho(r)$ according to $k(x, x') = \rho(r)$ where $r = |x - x'|$. The Wiener-Khintchine Theorem[1,10] can be used to show that a stationary correlation function can be described in terms of a spectral decomposition involving Fourier functions and a spectral density function, the latter a probability density $f(\xi)$. A proposed function $\rho(r)$ is an admissible correlation function if its associated spectral density $f(\xi)$ is non-negative everywhere.

For some commonly used univariate correlation functions, the associated spectral density functions can be evaluated explicitly. For example, the correlation functions

$$\rho(r) = e^{-\beta r^2}, \quad \rho(r) = e^{-\beta r}, \tag{23}$$

have spectral density functions such that

$$f(\xi) \propto e^{-\xi^2/4\beta}, \quad f(\xi) \propto \frac{1}{(\beta^2 + \xi^2)^{(n+1)/2}}, \quad \xi > 0, \tag{24}$$

respectively. The first of these correlation functions above is associated with the correlation kernel in (21) and the corresponding $f(\xi)$ is a Gaussian density.

Variance matrices associated with $\boldsymbol{x}$ derived from a correlation function ρ can be thought of as finite dimensional approximations of operators operating on function spaces and their eigenvalue decomposition reflects the spectral decomposition of their associated correlation function. In particular, their eigenvectors are similar to Fourier series basis functions and their eigenvalue distributions follows a similar behaviour to those of the correlation function. The eigenvectors represent increasing lack of smoothness (increasing frequency in the Fourier-like basis functions) as i increases. In this case, second penalising term in the objective function in (19) promotes explaining the response in terms of the smooth behaviour and while high frequency variation in the data vector $\boldsymbol{y}$ is explained in terms of the uncorrelated random effects represented by $\boldsymbol{\epsilon}$. The length scale parameter λ in (21) determines the rate at which the eigenvalues decreases to zero as i increases. As λ increases, the quicker the eigenvalues d_i^2 decrease to zero as i increases, resulting in a smoother approximation to the data.

For the correlation kernel in (21) and for approximately equally spaced $x_i - x_{i-1} = \Delta$, we expect that the eigenvalues d_i^2 of the associated variance matrix V follow a Gaussian density as in (24) of the form $d_i^2 \approx \alpha e^{-\beta(i\Delta)^2}$, for some α and β. We know that the eigenvalues of V must sum to the trace of V so that in this case $\sum_{i=1}^{m} d_i^2 = m$. If we also can approximate the first eigenvalue d_1^2 then we have two pieces of information to estimate α and β and hence the complete set of eigenvalues.

In order to estimate the leading eigenvalues of V, we use that fact that the associated eigenvectors are similar to low order Fourier series as functions of x_i and can therefore be approximated by Chebyshev polynomials or other appropriate basis functions. The general theory is as follows.

Suppose an eigenvector $\boldsymbol{u}$ of V can be expressed in terms of p basis functions $\boldsymbol{u} = G\boldsymbol{b}$ where G is the $m \times p$ matrix of basis functions. Then

$$V\boldsymbol{u} = VG\boldsymbol{b} = \lambda G\boldsymbol{b}, \quad G^\top VG\boldsymbol{a} = \lambda G^\top G\boldsymbol{a}.$$

If G has QR decomposition $G = Q_G R_G$ where Q_G is an $m \times p$ orthogonal matrix and R_G is a $p \times p$ upper-triangular matrix, then

$$R_G^\top Q_G^\top V Q_G R_G \boldsymbol{b} = \lambda G_1^\top G_1 \boldsymbol{b},$$

so that

$$Q_G^\top V Q_G \tilde{\boldsymbol{b}} = \lambda \tilde{\boldsymbol{b}}, \quad \tilde{\boldsymbol{b}} = R_G \boldsymbol{b}.$$

Thus, the eigenvalues of V can be approximated from the eigenvalues of the $p \times p$ matrix $Q_G^\top V Q_G$. Furthermore, if $\tilde{\boldsymbol{b}}$ is an eigenvector of $Q_G^\top V Q_G$, then $\boldsymbol{a} = Q_G \tilde{\boldsymbol{b}}$ is approximate eigenvector of V.

7.2. *Gaussian process models using reduced rank approximations*

The fact that the eigenvalues of the variance matrix associated with the kernel (21) die off exponentially means that for many problems the effective of rank of the associated variance matrix V is much less that the number of data points. As above, we can use $p \ll m$ Chebyshev polynomials to approximate the eigenvectors of V. If $Q_G^\top V Q_G = \tilde{U}\tilde{D}^2\tilde{U}^\top$, then

$V \approx Q_G \tilde{U} \tilde{D}^2 \tilde{U}^\top Q_G^\top$ and we can model[3] the data according to

$$\boldsymbol{y} = Q_Q \tilde{U} \tilde{D} \tilde{\boldsymbol{e}} + B\boldsymbol{a} + \boldsymbol{\epsilon}, \quad \tilde{\boldsymbol{e}} \in \mathrm{N}(\boldsymbol{0}, I), \quad \boldsymbol{\epsilon} \in \mathrm{N}(\boldsymbol{0}, \sigma_M^2 I),$$

where now $\tilde{\boldsymbol{e}}$ is a p-vector. With this approximating model the required calculations $O(m(p+n)^2)$ as opposed to $O(m^3)$ for the full model.

8. Numerical example: polynomial model associated with platinum resistance thermometers

In this example, we look at data generated according to a model

$$\boldsymbol{y} = B\boldsymbol{a} + \boldsymbol{\epsilon}, \quad \boldsymbol{\epsilon} \in \mathrm{N}(\boldsymbol{0}, \sigma_M^2)$$

where $B\boldsymbol{a}$ represents an order 10 (degree 9) polynomial. The polynomial arises as a reference function $W(T)$ that describes the resistance ratio[12] of a standard platinum resistance thermometer (PRT) as a function of temperature T in the range 0 °C to 961.78 °C, the later temperature the freezing point of silver. The polynomial $W(x)$ is graphed in figure 1 as a function of the transformed variable

$$x = x(T) = (T/K - 481 - k_0)/481, \quad k_0 = 273.15,$$

in the range corresponding to 0 °C to 660 °C, the latter the approximate freezing point of aluminium, in steps of 1 °C. The polynomial can be written as an order 10 Chebyshev polynomial with coefficients $a_1 = 4.4390$, $a_2 = 1.1876$, $a_3 = -0.0319$ and $(a_4, \ldots, a_{10})^\top = 10^{-5} \times (-5.9531, -10.5038, -0.4381, 0.7371, 0.1364, -0.0672, 0.0060)^\top$. Data vector $\boldsymbol{y}$ was generated according to the model

$$y_i = W(x_i) + \epsilon_i, \quad \epsilon_i \in \mathrm{N}(0, \sigma_M^2),$$

with $\sigma_M = 10^{-5}$. The graph also shows $W(x) - W_Q(x)$ where $W_Q(x)$ is a quadratic approximation to $W(x)$, along with $y_i - W_Q(x_i)$.

We have fitted the following four models to the data, described below.

Least squares (LS) using an order 10 Chebyshev polynomial. This model is the model we would use if we knew that the data arose from $W(x)$ and represents the maximum likelihood fit to the data, the best that can be achieved.

Tikhonov regularisation, (TR) using an order 50 polynomial expressed in terms of power series basis functions $f_j(x) = x^{j-1}$. As is well known, the use of power series basis can lead to severe problems with ill-conditioned matrices and loss of numerical accuracy. For this problem, it leads to a observation matrix that is numerically rank deficient with numerical rank 29 (instead of 50). In order to proceed, a Tikhonov regularisation was implemented, section 3, with κ chosen so that the effective number of degrees of freedom associated with the model was approximately 10, the order associated with $W(x)$.

Model average (MA) constructed from Chebyshev polynomial models of up to order 50, section 5. (Fitting large degree Chebyshev polynomials to data can be implemented without any numerical stability difficulties.) The weighting of the model predictions were based on the Bayes Information Criteria, section 5.3. The effective number of degrees of freedom associated with the average model was $n_d(MA) = 7.1$.

Gaussian process (GP) model using correlation kernel (21), section 7, along with an order three Chebyshev polynomial. The model involves three statistical parameters, σ_E, σ_M and length scale parameter λ. Point estimates of these parameters where found by maximising the marginalised posterior distribution $p(\sigma_E^2, \sigma_M^2, \lambda|\boldsymbol{y})$ to determine $\hat{\sigma}_E^2$, $\hat{\sigma}_M^2$ and $\hat{\lambda}$. The fitted model was then calculated using these estimates. The posterior distribution $p(\lambda|\hat{\sigma}_E^2, \hat{\sigma}_M^2, \boldsymbol{y})$ for λ given $\boldsymbol{y}$ and point estimates $\hat{\sigma}_E^2$ and $\hat{\sigma}_M^2$ is graphed in figure 2 along with the effective number of degrees of freedom $n_d(\lambda)$ as a function of λ, calculated using the eigenvalues of $V(\lambda)$. The effective number of degrees of freedom associated with the model specified by $\hat{\lambda}$ was $n_d(GP) = 8.0$. The bounds for $n_d(GP)$ were calculated from (17), firstly, using the eigenvalues for $V(\hat{\lambda})$ yielding $7.8 \leq n_d(GP) \leq 10.8$, and, secondly, using the approximate eigenvalues of $V(\hat{\lambda})$ determined using the strategy described in section 7.1, yielding $7.1 \leq n_d \leq 10.1$. That $n_d(GP)$ is nearer the lower bound reflects that fact the first three eigenvectors of $V(\hat{\lambda})$ can be approximated to some extent by order three polynomials. Figure 3 graphs the first five eigenvectors associated with the matrix $V(\hat{\lambda})$.

Figure 4 graphs $\hat{y}_i - W(x_i)$ where $\hat{y}_i$ is the model fit to data y_i according to four models above. The graph reflects for the most part good fits of the models to the data, bearing in mind that the random effects in the data were drawn from a Gaussian with standard deviation $\sigma_M = 10^{-5}$. Only the fit associated with the Tikhonov regularisation (TR) is poor. Although the effective number of degrees of freedom associated with the model is 10, the regularisation term is skewing the fit away from the data. Reducing the weighting parameter κ improves the fit but reducing it by too much leads to numerical difficulties. The model averaging and Gaussian process models determine estimates of the degrees of freedom that are consistent with the actual degrees of freedom associated with the model.

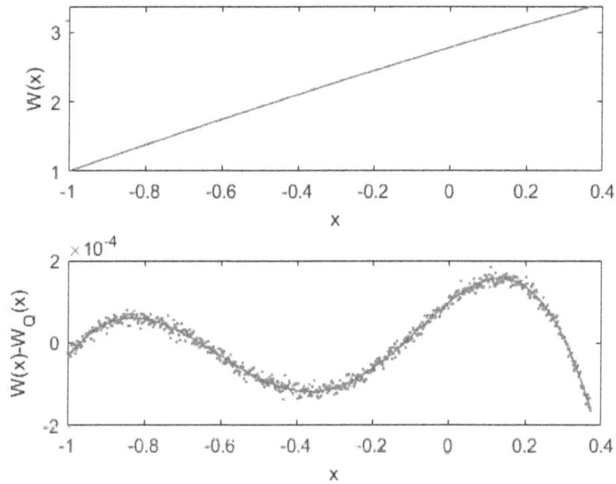

Fig. 1. Upper graph: an order 10 polynomial describing a reference resistance ratio $W(x)$ as a function of a transformed variable x. Lower graph: $W(x) - W_Q(x)$ where $W_Q(x)$ is a quadratic approximation to $W(x)$, along with $y_i - W_Q(x_i)$ where y_i was generated according to $y_i = W(x_i) + \epsilon_i$, $\epsilon_i \in \mathrm{N}(0, \sigma_M^2)$.

9. Concluding remarks

This paper has been concerned with calculating the effective number of degrees of freedom associated with a model and we have shown how it can be calculated for a range of models that are commonly used in metrology and more widely in science and engineering. The estimate of the effective degrees of freedom associated with a model can used to i) determine more efficient approximation schemes using models better tuned to the underlying physical system and ii) provide a guide as to how many measurements are required to characterise the behaviour of a system.

Acknowledgements

This work was supported by the UK's National Measurement System programme for Data Science and the European Metrology Programme for Inno-

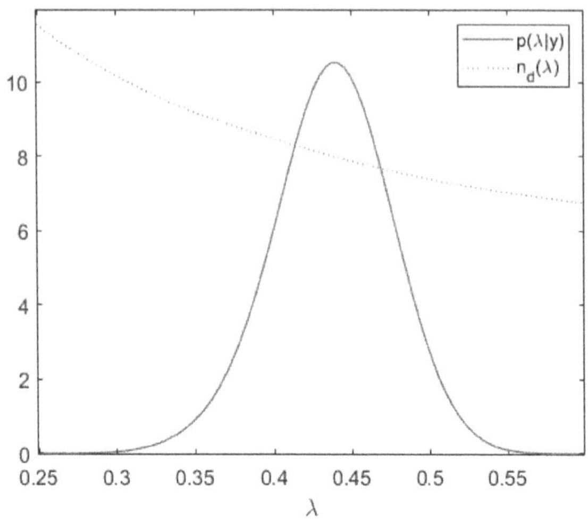

Fig. 2. The posterior distribution $p(\lambda|\hat{\sigma}_E^2, \hat{\sigma}_M^2, \boldsymbol{y})$ for λ given $\boldsymbol{y}$ and point estimates $\hat{\sigma}_E^2$, $\hat{\sigma}_M^2$ along with the effective number of degrees of freedom $n_d(\lambda)$ as a function of λ, calculated using the eigenvalues of $V(\lambda)$.

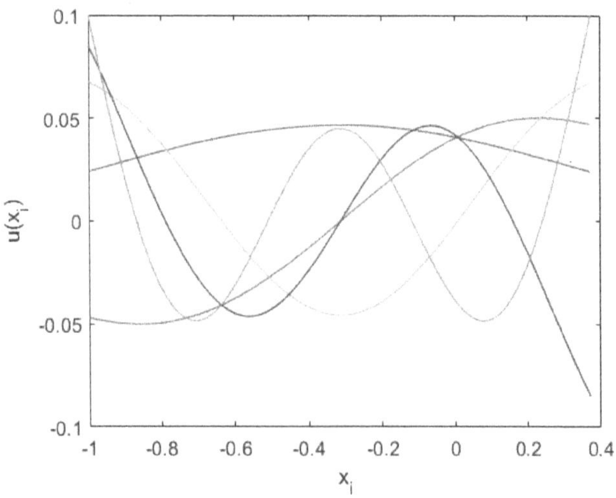

Fig. 3. The first five eigenvectors associated with the spatial correlation matrix $V(\hat{\lambda})$.

vation and Research (EMPIR) project 17IND12, Metrology for the Factory of the Future. The EMPIR programme is co-financed by the Participating

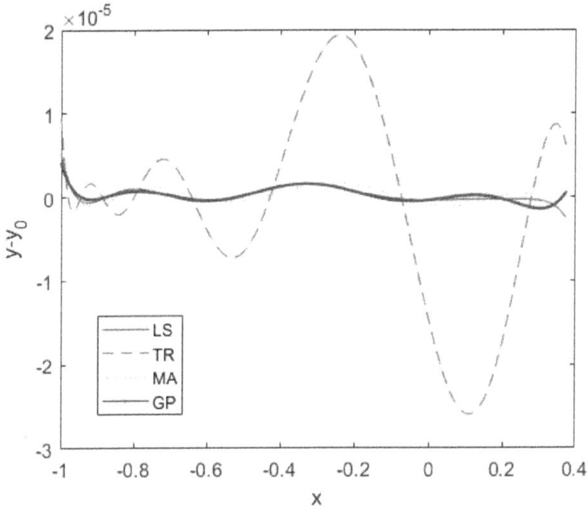

Fig. 4. The graphs of $\hat{y}_i - W(x_i)$ where $\hat{y}_i$ is the model fit to data y_i according to four models.

States and from the European Unions Horizon 2020 research and innovation programme. I thank my colleague Kavya Jagan for comments on an earlier draft of this paper.

References

1. P. Abrahamsen. *A Review of Gaussian Random Fields and Correlation Functions.* Norsk Regnesentral/Norwegian Computing Centre, 1997.
2. H. Akaike. A new look at the statistical model identification. *IEEE Transactions on Automatic Control*, 19:716–723, 1974.
3. A. B. Forbes. Empirical functions with pre-assigned correlation behaviour. In F. Pavese, W. Bremser, A. Chunovkina, N. Fischer, and A. B. Forbes, editors, *Advanced Mathematical and Computational Tools for Metrology X*, pages 17–28, Singapore, 2015. World Scientific.
4. G. E. Forsythe. Generation and use of orthogonal polynomials for data fitting with a digital computer. *SIAM Journal*, 5:74–88, 1957
5. G.H. Golub and C.F. Van Loan. *Matrix Computations.* Johns Hopkins University Press, Baltimore, fourth edition, 2013.
6. I. Guttman, D. Pena, and D. Redondas. A Bayesian approach for predicting with polynomial regression of unknown degree. *Technometrics*, 47(1), 2005.
7. T. Hastie, R. Tibshirani, and J. Friedman. *Elements of Statistical Learning.* Springer, New York, 2nd edition, 2011.
8. M. C. Kennedy and A. O'Hagan. Bayesian calibration of computer models. *J. Roy. Sat. Soc. B*, 64(3):425–464, 2001.

9. A. E. Raftery, D. Madigan, and J. A. Hoeting. Bayesian model averaging for linear regression. *Journal of the American Statistical Association*, 92:179–191, 1997.
10. C. E. Rasmussen and C. K. I. Williams. *Gaussian Processes for Machine Learning*. MIT Press, Cambridge, Mass., 2006.
11. G. Schwarz. Estimating the dimension of a model. *Annals of Statistics*, 6:461–464, 1978.
12. G. F. Strouse. Standard platinum resistance thermometer calibrations from the Ar TP to the Ag FP. Technical report, National Institute of Standards and Technology, Gaitersburg, January 2008.
13. X.-S. Yang and A. B. Forbes. Model and feature selection in metrology data approximation. In E. H. Georgoulis, A. Iske, and J. Levesley, editors, *Approximation Algorithms for Complex Systems*, pages 293–307, Heidelberg, 2010. Springer-Verlag.

F Pavese, A B Forbes, N F Zhang, A G Chunovkina (Eds):
Advanced Mathematical and Computational Tools in Metrology and Testing XII
Series on Advances in Mathematics for Applied Sciences, Vol. 90

https://doi.org/10.1142/9789811242380_0003

Bayesian signal analysis and parameter estimation – A tutorial review

G. A. Kyriazis
Instituto Nacional de Metrologia, Qualidade e Tecnologia (Inmetro),
Duque de Caxias, RJ 25250-020, Brazil
E-mail: gakyriazis@inmetro.gov.br
www.inmetro.gov.br

Bayesian parameter estimation techniques have been successfully applied to the analysis of acoustic and oceanographic chirped signals and of decaying sinusoidal signals in chemical material analysis employing nuclear magnetic resonance. It has been also applied to astronomy, radar, telecommunication, and audio testing and measurement. The method has been additionally applied to waveform metrology. The main objective of this tutorial review is to explain the essentials of the method so that it will be accessible to scientists and engineers involved in measurement and interested in disseminating and advancing best practices for measurement.

Keywords: Bayesian methods; Signal analysis; Tutorial review.

1. Introduction

Bayesian parameter estimation techniques have been successfully applied to the analysis of acoustic and oceanographic chirped signals [1] and of decaying sinusoidal signals in chemical material analysis employing nuclear magnetic resonance [2]-[6]. It has been also applied to astronomy [7], radar [8], telecommunication [9], and audio testing and measurement [10].

The method has been also applied to waveform metrology [11][12]. Here the objective is the analysis of deterministic signals contaminated with additive noise that are generated and measured for instrument calibration purposes. There is interest in estimating the signal parameters from the data sampled and comparing the estimates with the readings (settings) of the meters (sources) under calibration. The parameters are assumed not to vary over the interval of observation. This is reasonable for the stable signal generators used by metrologists.

This problem of inference has many applications in the calibration or performance evaluation of data converters, waveform recorders, phase angle meters, ac voltmeters, ac voltage calibrators, power and energy meters, power calibrators, signal generators, wideband signal analyzers, phasor measurement units, and power quality measuring instruments.

The main objective here is to explain the essentials of the work in [1][2][11][12] so that it will be accessible to scientists and engineers involved in measurement and interested in disseminating and advancing best practices for measurement. The typical reader will have a college degree in one of the sciences, or engineering, and will have had at least one introductory, undergraduate course on calculus, probability theory and matrix algebra.

After briefly reviewing Bayesian inference in section 2 and presenting the model in section 3, we start in section 4 with perhaps the simplest application: a dc signal. We study the common problem of fitting an unknown constant to a time series created by the readings of a digitizer used to sample the output of a dc voltage source. The conditions for regarding the average of the sampled data as the estimate of that constant are presented.

The discussion evolves in section 5 to a less simple signal: a sinusoidal signal with frequency known. We study the common problem of fitting a sinusoid to a time series created by the readings of a digitizer used to sample the output of a sinusoidal generator. The treatment is restricted in this introductory level to data uniform and coherently sampled so that the reader can easily compare the analysis with that of the previous section. The conditions for regarding the one-sided numerical Fourier transformation of the data evaluated solely at the known frequency as the estimate of the fitting parameters of the sinusoid are presented.

The problem is generalized in section 6 to arbitrary periodic signals with frequency known. The conditions for regarding the estimates computed from linear least-squares fitting of the data as the estimates of the harmonic amplitudes are presented.

The case where the frequency is unknown is discussed in section 7. It is shown that only the highest peak of the Lomb-Scargle periodogram is relevant for estimating the frequency of a sinusoidal signal, and that the uncertainty associated with the estimate is proportional to the noise standard deviation and inversely proportional to the square root of the curvature of the periodogram at its highest peak. The conditions for regarding the highest peak of the Lomb-Scargle periodogram as the estimate of the sinusoidal frequency are presented.

When the data show evidence for multiple nonharmonically related frequencies or complex behavior, the Lomb-Scargle periodogram may give misleading or incorrect results in light of more realistic models. In section 8, we finally apply the rules of probability theory to estimate the parameters of complex generalized signals.

The mathematical level increases gradually as the reader advances with the text. All integrations can be easily accessed from a table of integrals [13]. Matrix algebra was used in sections 6 and 8 to simplify the equations and make them resemble the scalar ones presented in sections 5 and 7, respectively. A good introduction to linear algebra is contained in [14].

Some simple simulation examples related to signal analysis are presented in section 9. The conclusions are drawn in section 10.

2. Bayesian inference

Bayesian statistics only requires the use of the rules of probability theory and interprets probability as providing a numerical description of the state of incomplete knowledge based on any kind of relevant information available.

A model for fitting the data is introduced. Since metrologists have cogent prior information about the signal waveform that is being generated and sampled for calibration purposes, it is here assumed that the model (and its order) is completely known except for the value of its parameters.

A sampling distribution with unknown location and scale parameters is assigned that describes the prior knowledge about the noise. The purpose of assigning a sampling distribution to the noise is to define the range of different possible variations of the data due to noise that one shall allow for in the inference, one of which will actually be realized. As a function of those parameters the sampling distribution is then termed the *likelihood* of the parameters given the data.

All participating parameters in the signal and noise models are regarded as unknown constants whose probability density functions (pdf) encode and convey states of (incomplete) knowledge about them. Metrologists generally know little (or wish to proceed as if they knew little) about the parameters of interest before the experiment. Thus there is a need for explicit rules for selecting *priors* to represent ignorance. Following [15]-[17], noninformative priors are used in this text: a uniform prior is used for unknown location parameters and Jeffreys' prior [16] is used for unknown scale parameters.

Noise and prior information are combined with Bayes' theorem, that is *posterior* $\propto$ *prior* $\times$ *likelihood* (the symbol $\propto$ means 'proportional to'), thus obtaining the joint posterior pdf for all the parameters. This pdf expresses all the information about the parameters that is contained in the prior information and data. If one is interested in a specific parameter, the others being "nuisance parameters", the marginal posterior pdf for this parameter is obtained by integrating the joint posterior pdf over the nuisance parameters. That is the end of the inference problem. The posterior pdf should be preferably plotted as it provides the most complete and detailed information about the unknown parameters.

Nevertheless, almost always the metrologist is interested just in the estimates of the parameter values and the associated uncertainties. This is particularly true when the number of parameters is large so that it is inconvenient to plot the several posterior densities and present them in a calibration certificate.

Assuming a quadratic loss function, the expectation of the posterior pdf (which for symmetric densities coincides with the mode) is taken as the estimate of the parameter. Having made this decision, the square root of the variance, that is the standard deviation, summarizes the typical spread of the pdf around the expectation, and thus is taken to be the standard uncertainty associated with the estimate.

We recommend the reader to consult [16]-[19] for an account of Bayes-Laplace approach to inference. A partial summary of Jaynes′ ideas with applications to metrology is contained in [20]. A good introduction to Bayesian statistics is available in [21].

3. Model

A digital sampling experiment is interpreted here as the act of sampling the voltage (dependent variable) of a signal at given instants of time t_i (independent variable) with a digitizer. Such an experiment is designed by choosing the values of t_i at which the signal is to be sampled. A total of N real samples is taken at time instants t_i, where $i = 0, \ldots, N - 1$. The problem is to fit a signal model $f(t_i)$ to a scatter plot of data $\{(t_0, y(t_0)) \ldots (t_{N-1}, y(t_{N-1}))\}$ in which each t_i is accurately known (assuming that the uncertainty associated with the time quantities due to sampling jitter and systematic time base error can be disregarded) and each digitizer reading $y(t_i)$ is subject to additive noise. Hence, it is assumed that the data can be modeled as

$$y(t_i) = f(t_i) + e(t_i), \tag{1}$$

where $y(t_i)$ is the sample value and $e(t_i)$ is the noise term at the digitizer output both at the time instant t_i. It is also assumed that the several $e(t_i)$ are normal and independently distributed (white noise) [22][23], each with mean zero and unknown common variance σ^2; that is $E[e(t_i)] = 0$, $E[e^2(t_i)] = \sigma^2$, and $E[e(t_i)e(t_j)] = 0$, $i \neq j$, where the operator $E[\cdot]$ denotes mathematical expectation.

Here it is implicitly assumed that: (a) the noise introduced by the sampling process within the digitizer is negligible compared with that present at the signal generator output and (b) the signal is sampled with a high-resolution digitizer so that the effect of quantization [24][25] can be disregarded.

All signal and noise parameters are regarded as unknown constants whose pdfs encode and convey states of (incomplete) knowledge about them. One is interested in estimating such parameters and assessing the uncertainties associated with the estimates. The question of interest here is: after getting the data, how accurately does the *single* data set we actually have determine the value of the parameters? We are not concerned with other data sets that might have been obtained but were not.

4. dc signal

This is perhaps the simplest situation. The signal is assumed not to vary over the interval of observation, that is

$$f(t_i) = C_0, \tag{2}$$

where C_0 is a dc component uncorrected for the systematic effects.

If the signal is known to be true, the difference between the data and the signal is just noise. The expression for the probability of obtaining the data, given the signal and noise models and their associated parameters, when seen solely as a function of such parameters, is the so-called likelihood function:

$$l\big(C_0, \sigma; y(t_0), \cdots, y(t_{N-1})\big) \propto \sigma^{-N} \exp\left\{-\frac{NQ(C_0)}{2\sigma^2}\right\}, \tag{3}$$

where

$$Q(C_0) \equiv \frac{1}{N}\sum_{i=0}^{N-1} y^2(t_i) - \frac{2C_0}{N}\sum_{i=0}^{N-1} y(t_i) + C_0^2. \tag{4}$$

If there is no cogent prior information about the parameters, a noninformative prior is assumed by taking C_0 and $\ln \sigma$ independent and uniformly distributed; that is $p(C_0, \sigma) \propto \sigma^{-1}$, with $-\infty < C_0 < \infty$ and $0 < \sigma < \infty$.

Noise and prior information are combined with Bayes' theorem, thus obtaining the joint posterior pdf for C_0 and σ. On combining (3) and the above prior, completing the square, lumping the terms that do not depend on C_0 and σ into the proportionality constant, the joint posterior pdf for C_0 and σ is

$$p\big(C_0, \sigma | y(t_0), \cdots, y(t_{N-1})\big) \propto \sigma^{-N-1} \exp\left\{-\frac{1}{2\sigma^2}\left[\nu s^2 + N(C_0 - h)^2\right]\right\}, \tag{5}$$

where $h = \Sigma y(t_i)/N$, $\nu s^2 = \Sigma y^2(t_i) - Nh^2$, and $\nu = N - 1$.

Integrating with respect to σ, the marginal posterior pdf for C_0 is

$$p(C_0|y(t_0),\cdots,y(t_{N-1})) \propto [\nu s^2 + N(C_0 - h)^2]^{-N/2}, \tag{6}$$

which is a Student t pdf [17] with mean h and variance $s^2\nu/(\nu-2)N$.

If instead, one integrates the joint posterior with respect to C_0, one obtains the marginal posterior pdf for σ, that is $p(\sigma|y(t_0), \ldots, y(t_{N-1})) \propto \sigma^{-\nu-1}\exp(-\nu s^2/2\sigma^2)$, which is in the form of an inverted gamma pdf with moment of order k given by $\Gamma((\nu-k)/2)\Gamma(\nu/2)^{-1}(\nu s^2/2)^{k/2}$ [2], where $\Gamma(\cdot)$ is the gamma function. One can use these moments to obtain the estimate of the noise variance at the digitizer output, that is $\hat{\sigma}^2 = (s^2\nu/(\nu-2))\,(1 \pm u)$, where $u = (2/(\nu-4))^{1/2}$. The estimation of the noise variance is of great interest. It indicates how good the data appear to be in the light of the model and can help one to decide whether to try a new model.

In the presence of white Gaussian noise, the functions h and νs^2 are sufficient to enable the posterior (6) to be derived without further reference to the data set. The function h summarizes all the information in the data that is relevant for estimating C_0. The estimate of C_0, uncorrected for the systematic effects, thus coincides with the data average.

To conclude, the estimate of C_0, uncorrected for the systematic effects, namely $h \pm s(\nu/(\nu-2)N)^{1/2}$, is optimal if the signal is dc and all the assumptions in section 3 are valid. If any of these conditions is not met, the data average may give misleading or simply incorrect results in light of more realistic models.

5. Sinusoidal signal with frequency known

Now consider a sinusoidal signal with null dc component, that is

$$f(t_i) = A_1\cos(\omega_0 t_i) + B_1\sin(\omega_0 t_i), \tag{7}$$

where A_1 and B_1 are the amplitudes and ω_0 is the (angular) frequency, all uncorrected for the systematic effects. It is here assumed that the frequency ω_0 of the *sampled* signal is known. It is also assumed that the data set has a zero mean value (any nonzero average value has been subtracted from the data).

The likelihood function is

$$l(A_1,B_1,\sigma;y(t_0),\cdots,y(t_{N-1})) \propto \sigma^{-N}\exp\left\{-\frac{NQ(A_1,B_1)}{2\sigma^2}\right\}, \tag{8}$$

where

$$Q(A_1,B_1) \equiv \frac{1}{N}\sum_{i=0}^{N-1} y^2(t_i) - \frac{2}{N}(A_1 g_1 + B_1 g_2) + \frac{1}{N}(A_1^2 a + B_1^2 b + 2A_1B_1c), \tag{9}$$

$g_1 = \Sigma\, y(t_i)\cos(\omega_0 t_i)$, $g_2 = \Sigma\, y(t_i)\sin(\omega_0 t_i)$, $a = \Sigma \cos^2(\omega_0 t_i)$, $b = \Sigma \sin^2(\omega_0 t_i)$ and $c = \Sigma \cos(\omega_0 t_i)\sin(\omega_0 t_i)$.

The quadratic term in the likelihood does not simplify. Although any multivariate Gaussian integral can be done (see next section), the key to being able to remove nuisance parameters easily, and above all, selectively, is that the likelihood factors into independent parts. This is exactly what happens when one adopts an orthogonal design for the experiment [26][27]. We describe one such design in the sequel.

Uniform sampling is typically employed in practice: a total of N samples is taken at time instants $t_i = it_{\text{samp}}$, where t_{samp} is the sampling interval. Aliasing is an undesirable consequence of uniform sampling and to avoid it t_{samp} is selected so that the sampling theorem [28][29] is satisfied, that is $t_{\text{samp}} < \pi/\omega_0$. The following approximations are valid for t_{samp} much smaller than the signal period and for large N [26]: $a \approx (N/2)(1 - \delta_1)$, $b \approx (N/2)(1 + \delta_1)$, $c \approx (N/2)\delta_2$, where $\delta_1 = (1/\eta_1)[(\cos\eta_1 - 1)\sin\eta_2 - \sin\eta_1\cos\eta_2]$, $\delta_2 = (1/\eta_1)[(\cos\eta_1 - 1)\cos\eta_1 - \sin\eta_1\sin\eta_2]$, $\eta_1 = 2\omega_0 N t_{\text{samp}}$ and $\eta_2 = 2\omega_0 t_0$.

Sampling is coherent if Nt_{samp} covers an integer number of signal periods, or in other words, if the sampling rate is an integer multiple of the signal frequency. It is implemented in practice with synchronous synthesizing and sampling systems [30]-[34]. In this case, $\omega_0 t_{\text{samp}} = 2\pi/N_T$, where N_T is the number of samples per period ($N_T > 2$ to avoid aliasing). If M_T is the number of periods in an acquisition, that is $N = N_T M_T$, one can easily see from the above approximations that $a = N/2$, $b = N/2$ and $c = 0$, so that (9) simplifies to

$$Q(A_1, B_1) \equiv \frac{1}{N}\sum_{i=0}^{N-1} y^2(t_i) - \frac{2}{N}(A_1 g_1 + B_1 g_2) + \frac{1}{2}(A_1^2 + B_1^2), \qquad (10)$$

where $g_1 = \Sigma\, y(it_{\text{samp}})\cos(2\pi i/N_T)$ and $g_2 = \Sigma\, y(it_{\text{samp}})\sin(2\pi i/N_T)$. Note that ω_0 in model (7) has dropped out from consideration in (10)[a].

A noninformative prior is assumed by taking A_1, B_1 and ln σ independent and uniformly distributed; that is $p(A_1, B_1, \sigma) \propto \sigma^{-1}$, with $-\infty < A_1 < \infty$, $-\infty < B_1 < \infty$ and $0 < \sigma < \infty$.

On combining (8), (10) and the above prior, completing the square, lumping the terms that do not depend on A_1, B_1 and σ into the proportionality constant, the joint posterior pdf for A_1, B_1 and σ is

$$p(A_1, B_1, \sigma|\boldsymbol{y}) \propto \sigma^{-N-1}\exp\left\{-\frac{1}{2\sigma^2}\left[\nu s^2 + \frac{N}{2}\left((A_1 - h_1)^2 + (B_1 - h_2)^2\right)\right]\right\}, \quad (11)$$

---

[a] When the digitizer readings are taken with a sampling rate synchronized to the frequency of the signal to be analyzed, the signal frequency is directly computed from the time base, that is $\omega_0 = 2\pi/N_T t_{\text{samp}}$. That is the reason why metrologists do prefer to deploy synchronous synthesizing and sampling systems. By so doing, they are able to simplify the estimation problem and improve the accuracy of the amplitude estimates.

where $\boldsymbol{y} = (y(t_0), \ldots, y(t_{N-1}))$ is the data set, $h_1 = (2/N)\Sigma\, y(it_{\text{samp}})\cos(2\pi i/N_{\text{T}})$, $h_2 = (2/N)\Sigma\, y(it_{\text{samp}})\sin(2\pi i/N_{\text{T}})$, $\nu s^2 = \Sigma\, y^2(it_{\text{samp}}) - N(h_1{}^2 + h_2{}^2)/2$, and $\nu = N - 2$.

Integrating the joint posterior pdf with respect to σ, the marginal posterior pdf for A_1 and B_1 is

$$p(A_1, B_1 | y(t_0), \cdots, y(t_{N-1})) \propto \left[\nu s^2 + \frac{N}{2}((A_1 - h_1)^2 + (B_1 - h_2)^2)\right]^{-N/2} \quad (12)$$

This is a bivariate Student t pdf [17] with mean vector (h_1, h_2) and diagonal covariance matrix with elements $2s^2\nu/(\nu - 2)N$.

If instead, one integrates the joint posterior with respect to A_1 and B_1, one obtains the marginal posterior pdf for σ, that is $p(\sigma\, |y(t_0), \ldots, y(t_{N-1})) \propto \sigma^{-\nu-1}\exp(-\nu s^2/2\sigma^2)$, which is in the form of an inverted gamma pdf. One can use the moments of this pdf (see related discussion in the previous section) to obtain the estimate of the noise variance at the digitizer output, that is $\hat{\sigma}^2 = (s^2\nu/(\nu - 2))\,(1 \pm u)$, where $u = (2/(\nu - 4))^{1/2}$.

In the presence of white Gaussian noise, the functions h_1, h_2 and νs^2 are sufficient to enable the posterior (12) to be derived without further reference to the data set. The functions h_1 and h_2 summarize all the information in the data that is relevant for estimating A_1 and B_1, respectively. The estimates of the fitting parameters of a sinusoidal signal without dc component, uncorrected for the systematic effects, thus coincide with the one-sided numerical Fourier transformation [35] of the data evaluated solely at the known frequency $2\pi/N_{\text{T}}$ rad. Probability theory is telling one to project all the N samples onto the sine and cosine components with frequency $2\pi/N_{\text{T}}$ rad. Note that this contrasts with the discrete Fourier transform (DFT) method [35]-[37], in which the N samples are mapped into N discrete frequencies $2\pi k/N$ rad, where $k = 0, \ldots, N - 1$ with N generally a power of two. Probability theory is telling one that there is no information in the data at those discrete frequencies defined by the DFT resolution except at those frequencies for which $k = M_{\text{T}}$ and $k = N - M_{\text{T}}$. Though one may still use the fast Fourier transform (FFT) algorithm [35]-[37] to save computation time, the functions h_1 and h_2 provide more accurate estimates for the amplitudes A_1 and B_1, respectively, since all the N samples are being used to compute the estimates at just one frequency.

To conclude, the estimates of the amplitudes A_1 and B_1, uncorrected for the systematic effects, namely $h_1 \pm s(2\nu/(\nu - 2)N)^{1/2}$ and $h_2 \pm s(2\nu/(\nu - 2)N)^{1/2}$, respectively, are optimal if the following conditions are met: (a) the signal contains a single frequency, (b) there is no dc component in the data, (c) the sampling is uniform and there is no aliasing, (d) the sampling is coherent, and (e) all the assumptions in section 3 are valid. If any of these conditions is not met, the functions h_1 and h_2 may give misleading or simply incorrect results in light of more realistic models.

The amplitudes are often reported in polar coordinates, namely the magnitude $a = (A_1^2 + B_1^2)^{1/2}$ and the phase angle $\theta = \text{atan2}\ (-B_1, A_1)$, where atan2 is the four quadrant inverse tangent function. It is worth evaluating the asymptotic behaviour of the posterior pdfs for a and θ when the sample size is large and the pdfs for A_1 and B_1 are independent (which happens to be the case for uniform and coherent sampling). The posterior pdf for a is then a Rician or noncentral Rayleigh [38][39]. For high signal-to-noise ratios (SNR), this pdf becomes approximately Gaussian with variance $2s^2/N$ in the region around its mean value $(h_1^2 + h_2^2)^{1/2}$. The posterior pdf for θ also becomes approximately Gaussian with variance $2s^2/N(h_1^2 + h_2^2)$ for high SNR and small phase displacements relative to the mean atan2 $(-h_2, h_1)$ [38][39]. Thus, the smaller the estimate of the magnitude is relative to the noise, the larger is the uncertainty associated with the estimate of the corresponding phase angle.

Therefore, in addition to the aforementioned conclusion, the estimates in polar coordinates, uncorrected for the systematic effects, namely $(h_1^2 + h_2^2)^{1/2} \pm s(2/N)^{1/2}$ for the magnitude and atan2 $(-h_2, h_1) \pm s(2/N\,(h_1^2 + h_2^2))^{1/2}$ for the phase angle, are optimal if the following additional conditions are met: (f) the data set is large and (g) the SNR is high.

6. Arbitrary periodic signal with frequency known

In this section we increase the number of harmonic functions in the model:

$$f(t_i) = C_0 + \sum_{j=1}^{m}\{A_j \cos(j\omega_0 t_i) + B_j \sin(j\omega_0 t_i)\}, \tag{13}$$

where C_0 is the dc component, A_j and B_j are the j-th amplitudes, $j\omega_0$ is the j-th harmonic of the known fundamental frequency ω_0 of the *sampled* signal, and m is the specified number of harmonic functions of the truncated trigonometric Fourier series. All parameters are uncorrected for the systematic effects.

Model (1) can be written in matrix notation as

$$\boldsymbol{y} = \boldsymbol{D}\boldsymbol{x} + \boldsymbol{e} \tag{14}$$

where $\boldsymbol{y} = (y(t_0), \ldots, y(t_{N-1}))'$ is here the data vector, $\boldsymbol{D}$ is the known $N \times (2m + 1)$ matrix with $(i, 0)$ element one and (i, l)-th element $\cos(l\omega_0 t_i)$ for $l = 1, \ldots, m$ and $\sin((l - m)\omega_0 t_i)$ for $l = m + 1, \ldots, 2m$, $\boldsymbol{x} = (C_0, A_1, \ldots, A_m, B_1, \ldots, B_m)'$ is the vector of fitting parameters, $\boldsymbol{e} = (e(t_0), \ldots, e(t_{N-1}))'$ is the vector of the noise terms at the digitizer output, and the prime (′) denotes the transpose. Here we consider that a single data set is available[b]. It is assumed that the elements of $\boldsymbol{e}$ are normal and independently distributed, each with mean zero and unknown common variance σ^2; that is, $E[\boldsymbol{e}] = \mathbf{0}$ and $E[\boldsymbol{ee}'] = \sigma^2\boldsymbol{I}_N$, where $\boldsymbol{I}_N$ is the identity matrix of order N.

The likelihood function is

$$l(\boldsymbol{x}, \sigma; \boldsymbol{y}) \propto \sigma^{-N}\exp\left\{-\frac{1}{2\sigma^2}(\boldsymbol{y} - \boldsymbol{Dx})'(\boldsymbol{y} - \boldsymbol{Dx})\right\}$$

$$\propto \sigma^{-N}\exp\left\{-\frac{1}{2\sigma^2}[\nu s^2 + (\boldsymbol{x} - \hat{\boldsymbol{x}})'\boldsymbol{D}'\boldsymbol{D}(\boldsymbol{x} - \hat{\boldsymbol{x}})]\right\}, \quad (15)$$

where $\nu = N - 2m - 1$ and

$$\hat{\boldsymbol{x}} = (\boldsymbol{D}'\boldsymbol{D})^{-1}\boldsymbol{D}'\boldsymbol{y}, \quad (16)$$

$$\nu s^2 = \boldsymbol{y}'\boldsymbol{y} - \boldsymbol{y}'\boldsymbol{D}\hat{\boldsymbol{x}}, \quad (17)$$

are sufficient statistics.

A noninformative prior is assumed in the analysis of the multiple regression model by taking the elements of $\boldsymbol{x}$ and $\ln \sigma$ independent and uniformly distributed; that is $p(\boldsymbol{x}, \sigma) \propto \sigma^{-1}$, with $-\infty < (\boldsymbol{x})_l < \infty$ and $0 < \sigma < \infty$, where the symbol $(\cdot)_l$ denotes the l-th element, $l = 0, \ldots, 2m$.

On combining (15) and the above prior, the joint posterior pdf for the parameters $\boldsymbol{x}$ and σ is

$$p(\boldsymbol{x}, \sigma|\boldsymbol{y}) \propto \sigma^{-N-1}\exp\left\{-\frac{1}{2\sigma^2}[\nu s^2 + (\boldsymbol{x} - \hat{\boldsymbol{x}})'\boldsymbol{D}'\boldsymbol{D}(\boldsymbol{x} - \hat{\boldsymbol{x}})]\right\}, \quad (18)$$

which can be factored into the product of two terms: the conditional posterior pdf for $\boldsymbol{x}$, given σ [that is $p(\boldsymbol{x} \mid \sigma, \boldsymbol{y})$], namely a multivariate normal pdf, and the marginal posterior pdf for σ, that is $p(\sigma \mid \boldsymbol{y}) \propto \sigma^{-\nu-1}\exp(-\nu s^2/2\sigma^2)$, which is in the form of an inverted gamma pdf. The latter can also be obtained upon integrating (18) over all elements of $\boldsymbol{x}$.

Integrating (18) with respect to σ, the marginal posterior pdf for the elements of $\boldsymbol{x}$ is

$$p(\boldsymbol{x}|\boldsymbol{y}) \propto [\nu s^2 + (\boldsymbol{x} - \hat{\boldsymbol{x}})'\boldsymbol{D}'\boldsymbol{D}(\boldsymbol{x} - \hat{\boldsymbol{x}})]^{-N/2}. \quad (19)$$

This is a multivariate Student t pdf [17] with ν degrees of freedom with mean vector $\hat{\boldsymbol{x}}$ and covariance matrix $s^2(\boldsymbol{D}'\boldsymbol{D})^{-1}\nu/(\nu - 2)$.

In the presence of white Gaussian noise, the functions $\hat{\boldsymbol{x}}$ and νs^2 are sufficient to enable the posterior (19) to be derived without further reference to

[b] The interested reader may consult [26][27] for a generalization of this problem to several data sets and for an analysis of the general case where the vector of fitting parameters is allowed to differ for each of those data sets.

the data set. The function $\hat{\boldsymbol{x}}$ summarizes all the information in the data that is relevant for estimating the elements of $\boldsymbol{x}$. It follows that the estimate of the vector of fitting parameters, uncorrected for the systematic effects, is $\hat{\boldsymbol{x}}$ with associated covariance matrix $s^2(\boldsymbol{D}'\boldsymbol{D})^{-1}\nu/(\nu-2)$. From the moments of the inverted gamma pdf (see section 4), it follows that the estimate of the noise variance at the digitizer output is $\hat{\sigma}^2 = (s^2\nu/(\nu-2))\,(1 \pm u)$, where $u = (2/(\nu-4))^{1/2}$.

Here we digress to remind the reader that we preferred in the previous section to adopt at the outset an orthogonal design to simplify calculations. Now we present the estimates for the general case where no orthogonal design is deployed: if $C_0 = 0$ and $m = 1$, then model (13) reduces to that in (7) and the estimates of A_1 and B_1 in (6) are $(bg_1 - cg_2)/(ab - c^2) \pm s(b\nu/(\nu-2)(ab-c^2))^{1/2}$ and $(ag_2 - cg_1)/(ab - c^2) \pm s(a\nu/(\nu-2)(ab-c^2))^{1/2}$, respectively, where $\nu s^2 = \Sigma y^2(t_i) - (bg_1^2 + ag_2^2 - 2cg_1g_2)/(ab - c^2)$, $g_1 = \Sigma\, y(t_i)\cos(\omega_0 t_i)$, $g_2 = \Sigma\, y(t_i)\sin(\omega_0 t_i)$, $a = \Sigma\cos^2(\omega_0 t_i)$, $b = \Sigma\sin^2(\omega_0 t_i)$, $c = \Sigma\cos(\omega_0 t_i)\sin(\omega_0 t_i)$, and $\nu = N-2$. The estimate of the noise variance at the digitizer output is $\hat{\sigma}^2 = (s^2\nu/(\nu-2))\,(1 \pm u)$, where $u = (2/(\nu-4))^{1/2}$. The reader may easily verify that all such estimates reduce to those derived in the previous section when the sampling is uniform and coherent.

The least-squares estimate of the vector of fitting parameters, uncorrected for the systematic effects, namely $\hat{\boldsymbol{x}}$ with associated covariance matrix $s^2(\boldsymbol{D}'\boldsymbol{D})^{-1}\nu/(\nu-2)$, is optimal if the following conditions are met: (a) the signal is periodic, (b) the signal frequency is known, (c) the signal is band-limited, and (d) all the assumptions in section 3 are valid. If any of these conditions is not met, the function $\hat{\boldsymbol{x}}$ may give misleading or simply incorrect results in light of more realistic models.

However, this estimate presents the following disadvantages: the matrix operations take too much computer time and memory, the elements of the vector of fitting parameters are generally correlated, and the estimates of fitting parameters are rather sensitive to the frequency accuracy.

An experiment design that yields estimates not presenting these disadvantages was explored in the previous section. It is to take the digitizer readings with a sampling rate synchronized to the frequency of the signal to be analyzed. But before studying its consequences, let us add a few remarks.

It is interesting to note that all the above analysis in this section is valid whether the sampling intervals are uniformly or nonuniformly spaced, and whether the sampling is coherent or not. Missing observations make no difference whatever to probability theory. Probability theory analyzes the data you actually have, regardless of whether the data are uniformly sampled or not.

Indeed, in Bayesian probability theory there is no such thing as a missing data problem [40]. In addition, all the above analysis is valid whether the signal has a dc component or not. If one does probability theory calculations using an additive constant in the model, probability theory will never lead one to subtract the average from the data as has been done in the last section (and as will be done in the next paragraph, and also in the next section). Rather, other statistics will appear that account for the presence of the constant in a consistent manner [40].

Let us now check what happens when a periodic signal without dc component is uniform and coherently sampled. It is here assumed that the data set has a zero mean value (any nonzero average value has been subtracted from the data). In this case, $\omega_0 t_{\text{samp}} = 2\pi/N_{\text{T}}$, where N_{T} is the number of samples per period ($N_{\text{T}} > 2m$ to avoid aliasing), $\boldsymbol{D}$ is the known $N \times 2m$ matrix with (i, l)-th element $\cos(2\pi i(l + 1)/N_{\text{T}})$ for $l = 0, \ldots, m - 1$ and $\sin(2\pi i(l - m + 1)/N_{\text{T}})$ for $l = m, \ldots, 2m - 1$, and $\boldsymbol{x} = (A_1, \ldots, A_m, B_1, \ldots, B_m)'$ is the vector of fitting parameters uncorrected for the systematic effects. Note that ω_0 has dropped out from consideration in models (13) and (14)[c]. If M_{T} is the number of periods in an acquisition, that is $N = N_{\text{T}}M_{\text{T}}$, one can easily show that the columns of $\boldsymbol{D}$ are now orthogonal and that their Euclidean norms are equal to $(N/2)^{1/2}$. Therefore, the design matrix $\boldsymbol{D}'\boldsymbol{D}$ is diagonal with all elements equal to $N/2$. The marginal posterior pdf for the elements of $\boldsymbol{x}$ then becomes

$$p(\boldsymbol{x}|\boldsymbol{y}) \propto \left[\nu s^2 + \frac{N}{2}\sum_{l=0}^{2m-1}((\boldsymbol{x})_l - (\boldsymbol{h})_l)^2\right]^{-N/2}, \tag{20}$$

where the symbol $(\cdot)_l$ denotes the l-th element ($l = 0, \ldots, 2m - 1$), $\boldsymbol{h} = 2\boldsymbol{D}'\boldsymbol{y}/N$ is the projection of the data onto the orthonormal model functions, $\nu s^2 = \boldsymbol{y}'\boldsymbol{y} - (N/2)\boldsymbol{h}'\boldsymbol{h}$, and $\nu = N - 2m$. This is a multivariate Student t pdf with mean vector $\boldsymbol{h}$ and covariance matrix $(2s^2\nu/(\nu - 2)N)\, \boldsymbol{I}_{2m}$, where $\boldsymbol{I}_{2m}$ is the identity matrix of order $2m$. Thus, the elements of $\boldsymbol{x}$ are uncorrelated. For $m = 1$, this expression reduces to that in (12) as should be expected.

In the presence of white Gaussian noise, the functions $\boldsymbol{h}$ and νs^2 are sufficient to enable the posterior (20) to be derived without further reference to the data set. The function $\boldsymbol{h}$ summarizes all the information in the data that is relevant for estimating the elements of $\boldsymbol{x}$. The estimates of the harmonic amplitudes of a periodic signal without dc component, uncorrected for the systematic effects, thus coincide with the one-sided numerical Fourier

---

[c] As mentioned in the previous section, when the digitizer readings are taken with a sampling rate synchronized to the frequency of the signal to be analyzed, the signal frequency is directly computed from the time base, that is $\omega_0 = 2\pi/N_{\text{T}}t_{\text{samp}}$.

transformation [35] of the data evaluated solely at the known frequencies $2\pi j/N_T$ rad, where j = 1, …, m. Probability theory is telling one to project all the N samples onto the sine and cosine components with frequencies $2\pi j/N_T$ rad, where j = 1, …, m. Note that this contrasts with the DFT method [35]-[37], in which the N samples are mapped into N discrete frequencies $2\pi k/N$ rad, where k = 0, …, N – 1 with N generally a power of two. Probability theory is telling one that there is no information in the data at those discrete frequencies defined by the DFT resolution except at those for which $k = jM_T$ and $k = N - jM_T$, where j = 1, …, m. Though one may still use the FFT algorithm [35]-[37] to save computation time, the function $\boldsymbol{h}$ provides more accurate estimates for the harmonic amplitudes of a periodic signal, since all the N samples are being used to compute the estimates at just m frequencies.

The estimate of the l-th harmonic amplitude, uncorrected for the systematic effects, namely $(\boldsymbol{h})_l \pm s(2\nu/(\nu-2)N)^{1/2}$, is optimal if the following conditions are met: (a) the signal is periodic, (b) the signal is band-limited, (c) there is no dc component in the data, (d) the sampling is uniform and there is no aliasing, (e) the sampling is coherent, and (f) all the assumptions in section 3 are valid. If any of these conditions is not met, the function $\boldsymbol{h}$ may give misleading or simply incorrect results in light of more realistic models.

Least squares always provide the "best" estimates by the mean-squared criterion of the harmonic amplitudes of a periodic signal with frequency known. The estimates computed from the one-sided numerical Fourier transformation of the data evaluated at the frequencies $2\pi j/N_T$ rad, where j = 1, …, m, coincide with those from least squares under uniform and coherent sampling, and under the absence of a dc component. The design matrix $\boldsymbol{D'D}$ then becomes orthogonal and the data are expanded on an orthonormal set of vectors. The expansion coefficients are just the projections of the data onto the expansion vectors. Under such conditions, the one-sided numerical Fourier transformation and the one-sided continuous Fourier transform are exactly the same within a scaling constant. Since the one-sided continuous Fourier transform of a band-limited periodic signal is exactly the same as the truncated trigonometric Fourier series within a scaling constant, we conclude that the least squares, the numerical Fourier transformation, the continuous Fourier transform and the Fourier series all coincide within scaling constants under uniform and coherent sampling of a band-limited periodic signal without dc component.

The remaining paragraphs of this section deal with a technical point. The reader may skip them on a first reading as they are not necessary for an understanding of the subsequent sections.

What about a given signal with two or more nonharmonically related frequencies? Is it possible to sample it uniform and coherently? The answer is yes if the signal results to be periodic and it will be so if all its frequencies are commensurable[d]. However, it is not straightforward, for instance, to use the DFT to measure interharmonics whose frequencies are not integer multiples of the fundamental. The frequency resolution with which the interharmonics can be measured depends on the number of signal periods used in the DFT. For example, if 10 signal periods of a 60-Hz signal are analyzed using the DFT, the discrete frequencies will be spaced at 6-Hz (60/10) intervals. To measure noninteger frequencies to a resolution of, for example, 0.01 Hz, 6000 stable signal periods would need to be uniform and coherently sampled, and analyzed using the DFT. This places a severe requirement on the stability of the source and may lead metrologists to doubt whether in fact the signal and noise parameters have been kept constant during the whole sampling.

As a further example, a carrier signal is typically modulated with a signal whose frequency is not commensurable with the carrier frequency. Modulated signals can be uniform and coherently sampled provided the carrier and modulation frequencies are commensurable and the signals are prepared in the time domain prior to locking. Since the DFT peaks of the carrier and modulating signals rarely coincide, measurements over many cycles are needed. For example, by carefully selecting the carrier and modulation frequencies one may ensure synchronization in most cases. Therefore, a 50-Hz signal modulated with 8.8 Hz (for instance, for flicker measurements) can be sampled over 500 periods of 50 Hz (or 88 periods of 8.8 Hz). The DFT resolution will be 0.1 Hz in this case.

Even so, despite all care, interharmonics and/or modulation frequencies may still fall between the discrete frequencies defined by the DFT resolution giving rise to serious measurement errors related to the so-called spectral leakage [37]. Although it is possible to use the DFT in those above cases and in many other similar ones, far fewer cycles can be used for the analysis by deploying the method to be described in section 8 which, incidentally, does not require the signal to be periodic (or its frequencies to be commensurable).

7. Arbitrary periodic signal with frequency unknown

Up to now only signal models that are linear in the unknowns have been discussed. Now consider a sinusoidal signal with null dc component,

---

[d] Two frequencies are commensurable when their ratio is a rational number.

$$f(t_i) = A_1 \cos(\omega t_i) + B_1 \sin(\omega t_i), \tag{21}$$

whose frequency ω is also unknown. All parameters are uncorrected for the systematic effects. It is here assumed that the data set has a zero mean value (any nonzero average value has been subtracted from the data).

The likelihood function is

$$l(A_1, B_1, \omega, \sigma; y(t_0), \cdots, y(t_{N-1})) \propto \sigma^{-N} \exp\left\{-\frac{Q(A_1,B_1,\omega)}{2\sigma^2}\right\}, \tag{22}$$

where

$$Q(A_1, B_1, \omega) \equiv \textstyle\sum_{i=0}^{N-1} y^2(t_i) - 2(A_1 g_1 + B_1 g_2) + A_1^2 a + B_1^2 b + 2A_1 B_1 c, \tag{23}$$

$g_1(\omega) = \Sigma\, y(t_i)\cos(\omega t_i)$, $g_2(\omega) = \Sigma\, y(t_i)\sin(\omega t_i)$, $a(\omega) = \Sigma \cos^2(\omega t_i)$, $b(\omega) = \Sigma \sin^2(\omega t_i)$ and $c(\omega) = \Sigma \cos(\omega t_i)\sin(\omega t_i)$.

A noninformative prior is assumed by taking A_1, B_1, ω and $\ln \sigma$ independent and uniformly distributed; that is $p(\omega, A_1, B_1, \sigma) \propto \sigma^{-1}$, with $-\infty < A_1 < \infty$, $-\infty < B_1 < \infty$, $-\infty < \omega < \infty$ and $0 < \sigma < \infty$.

Upon multiplying (22) by the above prior, the joint posterior pdf for A_1, B_1, ω and σ is

$$p(A_1, B_1, \omega, \sigma | y(t_0), \cdots, y(t_{N-1})) \propto \sigma^{-N-1} \exp\left\{-\frac{Q(A_1,B_1,\omega)}{2\sigma^2}\right\}, \tag{24}$$

Integrating with respect to A_1 and B_1, the conditional posterior pdf for ω and σ is

$$p(\omega, \sigma|\boldsymbol{y}) \propto \frac{\sigma^{-N+1}}{\sqrt{a(\omega)b(\omega) - c^2(\omega)}} \exp\left\{-\frac{1}{2\sigma^2}\left[\textstyle\sum_{i=0}^{N-1} y^2(t_i) - h^2(\omega)\right]\right\}, \tag{25}$$

where $\boldsymbol{y} = (y(t_0), \ldots, y(t_{N-1}))$ is the data set and

$$h^2(\omega) = \frac{b(\omega)g_1^2(\omega) + a(\omega)g_2^2(\omega) - 2c(\omega)g_1(\omega)g_2(\omega)}{a(\omega)b(\omega) - c^2(\omega)} \tag{26}$$

is numerically equal to the continuous Lomb-Scargle periodogram [40]-[43]. The frequency appearing in (24) is a continuous and unbounded parameter; it is neither in any way restricted to discrete values nor limited to the Nyquist critical frequency [28][29]. Probability theory is telling one that: (a) there is information in the data at frequencies between the discrete ones used in the DFT and (b) alias are real indications of the presence of a frequency and absolutely nothing in the data can tell one which is the correct frequency, only prior information can do that [40].

If there is no prior information available about the noise, then σ is a nuisance and can be eliminated by integrating (25) over σ, so that the marginal posterior pdf for ω is

$$p(\omega|y(t_0),\cdots,y(t_{N-1})) \propto \frac{1}{\sqrt{a(\omega)b(\omega)-c^2(\omega)}}[\sum_{i=0}^{N-1} y^2(t_i) - h^2(\omega)]^{(2-N)/2}. \quad (27)$$

This answers the question: "Conditional on the data, what is the probability that the sinusoidal frequency lies in the interval (ω, ω + dω) irrespective of the sinusoidal amplitude and phase?".

Assuming the sample size is large, one may summarize (27) by expanding the Lomb-Scargle periodogram about the maximum $h^2(\widehat{\omega})$, that is

$$h^2(\omega) = h^2(\widehat{\omega}) - \frac{\beta}{2}(\omega - \widehat{\omega})^2 + \cdots, \quad (28)$$

where

$$\beta \equiv \left.\frac{d^2h^2(\omega)}{d\omega^2}\right|_{\omega=\widehat{\omega}} > 0. \quad (29)$$

The Gaussian approximation is

$$p(\omega|y(t_0),\cdots,y(t_{N-1}),\sigma) \approx \left(\frac{2\beta}{\pi\sigma^2}\right)^{1/2} \exp\left\{-\frac{\beta(\omega-\widehat{\omega})^2}{2\sigma^2}\right\}, \quad (30)$$

from which the estimate of the frequency is $\widehat{\omega} \pm \sigma/\sqrt{\beta}$. When the sample size is large the noise standard deviation may be estimated from the data and the parameter be constrained to that value. Only the highest peak of the continuous Lomb-Scargle periodogram is relevant for estimating the frequency of a sinusoidal signal, and the uncertainty associated with the estimate is proportional to the noise standard deviation and inversely proportional to the square root of the curvature of the continuous periodogram at its highest peak. Note that one does not need to normalize (27) to estimate ω.

The joint posterior pdf for A_1, B_1, ω and σ, namely (24), is a function of ω and to find the posterior for A_1, B_1 and σ independently of it one must integrate over ω. An approximation may be obtained by factoring the joint posterior pdf into marginal and conditional pdfs. The posterior pdf for A_1, B_1 and σ is then an average of the conditional pdfs $p(A_1, B_1, \sigma| y(t_0), \ldots, t_{N-1}, \omega)$ weighted by the marginal posterior for ω [that is $p(\omega | y(t_0), \ldots, t_{N-1})$]. Thus if the frequency is very well determined by the data so that the posterior pdf (27) has a very sharp peak at $\omega = \widehat{\omega}$, then one may use the results already derived in the last section for sinusoidal signals with frequency known. The estimates of the amplitudes A_1 and B_1 are then $(bg_1 - cg_2)/(ab - c^2) \pm s(b\nu/(\nu - 2)(ab - c^2))^{1/2}$ and $(ag_2 - cg_1)/(ab - c^2) \pm s(a\nu/(\nu - 2)(ab - c^2))^{1/2}$, respectively, where $\nu s^2 = \Sigma y^2(t_i) - (bg_1^2 + ag_2^2 - 2cg_1g_2)/(ab - c^2)$, $g_1 = \Sigma\, y(t_i)\cos(\widehat{\omega}t_i)$, $g_2 = \Sigma\, y(t_i)\sin(\widehat{\omega}t_i)$, $a = \Sigma \cos^2(\widehat{\omega}t_i)$, $b = \Sigma \sin^2(\widehat{\omega}t_i)$, $c = \Sigma \cos(\widehat{\omega}t_i)\sin(\widehat{\omega}t_i)$ and $\nu = N - 2$. The estimate of the noise variance at the digitizer output is $\widehat{\sigma}^2 = (s^2\nu/(\nu - 2))\,(1 \pm u)$, where $u = (2/(\nu - 4))^{1/2}$.

To conclude, the estimates of the fitting parameters, uncorrected for the systematic effects, namely $\hat{\omega} \pm \hat{\sigma}/\sqrt{\beta}$, the highest peak in the Lomb-Scargle periodogram for the frequency ω, and $(bg_1 - cg_2)/(ab - c^2) \pm s(b\nu/(\nu - 2)(ab - c^2))^{1/2}$ and $(ag_2 - cg_1)/(ab - c^2) \pm s(a\nu/(\nu - 2)(ab - c^2))^{1/2}$ for the amplitudes A_1 and B_1, respectively, are optimal if the following conditions are met: (a) the data set is large, (b) the signal contains a single frequency, (c) there is no dc component in the data, and (d) all the assumptions in section 3 are valid.

The analysis showed that calculating the Lomb-Scargle periodogram is equivalent to postulating a simple sinusoidal model and then looking for the sinusoid that best fits the data. If any of those aforementioned conditions is not met, the Lomb-Scargle periodogram may give misleading or simply incorrect results in light of more realistic models. Obviously, when the signal has more structure than a sinusoid, the sinusoidal model cannot be fitted to the data well. Models which take on more of the characteristics of the real signal will do better (in the sense of more precise estimates) in determining the frequencies and other parameters. The model is generalized in the next section.

The remaining paragraphs of this section deal with a technical point. The reader may skip them on a first reading as they are not necessary for an understanding of the subsequent sections.

It is interesting to note that all the above analysis in this section is valid whether the sampling intervals are uniformly or nonuniformly spaced, and whether the sampling is coherent or not. In fact, in some cases, synchronization between source and digitizer is not possible (for instance, there may be no phase-locking facility associated with the signal source from which to gain synchronization). Let us see however what happens to the statistic (26) when the sampling is uniform and coherent. In this case, $a(\omega) = N/2$, $b(\omega) = N/2$ and $c(\omega) = 0$, and $h^2(\omega) = 2C(\omega)$, where $C(\omega) = (g_1^2(\omega) + g_2^2(\omega))/N$ is the continuous Shuster periodogram [44]. Or, in words, the Lomb-Scargle periodogram is exactly the same as the Schuster periodogram within a scaling constant when the sampling is uniform and coherent. Actually, one does not need to estimate the sinusoidal frequency under such sampling conditions. As noted in the previous two sections, when the digitizer readings are taken with a sampling rate synchronized to the frequency of the signal to be analyzed, the value of ω can be computed directly from the time base, that is $\omega = 2\pi/N_T t_{samp}$. That is the reason why metrologists do prefer to deploy synchronous synthesizing and sampling systems. By so doing, they are able to eliminate the signal frequency and concentrate only on the amplitudes, thereby simplifying the estimation problem and improving the accuracy of the amplitude estimates.

Anyway, let us investigate what happens if one insists on estimating the signal frequency from data uniform and coherently sampled. From (30), it is seen that almost all the posterior probability is concentrated in a narrow region of uncertainty at approximately the true value of ω. Actually, the maximum of the periodogram does not fall exactly at the true value of ω, for if it did so other than by chance, there would be an inherent contradiction in the method. If the most probable value of ω always agreed precisely with the true value, we should have an infallible method of determining ω precisely and this would contradict the very uncertainty which the posterior pdf (27) represents. In fact, if there were several data sets, and if one wanted to estimate ω conditionally on each of them at a time, the most probable value of ω would differ for each of those data sets due to the noise added to the signal. Of course, such differences would decrease as the SNR and/or the number of samples increases.

8. Complex generalized signals

Now we generalize further and consider an arbitrary signal with several unknown parameters, that is

$$f(t_i) = \sum_{l=0}^{m-1} B_l G_l(t, \boldsymbol{\omega}), \tag{31}$$

where B_l is the l-th amplitude and the model functions $G_l\,(t_i, \boldsymbol{\omega})$ are themselves functions of a set of r additional parameters which we label collectively $\boldsymbol{\omega}$. These parameters might be frequencies, chirp rates, decay times, phase displacements, delays, or any other quantities. All parameters are uncorrected for the systematic effects. The function $G_0\,(t_i, \boldsymbol{\omega}) = 1$ is used to model a dc component. It is assumed here that the signal is so complex that it is not reasonable to deploy synchronous synthesizing and sampling systems.

Model (1) can be written in matrix notation as

$$\boldsymbol{y} = \boldsymbol{D}(\boldsymbol{\omega})\boldsymbol{x} + \boldsymbol{e} \tag{32}$$

where $\boldsymbol{y} = (y(t_0), \ldots, y(t_{N-1}))'$ is the data vector, $\boldsymbol{D}(\boldsymbol{\omega})$ is the unknown $N \times m$ matrix with (i, l)-th element $G_l\,(t_i, \boldsymbol{\omega})$ for $l = 0, \ldots, m-1$, $\boldsymbol{x} = (B_0, B_1, \ldots, B_{m-1})'$ is the vector of amplitudes, and $\boldsymbol{e} = (e(t_0), \ldots, e(t_{N-1}))'$ is the vector of the noise terms at the digitizer output. It is assumed that the elements of $\boldsymbol{e}$ are normal and independently distributed, each with mean zero and unknown common variance σ^2; that is, $E[\boldsymbol{e}] = \boldsymbol{0}$ and $E[\boldsymbol{ee}'] = \sigma^2\boldsymbol{I}_N$, where $\boldsymbol{I}_N$ is the identity matrix of order N.

The likelihood function is

$$l(\boldsymbol{\omega}, \boldsymbol{x}, \sigma; \boldsymbol{y}) \propto \sigma^{-N} \exp\left\{-\frac{1}{2\sigma^2}(\boldsymbol{y} - \boldsymbol{D}(\boldsymbol{\omega})\boldsymbol{x})'(\boldsymbol{y} - \boldsymbol{D}(\boldsymbol{\omega})\boldsymbol{x})\right\}$$

$$\propto \sigma^{-N} \exp\left\{-\frac{1}{2\sigma^2}(\boldsymbol{y}'\boldsymbol{y} - 2\boldsymbol{x}'\boldsymbol{D}'(\boldsymbol{\omega})\boldsymbol{y} + \boldsymbol{x}'\boldsymbol{D}'(\boldsymbol{\omega})\boldsymbol{D}(\boldsymbol{\omega})\boldsymbol{x})\right\} \quad (33)$$

The design matrix $\boldsymbol{D}'(\boldsymbol{\omega})\boldsymbol{D}(\boldsymbol{\omega})$ is not diagonal and to derive the joint posterior pdf for $\boldsymbol{\omega}$, $\boldsymbol{x}$ and σ in a form which is easy to marginalize, one needs to diagonalize it.

Since $\boldsymbol{D}'(\boldsymbol{\omega})\boldsymbol{D}(\boldsymbol{\omega})$ is real symmetric, it can be factored as [14]

$$\boldsymbol{D}'(\boldsymbol{\omega})\boldsymbol{D}(\boldsymbol{\omega}) = \boldsymbol{Q}(\boldsymbol{\omega})\boldsymbol{\Lambda}(\boldsymbol{\omega})\boldsymbol{Q}'(\boldsymbol{\omega}), \quad (34)$$

where $\boldsymbol{Q}(\boldsymbol{\omega})$ is the eigenvector matrix and $\boldsymbol{\Lambda}(\boldsymbol{\omega})$ is the eigenvalue matrix[e]. Assuming that $\boldsymbol{D}'(\boldsymbol{\omega})\boldsymbol{D}(\boldsymbol{\omega})$ is positive definite, we may write [14]

$$\boldsymbol{D}'(\boldsymbol{\omega})\boldsymbol{D}(\boldsymbol{\omega}) = \left[\boldsymbol{Q}(\boldsymbol{\omega})\boldsymbol{\Lambda}^{1/2}(\boldsymbol{\omega})\right]\left[\boldsymbol{\Lambda}^{1/2}(\boldsymbol{\omega})\boldsymbol{Q}'(\boldsymbol{\omega})\right]. \quad (35)$$

Therefore, rearranging the terms,

$$\boldsymbol{\Lambda}^{-1/2}(\boldsymbol{\omega})\boldsymbol{Q}'(\boldsymbol{\omega})\boldsymbol{D}'(\boldsymbol{\omega})\boldsymbol{D}(\boldsymbol{\omega})\boldsymbol{Q}(\boldsymbol{\omega})\boldsymbol{\Lambda}^{-1/2}(\boldsymbol{\omega}) = \boldsymbol{I}_m, \quad (36)$$

where $\boldsymbol{I}_m$ is the identity matrix of order m, so that the matrix $\boldsymbol{W}(\boldsymbol{\omega})$ defined as

$$\boldsymbol{W}(\boldsymbol{\omega}) = \boldsymbol{D}(\boldsymbol{\omega})\boldsymbol{Q}(\boldsymbol{\omega})\boldsymbol{\Lambda}^{-1/2}(\boldsymbol{\omega}) \quad (37)$$

is a rectangular matrix with orthonormal columns.

Model (32) may now be rewritten as

$$\boldsymbol{y} = \boldsymbol{W}(\boldsymbol{\omega})\boldsymbol{z} + \boldsymbol{e}, \quad (38)$$

where $\boldsymbol{z}$ is the vector of amplitudes of the orthogonal model, that is $\boldsymbol{z} = (A_0, A_1, \ldots, A_{m-1})'$.

The likelihood function (32) is thus simplified,

$$l(\boldsymbol{\omega}, \boldsymbol{z}, \sigma; \boldsymbol{y}) \propto \sigma^{-N} \exp\left\{-\frac{1}{2\sigma^2}(\boldsymbol{y} - \boldsymbol{W}(\boldsymbol{\omega})\boldsymbol{z})'(\boldsymbol{y} - \boldsymbol{W}(\boldsymbol{\omega})\boldsymbol{z})\right\}$$

$$\propto \sigma^{-N} \exp\left\{-\frac{1}{2\sigma^2}(\boldsymbol{y}'\boldsymbol{y} - 2\boldsymbol{z}'\boldsymbol{h}(\boldsymbol{\omega}) + \boldsymbol{z}'\boldsymbol{z})\right\}, \quad (39)$$

where $\boldsymbol{h}(\boldsymbol{\omega}) = \boldsymbol{W}'(\boldsymbol{\omega})\boldsymbol{y}$ is the projection of the data onto the orthonormal model functions.

A noninformative prior is then assumed by taking the elements of $\boldsymbol{\omega}$, $\boldsymbol{z}$ and $\ln \sigma$ independent and uniformly distributed; that is $p(\boldsymbol{\omega}, \boldsymbol{z}, \sigma) \propto \sigma^{-1}$, with

---

[e] No specific knowledge about eigenvectors and eigenvalues is required here to apply intelligently the method. Codes for fast computing matrix operations are easily available as built-in functions of most currently available software packages.

$-\infty < (\boldsymbol{\omega})_p < \infty$, $-\infty < (\boldsymbol{z})_l < \infty$ and $0 < \sigma < \infty$, where the symbols $(\cdot)_p$ and $(\cdot)_l$ denote the p-th element ($p = 0, \ldots, r-1$) and the l-th element ($l = 0, \ldots, m-1$), respectively.

On combining (39) and the above prior, the joint posterior pdf for $\boldsymbol{\omega}$, $\boldsymbol{z}$ and σ is

$$p(\boldsymbol{\omega}, \mathbf{z}, \sigma|\mathbf{y}) \propto \sigma^{-N-1}\exp\left\{-\frac{1}{2\sigma^2}(\mathbf{y}'\mathbf{y} - 2\mathbf{z}'\boldsymbol{h}(\boldsymbol{\omega}) + \mathbf{z}'\mathbf{z})\right\} \tag{40}$$

and integrating with respect to the elements of $\boldsymbol{z}$, the marginal posterior pdf for $\boldsymbol{\omega}$ and σ is

$$p(\boldsymbol{\omega}, \sigma|\mathbf{y}) \propto \sigma^{-N+m-1}\exp\left\{-\frac{1}{2\sigma^2}\left(\mathbf{y}'\mathbf{y} - \boldsymbol{h}'(\boldsymbol{\omega})\boldsymbol{h}(\boldsymbol{\omega})\right)\right\}. \tag{41}$$

If σ is known, the posterior pdf for the elements of $\boldsymbol{\omega}$, conditional on the data and on our knowledge of σ, is

$$p(\boldsymbol{\omega}|\mathbf{y}, \sigma) \propto \exp\left\{\frac{\|\boldsymbol{h}(\boldsymbol{\omega})\|^2}{2\sigma^2}\right\}, \tag{42}$$

where the symbol $\|\cdot\|$ denotes the length of a vector [14]. Thus, the estimates of the parameters $\boldsymbol{\omega}$ of the signal $f(t_i)$ are those for which the length of the projection of the data onto the orthonormal model vectors is a maximum. The Bayesian nonlinear processing takes the noise level into account; wiggles below the noise level are almost certainly artifacts of the noise and are suppressed, while any peaks of $\|\boldsymbol{h}(\boldsymbol{\omega})\|^2$ that rise above the noise level will be strongly emphasized, indicating high probability of a parameter $\boldsymbol{\omega}$. As the ratio $\|\boldsymbol{h}(\boldsymbol{\omega})\|^2/\sigma^2$ is made larger and larger, the posterior pdf (42) changes into a sharply peaked density at its mode.

But if there is no prior information available about the noise, then σ is a nuisance and can be eliminated by also integrating (41) over σ, so that the marginal posterior pdf for $\boldsymbol{\omega}$ is

$$p(\boldsymbol{\omega}|\mathbf{y}) \propto \left[1 - \frac{\|\boldsymbol{h}(\boldsymbol{\omega})\|^2}{\|\mathbf{y}\|^2}\right]^{(m-N)/2}. \tag{43}$$

The function $\|\boldsymbol{h}(\boldsymbol{\omega})\|^2$ summarizes all the information in the data that is relevant for estimating the parameters $\boldsymbol{\omega}$ of the signal $f(t_i)$. It is a generalization of the continuous Lomb-Scargle periodogram to complex generalized models. This expression allows the following interpretation: estimating the parameters of interest $\boldsymbol{\omega}$ that are hidden in the model functions amounts essentially to finding the values of the $\boldsymbol{\omega}$ that permit $f(t_i)$ to make the closest possible fit to the data by the mean-square criterion. Probability theory is telling one that the most likely values of the $\boldsymbol{\omega}$ are those that permit a maximum amount of the mean-square data $||\boldsymbol{y}||^2$ to be accounted for by the systematic term, that is, those values that maximize $||\boldsymbol{h}(\boldsymbol{\omega})||^2$.

If one is interested in a specific parameter, for instance ω_r, the marginal posterior pdf for ω_r is obtained by integrating (43) over the other elements of $\boldsymbol{\omega}$.

One method of summarizing a multivariate density is based on the behavior of the density about its mode. First we note that (42) and (43) are not very different provided that $N \gg m$. Here it is necessary to interpret the parameter set $\boldsymbol{\omega}$ as a vector, namely $\boldsymbol{\omega} = (\omega_0, \omega_1, \omega_2, \ldots, \omega_{r-1})'$, $r \leq m$. Expanding $||\boldsymbol{h}(\boldsymbol{\omega})||^2$ in a second-order Taylor series about the mode $\hat{\boldsymbol{\omega}}$ of (43) we have

$$||\boldsymbol{h}(\boldsymbol{\omega})||^2 \approx ||\boldsymbol{h}(\hat{\boldsymbol{\omega}})||^2 - (\boldsymbol{\omega} - \hat{\boldsymbol{\omega}})'\boldsymbol{H}(\hat{\boldsymbol{\omega}})(\boldsymbol{\omega} - \hat{\boldsymbol{\omega}})/2, \tag{44}$$

where $\boldsymbol{H}(\hat{\boldsymbol{\omega}})$ is the Hessian matrix evaluated at the mode whose entries are

$$\left(\boldsymbol{H}(\hat{\boldsymbol{\omega}})\right)_{pq} = -\frac{1}{2}\frac{\partial^2 ||\boldsymbol{h}(\boldsymbol{\omega})||^2}{\partial(\boldsymbol{\omega})_p \partial(\boldsymbol{\omega})_q}\bigg|_{\boldsymbol{\omega}=\hat{\boldsymbol{\omega}}} > 0, \tag{45}$$

where the symbol $(\cdot)_{pq}$ denotes the (p, q)-th entry $(p, q = 0, 1, \ldots, r-1)$. By this expansion, the posterior pdf (43) is approximated by an r-dimensional normal pdf with mean $\hat{\boldsymbol{\omega}}$ and covariance matrix $\boldsymbol{\Sigma}_{\Omega} = \sigma^2 \boldsymbol{H}^{-1}(\hat{\boldsymbol{\omega}})$. When the sample size is large the noise variance may be estimated from the data and the parameter be constrained to that value. Numerical derivatives are needed to evaluate the Hessian matrix. To apply this Gaussian approximation, one needs to find the mode of (43). Note that one does not need to normalize (43) to estimate $\boldsymbol{\omega}$.

The joint posterior pdf (40) is a function of the $\boldsymbol{\omega}$ parameters and to find the posterior for $\boldsymbol{z}$ and σ independently of them one must integrate over all those $\boldsymbol{\omega}$ parameters. An approximation may be obtained by factoring the joint posterior pdf into marginal and conditional pdfs. The posterior pdf for $\boldsymbol{z}$ and σ is then an average of the conditional pdfs $p(\boldsymbol{z}, \sigma \mid \boldsymbol{\omega}, \boldsymbol{y})$ weighted by the marginal posterior for $\boldsymbol{\omega}$ [that is $p(\boldsymbol{\omega} \mid \boldsymbol{y})$]. Thus if the frequencies are very well determined by the data so that the posterior pdf (43) has a very sharp peak at $\boldsymbol{\omega} = \hat{\boldsymbol{\omega}}$, then

$$p(\boldsymbol{z}, \sigma|\boldsymbol{y}) \propto \sigma^{-N-1}\exp\left\{-\frac{1}{2\sigma^2}(\boldsymbol{y}'\boldsymbol{y} - 2\boldsymbol{z}'\boldsymbol{h}(\hat{\boldsymbol{\omega}}) + \boldsymbol{z}'\boldsymbol{z})\right\}. \tag{46}$$

Completing the square and factoring, one obtains: (a) the conditional posterior pdf for $\boldsymbol{z}$, given σ [that is $p(\boldsymbol{z} \mid \sigma, \boldsymbol{y})$], namely an m-dimensional normal pdf with mean $\boldsymbol{h}(\hat{\boldsymbol{\omega}})$ and covariance matrix $\boldsymbol{\Sigma}_Z = \sigma^2 \boldsymbol{I}_m$, where $\boldsymbol{I}_m$ is the identity matrix of order m, and (b) the marginal posterior pdf for σ $(0 < \sigma < \infty)$, that is $p(\sigma \mid \boldsymbol{y}) \propto \sigma^{-\nu-1}\exp(-\nu s^2/2\sigma^2)$, with $\nu s^2 = ||\boldsymbol{y}||^2 - ||\boldsymbol{h}(\hat{\boldsymbol{\omega}})||^2$ and $\nu = N - m$, which is in the form of an inverted gamma pdf. One can use the moments of the latter pdf (see section 4) to obtain the estimate of the noise variance at the digitizer output, that is $\hat{\sigma}^2 = (\nu s^2/(\nu - 2))\,(1 \pm u)$, where $u = (2/(\nu - 4))^{1/2}$.

To remove the σ dependence from the above-mentioned conditional posterior pdf for $\boldsymbol{z}$ one would need to multiply it by $p(\sigma \mid \boldsymbol{y})$ and integrate the product over σ. However, when the sample size is large the noise variance may be estimated from the data and the parameter be constrained to that value. The marginal posterior pdf for $\boldsymbol{z}$ is then approximated by an m-dimensional normal pdf with mean $\boldsymbol{h}(\hat{\boldsymbol{\omega}})$ and covariance matrix $\boldsymbol{\Sigma}_Z = \hat{\sigma}^2 \boldsymbol{I}_m$.

Such moments can be used to calculate those of the pdf for the $\boldsymbol{x}$ parameters in the nonorthogonal model. By inserting $\boldsymbol{W}(\hat{\boldsymbol{\omega}}) = \boldsymbol{D}(\hat{\boldsymbol{\omega}})\boldsymbol{Q}(\hat{\boldsymbol{\omega}})\boldsymbol{\Lambda}^{-1/2}(\hat{\boldsymbol{\omega}})$ in the orthogonal model, namely $\boldsymbol{y} = \boldsymbol{W}(\hat{\boldsymbol{\omega}})\boldsymbol{z} + \boldsymbol{e}$, and comparing with the nonorthogonal model, namely $\boldsymbol{y} = \boldsymbol{D}(\hat{\boldsymbol{\omega}})\boldsymbol{x} + \boldsymbol{e}$, yields $\boldsymbol{x} = \boldsymbol{Q}(\hat{\boldsymbol{\omega}})\boldsymbol{\Lambda}^{-1/2}(\hat{\boldsymbol{\omega}})\boldsymbol{z}$, which is a linear transformation. The posterior pdf for $\boldsymbol{x}$ is then an m-dimensional normal pdf with mean $\hat{\boldsymbol{x}} = \boldsymbol{Q}(\hat{\boldsymbol{\omega}})\boldsymbol{\Lambda}^{-1/2}(\hat{\boldsymbol{\omega}})\boldsymbol{h}(\hat{\boldsymbol{\omega}})$ and covariance matrix $\boldsymbol{\Sigma}_X = \hat{\sigma}^2 \boldsymbol{Q}(\hat{\boldsymbol{\omega}})\boldsymbol{\Lambda}^{-1}(\hat{\boldsymbol{\omega}})\boldsymbol{Q}'(\hat{\boldsymbol{\omega}})$. This follows immediately from the properties of the multivariate Gaussian distribution [38][39].

To conclude, the estimates of the fitting parameters, uncorrected for the systematic effects, namely $\hat{\boldsymbol{\omega}}$ with associated covariance matrix $\boldsymbol{\Sigma}_\Omega = \hat{\sigma}^2 \boldsymbol{H}^{-1}(\hat{\boldsymbol{\omega}})$ for the parameters $\boldsymbol{\omega}$, and $\hat{\boldsymbol{x}}$ with associated covariance matrix $\boldsymbol{\Sigma}_X = \hat{\sigma}^2 \boldsymbol{Q}(\hat{\boldsymbol{\omega}})\boldsymbol{\Lambda}^{-1}(\hat{\boldsymbol{\omega}})\boldsymbol{Q}'(\hat{\boldsymbol{\omega}})$ for the amplitudes, are optimal if the following conditions are met: (a) the number of data values is large compared with the number of model functions and (b) all the assumptions in section 3 are valid.

Harmonic parameters are often reported in polar coordinates, namely the vector of magnitudes $\boldsymbol{a}$ with j-th element $(\boldsymbol{a})_j = [(\boldsymbol{x})^2_{jc} + (\boldsymbol{x})^2_{js}]^{1/2}$ and the vector of phase angles $\boldsymbol{\theta}$ with j-th element $(\boldsymbol{\theta})_j = \text{atan2}\,[-(\boldsymbol{x})_{js}, (\boldsymbol{x})_{jc}]$, where the symbols $(\cdot)_{js}$ and $(\cdot)_{jc}$ denote the j-th sine and cosine amplitudes, respectively. It is worth evaluating the asymptotic behaviour of the posterior pdfs for $(\boldsymbol{a})_j$ and $(\boldsymbol{\theta})_j$ when the sample size is large and the pdfs for the $\boldsymbol{x}$ parameters can be assumed to be mutually independent. We know that the pdfs for the $\boldsymbol{z}$ parameters are mutually independent. But this may not be so for the parameters of the nonorthogonal model. However, we confirmed that if any two frequencies are not too close[f] then both $\boldsymbol{\Sigma}_\Omega$ and $\boldsymbol{\Sigma}_X$ are nearly diagonal. The posterior pdf for $(\boldsymbol{a})_j$ can then be approximated by a Rician or noncentral Rayleigh [38][39]. For high SNR, this pdf becomes approximately Gaussian with variance $(\boldsymbol{\Sigma}_X)_{jj}$ in the region around its mean value $[(\hat{\boldsymbol{x}})^2_{jc} + (\hat{\boldsymbol{x}})^2_{js}]^{1/2}$, where the symbol $(\cdot)_{jj}$ denotes the j-th diagonal entry. The posterior pdf for $(\boldsymbol{\theta})_j$ also becomes approximately Gaussian with variance $(\boldsymbol{\Sigma}_X)_{jj}\,[(\hat{\boldsymbol{x}})^2_{jc} + (\hat{\boldsymbol{x}})^2_{js}]^{-1}$ for high SNR and small phase angles relative to the mean atan2 $[-(\hat{\boldsymbol{x}})_{js}, (\hat{\boldsymbol{x}})_{jc}]$ [38][39]. Thus, the smaller the estimate of the

---

[f] Two frequencies ω_0 and ω_1 are regarded as too close when $|\omega_0 - \omega_1| < 2\pi/N$ rad.

magnitude is relative to the noise, the larger is the uncertainty associated with the estimate of the corresponding phase angle.

Therefore, in addition to the aforementioned conclusion, the estimates in polar coordinates, uncorrected for the systematic effects, namely $[(\hat{\boldsymbol{x}})^2_{jc} + (\hat{\boldsymbol{x}})^2_{js}]^{1/2} \pm [(\boldsymbol{\Sigma}\mathrm{x})_{jj}]^{1/2}$ for the j-th magnitude and $\mathrm{atan2}\,[-(\hat{\boldsymbol{x}})_{js}, (\hat{\boldsymbol{x}})_{jc}] \pm [(\boldsymbol{\Sigma}\mathrm{x})_{jj}]^{1/2}[(\hat{\boldsymbol{x}})^2_{jc} + (\hat{\boldsymbol{x}})^2_{js}]^{-1/2}$ for the j-th phase angle, are optimal if the following additional conditions are met: (c) the SNR is high, and (d) any two frequencies are not too close.

Nevertheless, the method of summarizing a multivariate density discussed above present a disadvantage: the matrix operations take too much computer time and memory space. Though probability theory allows us to eliminate nuisance parameters and concentrate only on the interesting ones, thus permitting a reduction in the dimensionality of the search algorithm with consequent time savings, computation time is expected to grow approximately as the square of the number of functions in the model. As mentioned in the discussion around (43), the Bayesian framework provides theoretical justification for the application of least squares methods. Some of the latter are much simpler and faster than the Bayesian approach [45]. They also require much less memory space. Nevertheless, given the ever increasing power of computers, computation time does not seem to be an impediment to widespread use of the Bayesian method discussed here. Additionally, the Bayesian framework allows the incorporation of prior information and simplifies the evaluation of the uncertainty associated with the estimates: since we are interested in extracting conclusions from the sole data available, the problem of determining the sampling distribution of an 'estimator' is eliminated.

9. Computer simulated examples

All the calculation steps presented in the last section have been summarized in the simple algorithm reported in [11]. Uniform (though not coherent) sampling is assumed and the total sampling time Nt_{samp} should cover at least one or two periods of the lowest frequency component present in the signal to ensure algorithm convergence. The pattern search algorithm introduced by Hooke and Jeeves [46] is used to find the posterior mode required for the Gaussian approximation. This nonlinear optimization algorithm was improved over the decades (see historical note in [47] describing the progress until 1994).

The examples discussed in this section assume that all parameters in the signal and noise models are constant, that the data can be modeled with trigonometric equations, and that there is no prior information about the model parameters (see [48]-[50] for applications to exponential decay models where prior information is available, [51] for a method to save computation time for complex models with highly informative priors, and [52], [53] for the situation where parameter values change with time). The main objective here is to provide the estimates of $\boldsymbol{\omega}$, $\boldsymbol{x}$, $\boldsymbol{a}$, $\boldsymbol{\theta}$ and σ, and their associated uncertainties. This is valid when the sample size is large and/or the SNR is high, as is typically the case in waveform metrology.

In order to save time, the method does not require numerical integration techniques, since all required integrations have been performed analytically in the last section. Numerical derivatives are also reduced to those needed to evaluate the Hessian matrix entries in (45). We use well known second order accurate finite difference formulas with small increment δ (a value of δ such as 0.000001 is low enough to approximate the derivative and high enough to avoid roundoff errors on the computer). The signal power spectral density is not estimated as this would require multidimensional integrations [54].

The method can be easily implemented with those commercial laboratory software packages routinely used by metrologists for controlling instrumentation and reporting measurement results. It was programmed with LabWindows/CVI v. 6.0 using the FORTRAN routines described in [2], [3].

The method was applied in all examples discussed in the following. The uncertainties associated with the estimates decrease with increasing SNR and/or number of samples. The first two examples with noise variance equal to 1 are presented to confirm results obtained in [2]. The noise variance of the remaining examples (except the last) is the order of magnitude of that typically found in calibration systems that deploy commercial signal generators. Marginal posteriors for the elements of $\boldsymbol{\omega}$ are additionally plotted for illustration in the last example. In this case, the integrations were performed with Simpson´s rule. The noise variance of this example was chosen for clear plotting considerations.

Additional examples on the Bayesian analysis of signal generator outputs from digital sampling data were reported in [11], [12], [55]-[59].

9.1. *Single frequency*

This example was discussed in [2]. We generated a set of 512 samples from

$$y[i] = 0.001 + \cos[0.3i + 1] + e[i],$$

where $e[i] \sim \mathcal{N}(0,1)$. The corresponding signal model is

$$f[i] = C_0 + A_1 \cos[\omega_0 i] + B_1 \sin[\omega_0 i].$$

There are two amplitudes (A_1 and B_1), one dc component (C_0), one frequency (ω_0) and one noise variance (σ^2). Probability theory has eliminated all parameters except the frequency. The pattern search algorithm was used to locate the maximum of the posterior pdf for ω_0 using any starting point within ±3% of the true frequency. The factor ρ and the "minimum" step size ε in [47] were set at 0.90 and $1{\cdot}10^{-8}$, respectively. The estimates are listed in Table 1.

Table 1. Parameter estimates for a sinusoidal signal.

Parameter	True value	Estimate	Uncertainty
ω_0	0.3	0.2996	0.0004
C_0	0.001	0.02	0.04
A_1	0.540302	0.62	0.06
B_1	−0.841471	−0.83	0.06
$(\boldsymbol{a})_1$	1	1.04	0.06
$(\boldsymbol{\theta})_1$	1	0.93	0.06
σ	1	0.97	0.06

Frequency and phase angle in radians. Amplitudes and noise standard deviation in arbitrary units.

Table 2. Parameter estimates for a signal containing two close frequencies.

Parameter	True value	Estimate	Uncertainty
ω_1	0.3	0.299	0.003
ω_2	0.307	0.309	0.003
σ	1	0.98	0.06

Frequencies in radians. Noise standard deviation in arbitrary unit.

The covariance matrix associated with the estimates of C_0, A_1 and B_1 was

$$\boldsymbol{\Sigma}_{\mathbf{X}} = \begin{bmatrix} 1.9 & -0.013 & -0.047 \\ -0.013 & 3.8 & -0.0029 \\ -0.047 & -0.0029 & 3.8 \end{bmatrix} \times 10^{-3}.$$

9.2. *Two close frequencies*

This example was also discussed in [2]. We generated a set of 512 samples from

$$y[i] = \cos[0.3i + 1] + \cos[0.307i + 2] + e[i],$$

where $e[i] \sim \mathcal{N}(0,1)$. The corresponding signal model is

$$f[i] = C_0 + A_1\cos[\omega_1 i] + B_1\sin[\omega_1 i] + A_2\cos[\omega_2 i] + B_2\sin[\omega_2 i].$$

This time series has two frequencies separated by approximately 0.6 bin in the DFT. One bin corresponds to $|\omega_0 - \omega_1| \approx 2\pi/512$ rad. These two frequencies cannot be distinguished with Fourier transform methods simply because there are no bins in between the two frequencies. There are four amplitudes (A_1, B_1, A_2

and B_2), two frequencies (ω_1 and ω_2), one dc component (C_0) and one noise variance (σ^2). Probability theory has eliminated six parameters except the two frequencies. The pattern search algorithm was used to locate the maximum of the posterior pdf for ω_1 and ω_2. The factor ρ and the "minimum" step size ε in [47] were set at 0.50 and $1\cdot10^{-8}$, respectively. The estimates are listed in Table 2.

As can be seen from the covariance matrix associated with the estimates of ω_1 and ω_2, namely

$$\boldsymbol{\Sigma}_\Omega = \begin{bmatrix} 6.1 & -5.5 \\ -5.5 & 5.9 \end{bmatrix} \times 10^{-6}\ \text{rad}^2,$$

the two frequencies are highly correlated.

9.3. *Two separate frequencies*

We generated a set of 512 samples from

$$y[i] = \cos[0.1i + 2] + \cos[0.3i + 1] + e[i],$$

where now $e[i] \sim \mathcal{N}(0, 10^{-6})$. This time series has two frequencies separated by 16 bins in the DFT. The estimates are listed in Table 3.

Table 3. Parameter estimates for a signal containing two separate frequencies.

Parameter	True value	Estimate	Uncertainty
ω_1	0.3	0.3000006	0.0000004
ω_2	0.1	0.0999998	0.0000004
C_0	0	0.00007	0.00004
A_1	0.540302	0.54032	0.00006
B_1	−0.841471	−0.84151	0.00006
A_2	−0.416147	−0.41598	0.00006
B_2	0.909297	0.90933	0.00006
$(\boldsymbol{a})_1$	1	1.00004	0.00006
$(\boldsymbol{a})_2$	1	0.99996	0.00006
$(\boldsymbol{\theta})_1$	1	1.00001	0.00006
$(\boldsymbol{\theta})_2$	2	1.99983	0.00006
σ	0.001	0.00097	0.00006

Frequencies and phase angles in radians. Amplitudes and noise standard deviation in arbitrary units.

As can be seen from the covariance matrix associated with the estimates of ω_1 and ω_2, namely

$$\boldsymbol{\Sigma}_\Omega = \begin{bmatrix} 1.7 & 0.018 \\ 0.0185 & 1.6 \end{bmatrix} \times 10^{-13}\ \text{rad}^2,$$

the two frequencies are almost uncorrelated in this case. The covariance matrix associated with the estimates of C_0, A_1, B_1, A_2 and B_2 was

$$\mathbf{\Sigma}_{\Omega} = \begin{bmatrix} 1.8 & -0.014 & -0.057 & -0.043 & -0.025 \\ -0.014 & 3.7 & -0.035 & -0.0038 & 0.0093 \\ -0.057 & -0.035 & 3.6 & -0.075 & -0.041 \\ -0.043 & -0.0038 & -0.075 & 3.7 & -0.041 \\ -0.025 & 0.0093 & -0.041 & -0.041 & 3.7 \end{bmatrix} \times 10^{-9}.$$

9.4. *Amplitude modulation*

We generated a set of 1000 samples from

$$y[i] = \cos[\omega_0 i] \cdot (1 + m \cos[\omega_1 i]) + e[i],$$

where ω_0 and ω_1 are respectively the carrier and modulation frequencies (see their true values in Table 4), m (= 0.1) is here the degree of modulation, and $e[i] \sim \mathcal{N}(0, 10^{-6})$.

Table 4. Parameter estimates for a signal modulated in amplitude.

Parameter	True value	Estimate	Uncertainty
ω_0	0.376991118	0.3769911	0.0000001
ω_1	0.188495559	0.188494	0.000003
C_0	0	0.00002	0.00003
$(\boldsymbol{a})_1$	1	0.99996	0.00006
$(\boldsymbol{a})_2$	0.05	0.04999	0.00006
$(\boldsymbol{a})_3$	0.05	0.04994	0.00006
$(\boldsymbol{\theta})_1$	0	0.00007	0.00006
$(\boldsymbol{\theta})_2$	0	−0.0004	0.0012
$(\boldsymbol{\theta})_3$	0	0.0002	0.0012
σ	0.001	0.00098	0.00004

Frequencies and phase angles in radians. Amplitudes and noise standard deviation in arbitrary units.

Using trigonometric identities, it is straightforward to show that the signal model is

$$f[i] = C_0 + A_0 \cos[\omega_0 i] + B_0 \sin[\omega_0 i] + C_1 \cos[(\omega_0 + \omega_1)i]$$
$$+ D_1 \sin[(\omega_0 + \omega_1)i] + E_1 \cos[(\omega_0 - \omega_1)i] + F_1 \sin[(\omega_0 - \omega_1)i],$$

where C_0 is the dc component, A_0 and B_0 are the carrier amplitudes, C_1, D_1, E_1 and F_1 are sideband amplitudes (proportional to the degree of modulation) corresponding to the modulation frequency ω_1 (ω_0 and ω_1 are not necessarily commensurable).

The true value of the vector of magnitudes is $\boldsymbol{a} = [C_0, (A_0^2 + B_0^2)^{1/2}, (C_1^2 + D_1^2)^{1/2}, (E_1^2 + F_1^2)^{1/2}]' = (0, 1, m/2, m/2)'$. The true value of the vector of phase angles $\boldsymbol{\theta}$ is the null vector. There are six amplitudes (or three polar magnitudes and three phase angles), one dc component, two frequencies and one noise variance. Probability theory has eliminated eight parameters except the two frequencies. The pattern search algorithm was used to locate the maximum of the posterior pdf for ω_0 and ω_1 using the starting guess $(\boldsymbol{\omega})_0^{(0)}$ within ±0.2% of the true value and the starting guess $(\boldsymbol{\omega})_1^{(0)}$ within ±0.3% of the true value. The optimization parameters ρ = 0.50 and ε = $1{\cdot}10^{-8}$ were used in [47]. Table 4 shows the estimates.

The covariance matrix associated with the estimates of ω_0 and ω_1 was

$$\boldsymbol{\Sigma}_\Omega = \begin{bmatrix} 0.012 & 0.000019 \\ 0.000019 & 9.6 \end{bmatrix} \times 10^{-12}\ \text{rad}^2.$$

9.5. *Chirp*

We generated a set of 31000 samples from

$$y(it_{\text{samp}}) = \cos(i\omega_0 t_{\text{samp}} + i^2\alpha_0 t_{\text{samp}}^2) + e(it_{\text{samp}}),$$

where ω_0 and α_0 are respectively the frequency and the chirp rate (see their true values in Table 5), t_{samp} (= 0.0000652 s) is the sampling interval and $e(it_{\text{samp}}) \sim \mathcal{N}(0, 10^{-6})$. The corresponding signal model is

$$y(it_{\text{samp}}) = C_0 + A_1\cos(i\omega_0 t_{\text{samp}} + i^2\alpha_0 t_{\text{samp}}^2) + B_1\sin(i\omega_0 t_{\text{samp}} + i^2\alpha_0 t_{\text{samp}}^2),$$

where C_0 is the dc component, and A_1 and B_1 are the amplitudes.

Table 5. Parameter estimates for a chirped signal.

Parameter	True value	Estimate	Uncertainty
ω_0	59.9876543210	59.987651	0.000009
α_0	0.9876543210	0.987656	0.000009
C_0	0	0.000002	0.000006
$(\boldsymbol{a})_1$	1	1.000002	0.000011
$(\boldsymbol{\theta})_1$	0	0.000007	0.000011
σ	0.001	0.001002	0.000008

Frequency in Hz. Chirp rate in Hz/s. Phase angle in radians. Amplitudes and noise standard deviation in arbitrary units.

There are two amplitudes, one dc component, one frequency, one chirp rate and one noise variance. Probability theory has eliminated four parameters except the frequency and the chirp rate. The pattern search algorithm was used to locate the maximum of the posterior pdf for ω_0 and α_0. The factor ρ and the "minimum" step size ε in [47] were set at 0.50 and $1{\cdot}10^{-8}$, respectively. The estimates are listed in Table 5.

The covariance matrix associated with the estimates of ω_0 and α_0 was

$$\boldsymbol{\Sigma}_\Omega = \begin{bmatrix} 7.7\times10^{-11}\ \text{Hz}^2 & 3.9\times10^{-16}\ \text{Hz}^2/\text{s} \\ 3.9\times10^{-16}\ \text{Hz}^2/\text{s} & 7.0\times10^{-11}\ \text{Hz}^2/\text{s}^2 \end{bmatrix},$$

so that the frequency and the chirp rate can be regarded as uncorrelated.

The covariance matrix associated with the estimates of C_0, A_1 and B_1 was

$$\boldsymbol{\Sigma}_{\mathrm{X}} = \begin{bmatrix} 1.9 & -0.013 & -0.047 \\ -0.013 & 3.8 & -0.0029 \\ -0.047 & -0.0029 & 3.8 \end{bmatrix} \times 10^{-3}.$$

9.6. *Single frequency with exponential decay*

We generated a set of 512 samples from

$$y[i] = 0.001 + \cos[0.3i + 1]e^{-i/100} + e[i],$$

where now $e[i] \sim \mathcal{N}(0,0.01)$. The corresponding signal model is

$$f[i] = C_0 + A_1 \cos[\omega_0 i]\, e^{-i/\tau} + B_1 \sin[\omega_0 i]\, e^{-i/\tau}.$$

There are two amplitudes (A_1 and B_1), one dc component (C_0), one frequency (ω_0), one decay time (τ), and one noise variance (σ^2). Probability theory has eliminated all parameters except the frequency and the decay time. The pattern search algorithm was used to locate the maximum of the posterior pdf for ω_0 and τ. The factor ρ and the "minimum" step size ε in [47] were set at 0.85 and $1{\cdot}10^{-8}$, respectively. The estimates are listed in Table 6.

Table 6. Parameter estimates for a sinusoidal signal with exponential decay.

Parameter	True value	Estimate	Uncertainty
ω_0	0.3	0.2996	0.0004
τ	100	95.4	4.0
C_0	0.001	0.0004	0.0046
A_1	0.540302	0.522	0.029
B_1	−0.841471	−0.885	0.029
$(\boldsymbol{a})_1$	1	1.028	0.029
$(\boldsymbol{\theta})_1$	1	1.038	0.029
σ	0.1	0.1027	0.0064

Frequency and phase angle in radians. Amplitudes, decay time and noise standard deviation in arbitrary units.

The covariance matrix associated with the estimates of τ and ω_0 was

$$\boldsymbol{\Sigma}_{\Omega} = \begin{bmatrix} 16 & -3.5 \times 10^{-5}\ \mathrm{rad} \\ -3.5 \times 10^{-5}\ \mathrm{rad} & 1.7 \times 10^{-7}\ \mathrm{rad}^2 \end{bmatrix}.$$

The posterior distributions for ω_0 and τ are plotted in Figs. 1 and 2, respectively.

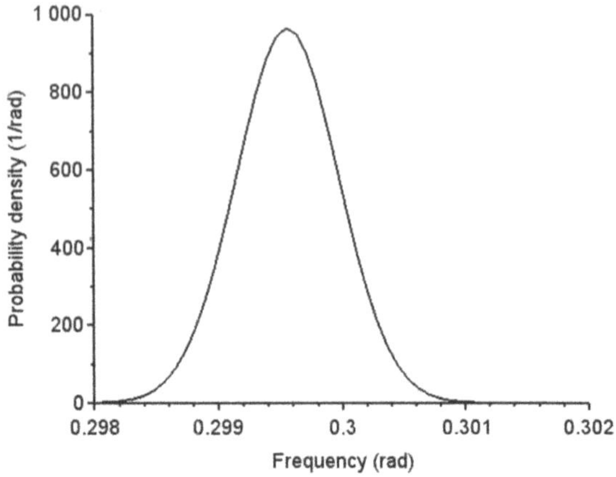

Fig. 1. Posterior distribution for the frequency.

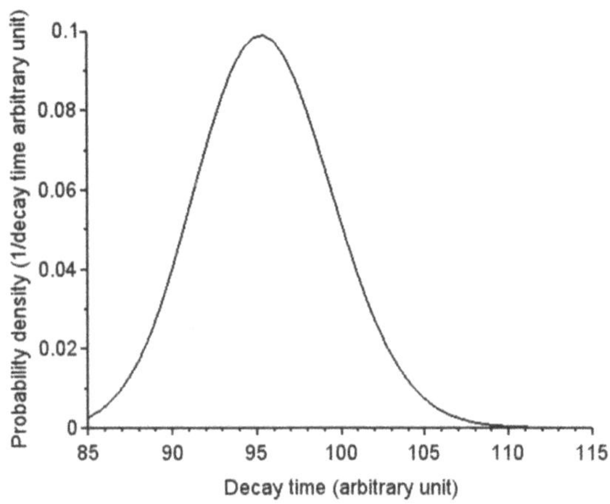

Fig. 2. Posterior distribution for the decay time.

10. Conclusions

Probability theory tells one how to analyze a particular signal optimally. As one has little prior information beyond the signal waveform, Bayesian estimates should not differ significantly from the estimates one would obtain from least squares. However, probability theory allows us to eliminate nuisance parameters and concentrate only on the interesting ones. This permits a reduction in the dimensionality of the search algorithm with consequent time savings. Additionally, the Bayesian framework allows the incorporation of prior information and simplifies the evaluation of the uncertainty associated with the estimates: since we are interested in extracting conclusions from the sole data available, the problem of determining the sampling distribution of an 'estimator' is eliminated.

As the metrologist knows in advance which signal has been generated for instrument calibration purposes, the question to be asked simply becomes "What is the evidence of such a specific signal waveform in these data?" The Bayesian framework then derives automatically the statistic that is best suited to answer the question and also indicates how it should be processed to obtain the estimates of signal parameters.

Statistics have been derived here for several particular signals under some conditions. For estimation of the amplitude of a dc signal, the statistic is the average of the data. For estimation of the fitting parameters of a sinusoidal signal without dc component from data uniform and coherently sampled, the statistic is the numerical Fourier transformation of the data evaluated at the known frequency. For estimation of the harmonic amplitudes of an arbitrary periodic signal with frequency known, the statistic is that resulting from the linear least-squares fitting of the data. This statistic reduces to the numerical Fourier transformation of the data evaluated at the known harmonic frequencies if the sampling is uniform and coherent and if the signal does not have a dc component.

For estimation of the frequency of a sinusoidal signal without dc component, the statistic is the continuous Lomb-Scargle periodogram. However, when the true signal deviates from a sinusoidal or periodic signal, for example, when there are multiple nonharmonically related sinusoids or the data contain a constant offset, then this statistic is never an optimal frequency estimator. It was shown that the statistic for the estimation of the frequencies of a complex generalized signal is the length of the projection of the data onto the orthonormal model functions corresponding to that complex generalized model.

The role of those synchronous synthesizing and sampling systems deployed by metrologists to simplify the estimation problem has been pointed out in many sections.

Some computer simulated examples have been also presented on the application to specific signals of the Bayesian method discussed here.

References

1. E. T. Jaynes, Bayesian spectrum and chirp analysis, in C. R. Smith and G. J. Erickson (eds.), *Maximum Entropy and Bayesian Spectrum Analysis and Estimation Problems* (D Reidel, 1987), pp. 1-29.

2. G. L. Bretthorst, Bayesian spectrum analysis and parameter estimation, in J. Berger, S. Fienberg, J. Gani, K. Krickenberg, and B. Singer (eds.), *Lecture Notes in Statistics* **48** (New York: Springer-Verlag, 1988).
3. G. L. Bretthorst, Excerpts from Bayesian spectrum analysis and parameter estimation, in G. J. Erickson and C. R. Smith (eds.), *Maximum-Entropy and Bayesian Methods in Science and Engineering* (Kluwer, Dordrecht, the Netherlands, 1988), pp. 75-145.
4. G. L. Bretthorst, Bayesian analysis of time-domain magnetic resonance signals. *J. Magn. Reson.* **79**, 369 (1988).
5. G. L. Bretthorst, Bayesian analysis I: parameter estimation. *J. Magn. Reson.* **88**, 533 (1990).
6. G. L. Bretthorst, Bayesian analysis III: examples relevant to NMR. *J. Magn. Reson.* **88,** 571 (1990).
7. P. Gregory and T. J. Loredo, A new method for detection of a periodic signal of unknown shape and period. *Astrophys. J.* **398**, 146 (1992).
8. P. M. Woodward, *Probability and Information Theory with Applications to Radar* (Pergamon Press, London, 1953).
9. P. M. Woodward and I. L. Davies, Information theory and inverse probability in telecommunication. *Proc. IEE., Pt. III* **99**(58), 37 (1952).
10. T. J. Loredo, Bayesian harmonic analysis for audio testing and measurement. In *111th Convention of the Audio Engineering Society*, (Sep 21-24, New York, NY, USA, 2001).
11. G. A. Kyriazis, A simple algorithm for Bayesian signal analysis with applications to metrology, *IEEE Trans. Instrum. Meas.* **60**, 2314 (2011).
12. G. A. Kyriazis, Bayesian inference in waveform metrology, in F. Pavese, M. Bär, J. –R. Filtz, A. B. Forbes, L. Pendrill, and K. Shirono (eds.), *Advanced Mathematical and Computational Tools in Metrology and Testing IX, Series in Advances in Mathematics for Applied Sciences* **84** (World Scientific, Singapore, 2012), pp. 232-243.
13. S. Lipschultz, M. Spiegel and J. Liu, *Mathematical Handbook of Formulas and Tables*, 4th ed. (Schaum´s Outline Series, 2012).
14. G. Strang, *Linear Algebra and Its Applications*, 3rd ed. (San Diego: Harcourt Brace Jovanovich, 1988).
15. E. T. Jaynes, Prior probabilities, *IEEE Trans. Sys. Sci. Cyber.* **4**, 227 (1968).
16. H. Jeffreys, *Theory of Probability*, 3rd ed. (London: Oxford University Press, 1961).
17. A. Zellner, *An Introduction to Bayesian Inference in Econometrics* (New York: John Wiley & Sons, 1971).

18. P. S. Laplace, *A Philosophical Essay on Probabilities* [unabridged and unaltered reprint of the Truscott and Emory translation] (New York: Dover Publications, 1951 [1814]).
19. E. T. Jaynes, *Probability Theory – The Logic of Science* (Cambridge: Cambridge University Press, 2003).
20. W. T. Estler, Measurement as inference: fundamental ideas, *Ann. CIRP* **48**(2), 611 (1999).
21. P. M. Lee, *Bayesian Statistics: an Introduction*, 4th ed. (Wiley, 2012).
22. G. A. Kyriazis, Uncorrelated Gaussian noise assumption in harmonic analysis with integrating digital sampling voltmeters, in *Conference on Precision Electromagnetic Measurements* (*CPEM*) (USA, 8-13 June, 2008).
23. D. R. White and J. F. Clare, Noise in measurements obtained by sampling. *Meas. Sci. Technol.* **3**, 1 (1992).
24. C. Elster, Evaluation of measurement uncertainty in the presence of combined random and analogue-to-digital conversion errors, *Meas. Sci. Technol.* **11**, 1359 (2000).
25. C. Elster and I. Lira, On the choice of a noninformative prior for Bayesian inference of discretized normal observations, *Comp. Stat.* **27**, 219 (2011).
26. G. A. Kyriazis and M. L. R. Campos, Bayesian inference of linear sine-fitting parameters from integrating digital voltmeter data, *Meas. Sci. Technol.* **15**, 337 (2004).
27. G. A. Kyriazis and M. L. R. Campos, Erratum, *Meas. Sci. Technol.*, **15**, 1947 (2004).
28. H. Nyquist, Certain topics in telegraph transmission theory, *Trans. AIEE*, **3**, 617 (1928).
29. H. Nyquist, Certain factors affecting telegraph speed. *Bell Syst. Tech. Jour.*, **3**, 324 (1924).
30. G. Ramm, H. Moser and A. Braun, A new scheme for generating and measuring active, reactive and apparent power at power frequencies with uncertainties of 2.5×10^{-6}, *IEEE Trans. Instrum. Meas.* **48**, 422 (1999).
31. W. G. Kürten Ihlenfeld, Traceability of ac voltage ratios and ac power by synchronous digital synthesis and sampling, PTB-E-76, (Physikalisch-Technische Bundesanstalt, Braunschweig, Oct. 2001).
32. W. G. Kürten Ihlenfeld, Maintenance and traceability of ac voltages by synchronous digital synthesis and sampling, PTB-E-75, (Physikalisch-Technische Bundesanstalt, Braunschweig, Aug. 2001).
33. S. Svensson, Power Measurement Techniques for Non-sinusoidal Conditions, Ph.D. Thesis, Department of Electric Power Engineering, Chalmers University of Technology, Göteborg, Sweden (1999).

34. A. Ferrero and R. Ottoboni, High-accuracy Fourier analysis based on synchronous sampling techniques, *IEEE Trans. Instrum. Meas.* **41**, 780 (1992).
35. R. N. Bracewell, *The Fourier Transform and Its Applications*, 2nd ed., revised (McGraw-Hill, 1986).
36. S. W. Smith, *The scientist and engineers guide to digital signal processing*, URL http://www.dspguide.com/
37. E. Brigham, *The Fast Fourier Transform and Its Applications* (Prentice Hall, 1974).
38. A. Whalen, *Detection of Signals in Noise* (Academic Press, 1971).
39. D. Middleton, *An Introduction to Statistical Communication Theory* (IEEE Press: New Jersey, 1996), reprint, originally published: New York, McGraw-Hill, 1960.
40. G. L. Bretthorst, Nonuniform sampling: bandwidth and aliasing, in G. J. Erickson (ed.), *Maximum Entropy and Bayesian Methods in Science and Engineering* (Kluwer, 2000).
41. N. R. Lomb, Least-squares frequency analysis of unevenly spaced data, *Astrophys. Space Sci.* **39**, 447 (1976).
42. J. D. Scargle, Studies in astronomical time series analysis II. Statistical aspects of spectral analysis of unevenly spaced data. *Astrophys. Jour.* **263**, 835 (1982).
43. J. D. Scargle, Studies in astronomical time series analysis III. Fourier transforms, autocorrelation and cross-correlation functions of unevenly spaced data. *Astrophys. Jour.* **343**, 874 (1989).
44. A. Shuster, The periodogram and its optical analogy, *Proceedings of the Royal Society of London* **77**, 136 (1905).
45. G. A. Kyriazis, P. M. Ramos and A. Cruz Serra, Bayesian and least squares algorithms for estimating signal harmonics: a comparative study, *Measurement* **45**, 2203 (2012).
46. R. Hooke and T. A. Jeeves, Direct search solution of numerical and statistical problems, *J. Assoc. Comp. Mach.* **8**, 212 (1961).
47. M. G. Johnson, Nonlinear optimization using the algorithm of Hooke and Jeeves (1994). URL http://www.netlib.org/opt/hooke.c
48. G. L. Bretthorst, How accurately can parameters from exponential models be estimated? A Bayesian review, *Concepts in Magnetic Resonance*, **27A**, 73 (2005).
49. C. Elster and G. Wübbeler, Bayesian regression versus application of least-squares – an example, *Metrologia* **53**, S10 (2016).

50. G. A. Kyriazis, Bayesian and least-squares approaches for estimating parameters of decay processes, *Measurement* **134**, 218 (2019).
51. G. A. Kyriazis, A Cartesian method to improve the results and save computation time in Bayesian signal analysis, in F. Pavese, W. Bremser, A. G. Chunovkina, N. Fischer and A. B. Forbes (eds.), *Advanced Mathematical and Computational Tools in Metrology and Testing X, Series in Advances in Mathematics for Applied Sciences* **86** (World Scientific, Singapore, 2015), pp. 229-240.
52. G. A. Kyriazis, M. A. F. Martins and R. A. Kalid, Bayesian recursive estimation of linear dynamic system states from measurement information, *Measurement* **45**, 1558 (2012).
53. W. G. Kürten Ihlenfeld, G. A. Kyriazis and R. P. Landim, Kalman filtering applied to the calibration of a high-resolution analog-to-digital converter, *Journal of Physics: Conference Series* (*Online*) **575**, 012013 (2015).
54. P. M. Djuric, Bayesian spectrum estimation of harmonic signals, *IEEE Signal Processing Letters* **2**, 213 (1995).
55. G. A. Kyriazis, Comparison of two algorithms for accurate harmonic analysis of periodic arbitrary signals using nonsynchronous digital sampling, in *Conference on Precision Electromagnetic Measurements* (*CPEM*), (Korea, June 13-18, 2010).
56. G. A. Kyriazis, Bayesian estimation of amplitude-modulated signal parameters from digital sampling data, in *Conference on Precision Electromagnetic Measurements* (*CPEM*), (Korea, June 13-18, 2010).
57. G. A. Kyriazis, Bayesian estimation of interharmonic and fluctuating harmonic parameters from digital sampling data, in *Conference on Precision Electromagnetic Measurements* (*CPEM*), (Korea, June 13-18, 2010).
58. G. A. Kyriazis, Estimating parameters of complex modulated signals from prior information about their arbitrary waveform components, *IEEE Trans. Instrum. Meas.* **62**, 1681 (2013).
59. G. A. Kyriazis, W. G. Kürten Ihlenfeld and R. P. Landim, Estimating parameters of combined AM and PM signals using prior information, *IEEE Trans. Instrum. Meas.* **64**, 1760 (2015).

F Pavese, A B Forbes, N F Zhang, A G Chunovkina (Eds):
Advanced Mathematical and Computational Tools in Metrology and Testing XII
Series on Advances in Mathematics for Applied Sciences, Vol. 90

https://doi.org/10.1142/9789811242380_0004

Review on questionable consistency issues between the CGPM Resolution 1 on revised SI (2018) and the 9th BIPM SI Brochure (2019)

F. Pavese
Formerly at CNR-IMGC (then INRiM),
Torino, 10135, Italy
E-mail: frpavese@gmail.com

One of the major issues debated during the preparation of the 2018 Resolution 1 of the CGPM concerned the switch from the use of physical states to the use of physical constants—in both cases experiments are involved. The differences in the necessary competence and in the implementation difficulties are very considerable. As a consequence, two positions have emerged: the first, now implemented, does not recognise a two-frames structure to the revised SI and does not consider necessary to locally implement any realisation of the constants. This is the BIPM position, now implemented in the text of the BIPM 9th Edition of the SI Brochure. The second position asserts two frames in a revised SI structure, and requires a demonstration that the stipulated values are effectively locally valid. The aim of this paper is to bring evidence that, as a consequence, the last SI Brochure edition contains the above and other interpretations of the Resolution 1, the only official reference document, that are in contradiction with the prescriptions, explicit or implicit, of the latter.

Keywords: 9th SI Brochure; CGPM 2018 1st Resolution; SI-2018.

1. Introduction

One of the major issues arising during the preparation of the SI revision (rSI-2018 in the following), then promulgated by the 2018 Resolution 1 (R1-2018 in the following), [1] of the CGPM, concerned some consequences of the switch from the use of physical states—that can, and have to, be measured—to physical constants—that can, and have to, included in experiments implementing an equation in which they are proportionality constants.[1]

Two positions had emerged: the first does not recognise a two-frames structure to the revised SI and does not consider necessary to locally implement a realisation of the stipulated constants; the second position asserts two frames in revised SI structure, and requires a demonstration that the stipulated values

[1] The term "proportionality constant" is not universally accepted for all the constants in question. [2]

are actually locally valid, as a necessary condition to validate the consistency of the local realisations with each other in the rSI-2018.

The former is the BIPM position, now implemented in the text of the BIPM 9th Edition of the SI Brochure (SI9). In addition, the aim of this paper is to bring evidence that the last SI Brochure contains other interpretations of the R1-2018—which is the only official reference document—that are in contradiction with the prescriptions, explicit or implicit, of the latter.

The paper will start its analysis where the recent paper from Johansson [2] ends, in its excellent demonstration, performed with the tools of philosophy of science, of internal logical and substantial problems of the 9th Brochure (SI9) [3] in spelling out what the BIPM considers the implementation of the R1-2018. The present paper ambition will instead be to detect and illustrate the departures of the 9th Brochure from the letter and the spirit of the R1-2018, from a strict metrological viewpoint, but complementing, without contradictions, the statements in [2], where the metrologists, but also the philosophers of science, are directed for a more complete picture of the present situation.

A short overview is useful to correctly understand some highly important consequences entailed by the new SI definition, from which the new definition of the (same) *base units* depends and becomes modified with respect to the previous one, the SI-1960,[2] that eventually was simply defining seven base units, some containing stipulated numerical value of physical states.

The revision of the SI definition in the R1-2018 consists of a single period, stating that the rSI "... *is the system of units in which* ..."—e.g., for the Boltzmann constant, *k*,—"*the Boltzmann constant k is* 1.380 649 10^{-23} J/K ...", and similarly for all seven constants. This peculiarity comes very clearly in the R1-2018, and concerns all seven constants.

Then, a *distinct* definition of the seven *base units* (for seven quantities) is *not* part of the rSI definition, but appears only as the Appendix 3 to the R1-2018,[3] after an Appendix 1 having formulated in [1]: "*Abrogation of former definitions of the base units*" and an Appendix 2 (see later). These Appendices start with the sentence "*It follows from the new definition of the SI described above* ...", which is a confirmation that the *new* base units—though having the same name and relating to the same quantities— *are not anymore* the SI definition they *were* in the SI-1960 (and in the related 8th Brochure (SI8)). [4]

---

[2] In fact, the promulgation process of the previous SI has been quite complicated, starting in 1948, with an important step in 1954 and ended only in 1960, but here is indicated for convenience as SI-1960.

[3] It has to be noted that the term "base unit" is supplied with no specific information on the SI9.

Appendix 2 specifies the level of uncertainty now associated to the numerical values stipulated exact in the previous SI definition, i.e., the degree of the *magnitude continuity* (see later) of the new with respect to the SI-1960 units for the same base quantities.

In conclusion, it is uncontroversial that the *new* base unit definitions in Appendix 3 are *not* part of the rSI-2018 definition, but are *based* on the numerical values of the constants stipulated in the rSI-2018—e.g., for the kelvin by referring to the Boltzmann constant as follows: "*the SI unit of thermodynamic temperature ... defined by taking the fixed numerical value of the Boltzmann constant k to be* 1.380 649 10^{-23} *when expressed in the unit* J K^{-1} ..." [3]. [4]

This paper will identify and illustrate the following discrepancies between the SI9 and the R1-2018 promulgating the rSI-2018—the Section number is indicated:

2) The lack of recognition of the two-step structure of the rSI-2018, bringing to a logically inconsistent formulation, including its definition (SI9, p. 125).

3) The lack of correct information about the reasons for the choice of the same base units (with a different definition) with respect to the SI-1960 and the SI8 (SI9, p. 129);

4) The lack of explicit recognition of the consequences of the fact that the rSI-2018 definition only concerns the stipulation of the seven constants, and that the definitions of the units are not part of it, but only refer to the rSI-2018 definition with a mandatory condition: that, for each constant, there is a *continuity condition* that must be satisfied in order to ensure that it allows to obtain the stipulated numerical value;

5) The consequent need to demonstrate that the locally-realised units satisfy the following requirement: to be consistent with the stipulated numerical values of the relevant constants;

6) An ambiguous distinction between *primary* realizations and *mise en pratique* realizations (SI9, p. 135);

7) The lack of information about the consequences of the circularity in the definition of some base units, and about the actual possibility to realize the Planck constant h according to its definition (SI9, p. 126) (after [2].

---

[4] That is somewhat odd, being the SI the "Système International des Unités", but the present definition instead only stipulates *the numerical values of the constants* that must be realized by the same measurement units, after having abrogated the previous definitions—though the rSI-2018 is *not* explicitly representing a mere revision of the SI-1960 (in this respect, see [5])

Finally, an Appendix here contains an analysis about the correct calculation and different representations of the numerical values of the *derived constants* depending on some of the newly stipulated ones.

2. Inconsistency of the formulation of the rSI-2018 definition

The definition is reported in the Introduction. It is to be considered as a new definition, because the SI-1960 did only define the seven base units, though later, for length was already based on the constant *c*. [5]

However, it looks like the chosen spelling overlooks the fact that, by indicating the numerical value of each constant, the need arises to also indicate the *measurement units* used to obtain that value, and both are the SI-1960 ones that are retained (except now for their present stipulated value). However, as to the rSI-2018, the (independent) formulation of the new unit definitions comes only after the abrogation of the SI-1960 ones (Appendix 1 to the R1-2018) and as a subordinate to the rSI-2018 definition. Thus, the units definitions indicated in the R1-2018 are logically meaningless.

On the other hand, how to make the necessary connection to the SI-1960 without losing the claimed independence of the rSI-2018? This can be done only via *two alternatives*, both already suggested before the rSI-2018 promulgation:

A) By considering the rSI-2018 not a brand new one, but a *revision of some parts* of the SI-1960, e.g., as formulated in [5] (consult it for the full proposal):
(start)

> "... the practical system of measurement units, the International System of Unities, SI, as defined by the Resolution 6 of the 10th CGPM Meeting (1960) and subsequent amendments, is further amended as follows.
> For the aim of the SI, the *numerical values* of the following constant quantities have no uncertainty:
> – for the frequency of the non perturbed radiation corresponding to the transition between the two hyperfine levels of the fundamental state of the caesium 133 atom $\Delta\nu_{Cs}$ it is 9 192 631 770,

[5] In the SI9, the Resolution 1 translated into the following sentence: "*The definition of the SI units is established in terms of a set of seven defining constants. The complete system of units can be derived from the fixed values of these defining constants, expressed in the units of the SI. These seven defining constants are the most fundamental feature of the definition of the entire system of units. These particular constants were chosen after having been identified as being the best choice, taking into account the previous definition of the SI, which was based on seven base units, and progress in science.*" (SI9, p 125, 1.1 "*The SI and the defining constants*"). Not as sharp and unambiguous as CGPM Resolution 1, and Clause 1.3 in SI9 "*Implementation of the SI*" only refers to the SI units.

– for the speed of light in vacuum c it is 299 792 458,
– for the Planck constant h it is 6.626 XX $\times 10^{-34}$, …" [*similarly for the rest*]
"…The units of the practical system of measurement units remain in place, but the definition of the seven base units are repealed and the following equivalent definitions are adopted:
- The duration of the second is equal to the total duration of the number of periods corresponding to the numerical value of the adopted frequency of the radiation $\Delta\nu_{Cs}$ of the caesium 133 atom; …" [*similarly for the rest*]

(end)

B) By recognizing that the full revision of the SI-1960 is made by means of *two* logical and operative steps, e.g., as formulated in [6–8] (consult them for the full proposal). A full conceptual roadmap was made according to the formulation in [6–8]:

(start)

"… by construing a system of units according to a two-stage structure that explicitly includes a fundamental system and a conventional one:

• both a *system of quantities* and a set of constants corresponding to the dimensions of the base of the system are assumed; this represents the fundamental system where the numerical value of each constant is 1;

• a *conventional system* is then considered linked to the above fundamental system, where it is admitted that the numerical values of the constants can have values different from 1, assigned according to the best available present knowledge so that in the change to the new system the units maintain their values as expected according to the principle of continuity.

In the "Global Explicit Constant Definition" (GCD) required for the "New SI", the units are defined by taking as referent a set $\{C_i\}$ of quantities modelled as constants, and taking a given numerical value k_i for each C_i. Each unit $[Q]$ is then defined as:

$$[Q] = \Pi_i(C_i / k_i)\, n_i. \qquad (1)$$

Step A: The *SI-fundamental* is the coherent system of units in which C1 is the fundamental-unit of frequency, C2 is the fundamental-unit of speed and C3 is the fundamental-unit of action, ….

Step B: The *SI-conventional* is the coherent system of units in which

the unit of time (duration) is the second, symbol s, the unit of length is the metre, symbol m, and the unit of mass is the kilogram, symbol kg;
the frequency $\Delta\nu_{Cs}$ is k_1 s^{-1}, the speed *c* is k^2 m s^{-1}, the action *h* is k_3 kg m^2 s^{-2}, …
Then, the numerical values of the k_i should be specified in a relevant section of the SI-2018 and of the SI9."

(end)

The inconsistency in the definition formulation (Section 2) comes as a consequence of the lack of this distinction.

Notice that frame B is obviously un-avoidable, being the true aim (see Footnote 4) of the SI—to define the measurement units—but it is based on the former. In addition, it must ensure the *continuity* of the unit magnitudes with those of the former-SI (and *consistency* with the ISQ-ISO); however, this fact is not indicated in the SI9 to be part of *the definition* of the revised SI—in fact, it is put in [1] in Appendix 3 (see above and in Section 3 below).

3. Lack of correct information about the reasons for the chosen base units with respect to the SI-1960

As said in Section 2, the units indicated in the rSI-2018 definition for the constants are taken from SI-1960 without any comment or previous indication.

Thus, since for the new definition of the same units the rSI-2018 indicates a new mandatory condition—that each new standard must ensure that it allows to obtain the stipulated numerical value of a (single) reference constant—and the previous unit definitions are abrogated (Appendix 1 of the R1-2018), the new definitions of those units are implicitly determined by a kind of tautology: if the magnitude continuity principle (see Section 4) is respected, the stipulated numbers are consistent with the SI-1960 units, and the "new" rSI-2018 definitions of the units only reflect that information. In fact, those numerical values are obtained using the previous SI.

Therefore, the sentences in the SI9 telling

"*This description in terms of base and derived units is maintained in the present definition of the SI, but has been reformulated as a consequence of adoption of the defining constants*" (SI9, p 124), and

"*Defining the SI by fixing the numerical values of seven defining constants has the effect that this distinction is, in principle, not needed, since all units, base as well as derived units, may be constructed directly from the defining constants. Nevertheless, the concept of base and derived units is maintained because it is useful and historically well established*" (SI9, p 129), are incorrect.

However, the rSI-2018 was *obliged* to keep intact the previous system of units, since only in that sense, the definition of the units are not a circular reasoning.

The rSI-2018 new unit definitions do *not* indicate anymore a *definitional method* for their realization, so leaving full freedom to choose one. However, one can deduce from the above that, among the various exemplifications, the *previously* used methods *remain valid*, having provided evidence that they allowed to obtain the stipulated values (obviously within their declared uncertainties). The same situation is concerning the constants, where the theoretical methods/experiments having allowed to obtain the values then stipulated are by default also valid in future with higher confidence (within the then-obtained uncertainties).

However, no information is provided in the SI9 about the fact that the new base units—that looks not to be anymore the ones having defined the SI-1960—have not only the same name and are relating to the same quantities, but are actually *the same in all respect*, except for the lack of restrictions concerning their realizations).

4. Lack of illustration of the dilemma between units magnitude continuity or significant discontinuity with respect to those of the SI-1960, and reasons for choosing the first

Stipulated values must be used in any SI definition, since it is not admitted that they may be considered still uncertain like the experimental numerical values they take the origin from. It means that digits following the expressed ones by the stipulated number are a periodic zero.

Therefore, let us start from a recall about the way the stipulated numerical values originate and are expressed in metrology. In that frame, the stipulation process, which is based on the current "best" experimental value and associated uncertainty, should in turn also *avoid stipulated values not consistent* with the original experimental values. This requirement involves the need that two conditions are satisfied:

(i) to avoid the stipulate digits of the experimental data *affected by uncertainty* (directly or within the interval determined by the expanded uncertainty);

(ii) to use rounding instead of truncation for the last expressed digit, the experimental values having normally a larger number of expressed digits.

The CGPM based its stipulation on the original experimental data and their associated uncertainties according to the CIPM [9] (and the CCU) expressed preference "*for the minimum number of digits* [of the stipulated value] *for each defining constant h, e, k, and N_A of the revised SI that yields consistency factors equal to* 1 [exactly] within *their uncertainties*".

Table 1 shows both the resulting CODATA/CGPM adjusted/stipulated numerical values and, in the last column, the representation of what will be called here the "exactly known number", meaning that their expressed digits are not affected by the original uncertainty. The CODATA stipulations, presented here in the second column of Table 1, are *different from the values in the third column* (stipulated by the CGPM).

Furthermore, should the "expanded uncertainty", *U*, be used—commonly advised be more of standard use in science— also for the stipulation of the numerical values, the uncertainties of the numerical values in the second column of Table 1 would be multiplied by ≈2.

Table 1. Different ways of treating the digits with the same numerical values for the four new constants.

Constant numerical value	CODATA 2017 ($k = 1$) [10]	CGPM stipulated [a]	Exactly known number [b]
$\{h\} \times 10^{34}$	6.626 070 150(69), $u_{rel} = 1.1 \times 10^{-8}$	6.626 070 15, $u_{rel} = 7.5 \times 10^{-10}$	6.626 070 (1), $u_{rel} = 2.3 \times 10^{-8}$
$\{e\} \times 10^{19}$	1.602 176 6341(93), $u_{rel} = 0.6 \times 10^{-8}$	1.602 176 634, $u_{rel} = 3.1 \times 10^{-10}$	1.602 176 6 (3), $u_{rel} = 4.0 \times 10^{-7}$
$\{N_A\} \times 10^{-23}$	6.022 140 758(62), $u_{rel} = 1.1 \times 10^{-8}$	6.022 140 76, $u_{rel} = 8.3 \times 10^{-10}$	6.022 140 (8), $u_{rel} = 1.3 \times 10^{-7}$
$\{k\} \times 10^{23}$	1.380 649 03(51), $u_{rel} = 7.4 \times 10^{-7}$	1.380 649, $u_{rel} = 3.6 \times 10^{-7}$	1.380 6 (5), $u_{rel} = 3.5 \times 10^{-5}$

[a] The smaller case digits in the proposed stipulations are used here for those affected by the uncertainty interval in the previous column.

[b] Here "exact" means unaffected by the original experimental uncertainty but not stipulated. The added smaller case digit in parentheses is not exact, being affected by the CODATA uncertainty interval indicated in the previous column: it is reported because the rounding up, different from truncation, would affect the exact numerical value of N_A and *k*.

In this case, the above divergence would not change for any of the four constants if the "exactly known numbers" were not rounded. In the latter case, when instead rounded, *k* and N_A would "miss" one further digit.

However, there is *another basic principle*, explicitly spelled out within the CIPM/CCU rules when they speak of "consistency factors", which *must also be respected* in stipulation: the "continuity principle" (see in rSI-2018 and SI9). Values of the "consistency factors" can be computed from the stipulated values reported in Table 1 and are reported here in the equations below—in square brackets the relative level of continuity indicated in Appendix 2 of the R1-2018:

$$[m(\mathrm{K})/(\mathrm{kg})_{\mathrm{rev}}]/1 = 1.000\,000\,001(10)\ [1\times10^{-8}] \quad (2)$$
$$[\mu_0/(\mathrm{H\,m^{-1}})_{\mathrm{rev}}]/(4\pi\times10^{-7}) = 1.000\,000\,000\,20(23)\ [2.3\times10^{-10}] \quad (3)$$
$$[M(^{12}\mathrm{C})/(\mathrm{kg\,mol^{-1}})_{\mathrm{rev}}]/0.012 = 1.000\,000\,000\,37(45)\ [4.5\times10^{-10}] \quad (4)$$
$$[T_{\mathrm{TPW}}/(\mathrm{K})_{\mathrm{rev}}]/273.16 = 1.000\,000\,01(37)\ [3.7\times10^{-7}]. \quad (5)$$

These factors are considered to correspond to the CIPM's indicated criteria. However, though the above equations are the direct consequence of using all CODATA stipulated digits, they are *conflicting with the continuity principle*.

The conflict occurs when (see Table 1) the principle, though rather obvious, is violated that one should not stipulate a number whose (last) digit(s) are not *firmly* confirmed by the experiments. Could one stipulate a number having, say, three digits more than the experimental uncertainty level? That is what is happening. The fact that the uncertainty is eventually dropped in stipulation is *entirely irrelevant*: uncertainty means that the digit(s) affected by it could *presently be different* from the stipulated one(s) when using different statistical tools.

Therefore, a *basic dilemma* arises (apparently undetected, certainly unpublished and not explicitly resolved) for the constants in the above situation:

(0) **Accepting constant stipulated values exceeding the present experimental accuracy, but preserving unit magnitude continuity**,

or,

(1) **Requiring constant stipulated values to conform to present experimental accuracy, but inducing a unit magnitude discontinuity (small but significant)**.

In fact, the founding experimental results, despite their conspicuous five-time uncertainty lowering in a decade, still do *not support the firm continuity* of the units' magnitudes. To obtain the desired continuity, one is obliged to "guess" that the last digit(s) affected by uncertainty are correct.

Since the issue of continuity is strictly one of the *most basic features* of the metrology regulatory context, one may justify considering correct what the BIPM did as just sufficient to ensure continuity. However, the SI9 does not contain an explicit illustration of the problem nor a declaration justifying the choice in this respect, including the fact that the experimental data are *supporting only two digits less*.

5. Need to *demonstrate* that the locally-realised units satisfy the requirement to be consistent with the stipulated numerical values of the relevant constants

Every NMI is at the top of its *national* traceability chain, so it must demonstrate the claim that its SI unit realisation satisfies the *new condition* to be "such that" their realized units are consistent with (or brings to) the stipulated values of the constants, obviously within the assigned uncertainty of their realisation.

This condition is not automatically fulfilled, but it can only be fulfilled by means of *experimental demonstrations* of the correct realisation of the new units, based, first of all, on an independent *experimental realisation* of the relevant constant(s) using *absolute primary* methods. [12]

The latter has to be performed in a way similar to what has been done so far, e.g., for the realisation of the triple point of water (TPW) according to the previous definition of the kelvin. [6]

It means that experimenters will now do their best to be confident in the assumption that their obtained numerical values of the relevant constants are consistent with the *stipulated* ones (in an experiment only consistency can be obtained, not the exactness of the value).

It will be exactly what some of them did before May 20, 2019 when their own measurements were performed, leading in the case of the elementary charge, to values of *e* intended to be used in the database eventually having established the numerical "best value" of the constant, then adopted in the revised SI—but with a basic difference.

Now, e.g., for the numerical value of the TPW, they *have to, and only can, assign* as the numerical result of their realisation the *stipulated* value of the Boltzmann constant, *k*, as they did assign in the past the stipulated value 273.16 K to their own realisation of the TPW. No means exist within a single laboratory to test the assumption that the local realisation of the units comply with the new requirement—that the corresponding value of *k* is the stipulated one—and within the aimed local uncertainty. These can be called *local*

---

[6] Here the kelvin and the TPW are used as an example.

realisations of the SI definition or, correspondingly, *local realization* of the ampere, etc… and the check can only be performed, as usual, by means of *inter-laboratory comparisons*.

In fact, as the equivalent of a Key Comparison (KC), e.g., of k_{local}, it should closely follow the former practice for checking the realizations of the previous standards [13]. This kind of KC will allow for multiple realization methods in order to find the *degree of equivalence* (DoE) of each local absolute primary realization. However, only differences between realisations can be obtained from a KC: it remains to be decided to which laboratory, if any, the *stipulated* value should be attributed by default (perhaps to BIPM as the international reference laboratory). That will be necessary to allow performing corrections of k_{local} in the example, if corrections will be allowed. These results may also lead to the possibility of Significant Unresolved Deviations (SUD) occurring (e.g., for the realisations of the Planck constant, *h*, where discrepant values were obtained).

One peculiar and *exclusive group of local realisations of absolute* primary methods (not specified in the SI9) concerns today those laboratories that *already realized* a specific constant before September 2017. Some of them were used by CODATA to recommend a numerical value of that constant, which was eventually adopted and fixed by the CGPM. [1] For them the local value of the constant realised by them is *already known*, therefore also the DoE is known of their realisations with respect to the *stipulated value* (contrarily to the case of a KC)—see also end of Section 3 about that. Should the DoE lack consistency with the latter, they should broaden their uncertainty, as is normally prescribed in KCs for the SUDs.

However, a peculiar difficulty remains: which are the SI units responsible for a SUD on the *local* constant realisation? *In the case of electrical current, not necessarily only the unit coulomb, because e is multidimensional*. It is the same also for other units.

Therefore, performing the correction to the *magnitude of each relevant unit (or an experimental quantity)* is not a trivial task and certainly requires more pondering.

The obligations arising from the status of the two-frames new definition (see Section 3) are *not explicitly indicated*. In addition (see also Table 2):

• If frame A) constitutes the *top of the international traceability chain*, the revised system may lead to a hierarchical situation illustrated later, if not correctly applied;

• If frame A) is *not* its top, the function of the constants, the deriving obligations, and the local implementations, are *not explicitly formulated*: the status of frame (B) is *not explicitly indicated.*

Table 2. Metrological traceability chain according to the SI (example: length).

Traceability Level	Former-S I	Revised-SI
Top	Definitional method. *Method using "distance" and "time interval"* [a]	No definitional method. "New condition": to reproduce the stipulated constant(s) exact value
−1	"Mise en pratique". *Other method(s):* *using frequency and period*	Primary methods: *c_0 and t explicit in the model*
−2	Secondary methods. *Other method(s):* *stabilised laser*	" Mise en pratique". *Other method(s):* *using frequency and period*
−3	Workshop methods. *Gauge blocks*	Secondary methods *Other method(s):* *stabilised laser*
−4 ...	Lower ranks ...	Workshop methods: *gauge blocks, ...*

[a] This requirement was not always respected in the traceability chain, using instead frequency and period.

The position of who denies the need of the local consistency proof is generally justified by the fact that the value of the constants is fixed after stipulation.

This fact is actually not at all new. Already in the past any numerical value included in a unit definition was necessarily stipulated. An example is the kelvin, where the triple point of water, its definition before 20 May 2019, was fixed to 273.16 K. Another example is the definition of the metre, involving the fixed valuc of the velocity of light. However, when the previous SI required to demonstrate that an NMI had complied to that condition, this only meant that any new realisation had to be assigned *by default* that value, while the truth of that assignment had to be confirmed by a key comparison (KC) and any significant difference found had to be taken into account. Also in that case, no Laboratory was supposed to be assigned the correct value by law, because no hierarchy did apply.

This will happen exactly in the same way for the new constants. The *new condition* (see Table 2) is that 'the revised SI is "such that" the stipulated values must apply': it *cannot* be satisfied *automatically, or necessarily, or* anywhere among the NMIs. Therefore it has to be *proved*: only the realisation of at least one primary method can do that. Only a few NMIs (i.e., Nations in the World) are *already* in the position to implement it, as noted before, since they realised constants for some quantities in their studies that formed the database used for the final stipulation of the value. They know the differences of their numerical values with each other prior stipulation, so they know their differences with respect to the stipulated value and would not require checking it by means of KCs. In principle they could even act as "reference" in future KCs.

However, this privilege looks quite unfair with respect to the rest of the NMIs: it would lead to a hierarchy that never existed before and would be contrary to the Treaty of the Metre.

The only way to escape that situation and that requirement (very heavy to perform), would be to admit that the top of the metrological traceability chain does not reside in those local realisations of the relevant constants, but at the (hierarchically) lower level of the practical realisation of the units. However, that would mean that it should not be anymore necessary to prove that NMI's standards comply with the SI definition, quite unusual, and again would establish a hierarchy among NMIs which was never necessary and never admitted before.

In all instances no guidance is presently available in this respect from BIPM documents.

6. Ambiguous distinction between "primary" realizations and "mise en pratique" realizations of a unit

No instructions on the realization of the condition that a unit must be compatible with the stipulated value of the relevant constant are included in the SI9, as a consequence of the BIPM position illustrated above in Section 2. In Section 3 the expression "*primary* method" was used without full explanation. It is introduced on p. 135 clause 2.3.2 "*Practical realization of SI units*" in [3]: "The highest-level experimental methods used for the realization of units using the equations of physics are known as *primary methods*. The essential characteristic of a primary method is that it allows a quantity to be measured in a particular unit by using only measurements of quantities that do not involve that unit.".

The term "mise en pratique" (MeP), term introduced during the previous version of the SI, is used only in SI9 and only on page 125, clause 1.1 "*A variety of experimental methods described by the CIPM Consultative Committees may be used to realize the definitions. Descriptions of these realizations are also referred to as 'mises en pratique'* " (MeP), but is deferred to the "practical realizations" decided by the BIPM Comités Consultatifs (CCs) (SI9, Appendix 2, actually available only from each CC [14]).

Therefore, note that the use of *primary methods* for determining the numerical value of a constant is precisely the reason why these methods *cannot be included* in a "mise en pratique"—unless their implementation is sufficiently simplified. This could be done, e.g. by using them in the "*relative mode*" (see below) thereby making them of practical use (in the sense of the methods of the MeP, as the ITS-90 for temperature).

These limited and vague indications brought to *incorrect instructions* to the BIPM CCs about the realization of the units. For example, taking the document from the CC for Thermometry (CCT), MeP-K_2019, [12] under the Chapter 3 "*Nomenclature*", one finds the following definitions:

"***Primary thermometry****: is carried out using a thermometer based on a well-understood physical system, for which the equation of state describing the relation between thermodynamic temperature T and other independent quantities, such as the ideal gas law or Planck's equation, can be expressed explicitly without unknown or significantly temperature-dependent constants. Thermodynamic temperature can be obtained by measuring the independent quantities*… .

Absolute primary thermometry*: allows to measure thermodynamic temperature directly in terms of the definition of the base unit kelvin, i.e. the defined numerical value of the Boltzmann constant. No reference is made to any temperature fixed point (n* = 0*, n = number of points) and all other parameters specified in the equation of state are measured or otherwise determined*… .

Relative primary thermometry*: allows to measure thermodynamic temperature indirectly using a specified equation of state, with one or more key-parameter values determined from temperature fixed points (n >* 0*), for which values for the thermodynamic temperature T and their uncertainties are known a priori from previous absolute or relative primary thermometry*… .

Defined temperature scales*: allow to assign temperature values, determined by primary thermometry, to a series of naturally occurring and highly reproducible states (e.g., the freezing and triple points of pure substances)*… ."

They were all already included in the MeP of the kelvin unit defined on SI-1960 by means of the triple point of water (TPW), having thermodynamic T = 273.16 K exact—by means of he numerical value of the Boltzmann constant in rSI-2018. In the past, the MeP-K was assumed to be able to include also thermodynamic measurements, similar to the two cases of the absolute or relative primary methods, but under a very different situation with respect to the present one—but that possibility was never implemented.

In fact, in the absolute-mode case, one had to measure T in a physical equation (see below Section 7.1) where all the parameters of the model, except, T were measurable with sufficiently low uncertainty, and the measurements had to include the TTPW, whose temperature was fixed. In the rSI-2018, instead, one needs to measure T in a physical equation where k is fixed at the stipulated value and all other parameters are measurable. That brings one back to the situation illustrated in Section 3, where no check is possible for any single experimenter for the correctness of his/her assumption in that particular case (as said there, a situation similar when previously the TPW had a fixed value). However, while it was relatively simple to ascertain if the used device for measuring T_{TPW} was accurate, now no such guidance is provided, indicating how to ascertain that k effectively has the stipulated fixed value in that occurrence (see Section 5 above).

In addition, according to the need to make a check for the latter issue, as illustrated in Section 5, the absolute primary methods are just those, and only those, that can allow a check, followed by a KC, of the compliance of the local rSI-2018 units to the condition set by the stipulation of the numerical values of the seven constants. Therefore, their inclusion in the MeP is, by definition, totally inappropriate. Their use do not concern (only) the unit definitions, but the required match with the stipulated values of the constants, which is not automatically achieved, so also not would be the necessary international traceability of each NMI.

7. Lack of information about the consequences of the circularity in the definition of some base units, in particular about the possibility to use in practice the Planck constant, *h*

The logical circularity issue of the rSI-2018 definitions of the units is fully treated in [2] and will not be illustrated here. The existence of this problem is not recognized in the SI9.

In the following only the most intriguing problems are illustrated.

7.1. *Case of the temperature unit*

In Fig. 1 the flow chart is shown for a planned measurement of temperature with a primary method, e.g., by using the indicated ideal-gas thermodynamic equation.

A primary method makes use of an equation where the constant on which the unit of the quantity to be measured, here thermodynamic temperature, T, must explicitly appear. Other quantities are measured for this purpose, and the figure shows that which other constants that defines the units of these other quantities should be implied. Each of them must be assigned the stipulated value.

The equation in Fig. 1 is not the measurement equation, which is much more complex and involves also other instrumental quantities, namely of the electrical category, thus involving also one further constant, the electrical charge, *e*. There the *ideal*, or *reference model* is shown shared by the Community of metrologists and physicists. For more details on the issue of the model, outside the scope of this paper, see [15].

In Fig. 1 the quantities involved are only indicated by means of the SI units involved, after their numerical value indicated in Greek letters. The Boltzmann constant, *k*, is the one defining the kelvin.

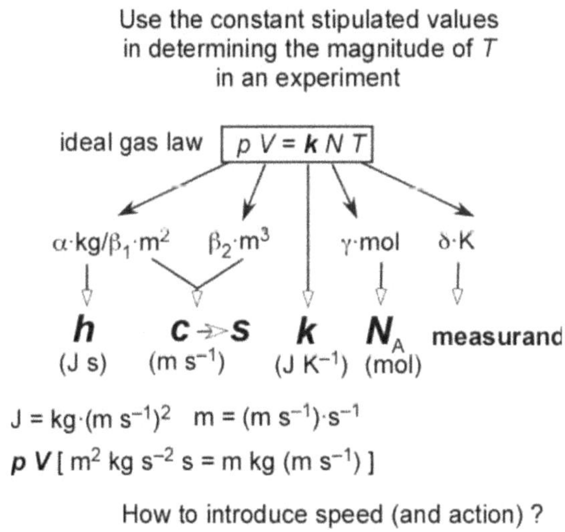

Fig. 1. Flow chart of an experiment using primary method for the determination of thermodynamic temperature *T* in the kelvin unit.

One can easily see that the length unit requires both the speed of light and the second, and that the energy unit (joule) is a derived unit requiring, in particular the kilogram that is defined by the Planck constant, *h*. The mole has no interaction with any other quantity.

The Planck constant is directly required by the measurement of pressure, p = force/surface (= mass × (acceleration/surface). However, as already indicate in [2] no equation is available for expressing mass in terms of *h*. Thus, there is no way to introduce into the calculations the stipulated numerical value of *h*, $\{h\}$, which must be explicit in a primary method equation. This is contrasting the rules for the use of any primary method and there is no guidance about a possible solution. In fact the BIPM CC of mass (CCM) has already expressed the need to resort to a KC, [16] the only one CC having recognized so far that general need. *

7.2. *Case of the mass unit*

The official primary source is the mass *mise en pratique* document, MeP-kg (2019), a CCM document, where primary methods are defined and indicated. [17] In it, the traceability chain for the kilogram is shown in Fig. 2.

According to [17], "the Planck constant is exactly $h = 6.626\ 070\ 15 \times 10^{-34}$ Js. This numerical value of h defines the unit joule second in the SI and, in combination with the SI second and metre, defines the kilogram. The second and metre are themselves defined by exact values of the hyperfine transition frequency $\Delta\nu_{Cs}$ of the caesium 133 atom and the speed of light in vacuum c. The numerical value of h given in the definition of the kilogram has ensured the continuity of the unit of mass with the previous definition", where the continuity degree is given here in the Eq. 2 above.

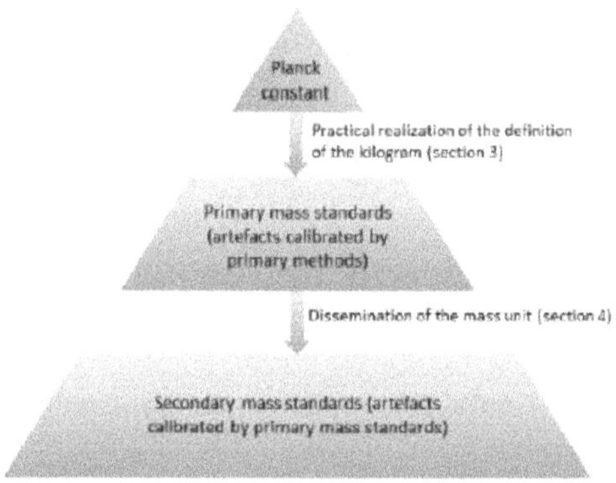

Fig. 2. Traceability chain for the realization of the kilogram according rSI-2018. [17]

* The first KC was then performed in 2019 [32], where it is indicated: "The objectives were to determine the level of agreement between realizations of the kilogram using Kibble and joule balances and the X-ray crystal density (XRCD) method and to provide input for the calculation of the first "consensus value" of the kilogram". The result has been: "The key comparison reference value … has a deviation of -0.0188 mg with respect to the mass unit maintained by the BIPM working standards, with a standard uncertainty of 0.0075 mg". One participant's result was not compatible for $k = 1$, all were for $k = 2$. "The BIPM working standards, traceable to the Planck constant through the IPK, and the "reference value" were consistent within $k = 1$. To be noted that the latter, "$u(\Delta m_{h(IPK)}$, … is dominated by the uncertainty of the mass of the IPK in terms of the Planck constant, 0.010 mg".

Two primary methods are indicated and illustrated in [17]: the Kibble balance and the silicon density method (the latter arising from the so-called "Avogadro Project" [18]).

The problem with the Kibble balance has already been clearly illustrated in [2], critics that are shared here: basically, it cannot be considered as primary, *not* because it is a material device (as the IPK was) *but* because it requires an empirical calibration, so, in principle it can be suitable only for measuring mass ratios.

About the "Avogadro method", it is correctly indicated: "*The measurements described here were first used to determine the value of the Avogadro constant* N_A, *which is defined as the number of elementary entities per mole of substance. An accurate measurement of* N_A *was an essential contribution on the road to redefining the kilogram in 2018.*". Then, it continues: "*Today, however, the numerical value of* N_A *is exactly defined when expressed in the SI unit* mol^{-1} *thus making the definition of the mole independent of the kilogram*". The fact that the mol is defined through its reciprocal, brings to an inconsistent situation already discussed in [2] for the second, with undesirable consequences also in this case because, as for the second, the defined number cannot express a count (of entities here, of periods for the second), being a generic real number. [7]

In addition, the founding equation, $m_s = N\, m(^{28}Si)$, [8] "is well known [to be] not exact because the right-hand side is reduced by the mass equivalent, E/c^2, of the total binding energy E of the atoms in the crystal, where c is the speed of light in vacuum" resorts to an equation for which there is no demonstration in current physics of the validity in this case (the fact that an insignificant correction is needed, is irrelevant). Thus:

$$m_s = h\, N\, (m(^{28}Si)/h), \qquad (6)$$

where "$m(^{28}Si)/h$ is a [new ?!] constant of nature" and the mass $m(^{28}Si)$ is that of an element's isotope, ^{28}Si. The ^{28}Si in the device used in the Avogadro measurements is very pure ^{28}Si, but *not* ideally pure, so that an *empirical*

[7] Note that the SI9 does not consider possible to have quantities that are counts (see also [2]).

[8] "CODATA value for the Si{220} lattice spacing d_{220} = 192.015 571 4×10^{-12} m, with a standard uncertainty of 0.000 003 2×10^{-12} m, (i.e. $\Delta d/d$ = 1.67×10^{-8}) at a temperature of 22.5 °C in vacuum. This is the lattice spacing of an ideal single crystal of natural-isotopically un-doped silicon that is free of impurities and imperfections." (from MeP-kg [17], see Section 6.3)

correction is needed. Furthermore, in contrast to the general statement about the use of constants [3], here a *particular* element is referred to, as for the time unit (^{133}Cs). Finally, also in this case a device (a silicon sphere, subject to boundary definition and to wearing in time as for the IPK, but without present experience about that) is used for implementing the method.

7.3. *Case of the length unit*

In addition to the logical problems arising from the fact that length is now defined my means of the concept of speed (and so needed also time, a solution a bit surprising but already in force since 1958), we propose here an analysis based on the official primary source, the length "mise en pratique", MeP-m (2019), a CCL document where primary methods are defined and indicated. [19]

In this case the founding equation is known:

$$l = c\,\Delta t, \tag{7}$$

where l is the "path travelled" and is the Δt is the "time duration", assumed to be, respectively, the "length" and the "time" quantities, and c is the "speed of light in vacuum" connecting them through the relationship whose units are ms^{-1}, where [m] is the unit of length and [s^{-1}] is the unit second according to its present definition.

The MeP-m states: "… *the definition of the metre can be realized in practice by one of the following primary methods: a) by direct measurement of light travelling time, b) by indirect measurement of light travelling time.*": only type (a) can strictly be considered consistent with the definition and its Eq. 7 above.

Then a note indicates difficulties in using type a) **method: not-in-vacuum**, "*…in most cases the realization of the length unit is performed under atmospheric pressure. Then the exact value of the influence of the air on the speed of light is of major importance. Therefore, a distinction has to be made between c, the speed of light in vacuum and c', the speed of light in general. Under atmospheric pressure, the air refractive index reduces the speed of light (c' = c/n) with a relative effect of the order of* 3×10^{-4} *corresponding to* 300 μm *per metre of measured length.", so a correction is needed; wavelength-used, "in the case of modulated light it is important to consider the group refractive index of air* n_g *instead of the (phase-) refractive index, n. For example, for green light* ($\lambda \approx 520$ nm) $n_g - n$ *is approximately* 10^{-5}*, i.e. considering instead of* n_g *causes an additional error of* 10 μm *per metre. This difference is significant and*

*comparable in size to the variation of the phase refractive index of air over the entire range of visible light: n(*380 nm*)* – *n(*780 nm*)* = 9×10^{-6}".
The Note concludes: "*Thus, determination of the exact speed of light to use in equation* [Eq. 7 here] *is a significant consideration in realizing the metre through primary methods*".

For such a determination, two are the basic parameters:

Frequency of the light. "*Knowledge of the frequency of the light, f, is an essential requirement for the realization of the unit of length. It provides the scaling factor between a measured phase difference and the length that is realized by interferometry. Often, the value of the so called 'vacuum wavelength', which describes the distance between the wavefronts in vacuum under idealized conditions* ($\lambda_0 = c/f$), *is stated instead of the frequency. For highest demands on the accuracy of the light frequency, a light source could be synchronized to the primary frequency standards by an appropriate technique.*";

Phase difference, $\Delta\varphi$. It can be obtained "*using length measuring interferometers is mostly performed in air. The presence of air reduces the speed of the light to c/n and the wavelength to* $\lambda = \lambda_0/n$. *Reduction of the speed of light leads to enlargement of the measured phase difference, i.e. the same length is realized in vacuum and in air according to Eq. 7. The actual amount of the air refractive index is dependent on the air parameters and the wavelength.*"

The above conditions involve critical *empirical* corrections.

7.4. *Measurement equations of primary methods*

All the above-reported equations are for ideal situations agreed by the scientific Community. The *measurement equations* used in each *specific* experiments are much more complex, as they must include all the influence quantities that arise, not only from technical reasons due to the experimental apparatus, but also from actual (and contingent) deviation from ideality of the physical quantities involved in that specific measurement. The reader is directed, e.g., to [15], for a full illustration.

In other words, those equations must include what are commonly called "corrections", to *take into account deviations from reference states or ideal conditions*. They are (necessarily) allowed also in the implementation of primary methods, though they can occasionally require the use of the same quantity under realization, typically temperature in specific sites of the equipment in Fig. 1. They are not assumed to affect by circularity the final results: in case of detected relevance, special measurement procedures are to be used to compensate for their possible effect on the final results.

As to the unavoidable overall uncertainty of the final result, it comes from estimate of a single component (the uncertainty budget) for each influence quantity, in order to build up the total uncertainty budget, with the indication of the type of evaluation of the data dispersion (for probabilities, their "distribution", not necessarily only Gaussian or uniform). No component is allowed to miss that information, unless it comes with a stipulated number or its uncertainty is declared irrelevant in respect to the aimed overall uncertainty.

8. Conclusions

The paper analysed the contents of the SI9 (2019) Brochure and compared them to the formulation of the R1-2018, finding some critical divergences of the SI9 from the CGPM R1-2018. In addition, several omissions where found of explication and guidelines about the *local* application of the statements in the R1-2019, which may prevent the users of the rSI-2018 from understanding its rational and from correctly implement their actions as required by the R1-2018.

In principle, the definition of the rSI-2019,

> the SI "… *is the system of units in which* …" the value of certain physical constants have a stipulated fixed numerical value when the measurement units are the ones indicated in the definition,

could be translated into the previous SI definition as follows, e.g. for length:

> the SI is the system of units in which the value of speed of light in vacuum is 299 792 458 m s^{-1} exactly (notice: the numerical value is identical).

Why this would *not be* possible?

The basic reason is that the SI-1960 had *first to define the units* of the relevant quantities (e.g., length), which may have been different in the SI preceding times—or not defined as base units—not only the magnitudes of some involved physical constants (or of some physical states for other units like the triple point of water for the kelvin, the unit of temperature).

In the case of the rSI-2019, the units indicated in their definitions *are the same* as in the SI-1960 with modified definitions, but *with magnitudes within the degree of continuity* indicated in the rSI-2018. Only the formal spelling of the unit definition in the Appendix 3 of the R1-2018 (notice, not in the rSI-2018) definition) is different, by basically only *changing the reference fixed value* used, now physical constants only (the same as before for length). The further wording is only supposed to avoid circularity—with no success. This confirms that the rSI-2018 is, in fact, *not* a new definition, but a *revision* of the SI-1960. [5] However, this is not made clear in the SI9. Furthermore, that kind of change, while avoiding the indication of definitional methods for the realization of the units, in principle strongly increases the possible non-uniqueness of the different "primary" methods, among them and when they are *local* realized: that issue should be explained and treated accordingly.

References

1. CGPM, Resolution 1 of the 26th Meeting, 16 November 2018, BIPM, Sèvres, https://www.bipm.org/en/CGPM/db/26.
2. I. Johansson, Quantities as metrical coordinative definitions and as counts on some definitional structures in the New SI Brochure, *Journal for General Philosophy of Science* **52** (2021) 407–429, https://doi.org/10.1007/s10838-020-09545-z
3. 9th SI Brochure, BIPM, Sèvres, 2019. https://www.bipm.org/en/publications/si-brochure/.
4. 8th SI Brochure, BIPM, Sèvres, 2006.
5. F. Pavese, A possible draft of the CGPM Resolution for the revised SI, compared with the CCU last draft of the 9th SI Brochure, *Measurement* **114**, 478–483 (2018)
6. L. Mari, F. Pavese, A conceptual roadmap for setting up a system of units in the 'New SI' context, arXiv Physics.ed-ph 1604.00982, pp. 6; *Journal of Physics Conf. Series* **772**, (2016) 012011, doi:10.1088/1742-6596/772/1/012011
7. F. Pavese, Some conceptual problems in the CCU 2016 draft of the New SI, *Journal of Physiscs Conf. Series* **772**, (2016) 012011, doi:10.1088/1742-6596/772/1/012011
8. L. Mari, P. Blattner, F. Pavese, Improving the understandability of the next edition of the International System of Units (SI) by focusing on its conceptual structure, *Measurement* **101**, 200–205 (2017)
9. CIPM, Decision CIPM/105-15, 2016, https://www.bipm.org/en/committees/cipm/meeting/105.html and CCU Decisions 2016, BIPM, https://www.bipm.org/cc/AllowedDocuments.jsp
10. D.B. Newell, F. Cabiati, J. Fischer, K. Fujii, S.G. Karshenboim, H.S. Margolis, E. de Mirandes, P.J. Mohr, F. Nez, K. Pachucki, T.J. Quinn, B.N. Taylor, M. Wang, B. Wood, Z.H. Zhang, The CODATA 2017 Values of h, e, k, and N_A for the Revision of the SI, *Metrologia* **55**, 29 (2018)
11. P.J. Mohr, D.B. Newell, B.N. Taylor, E. Tiesinga, Data and analysis for the CODATA 2017 special fundamental constants adjustment for the revision of the SI, *Metrologia* **55**, 125–146 (2018)
12. BIPM, CCT, Mise en pratique for the definition of the kelvin in the SI (MeP-K), SI Brochure, 9th edition (2019) Appendix 2, https://www.bipm.org/utils/en/pdf/si-mep/SI-App2-kelvin.pdf
13. BIPM, https://www.bipm.org/en/cipm-mra/cipm-mra-documents/

14. BIPM, https://www.bipm.org/utils/en/pdf/si-mep/
15. F. Pavese, On the classification into random and systematic effects, in *Mathematical and Computational Tools in Metrology and Testing XI*, Series on Advances in Mathematics for Applied Sciences vol. 89, © 2018 World Scientific Publishing Company, pp. 58-69
16. G. Wübbeler, H. Bettin, F. Härtig, C. Elster, Maintaining consensus for the redefined kiligram, Metrologia **55** 722–730 (2018)
17. BIPM, CCM, Mise en pratique for the definition of the kelvin in the SI (MeP-K), SI Brochure, 9th edition (2019) Appendix 2, https://www.bipm.org/utils/en/pdf/si-mep/SI-App2-kilogram.pdf
18. BIPM, The Avogadro constant and the revision of the SI, https://www.bipm.org/en/bipm/mass/avogadro/
19. BIPM, CCT, Mise en pratique for the definition of the ampere in the SI (MeP-K), SI Brochure, 9th edition (2019) Appendix 2, https://www.bipm.org/utils/en/pdf/si-mep/SI-App2-ampere.pdf
20. P.J. Mohr, D.B. Newell, B.N. Taylor, CODATA recommended values of the fundamental physical constants: 2014, (2015) arXiv:1507.07956v1, *Rev. Modern Physics* **88**, 1–73 (2016)
21. F. Pavese, On the revised SI, specifically on the numerical value of the Planck constant, *Acta IMEKO* **7**, (4) 95–101 (2018)
22. F. Pavese, The New SI: Some critical features and a critical review of the CCU 2016 Draft of the SI brochure, *Measurement* **98**, 325–338 (2017)
23. F. Pavese, The New SI and the CODATA recommended values of the fundamental constants 2017 compared with 2014, with a Comment to 'Possolo *et al.*, *Metrologia* **55**, 29 (2018)', ArXiv physics.data-an 1512.03668v5 (2018)
24. E.R. Cohen, B.N. Taylor, The 1973 Least-Squares Adjustment of the Fundamental Constants, *J. Phys. Chem. Ref. Data* **2**, 663–734 (1973)
25. F. Pavese, The New SI and the 'CODATA recommended values of the fundamental constants 2014 (on arXiv:at-phys-1507.07956)', ArXiv physics.data-an 1512.03668, August 2016, pp. 5
26. T.G. Blaney, C.C. Bradley, G.J. Edwards, D.J.E. Knight, W.R.C. Eowley, K.C. Shotton, P.T. Woods, *Ann. N.Y. Acad. Sci.* **251**, 46 (1974)
27. F. Pavese, Musing on use and misuse of numerical data in science, St. Petersburg Seminar, 2018. *Measurement Techniques* (2019) (June, in Russian N. 2 11–14; N.3 in English)
28. https://keisan.casio.com/calculator
29. CODATA, 2018 CODATA recommended values of the fundamental constants of physics and chemistry, *NIST SP 959* (June 2019). For an extensive constants list: https://physics.nist.gov/cuu/Constants/index.html
30. http://www.javascripter.net/math/calculators/100digitbigintcalculator.htm
31. https://www.calculator.net/big-number-calculator.html
32. BIPM, Report on the CCM key comparison of kilogram realizations CCM.M-K8.2019, Final Report, Sèvres, KCDB 2019.

Appendix

On the new numerical values of derived constants

The values of the constants stipulated and adopted by the Conférence Générale des Poids et Measures (CGPM) in October 2018 for the revised Système International des Unités (SI), in force since May 20 2019, [1] have been the values recommended by the CODATA in their analysis and 2017 special adjustment of the physical constants. [11]

The CODATA also provides at regular intervals the recommended values of a much larger set of other important physical constants (the last so far in 2014 [20]—the special 2017 adjustment only concerned the constants of SI interest). In addition, as already done in 1983 for the speed of light *in vacuum*, c (also often indicated with c_0), [9] and previously for μ_0, CODATA recommends that, also in computing the values of the constants whose definition equations contain some of the constants stipulated in the SI, the stipulated SI values are used. This decision implicitly assumes that the numerical values recommended for all constants are using the SI as the system of units.

This has also been the case for the recent 2018 adjustment of those constants of the set depending on the seven constants used by the seven base units of the SI. In Table 1 the 2018 SI values of these constants [1] are reported: stipulation means that these values are by consensus assumed to be exact, so no uncertainty is indicated and the non-indicated digits following the lowest one are all zero (the numbers being normally real ones). See [8, 21] for (un-solved) problems in selecting the correct number of digits in stipulation, related to the original experimental uncertainty.

The use of stipulated numbers in algebraic operations places, for the first time, (unusual) problems in correctly representing the resulting numerical value.

An illustration of these problems, with a proposed solution, follows.

In Table 3 a recall, convenient in the following, of the new stipulated values of the SI constants is reported, with also the previous CODATA recommended values since 2010 and the resulting change. [8, 21-23]

The corresponding new units, of which these constants are now the bases of the definition, have changed in magnitude as in Eq. 2–5 above. [1]

---

[9] See in [24, 25] the fact that the 1983 stipulated value of c was not the resulting one from the CODATA 1973 adjustment (299 792 456.2 ± 1.1) m s^{-1}, but a value suggested by the BIPM. In addition, the subsequent 1974 value from [26] (299 792 459.0(0.8) m s^{-1}) was not taken into consideration in 1983. Finally, it never happened until now that the stipulated value had to be adjourned, according to the statement in [24] "provided that the data upon which it is based are not subsequently proved to be in error", because it has not been possible so far to perform any independent measurement with sufficiently low uncertainty after the 1983 stipulation.

These changes ensure a new magnitude equal to that of the previous SI compatible with the current best experimental uncertainties, though this fact is not supported by the uncertainties of the 2017 experimental data used for the stipulations. [21]

Table 3. Change in numerical value of the CODATA adjustments 2010-2017 (stipulated in 2018) for the constants involved in the revised SI definition.

Constant	CODATA	Numerical value *	u / relative $\times 10^7$	Change ‡	Total shift
Planck	2006	6.626 069 0	3.3/0.50	—	
$h \times 10^{34}$	2010	6.626 069 6	3.5/0.53	$+6 \times 10^{-7}$	
	2014	6.626 070 04 [a]	0.8/0.12	$+4.4 \times 10^{-7}$	$+10.4 \times 10^{-7}$
	2017	6.626 070 15 [a]	0.7/0.11	*$+1.1 \times 10^{-7}$*	$+11.5 \times 10^{-7}$
SI stipulated	2018	**6.626 070 15**|—		$+0.17 \times 10^{-7}$	
Electron	2006	1.602 176 49	0.4/0.23	—	
charge	2010	1.602 176 57	0.35/0.19	$+8 \times 10^{-8}$	
$e \times 10^{19}$	2014	1.602 176 62	0.098/0.06	$+5 \times 10^{-8}$	$+1.3 \times 10^{-7}$
	2017	1.602 176 634 [b]	0.08/0.05	*$+1.4 \times 10^{-8}$*	$+1.44 \times 10^{-7}$
SI stipulated	2018	**1.602 176 234**|—		$+0.87 \times 10^{-8}$	
Avogadro	2006	6.022 141 8	3.0/0.50	—	
$N_A \times 10^{19}$	2010	6.022 141 3	2.7/0.45	-5×10^{-7}	
	2014	6.022 140 8 [c]	0.74/0.12	-4×10^{-7}	-9×10^{-7}
	2017	6.022 140 76 [c]	0.6/0.10	*$+1 \times 10^{-7}$*	-8×10^{-7}
SI stipulated	2018	**6.022 140 76**|—		-0.17×10^{-7}	
Boltzmann	2006	1.380 650 4 [d]	24/17	—	
$k \times 10^{23}$	2010	1.380 648 8 [d]	13/9.4	-1.6×10^{-6}	
	2014	1.380 648 5	7.9/5.8	-0.3×10^{-6}	-1.9×10^{-6}
	2017	1.380649 [e]	5/3.6	$+0.5 \times 10^{-6}$	-1.4×10^{-6}
SI stipulated	2018	**1.380 649**|—		$+0.36 \times 10^{-6}$	
Light speed c					
SI stipulated	1983	**299 792 458**|—	—	—	—

* The CODATA reports standard uncertainty u. The smaller-case digits, obtained from the CODATA two-digit uncertainty format, are those affected by *uncertainty*—except for Boltzmann constant k, see note c) and e). The (*new*) symbol |— [27] indicates a *stipulated* number, otherwise not identifiable.

‡ Changes exceeding the final uncertainty are shown in italics.

[a] The 2014 CODATA outcome is 6.626 070 040(81), therefore the numerical value can be as low as 6.626 069 959, thus affecting also the preceding digit. Similarly for the 2017 one, the CODATA outcome is 6.626 070 150(69). Consequently the upper bound of the 2017 interval is 6.626 070 121, and the lower bound of the 2017 interval in 6.626 070 081, *not significantly overlapping*.

[b] The 2017 CODATA outcome is 1.602 176 6341(83), therefore the numerical value can be as low as 1.602 176 6258 and as high as 1.602 176 6424, thus affecting also the preceding digit.

[c] The 2014 CODATA outcome is 6.022 140 857(74), therefore the numerical value can be between 6.022 140 783 and 6.022 140 931, thus affecting the preceding digit. The analysis is similar for the 2017 one, where the CODATA outcome is 6.022 140 758(62).

[d] Two digits are shown because the rounding affects also the preceding digit.

[e] The CODATA 2017 outcome is 1.380 649 03(51), thus the rounding does not include uncertain digits (the only occurrence in the Table). However, the numerical value can be as low as 1.380 648 50, thus in fact affecting the last digit.

These changes ensure a new magnitude equal to that of the previous SI compatible with the current best experimental uncertainties, though this fact is not supported by the uncertainties of the 2017 experimental data used for the stipulations. [21]

In Table 4, the changes are reported in the numerical values of the CODATA adjustments 2010-2017 for derived constants.

They are the ones that contain in their definition equation one or more of the constants newly stipulated by the revised SI. For the 2018 numerical values, both "stipulated/non-stipulated" values, computed by using a 100-digit computer, [28].

Three different considerations arise:

a) Being the stipulated values in Table 3 defined exact, any algebraic operation involving them is out coming an also exact value, whose digits must all be shown. No rounding or truncation is admitted. For expressions including only products the resulting number always has a finite number of decimal digits. That situation is shown in Table 4, e.g., for R and F for the "stipulated" value, not by CODATA [29] value for R (F is not listed by CODATA).

b) That rule is valid for all stipulated constants in Table 4, but is masked for the ones having π in the definition, because that mathematical constant π is an irrational number—so is the result of the algebraic equation. In the "stipulated" values, the dots have that meaning only. On the contrary, in the CODATA [29] values the meaning is mixed, being used also when no π appear in the equation, so being ambiguous.

In the case when π appears, the only possibility would be to compute exact a constant $C/\pi(x)$, where x is the possible power indicated in the equation: e.g., $h/\pi = (1/2) \times 6.626\ 070\ 15$, i.e., $h/2\pi = 3.313\ 035\ 075 \times 1/\pi$.

Table 4. Change in numerical value for *derived constants* 2006-2018.

Constant	Year	Numerical value {Q}	Change *
$\hbar \times 10^{34}$ =	2010	1.054 571 726(47)	—
$h/2\pi$ [J s]	2018	1.054 571 817 … (CODATA [29])	*8.6 ×10⁻⁸*
stipulated	2018	**1. 054 571 817 *646 156* …** [a]	***6.2 ×10⁻¹⁰***
$\alpha \times 10^{3}$ =	2010	7.297 352 569 8(24)	—
$\mu_0 e^2 c/2h$	2014	7.297 352 566 4(17)	*−4.7 ×10⁻¹⁰*
not stipulated [b]	2018	—	—
CODATA		7.297 352 569 3(11)	
$R = kN_A$	2010	8.314 462 1(75)	—
[J K⁻¹ mol⁻¹]	2014	8.314 459 8(48)	−2.8 ×10⁻⁷
stipulated	2018	**8.314 462 618 *153 24*** [c]	**3.4 ×10⁻⁷**
CODATA		***8.314 462 618 …***	
$K_J \times 10^{14}$ =	2010	4.835 978 70(11)	—
$2e/h$ [Hz V⁻¹]	2014	4.835 978 525(30)	*−3.6 ×10⁻⁸*
stipulated	2018	**4.835 978 484 *169 836* …** [d]	—
CODATA		***4.835 978 484 …***	
$R_K \times 10^{4}$ =	2010	2.581 280 744 34(84)	—
h/e^2 [Ω]	2014	2.581 280 745 55(59)	*4.7 ×10⁻¹⁰*
stipulated	2018	**2.581 280 745 *930 45* …** [d]	—
CODATA		***2.581 280 745 …***	
$F = N_A\ e$	2010	96 485.336 5(21)	—
[C mol⁻¹]	2014	96 485.332 89(59)	*−3,7 ×10⁻⁸*
stipulated	2018	**96 485.332 123 310 018 4** [c]	**−8 ×10⁻⁸**
$a_0 \times 10^{11}$ =	2010	0.529 177 210 92(17)	—
$\varepsilon_0 h^2/\pi m_e e^2$ [m]	2014	0.529 177 210 67(12)	*−4.7 ×10⁻¹⁰*
not stipulated [b]	2018	—	—
CODATA		0.529 177 210 903(80)	
$R_\infty \times 10^{-7}$ =	2010	1.097 373 156 8539(55)	—
$\alpha^2 m_e c/2h$ [m⁻¹]	2014	1.097 373 156 8508(65)	−2.8 ×10⁻¹²
not stipulated [b]	2018	—	—
CODATA		1.097 373 156 8160(21)	
$\sigma \times 10^{8}$ =	2010	5.670 373(21)	—
$2\pi^5 k^4/15h^3c^2$	2014	5.670 367(13)	−1 ×10⁻⁶
[W m⁻² K⁻⁴]			
stipulated	2018	**5.670 374 419 184 429 45…** [a,d]	**1.3 ×10⁻⁶**
$\sigma\pi^5 \times 10^{8}$	(2018)	[d] *See text (b) below*	—
CODATA		***5.670 374 419 …***	
$c_1 \times 10^{16}$ =	2010	3.741 771 53(17)	—
$2\pi hc^2$]W m²]	2014	3.741 771 790(46)	7 ×10⁻⁸
stipulated	2018	**3.741 771 852 192 758 011 367 155 5559…** [f,g]	**8.3 ×10⁻⁸**
$c_1/\pi \times 10^{16}$ [e]			
stipulated	(2018)	***11.910 429 723 971 884 140 794 8920*** [c]	—
CODATA		***3.741 771 852 … (See text)***	
$c_2 \times 10^{2}$ =	2010	1.438 7770(13)	—
hc/k [m K]	2014	1.438 777 36(83)	2.5 ×10⁻⁷
stipulated	2018	**1.438 776 877 50…** [h,d,c]	**3.4 ×10⁻⁷**
CODATA		***1.438 776 877 …***	

*Changes exceeding the final uncertainty are shown in italics.

[a] Dots for more digits because the exact number is irrational due to π . See also note c.

[b] Not stipulated because the defining equation contains μ_0= 1.256 637 0614(3) 10^{-6} H m^{-1} or ε_0 = 8.854 187 818(2) F m^{-1}, no more exact in the revised SI.

[c] See text for the reason of so many digits.

[d] Number almost certainly irrational (check made for up to 100 digits [28]).

[e] Stipulated but rational number (full digits displayed, see text).

[f] Using π = 3.141 592 653 589 793 238 462 643 383 279 502 884 1971.

[g] A number of digits is shown higher than the one of the exact stipulated value of $c_1\pi$.

[h] Using up to 100 digits [29]: c_2 = 1.438 776 877 503 933 8021 4667 … .

The case of the constant σ is a special one, where the above possibility is overcome by the fact that σ is in itself an irrational number, arising from the presence of a ratio, so that the dots cannot be avoided.

The case of the constant c_1, instead, allows the application of the above method, because it does not contain π. Three online 100-digits calculators have been used:

(i) by using the calculator [28], we got all results, including for c_1, equal to the CODATA one [29];

ii) for $c_1\pi$, by using [28, 30, 31], we obtain the same stipulated value:

$c_1\pi \times 1016 = 2hc^2 = 11.910\ 429\ 723\ 971\ 884\ 140\ 794\ 8920$ **exact**.

Then one can compute c_1 back, by dividing the above value by π (using the same number of digits indicated in Note f of Table 4), only with [28]. However, with the two others calculators (both using a Java method) we got instead: 3.791 207 529 837 527 785 333 809 728 404, different from the one in Table 4 and incorrect.

c) In the case of the constant c_2 only, though no π is present in the definition, the dots also mean a probable irrational number, as it can happen for algebraic expressions including ratios.

Therefore, computations of the correct total digits must be performed with a sufficiently precise computer. That possibility was checked here up to a maximum number of 100+ digits. [28]

Here below, examples are shown of what it may happen, instead, by use the popular MS Excel on a common 64 bit PC. Let us take as example:

—for a **product** the Gas Constant, $R = N_A k$, where N_A = 6.022 140 76 ×1023 mol–1 and k = 1.380 649 ×10–23 J K, exactly.

Thus one has to multiply the two finite-digit numbers, and all the resulting digits must be retained and should be explicit, since also R is now stipulated:

When using a 50-digit max computer one gets: 8.314 462 618 153 24 J K mol^{-1}.

When using MS Excel one gets: R = 8.314 462 618 153 24 J K mol^{-1}, correct, but … taking instead $F = N_A\ e$, where $N_A = 6.022\ 140\ 76 \times 10^{23}\ \text{mol}^{-1}$ and e = 1.602 176 634 C, exact,

and by using again MS Excel with 20 decimals, instead of 96 485.332 123 310 018 4 exact, one gets 96 485.332 123 310 (zero periodic), truncated with respect to the exact value (difference 2×10^{-12}).

— for a **ratio**, the Second Radiation Constant, $c_2 = hc_0/k$, where $h = 6.626\ 070\ 15 \times 10^{-34}$ J s, $c_0 = 299\ 792\ 458$ m s^{-1}, $k = 1.380\ 649 \times 10^{-23}$ J K, exactly.

Using a 50-digit max computation one gets: 0.014 387 768 775 039 338 021… m K (probably irrational).

Using MS Excel one gets: $c_2 = 0.014\ 387\ 768\ 775\ 039\ 30$ m K (difference 2×10^{-15}).

In conclusion, the stipulation of the values of more constants by the SI since 2018 places a new and unusual problem in the exact calculation of several derived constants, when the r*elative* uncertainty must be higher than about $u_r = 2\times10^{-12}$, which is the precision that common users can obtain, at best, from their PCs, using, e.g. MS Excel for the calculation from the SI-defined stipulated constants. Therefore, also MS Excel can be used if a relative precision up to about 10^{-10} is sufficient for own calculations.

However, it always stands the fact that the result of an algebraic operation involving defined numbers is itself a defined one, not allowing for truncation or rounding, unless specifically indicated to be sufficient to most users.

The above fact is masked for some of them by the fact that the defining equation contains a ratio ending in an irrational number, or contains the irrational constant π, or because the equation. In the latter case, it would be possible to report all exact stipulated digits if the numerical value of the constant divided by π are also reported, where dots (…) would not be necessary, unless the resulting number would be assumed to be an irrational one (like σ and c_2). Only the actual constants including π will anyway need the dots.

In that way, except in the latter occurrences, it would possible to end the indication of the numerical values at the last exact digit, where the total number of them may be quite higher than usual.

Finally, in order to get the stipulated exact numbers, in some cases the computation of the correct total number of digits must be performed by sufficiently-precise computing.

F Pavese, A B Forbes, N F Zhang, A G Chunovkina (Eds.):
Advanced Mathematical and Computational Tools in Metrology and Testing XII
Series on Advances in Mathematics for Applied Sciences, Vol. 90

https://doi.org/10.1142/9789811242380_0005

Statistical and computational tools for metrologists

S. Stoudt

Smith College,
Statistical and Data Sciences Program,
Northampton, MA 01063, USA
E-mail: sstoudt@smith.edu

A. Possolo

Statistical Engineering Division,
National Institute of Standards and Technology,
Gaithersburg, MD 20899-8980, USA
E-mail: antonio.possolo@nist.gov

Metrologists are increasingly being faced with challenges in statistical data analysis and modeling, data reduction, and uncertainty evaluation, that require an ever more demanding and comprehensive analytical and computational toolkit as well as a strategy for communication of more complex results. For example, conventional assumptions of Gaussian (or normal) measurement errors may not apply, which then necessitates alternative procedures for uncertainty evaluation.

This contribution, aimed at metrologists whose specialized knowledge is in a particular area of science, and whose prior study of topics in probability or statistics will have been merely introductory, provides illustrative examples and suggestions for self-study. These examples aim to empower metrologists to attain a working level of concepts, statistical methods, and computational techniques from these particular areas, to become self-sufficient in the statistical analysis and modeling of their measurement data, and to feel comfortable evaluating, propagating, and communicating the associated measurement uncertainty.

The contribution also addresses modern computational requirements in measurement science. Since it is becoming clear to many metrologists that tools like Microsoft *Excel*, Libreoffice *Calc*, or Apple's *Numbers* often are insufficiently flexible to address emerging needs, or simply fail to provide required specialized tools, this contribution includes accompanying *R* code with detailed explanations that will guide scientists through the use of a new computing tool.

Keywords: measurement; statistical model; measurement uncertainty; statistical computing; data science; missing data; probability; statistics; heterogeneity; interlaboratory study; calibration.

1. Introduction

Contemporary measurement is complex and nuanced, typically requiring appropriate statistical modeling and flexible computing. An appropriate model represents the relationship between the observations and the object of measurement sufficiently accurately to be fit for purpose. In the context of measurement science, this purpose may be deciding whether the mass fraction of arsenic in a shipment of rice exceeds a regulatory threshold, or certifying the concentration of a vitamin D metabolite in a reference material. Flexible computing involves the ability to do data reductions reproducibly with an appropriate model, which often calls for custom computational procedures.

This contribution reviews and illustrates the application of tools that facilitate the development of appropriate models and the development and implementation of custom software solutions for data reductions, while emphasizing the statistical aspects and the accessible communication of both.

In Section 2 we consider a common task in measurement science: the reduction of a set of replicated determinations of a measurand and the evaluation of the associated uncertainty. We illustrate how consistency with common assumptions may be verified and propose alternatives for what to do when such assumptions appear untenable.

Section 3 is dedicated to the construction of calibration functions, demonstrating how the best among several reasonable candidates may be selected. Once a function is selected we show how it may be fitted to the calibration data and how its inverse (analysis function) may be used to translate instrumental indications into estimates of the measurand. We conclude the section by showing how the nonparametric bootstrap may be used to characterize the uncertainty of the outputs produced by the analysis function, and to do so without resorting to formulas whose domain of validity is rather limited.

Section 4 discusses model selection in the context of interlaboratory studies and meta-analysis and illustrates how an unconventional random effects model may be fitted and the uncertainty of its outputs evaluated.

Section 5 reviews conclusions and offers practical recommendations for model building, validation, and fitting to data, highlighting how a flexible programming environment for data analysis goes a long way toward empowering metrologists to carry out data reductions that are well-tuned to the data in every instance of measurement.

Bell [1] and Hall and White [2] provide excellent, very accessible introductions to the evaluation of measurement uncertainty based on explicit measurement models. Possolo and Meija [3] offer an eclectic re-introduction to measurement uncertainty, with examples of statistical models, their computational implementation, and associated uncertainty evaluations, from many different areas of science and technology.

2. Replicated Observations

Many measurements, such as those made in the development of reference materials or to control the quality of manufactured food products involve some replication, and therefore pose the problem of how to combine replicates into an estimate of the measurand. In many cases, an additive model is adopted, which expresses the observations as $x_i = \xi + \varepsilon_i$, for replications $i = 1, \ldots, m$, where ξ denotes the true value of the measurand, and the $\{\varepsilon_i\}$ are non-observable measurement errors. Typically, the number, m, of replicates is small, say between two and five.

The principal factors that determine how the observations should be combined are: (i) the assumptions made about the nature of the measurement errors; and (ii) the choice of optimality criterion that the estimate of ξ should satisfy.

The most common version of (i) regards the $\{\varepsilon_i\}$ as realizations of independent Gaussian random variables with mean 0 and the same standard deviation σ. Many out-of-the-box computational approaches make this normality assumption for mathematical convenience, but in practice, errors with heavier tails may arise.

The usual choice made in (ii) is that the estimate $\widehat{\xi}$ should minimize mean squared error, which equals the squared bias of $\widehat{\xi}$ plus its variance. In these circumstances, the optimal estimate is the arithmetic average, $\overline{x}$. This approach leads to unbiased estimates, but there are cases where one might want to use a regularized optimality criterion, sacrificing some bias to decrease variance. Minimizing a criterion other than mean squared error may be optimal in some situations: for example, when the measurement errors have a Laplace distribution, the maximum likelihood solution involves minimizing the sum of the absolute values of the errors.

If the Gaussian assumption made in (i) is invalid, or a criterion other than mean squared error is adopted for (ii), then the best estimate need not be $\overline{x}$. If the replicates may have heavier tails than a Gaussian distribution, yet still may safely be assumed to be a sample from a symmetric probability distribution, then the Hodges-Lehmann estimator is much to be preferred over the arithmetic average because it extracts the information from the data almost as efficiently as the arithmetic average in the Gaussian scenario, and can vastly outperform the arithmetic average when the Gaussian assumption is not met.

Figure 1 lists determinations of the mass fraction of silica (SiO_2) in NIST Standard Reference Material 690 Iron Ore Concentrate (Canada), made using X-ray fluorescence, and the R code used to test the data for Gaussian shape and for symmetry, and then to compute the Hodges-Lehmann estimate, 3.678 cg/g and the associated standard uncertainty, 0.005 cg/g. (NOTE: The value of the mass fraction of silica assigned to this reference material is based on these and on other determinations.)

The Shapiro-Wilk test has the most power to detect a deviation from Gaussian shape when it truly exists, followed by the Anderson-Darling test [4]. If the resulting p-value from either of these tests is less than a predetermined significance value, typically chosen to be 0.05, we reject the null hypothesis that the data are a sample from a Gaussian distribution. Both these tests yield p-values of less than 0.007, thus suggesting that the Gaussian model is inappropriate for these data.

Although we have rejected the null hypothesis that the data come from a Gaussian distribution in this case, we can test if the distribution is at least symmetric. [5] developed a test that has greater power to detect deviations from symmetry than other tests of symmetry, so we use that approach here. This test is nonparametric and relies on a bootstrap procedure to calculate the p-value. We will give more details about nonparametric bootstrap procedures in Section 3, but now will point out that this approach is very appealing and widely applicable because it does not make any distributional assumptions about the data. This is important in our case as we have rejected the hypothesis that our data comes form a Gaussian distribution. If the resulting p-value from this symmetry test is less than our significance level, then we reject the null hypothesis that the distribution is symmetric.

For the CRF determinations of silica in NIST SRM 690, the test yields p-value 0.87, hence we do not reject the hypothesis of symmetry and proceed as if the data indeed are a sample from a symmetric distribution. In these circumstances we can use the Hodges-Lehmann procedure to estimate the true mass fraction ξ because it requires only that the data be like a sample from a symmetric distribution. To obtain this estimate and its associated uncertainty we can invert the Wilcoxon signed-rank test [6], as illustrated in Figure 1.

```
## Replicated determinations of mass fraction of silica in NIST
## SRM 690, expressed as percentages (cg/g)
w = c(3.7019, 3.6784, 3.6886, 3.6753, 3.683, 3.6629, 3.6672, 3.6642,
      3.5793, 3.6713, 3.7353, 3.6735, 3.6578, 3.6888, 3.6916, 3.693)

## Assessing conformity with Gaussian model
library(nortest)
ad.test(w)$p.value
## Assessing symmetry
library(symmetry)
symmetry_test(w, stat="MGG", bootstrap=TRUE, B=10000)$p.value

## Hodges-Lehmann estimate, and proxies for standard uncertainty,
## expanded uncertainty U95, and effective coverage factor k95
via inversion of Wilcoxon signed-rank test
w68 = wilcox.test(w, exact=TRUE, conf.int=TRUE, conf.level=0.68)
w95 = wilcox.test(w, exact=TRUE, conf.int=TRUE, conf.level=0.95)
w.x = w68$estimate; names(w.x) = NULL
w.u = abs(diff(w68$conf.int))/2
w.U95 = abs(diff(w95$conf.int))/2
w.k95 = w.U95 / w.u
c(w=w.x, uw=w.u, U95w=w.U95, k95=w.k95)
```

Fig. 1. Measured values and R code to test observations for Gaussian shape and symmetry, and then to compute the Hodges-Lehmann estimate (3.678 cg/g) of the mass fraction of silica in NIST SRM 690, the associated standard uncertainty (0.005 cg/g), expanded uncertainty for 95 % coverage (0.011 cg/g), and implied coverage factor $k_{95\,\%} = 2.215$.

The combination of non-Gaussianity and symmetry may lead us to choose to model the observations as a rescaled and shifted Student's t distribution with median ξ, scale σ, and ν degrees of freedom — all to be estimated. In this case we will need to use a more general optimality criterion, namely maximum likelihood, to estimate the mass fraction of silica in the material. Given the assumed structure of the Student's t distribution, we want to know which parameters maximize the likelihood of seeing the data we observed. When we fit the model using this criterion, by executing the R code in Figure 2, we get an estimate of 3.678 mg/kg, with approximate standard uncertainty 0.005 mg/kg. The maximum likelihood estimate of ν is 1.9 ± 1.0, thus suggesting that the distribution of the measurement errors has very heavy tails.

3. Calibration

Calibration involves determining a relationship between values of a measurand $\{c_j\}$, typically realized in standards of reference indexed by j, and

```
## Replicated determinations of mass fraction of silica in NIST
## SRM 690, expressed as percentages (cg/g)
w = c(3.7019, 3.6784, 3.6886, 3.6753, 3.683, 3.6629, 3.6672, 3.6642,
      3.5793, 3.6713, 3.7353, 3.6735, 3.6578, 3.6888, 3.6916, 3.693)

## Negative loglikelihood function for the rescaled and shifted
## Student's t distribution
negloglik.student = function (mu, sigma, nu, x,
                              tol=sqrt(.Machine$double.eps))
{
  if (any(c(sigma, nu) < tol)) { return(Inf)
    } else { return(-1*sum(dt((x-mu)/sigma, df=nu, log=TRUE) -
                        log(sigma))) }
}
## Computation of maximum likelihood estimates of the median,
## scale, and number of degrees of freedom of the rescaled and
## shifted Student's t distribution, and approximate standard
## uncertainties for these estimates
require(bbmle)
student.mle = mle2(negloglik.student, data=list(x=w),
                   start=list(mu=median(w), sigma=mad(w), nu=30),
                   skip.hessian=FALSE,
                   optimizer="optim", method="Nelder-Mead")
summary(student.mle)
```

Fig. 2. Measured values and R code to fit a rescaled and shifted Student's t distribution to the determinations of the mass fraction of silica in NIST SRM 690 using the method of maximum likelihood.

the responses $\{r_{ij}\}$ that a measuring instrument produces when exposed to those standards across replicates indexed by i. This relationship is often described by a *calibration function* that maps values of the measurand to values of the instrumental response.

Figure 3 depicts the calibration data listed in Table 1 in blue, and a calibration function in red, for an ion chromatography system that uses a glassy carbon electrode modified by electrochemical deposition of palladium particles as a detector [7] [8].

Three polynomial calibration functions were considered, of the form $r_{ij} = \varphi(c_j)+\varepsilon_{ij}$, where $i = 1, 2, 3$ denotes the replicate, $j = 1, \ldots, 8$ denotes the value of concentration of the measurand, and the $\{\varepsilon_{ij}\}$ represent measurement errors. The candidate polynomial φ was of first, second, or third degree, with 2, 3, or 4 coefficients, respectively. The model assumes that the uncertainties surrounding the concentrations $\{c_j\}$ are negligible by comparison with the uncertainties surrounding the instrumental responses $\{r_{ij}\}$, although there are approaches to handle the case where the uncertainties are of a similar magnitude [9].

Table 2 lists the values that the Akaike Information Criterion (AIC) and the Bayesian Information Criterion (BIC) take for each of these models [10]. AIC and BIC are measures of the adequacy of a model to a dataset, taking into account both goodness-of-fit between data and model and the complexity of the model. They strike a balance between how well the model fits the data and the number of adjustable parameters involved in the fit. The Bayesian Information Criterion penalizes the number of parameters more heavily than the Akaike Information Criterion, so it will in general favor less complex models. Both are designed so that the smaller the value of AIC or BIC, the more adequate the model. In this case, the two criteria agree, suggesting the second degree polynomial, which is also depicted in Figure 3. The selected model is of the form $\widehat{r}_j = \beta_0 + \beta_1 c_j + \beta_2 c_j^2$, where the $\{\widehat{r}_j\}$ are the fitted responses at the different values of the concentration.

The fact that the points in the QQ-plot in Figure 3 deviate markedly from a straight line, suggests that the residuals corresponding to a least squares fit are not normally distributed as required, so an alternative to least squares may be in order. Table 2 lists the estimates of the coefficients for the calibration function and associated uncertainties both for ordinary least squares (OLS) and for robust regression, an alternative approach when OLS residuals are heavy-tailed, as implemented in R function `lmrob` defined in package `robustbase`.

Table 1. Calibration data for the determination of the sulfide anion in flow injection analysis using a glassy carbon electrode detector that has been modified by electrochemical deposition of palladium particles [7, Table 4].

RUN 1		RUN 2		RUN 3	
c/µmol/L	r/µA	c/µmol/L	r/µA	c/µmol/L	r/µA
0.88	0.170	0.88	0.211	0.88	0.179
1.77	0.341	1.77	0.349	1.77	0.353
3.55	0.663	3.55	0.684	3.55	0.656
7.10	1.37	7.10	1.34	7.10	1.30
14.2	2.60	14.2	2.58	14.2	2.64
28.4	5.29	28.4	5.24	28.4	5.26
56.8	10.83	56.8	11.10	56.8	11.05
81.2	15.85	81.2	15.75	81.2	15.94

However, to be able to use the instrument to make measurements, the inverse relationship is needed, which produces an estimate of the value of

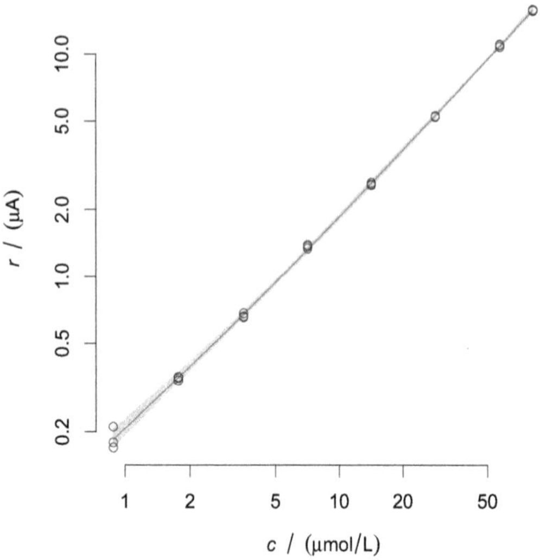

(a)

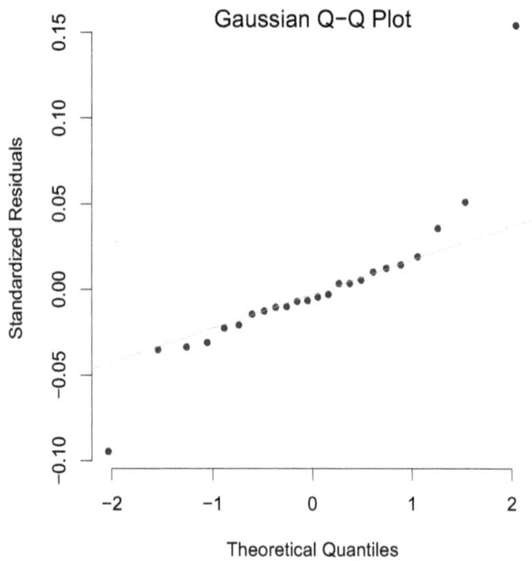

(b)

Fig. 3. (a) Triplicate instrumental responses (r) for each value of amount concentration (c), depicted on logarithmic scales, and the model (lighter line in (a)) deemed best according to the Bayesian Information Criterion (BIC), with associated, 95 % coverage (gray) band computed using the nonparametric bootstrap. (b) QQ plot of standardized residuals $\{\log r_{ij} - \log \widehat{r}_j\}$, with (gray) trend line determined by the 25th and 75th percentiles of the theoretical quantiles and of the standardized residuals, indicates heavier tails than the Gaussian model contemplates, and suggests that a robust alternative may be preferable to conventional least squares.

Table 2. LEFT PANEL: Model selection criteria for the ion chromatographic system used to measure sulfide anions. RIGHT PANEL: Coefficients of 2nd degree polynomial fitted either via ordinary least squares (LM), or via robust regression (LMROB).

MODEL SELECTION		
DEGREE	AIC	BIC
1st	−70.3	−66.8
2nd	−81.9	−77.1
3rd	−79.9	−74.0

COEFFICIENTS OF 2ND DEGREE POLYNOMIAL				
	LM		LMROB	
$\widehat{\beta}_0$	−1.57	±0.02	−1.61	±0.02
$\widehat{\beta}_1$	0.913	±0.019	0.941	±0.019
$\widehat{\beta}_2$	0.0168	±0.0042	0.0119	±0.0038

the concentration given an observed instrumental response. In chemistry, this is usually called the *analysis function* [11].

For example, suppose that a solution whose sulfide concentration is unknown produces the instrumental reading $r^* = 0.5\,\mu\text{A}$, and that one wishes to use the analysis function that corresponds to the robust calibration function to determine the unknown concentration c^*.

This involves finding c^* such that $\widehat{\beta}_2(\log c^*)^2 + \widehat{\beta}_1(\log c^*) + \widehat{\beta}_0 - \log r^* = 0$ and c^* lies between $0.88\,\mu\text{mol/L}$ and $81.2\,\mu\text{mol/L}$, where $\widehat{\beta}_0$, $\widehat{\beta}_1$, and $\widehat{\beta}_2$ are as listed on the right-hand side of Table 2, under "LMROB". The roots of the equation above are

$$\frac{-\widehat{\beta}_1 \pm \sqrt{\widehat{\beta}_1^2 - 4\widehat{\beta}_2(\widehat{\beta}_0 - \log(r^*))}}{2\widehat{\beta}_2} = -80.2 \text{ and } 0.961,$$

therefore $c^* = \exp(0.961)\,\mu\text{mol/L} = 2.61\,\mu\text{mol/L}$.

If a polynomial of degree higher than 2 would have been found to be the best model and then selected, we would not have been able to use a simple algebraic expression for the roots. Instead, we would have had to use a numerical root-finding procedure, for example R function `polyroot`, or an approximate solution, like an interpolating spline [12].

The uncertainty evaluation is done by application of the nonparametric bootstrap, using the R code in Figure 4, and yields a 95 % coverage interval ranging from $2.39\,\mu\text{mol/L}$ to $2.67\,\mu\text{mol/L}$ for the true amount concentration that corresponds to instrumental response $r^* = 0.5\,\mu\text{A}$.

The bootstrap [13] serves to characterize the sampling uncertainty of our estimate of the true concentration. The nonparametric version of the bootstrap does not make any assumptions about the underlying data distribution, but instead creates new pseudo-samples by sampling with replacement from our original data. We fit the same calibration function (of degree 2) to each of these pseudo-samples, invert the calibration function

```
## rSTAR is the value of the instrumental response for which an
## estimate of concentration is sought. The bootstrap procedure
## yields K estimates, which are stored in vector cSTAR
rSTAR = 0.5; K = 25000; cSTAR = numeric(K); nz = nrow(z)
for (k in 1:K)
{
  ## Indices of the data points selected for the current
  ## bootstrap sample: the same index may appear more than once
  ## in the vector i because sampling is with replacement
  i = sample(1:nz, size=nz, replace=TRUE)
  ## Fit robust regression line to bootstrap sample
  zi.lmrob = lmrob(logr~poly(logc, degree=2, raw=TRUE), data=z[i,],
    control=lmrob.control(k.max=500, max.it=1000, maxit.scale=5000))
  betai = coef(zi.lmrob)
  ## Results are exponentiated because the fits are for the
  ## logarithms of the instrumental responses and of the concentrations
  cRoots = exp(Re(-betai[2] + c(-1,1)*sqrt(betai[2]^2 -
                 4*betai[3]*(betai[1]-log(rSTAR))))/(2*betai[3]))
  cSTAR[k] = cRoots[(0.88 < cRoots) & (cRoots < 81.2)]
}
## Endpoints of 95 % coverage interval for concentration
## corresponding to instrumental reading of 0.5
quantile(cSTAR, probs=c(0.025, 0.975))
```

Fig. 4. This application of the nonparametric bootstrap assumes that the logarithms of the concentrations and instrumental responses in Table 1 are in columns labeled `logc` and `logr` of data-frame `z`.

to compute the analysis function, and predict the concentration associated with the instrumental response of interest. We then use suitable quantiles of these predictions of the concentration of interest to define a confidence interval for our original estimate.

4. Consensus Building Model

Combining measurement results for the same measurand to agree on a consensus value is a common task in measurement science, which arises in many different settings. For example, we may want to assign a value to a property of a reference material that has been measured using several methods, or by different laboratories working independently of one another; reduce uncertainty and broaden the evidentiary basis for a conclusion, via meta-analyses of multiple studies; or establish equivalence of measurement capabilities for national metrology institutes, in key comparisons.

Executing this task involves modeling the measurement results so that they can be naturally combined and developing a procedure for data

```
## Measured values and associated standard uncertainties
## reported by 17 laboratories that measured the amount fraction
## of mercury in kudzu (mg/kg)
lab = c("A", "B", "C", "D", "E", "F", "G", "H", "I",
        "J", "K", "L", "M", "N", "O", "P", "Q")
w = c(0.0246, 0.0177, 0.0224, 0.0102, 0.0075, 0.0140,
      0.0258, 0.0057, 0.0025, 0.0066, 0.0059, 0.0109,
      0.0088, 0.0059, 0.0059, 0.0157, 0.0066) ## mg/kg
u = c(0.0021, 0.0030, 0.0006, 0.0028, 0.0015, 0.0035,
      0.0005, 0.0011, 0.0012, 0.0002, 0.0001, 0.0015,
      0.0002, 0.0006, 0.0011, 0.0006, 0.0003) ## mg/kg

## Assessing mutual consistency of measurement results
## using Cochran's Q test
require(metafor)
summary(rma(yi=w, sei=u, method="DL"))

## Assessing symmetry and shape of distribution of the
## roughly standardized measured values
require(symmetry)
symmetry_test((w-median(w))/u, stat="MGG", bootstrap=TRUE, B=500000)
require(nortest)
ad.test((w-median(w))/u)

## Fitting Bayesian model with Laplace laboratory effects and
## Gaussian measurement errors: nu is the vector of numbers of
## degrees of freedom supporting the u's (all equal to 2)
library(rstan)
rstan_options(auto_write = TRUE)
n = length(w); nu = rep(2,n)
Hg.Data = list(N=n, w=w, u=u, nu=nu, gamma=mad(w), delta=median(u))
Hg.Fit = stan(file="LaplaceGaussian.stan", data=Hg.Data,
              control= list(adapt_delta=0.975),
              warmup=75000, iter=750000, chains=4, cores=4, thin=25)
Hg.Fit.post = extract(Hg.Fit)
c(omegaTILDE=mean(Hg.Fit.post$omega),
  omegaTILDE.u=sd(Hg.Fit.post$omega))
```

Fig. 5. Measured values, Type A evaluation of standard uncertainties, and numbers of degrees of freedom that the standard uncertainties are based on, followed by the commands that fit the Bayesian hierarchical model with Laplace random effects, which assume that a file `LaplaceGaussian.stan`, whose contents are listed in Figure 6, exists in the current R working directory.

reduction and uncertainty evaluations that is consistent with the model and whose results are fit for their intended purpose.

The measurement results for the mass fraction of mercury in kudzu that are listed in Figure 5 were obtained in an interlaboratory study whose results were part of the data used to assign value to NIST SRM 3268 *Pueraria montana* var. *Lobata* (Kudzu) Extract. Each measured value is the average of three determinations, and the reported uncertainty is the result of the common Type A evaluation, as the standard error of the average, as described in clause 4.2.3 of [14].

We will use a random effects model of the form $w_j = \omega + \lambda_j + \varepsilon_j$ for $j = 1, \ldots, 17$, where w_j denotes the mass fraction of mercury measured by laboratory j, λ_j is the effect of laboratory j, ε_j is the corresponding measurement error, and ω is the true mass fraction of mercury in the material, which the consensus value seeks to estimate. A random effects model, as opposed to a fixed effects model, is appropriate as we expect structure in the uncertainty to come from the various laboratories, but we are not necessarily interested in estimating each laboratory effect.

This model will be fitted to data comprising 17 triplets, one per participating laboratory: $\{(w_j, u_j, \nu_j)\}$, with the $\{\nu_j\}$, all equal to $3 - 1 = 2$, denoting the numbers of degrees of freedom that the laboratory-specific standard uncertainties $\{u_j\}$ are based on. An out-of-the-box random effects model typically assumes that the random effects $\{\lambda_j\}$ are a sample from a Gaussian distribution with mean 0 and standard deviation τ. We will test this assumption as part of our analysis.

While fitting this random effects model using the so-called DerSimonian-Laird procedure [15], as implemented in the R function `rma` (with argument `method = "DL"`) defined in the package `metafor` [16], we can assess whether the measurement results appear to be mutually consistent (or homogeneous), or not, using the common chi-squared test of homogeneity, which is Cochran's Q test [17]. This test determines whether the values measured by the different laboratories are more variable than their reported uncertainties suggest that they should be: or, in other words, whether $\tau > 0$.

This test rejects the hypothesis of mutual consistency decisively with a very small p-value reported. The R code in Figure 5 also produces $\tau = 0.0048\,\text{mg/kg}$ as estimate of *dark uncertainty* [18], which happens to be 4.3 times larger than the median of the $\{u_j\}$. This "dark uncertainty" is an uncertainty component that is not accounted for in the individual reported uncertainties but becomes apparent when intercomparing the measured values. Accordingly, we shall seek to estimate this "dark uncertainty" and to propagate it to the final results.

Next we ask whether the roughly standardized measured values, $\{(w_j - M)/u_j\}$, where M denotes the median of the $\{w_j\}$, look like a sample from

```
data {
  int < lower = 1 > N; // Number of labs
  real gamma;          // Prior median of dark uncertainty tau
  real delta;          // Prior median of {sigma[j]}
  real w[N];           // Measured values
  real u[N];           // Standard uncertainties
  real nu[N]; }        // Numbers of degrees of freedom

transformed data{
  real u2[N];
  for (j in 1:N) {u2[j] = u[j]^2;} }

parameters {
  real omega;
  real < lower = 0 > tau;
  real theta[N];
  real < lower = 0 > sigma[N]; }

model {
  omega ~ normal(0, 100000); // Prior for omega
  // Half-Cauchy prior for tau with median gamma
  tau ~ cauchy(0, gamma);
  // Random effects {theta[j]-omega}
  // Division by sqrt(2) makes tau the prior SD
  theta ~ double_exponential(omega, tau/sqrt(2));
  // Half-Cauchy prior for sigma[j] with median delta
  for (j in 1:N) sigma[j] ~ cauchy(0, delta);
  // Likelihood for u2[j]
  for (j in 1:N) {u2[j] ~ gamma(nu[j]/2,
                               nu[j]/(2*(sigma[j]^2)));}
  // Likelihood for w[j]
  for (j in 1:N) w[j] ~ normal(theta[j], sigma[j]); }
  // Must end with empty line
```

Fig. 6. Stan code used to fit the Bayesian hierarchical model with Laplace random effects and Gaussian measurement errors to the results of the interlaboratory study of the mass fraction of mercury in kudzu.

a symmetrical distribution, and whether this distribution may be Gaussian. We answer the first question using the test proposed in Ref. 5, which yields a large p-value, hence not ruling out symmetry, and we answer the second using the Anderson-Darling [19] test, which yields a p-value of 0.03, thus rejecting the null hypothesis that the Gaussian distribution is an appropriate model for the laboratory effects $\{\lambda_j\}$.

The plot in Figure 7 suggests that a distribution with tails heavier than the Gaussian tails may be appropriate. In this conformity, we will model

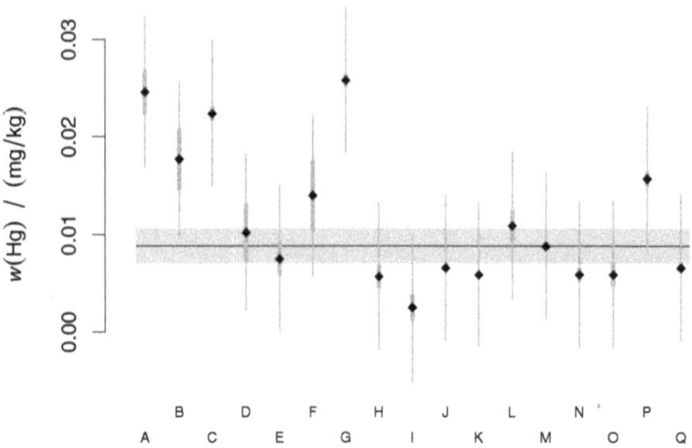

Fig. 7. Measured values (black diamonds), reported standard uncertainties (thick, vertical gray segments), standard uncertainties including contribution from dark uncertainty (thin, vertical gray segments), consensus value (horizontal, black line), and 1-σ coverage band (gray) for the true values of the consensus value.

the random effects $\{\lambda_j\}$ as a sample from a Laplace distribution, another alternative distribution that has fatter tails than a Gaussian, with mean 0 and standard deviation τ, and the measurement errors $\{\varepsilon_j\}$ as outcomes of independent, Gaussian random variables with mean 0 and laboratory-specific standard deviations $\{\sigma_j\}$. Furthermore, we will employ a Bayesian approach, adopting a vague prior distribution for the true mass fraction of mercury, ω, and weakly informative, half-Cauchy prior distributions for the dark uncertainty τ and the $\{\sigma_j\}$, as detailed in the Stan code in Figure 6.

Figure shows the results from this model fit. We find a consensus value for the true mass fraction of mercury in kudzu of 0.0088 mg/kg with standard uncertainty 0.0017 mg/kg. This uncertainty evaluation takes into consideration both the reported standard uncertainties (denoted as thick gray vertical lines), but also the uncertainty contribution from the dark uncertainty (the combined effect of reported and dark uncertainties is depicted by the thinner and longer gray vertical lines). Figure 8 shows us the posterior distribution of both the consensus value and of the value of the dark uncertainty. The preferred Bayesian estimate of this dark uncertainty is the median of the posterior distribution of τ because this distribution is markedly skewed to the right: 0.0075 mg/kg.

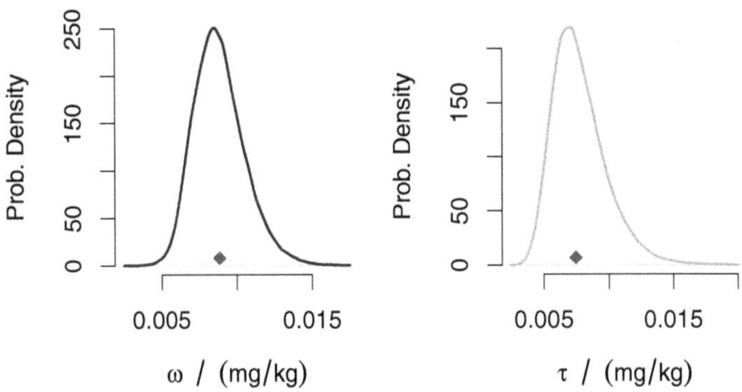

Fig. 8. LEFT PANEL: Posterior density of the mass fraction of mercury in kudzu. RIGHT PANEL: Posterior density of the dark uncertainty.

5. Conclusions and Recommendations

Since rigorous measurement is intrinsically demanding, often requiring substantial time and resources, it is only fitting that the reductions of data most often required in contemporary metrology be undertaken with comparable care, attention to the particulars of the situation, and mindfullness of the purposes that the results are intended for.

Explicit statistical models for measurement data, rather than just *ad hoc* recipes, should be developed and validated in all cases. If something is worth measuring, then it should be measured as well as the application demands, and assuring measurement quality involves the ability to scrutinize and to evaluate critically how the measurement data relate to the measurand.

The adequacy of the model to the data should be established using proven diagnostics, and the choice of manner of fitting an appropriate model to the data should be guided by principles of statistical estimation while considering the particulars of the data and how the results will be used.

In this article we have provided examples of three different types of measurement problems that metrologists may face in their day-to-day work: reducing replicated determinations of a measurand, constructing calibration functions and their inverses, the analysis functions, and selecting models for interlaboratory studies or to blend measurement results obtained by different methods, all accompanied by evaluation of the associated uncertainties.

In each real-world case we present a common statistical model, provide statistical tests and diagnostic plots to assess whether the models assumptions appear to be met, and provide an alternative, more customized statistical model to use when the requisite assumptions are not met. We have presented R code to execute all of these tasks, so that a practitioner can follow along with their own data.

By explaining the choices made when selecting models and performing data reductions for common measurement problems, providing a clear explanation of the required assumptions, and how to diagnose deviations from those assumptions, we have given metrologists a guide that they can refer to as they build and fit models for their own data. In addition, by providing concrete computational steps for model fitting and assessment, even in cases where methods have to be adapted to overcome unmet assumptions, we hope that metrologists will feel empowered to move beyond out-of-the-box approaches for their most nuanced problems.

Acknowledgments

The authors are much indebted to Dr. John Sieber (NIST) for sharing the XRF determinations of the mass fraction of silica in NIST SRM 690, and to Dr. Catherine Rimmer and Dr. Laura Wood (NIST) for allowing us to use the determinations of the mass fraction of mercury in kudzu, which were made during the development of NIST SRM 3268.

References

[1] S. Bell, *A Beginner's Guide to Uncertainty of Measurement*, Measurement Good Practice Guide, Vol. 11 (Issue 2) (National Physical Laboratory, Teddington, Middlesex, United Kingdom, 1999), Amendments March 2001.

[2] B. D. Hall and D. R. White, *An Introduction to Measurement Uncertainty* (Measurement Standards Laboratory of New Zealand, Lower Hutt, New Zealand, 2018), Also available from www.lulu.com.

[3] A. Possolo and J. Meija, *Measurement Uncertainty: A Reintroduction* (Sistema Interamericano de Metrologia, Montevideo, Uruguay).

[4] N. M. Razali and Y. B. Wah, Power comparisons of Shapiro-Wilk, Kolmogorov-Smirnov, Lilliefors and Anderson-Darling tests, *Journal of Statistical Modeling and Analytics* **2**, 21 (2011).

[5] W. Miao, Y. R. Gel and J. L. Gastwirth, A new test of symmetry about an unknown median, in *Random Walk, Sequential Analysis and Related Topics: A Festschrift in Honor of Yuan-Shih Chow*, eds. A. C. Hsiung, Z. Ying and C.-H. Zhang (World Scientific, Singapore, 2006) pp. 199–214.

[6] M. Hollander, D. A. Wolfe and E. Chicken, *Nonparametric Statistical Methods*, 3rd edn. (John Wiley & Sons, Hoboken, NJ, 2014).

[7] E. Desimoni, A program for the weighted linear least-squares regression of unbalanced response arrays, *Analyst* **124**, 1191 (1999).

[8] I. G. Casella, M. R. Guascito and E. Desimoni, Sulfide measurements by flow injection analysis and ion chromatography with electrochemical detection, *Analytica Chimica Acta* **409**, 27 (2000).

[9] T. Bartel, S. Stoudt and A. Possolo, Force calibration using errors-in-variables regression and Monte Carlo uncertainty evaluation, *Metrologia* **53**, 965 (2016).

[10] K. P. Burnham and D. R. Anderson, Multimodel inference: Understanding AIC and BIC in model selection, *Sociological Methods & Research* **33**, 261 (November 2004).

[11] ISO, *Gas analysis — Comparison methods for determining and checking the composition of calibration gas mixtures* (International Organization for Standardization (ISO), Geneva, Switzerland, 2001), International Standard ISO 6143:2001(E).

[12] Y. N. Subbotin, Spline interpolation, in *Encyclopedia of Mathematics*, ed. U. Rehmann et al. (European Mathematical Society, Berlin, Germany, 2011) http://encyclopediaofmath.org/index.php?title=Spline_interpolation&oldid=11892.

[13] B. Efron and R. J. Tibshirani, *An Introduction to the Bootstrap* (Chapman & Hall, London, UK, 1993).

[14] Joint Committee for Guides in Metrology (JCGM), *Evaluation of measurement data — Guide to the expression of uncertainty in measurement* (International Bureau of Weights and Measures (BIPM), Sèvres, France, 2008), BIPM, IEC, IFCC, ILAC, ISO, IUPAC, IUPAP and OIML, JCGM 100:2008, GUM 1995 with minor corrections.

[15] R. DerSimonian and N. Laird, Meta-analysis in clinical trials, *Controlled Clinical Trials* **7**, 177 (September 1986).

[16] W. Viechtbauer, Conducting meta-analyses in R with the metafor package, *Journal of Statistical Software* **36**, 1 (2010).

[17] W. G. Cochran, The combination of estimates from different experiments, *Biometrics* **10**, 101 (March 1954).

[18] M. Thompson and S. L. R. Ellison, Dark uncertainty, *Accreditation and Quality Assurance* **16**, 483 (October 2011).

[19] T. W. Anderson and D. A. Darling, Asymptotic theory of certain "goodness-of-fit" criteria based on stochastic processes, *Annals of Mathematical Statistics* **23**, 193 (1952).

F Pavese, A B Forbes, N F Zhang, A G Chunovkina (Eds.):
Advanced Mathematical and Computational Tools in Metrology and Testing XII
Series on Advances in Mathematics for Applied Sciences, Vol. 90

https://doi.org/10.1142/9789811242380_0006

Measurement results and fuzzy models

Reinhard Viertl

Institute of Statistics and MME, Vienna University of Technology,
Wiedner Hauptstrasse 7/105, 1040 Wien, Austria
Phone: +4366460588 7777,
E-mail: R.Viertl@tuwien.ac.at

Measurements are basic for all quantitative science and for many human activities. Usually numbers are considered to be basic for the description of measurement results of one-dimensional continuous quantities. By the limited accuracy of every measurement equipment the result of one measurement of a continuous quantity is not a number but a more or less vague result. This unavoidable uncertainty has to be analyzed in order to obtain realistic results. Using the concept of so-called *fuzzy numbers*, which are special fuzzy subsets of the set of real numbers, helps to give a more realistic description of the uncertainty of individual measurement results. Consequences of this description for the analysis of measurement data are described in this contribution.

Keywords: Fuzzy number, measurement result, uncertainty, statistical analysis

1. Introduction

When Galilei wrote "Measure everything which is measurable, and the non-measurable make measurable", this had strong influence on the development of science, and still it is governing all kinds of quantitative scientific work. So-called "exact sciences" are based on the possibility of measurement of continuous quantities like time, length, volume, mass etc.

Despite of all problems concerning the definition of measurement units, a practical measurement result of a continuous variable can never be a real number because it is impossible to know all - infinitely many - decimals. Real measurement equipments for continuous quantities produce "numbers" with finitely many digits. These are intervals of real numbers. Using fuzzy numbers is a more general method to model measurement uncertainty than rounding and truncation.

2. What is the result of a measurement?

Individual measurement results are connected with different kinds of uncertainty. The most important are errors and vagueness (fuzziness).

Whereas errors (systematic and random) are properly analysed by statistical methods, vagueness is not a statistical uncertainty, it is the unavoidable uncertainty coming from the impossibility to know the resulting numbers from measurement equipments without uncertainty. Vagueness has nothing to do with Heisenberg's uncertainty principle. Concerning the mathematical description of uncertainties and measurements compare the references[1], [2], and[4] .

Real measurement equipments for continuous quantities produce "numbers" with finitely many digits. Some digital measurement equipments produce too many digits, which are incompatible with the measurement accuracy of the mesurement equipments.

Therefore a more general concept than numbers is necessary to model measurement results. A first step is to use intervals $[a, b]$ of real numbers to describe the measurement result of a one-dimensional continuous quantity. The analysis method of this kind of data, i. e. $x_i^\star \hat{=} [a_i, b_i]$, is *interval mathematics*.

But many scientific data are of more general form, for example results of mass spectrography, color intensity pictures of radar equipments, and records of pointers.

For further explanations the following definition is needed:

Definition 2.1. For classical sets M, subsets $A \subseteq M$ are characterized by their *indicator function* $I_A(\cdot)$, which is defined by

$$I_A(x) := \begin{Bmatrix} 1 \text{ for} & x \in A \\ 0 \text{ for} & x \notin A \end{Bmatrix} \qquad \forall \quad x \in M.$$

Numbers x_i as well as intervals $[a, b]$ are characterized by their indicator functions $I_{\{x_i\}}(\cdot)$ and $I_{[a_i,b_i]}(\cdot)$ respectively. These both types of data are subsets of the set of real numbers $\mathbb{R}$.

In 1951 K. Menger published the idea of generalizing indicator functions of subsets and created the term *ensembles flous* for generalized subsets. Later such generalized subsets were called *fuzzy sets*. This concept, applied to $\mathbb{R}$ and specialized, is a suitable basis for the description of measurement results of one-dimensional continuous quantities.

3. Fuzzy numbers

A *fuzzy number* $x^\star$ is defined by its so-called *characterizing function.*

Definition 3.1. A characterizing function $\xi(\cdot)$ is a real function obeying the following:

(1) $0 \leq \xi(x) \leq 1 \qquad \forall\ x \in \mathbb{R}$
(2) $\text{support}[\xi(\cdot)] := \{x \in \mathbb{R} : \xi(x) > 0\}$ is a bounded set
(3) $\forall\ \delta \in (0,1]$ the so-called δ-cut $C_\delta\Big[\xi(\cdot)\Big]$ defined by $C_\delta\Big[\xi(\cdot)\Big] := \{x \in \mathbb{R} \colon \xi(x) \geq \delta\}$ is non-empty, and a finite union of compact intervals, i. e.

$$C_\delta\Big[\xi(\cdot)\Big] = \bigcup_{j=1}^{m_\delta} \Big[a_{\delta,j}, b_{\delta,j}\Big]$$

This definition contains the classical situation of real numbers x_0 as special case (in this case a one-point closed interval $[x_0, x_0] = C_\delta\left[I_{\{x_0\}}(\cdot)\right]$ is obtained).

Examples of characterizing functions are given in figure 1.

Fig. 1. Examples of characterizing functions of fuzzy numbers

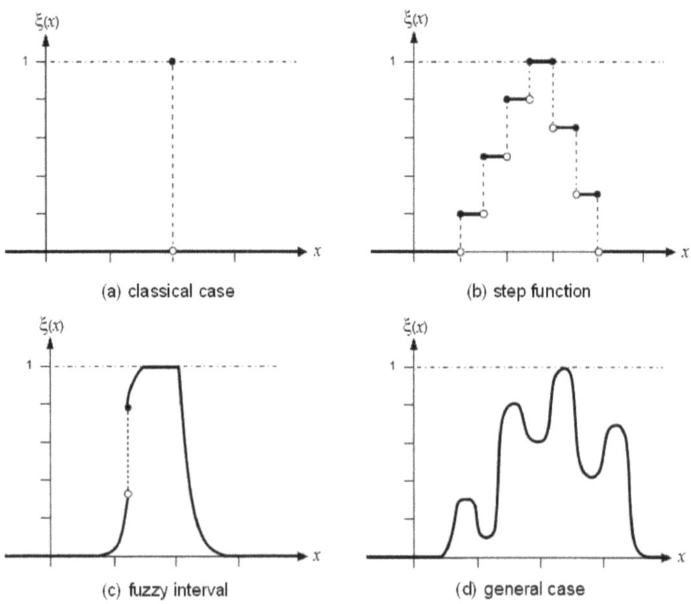

Remark 3.1. Characterizing functions are special forms of membership functions of fuzzy subsets of $\mathbb{R}$.

A main problem is how to obtain the characterizing function of a measurement. There is no general methodology for that but some methods for different application areas are available.

For digital displays on measurement instruments the corresponding function is the indicator function of the interval obtained from the reading x_i by

$$\xi(\cdot) = I_{[x_{i,0},\ x_{i,1}]}(\cdot)$$

where $x_{i,0}$ is the real number obtained from x_i by letting all decimals after the last available digit to be 0, and $x_{i,1}$ is the real number obtained by setting all decimals after the last available digit of x_i equal to 9.

For analog equipments the resulting measurements can be on meters or oscillographs. Meters as well as oscillographs produce color intensity pictures. From such pictures characterizing functions can be obtained using the color intensity along the real axis for one-dimensional quantities. For light points with light intensitiy function $h(\cdot)$ the characterizing function $\xi(\cdot)$ of the "data point" is obtained by

$$\xi(x) := \frac{h(x)}{\max\{h(x) \colon x \in \mathbb{R}\}} \qquad \forall \quad x \in \mathbb{R}.$$

The values $h(x)$ are precise numbers in case of light points on computer screens.

In figure 2 an example of the construction of the characterizing function of a non-precise data point is given.

In case of data given as boundaries of color intensity pictures the color intensity function $g(\cdot)$ can be used to obtain the characterizing function $\xi(\cdot)$ of the non-precise data point $x^\star$. In order to obtain $\xi(\cdot)$ the absolute value of the derivative

$$g'(x) = \frac{d}{dx}g(x)$$

is used, i. e. $|g'(x)| \quad \forall \quad x \in \mathbb{R}$.

The characterizing function $\xi(\cdot)$ is obtained by normalizing $|g'(x)|$:

$$\xi(x) = \frac{|g'(x)|}{\max\{|g'(x)| \colon x \in \mathbb{R}\}} \qquad \forall \quad x \in \mathbb{R}$$

The situation is depicted in figure 3. Rescaling the ordinate of the third graph in figure 3 yields the characterizing function $\xi(\cdot)$.

Fig. 2. Color intensity function of a light point and its characterizing function

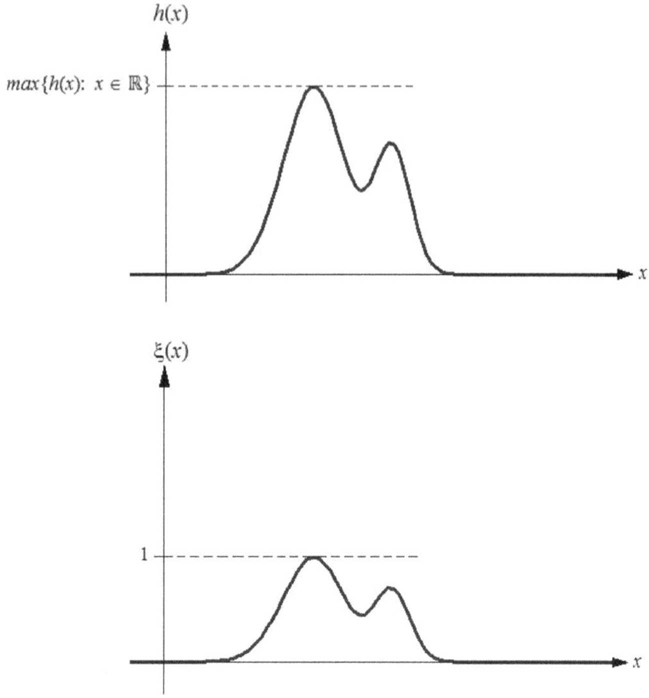

4. Fuzzy vectors

For continuous k-dimensional quantities $\underline{x} = (x_1, \cdots, x_k)$ real measurement results are again more or less non-precise. An informative example is the observation of the position of a ship on a radar screen.

To describe the result of an observation of a k-dimensional quantity so-called *fuzzy vectors* are the best mathematical model. Fuzzy vectors $\underline{x}^\star$ are defined by their so-called *vector-characterizing function* $\zeta(\cdot, \cdots, \cdot)$.

Definition 4.1. A vector-characterizing function $\zeta(\cdot, \cdots, \cdot)$ is a real valued function of k real variables $x_1, \cdots, x_k$ obeying the following:

($\underline{1}$) $0 \leq \zeta(x_1, \cdots, x_k) \leq 1 \qquad \forall \quad (x_1, \cdots, x_k) \in I\!R^k$

Fig. 3. Color intensity function of a fuzzy boundary and corresponding characterizing function

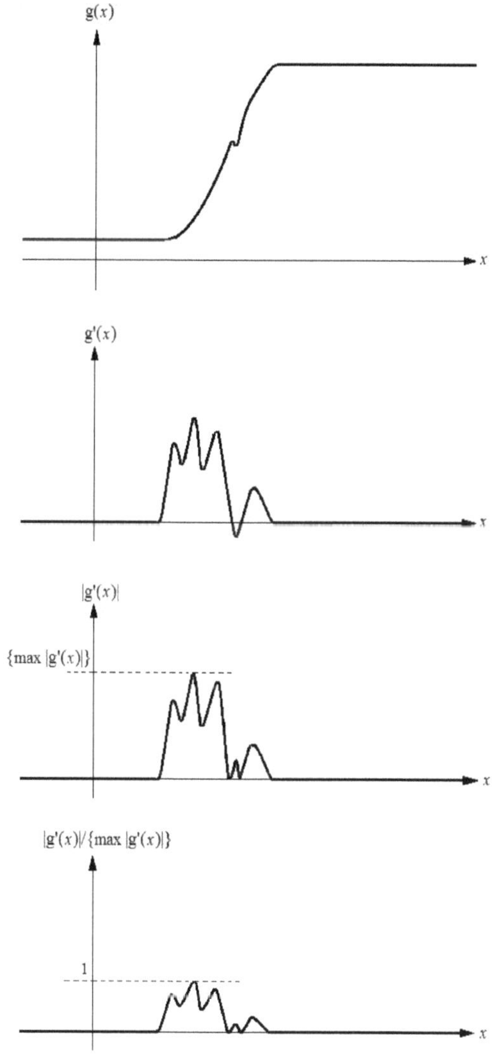

(2) support $[\zeta(\cdot,\cdots,\cdot)]$ is a bounded subset of $\mathbb{R}^k$

(3) $\forall\ \delta \in (0,1]$ the so-called δ-cut $C_\delta\,[\zeta(\cdot,\cdots,\cdot)]$ defined by

$$C_\delta\,[\zeta(\cdot,\cdots,\cdot)] := \left\{(x_1,\cdots,x_k) \in \mathbb{R}^k : \zeta(x_1,\cdots,x_k) \geq \delta\right\}$$

is a non-empty finite union of closed bounded and simply connected subsets of $\mathbb{R}^k$

An example of a vector-characterizing function is given in figure 4.

Fig. 4. Vector-characterizing function

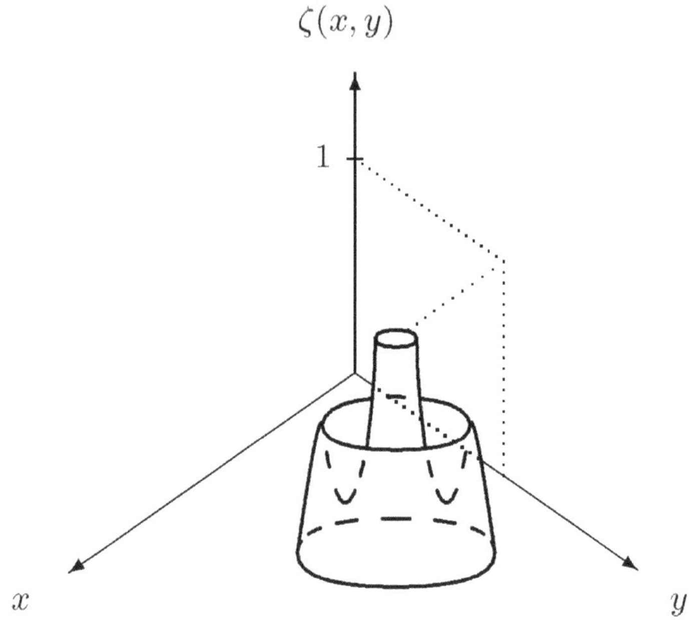

A problem is how to obtain the vector-characterizing function of a non-precise k-dimensional data point $\underline{x}^\star = (x_1, \cdots, x_k)^\star$.

The vector-characterizing function of a two-dimensional light point on a radar screen is obtained in the following way:
Let $h(x, y)$ be the light intensity of the non-precise light point, i. e.

$$h(x, y) \qquad \forall \quad (x, y) \in \mathbb{R}^2.$$

Then the vector-characterizing function $\zeta(\cdot, \cdot)$ of the corresponding fuzzy vector $\underline{x}^\star = (x, y)^\star$ is given by its values

$$\zeta(x, y) = \frac{h(x, y)}{\max\left\{h(x, y) \colon (x, y) \in \mathbb{R}^2\right\}} \qquad \forall \quad (x, y) \in \mathbb{R}^2.$$

The method above to obtain the vector-characterizing function of a non-precise observation can be used for all data which are given in form of color intensity pictures or graphs.

Remark 4.1. Contrary to classical models for vector quantities, where the vector $\underline{x} = (x_1, \cdots, x_k)$ is determined by its components $x_1, \cdots, x_k$, the vector-characterizing function of a fuzzy vector $\underline{x}^\star$ is not trivially determined by the characterizing functions $\xi_i(\cdot)$ of the non-precise coordinates $x_i^\star, i = 1(1)k$.

But there are so-called *t-norms (triangular norms)* to combine fuzzy numbers $x_1^\star, \cdots, x_k^\star$ in order to obtain a fuzzy vector $\underline{x}^\star = (x_1, \cdots, x_k)^\star$.

For details compare section 8 and[5].

The combination of fuzzy numbers is also necessary to generalize arithmetic operations for real numbers to the situation of fuzzy numbers. This generalization has to be in such a way that interval arithmetics is a special case.

The most important combination method of fuzzy numbers in order to obtain a fuzzy vector is thereforeby the so-called minimum-t-norm, i.e. the following:

$$\zeta(x_1, \cdots, x_k) := \min\{\xi_1(x_1), \cdots, \xi_k(x_k)\} \qquad \forall \quad (x_1, \cdots, x_k) \in \mathbb{R}^k$$

5. Arithmetic operations for fuzzy numbers

In order to analyze multiple measurement results of a continuous quantity - which are always more or less fuzzy - the generalization of mathematical calculations to the situation of fuzzy numbers is necessary. This is possible applying the so-called *extension principle* from the theory of fuzzy sets (compare[3]). This principle generalizes classical functions $f : \mathbb{R}^k \to \mathbb{R}$ to the situation of fuzzy argument values $\underline{x}^\star$ in $\mathbb{R}^k$: Let $\zeta(\cdot)$ be the vector-characterizing function of $\underline{x}^\star$. The value $f(\underline{x}^\star)$ is a fuzzy element $y^\star$ of $\mathbb{R}$ whose membership function $\eta(\cdot)$ is defined by its values $\eta(y)$ in the following way:

$$\eta(y) = \left\{ \begin{array}{ll} \sup\left\{\zeta(\underline{x}) : \underline{x} \in \mathbb{R}^k, f(\underline{x}) = y\right\} & \text{if } f^{-1}(\{y\}) \neq \emptyset \\ 0 & \text{if } f^{-1}(\{y\}) = \emptyset \end{array} \right\} \quad \forall\, y \in \mathbb{R}$$

For calculations based on fuzzy data it is necessary to generalize algebraic operations. This is possible based on the extension principle and the *minimum-t-norm*.

5.1. *Translation of a fuzzy number*

Let $x^\star$ be a fuzzy number with characterizing function $\xi(\cdot)$, and $c \in \mathbb{R}$, then the fuzzy number $x^\star + c$ is defined by its characterizing function $\eta(\cdot)$ whose values are given by

$$\eta(x) = \xi(x - c) \qquad \forall \quad x \in \mathbb{R}.$$

Remark 5.1. This is an application of the extension principle. Moreover it is a generalization of interval arithmetics.

5.2. *Scalar multiplication of fuzzy numbers*

Let $x^\star$ be a fuzzy number with characterizing function $\xi(\cdot)$ and $c \in \mathbb{R} \backslash \{0\}$, then the characterizing function $\eta(\cdot)$ of the fuzzy number $c \cdot x^\star$ is given by its values

$$\eta(x) = \xi\left(\frac{x}{c}\right) \qquad \forall \quad x \in \mathbb{R}.$$

Remark 5.2. $\eta(\cdot)$ is a characterizing function, because $0 \leq \eta(x) \leq 1$ $\forall\ x \in \mathbb{R}$, and support$\left[\eta(\cdot)\right]$ is bounded. Moreover all δ-cuts $C_\delta\left[\eta(\cdot)\right]$ are non-empty by $C_\delta\left[\eta(\cdot)\right] = \{x \in \mathbb{R} \colon \eta(x) \geq \delta\} = \left\{x \in \mathbb{R} \colon \xi\left(\frac{x}{c}\right) \geq \delta\right\} \neq \emptyset$ $\forall \quad \delta \in (0, 1]$.

5.3. *Sum of fuzzy numbers*

Let $x_1^\star, \cdots, x_n^\star$ be fuzzy numbers with corresponding characterizing functions $\xi_1(\cdot), \cdots, \xi_n(\cdot)$. Then the sum $x_1^\star \oplus x_2^\star \oplus \cdots \oplus x_n^\star$, also denoted by $\bigoplus\limits_{i=1}^{n} x_i^\star$, has characterizing function $\eta(\cdot)$ whose values, using the notation $\underline{x} = \left(x_1, \cdots, x_n\right) \in \mathbb{R}^n$, are given by

$$\eta(y) = \sup\left\{ \min\left\{\xi_1(x_1), \cdots, \xi_n(x_n)\right\} \colon \underline{x} \in \mathbb{R}^n, \sum_{i=1}^{n} x_i = y\right\} \quad \forall \quad y \in \mathbb{R}.$$

This is the application of the extension principle to the function $f(x_1, \cdots, x_n) = \sum_{i=1}^{n} x_i$.

Remark 5.3. The sum of fuzzy numbers is again a fuzzy number. For the proof compare[5].

5.4. *Difference of fuzzy numbers*

The difference $x^\star \ominus y^\star$ of two fuzzy numbers with corresponding characterizing functions $\xi(\cdot)$ and $\psi(\cdot)$ is the fuzzy number whose characterizing function $\eta(\cdot)$ is given by

$$\eta(z) = \sup\left\{ \min\{\xi(x), \psi(y)\} : (x, y) \in \mathbb{R}^2,\ x - y = z \right\} \quad \forall \quad z \in \mathbb{R}.$$

5.5. *Product of fuzzy numbers*

The product $x^\star \odot y^\star$ of two fuzzy numbers with corresponding characterizing functions $\xi(\cdot)$ and $\psi(\cdot)$ is the fuzzy number whose characterizing function $\eta(\cdot)$ is given by

$$\eta(z) = \sup\left\{ \min\{\xi(x), \psi(y)\} : (x, y) \in \mathbb{R}^2,\ x \cdot y = z \right\} \quad \forall \quad z \in \mathbb{R}.$$

Remark 5.4. There is a general theorem which assures that the result of the generalized algebraic operations yield fuzzy numbers.

Theorem 5.1. *Let $\underline{x}^\star$ be a k-dimensional fuzzy vector and $f : \mathbb{R}^k \to \mathbb{R}$ a continuous function. Then the generalized value $f(\underline{x}^\star)$, defined by the extension principle is a fuzzy number.*

For the proof compare[5].

6. Variation and vagueness of measurement results

Many observed measurements of one-dimensional quantities X exhibit variation as well as imprecision. Whereas variation can be modelled by stochastic models $P(\cdot \mid \theta), \theta \in \Theta$, vagueness of individual measurement results is modelled by fuzzy numbers.

In order to estimate so-called *parameters* θ of stochastic models, functions of observations $x_1, \cdots, x_n$ of X are essential, i. e. $g(x_1, \cdots, x_n)$. In case of stochastic quantities X the observations $x_1, \cdots, x_n$ are called *sample* of X, and measurable functions of $x_1, \cdots, x_n$ are called *statistics.*

The most important statistics are the *sample mean* $\overline{x_n} = g_1(x_1, \cdots, x_n) = \frac{1}{n}\sum_{i=1}^{n} x_i$ and the *sample variance*

$$s_n^2 = g_2(x_1, \cdots, x_n) = \frac{1}{n-1}\sum_{i=1}^{n}(x_i - \overline{x_n})^2.$$

For fuzzy samples $x_1^\star, \cdots, x_n^\star$ the resulting fuzzy values $g(x_1^\star, \cdots, x_n^\star)$ are obtained by application of the extension principle from section 5.

In case the measurement results $x_i^\star$ have characterizing function $\xi_i(\cdot)$, for which all δ-cuts are compact intervals - in this case the $x_i^\star$ are called *fuzzy intervals* - the values $g(x_1^\star, \cdots, x_n^\star)$ for continuous function $g(\cdot, \cdots, \cdot)$ are again fuzzy intervals and the following theorem holds.

Theorem 6.1. *Let $\underline{x}^\star$ be a k-dimensional fuzzy vector for which all δ-cuts $C_\delta[\underline{x}^\star]$ are star-shaped compact subsets, and $g(\cdot, \cdots, \cdot)$ a continuous function $g: I\!R^k \to I\!R$. Then the generalized value $y^\star = g(\underline{x}^\star)$ is a fuzzy interval whose δ-cuts are given by*

$$C_\delta[y^\star] = \left[\min\left\{g(\underline{x}) : \underline{x} \in C_\delta[\underline{x}^\star]\right\}, \max\left\{g(\underline{x}) : \underline{x} \in C_\delta[\underline{x}^\star]\right\}\right] \quad \forall\, \delta \in (0,1].$$

For the proof compare[5].

Remark 6.1. By the representation lemma for membership functions $\zeta(\cdot)$ of fuzzy subsets of an arbitrary set M, i. e.

$$\zeta(x) = \max\left\{\delta \cdot I_{C_\delta[\zeta(\cdot)]}(x) : \delta \in [0,1]\right\} \quad \forall \quad x \in M,$$

the characterizing function of $y^\star = g(\underline{x}^\star)$ is determined.

For details compare[3].

7. Analysis of measurement data

Considering stochastic models $X \sim P(\cdot \mid \theta), \theta \in \Theta$ for variation with observation space M_X, and fuzziness of observations, the generalization of statistical analysis is necessary.

Now it is possible to generalize *estimators* $\psi(x_1, \cdots, x_n)$ of parameters θ in stochastic models to the situation of fuzzy samples $x_1^\star, \cdots, x_n^\star$, applying the extension principle to the combined fuzzy sample (compare[3]).

Another important concept in the analysis of measurement data are *confidence intervals* for parameters θ. For samples $x_1, \cdots, x_n$ a confidence function $\kappa(X_1, \cdots, X_n)$ is a function which assigns to every sample $x_1, \cdots, x_n$ a subset $\kappa(x_1, \cdots, x_n)$ of the parameter space Θ, such that

$$\text{Prob}\Big[\theta \in \kappa\big[X_1, \cdots, X_n\big)\Big] = 1 - \alpha \qquad \forall \quad \theta \in \Theta,$$

where $1 - \alpha$ is denoted as *coverage probability.*

For fuzzy samples $x_1^\star, \cdots, x_n^\star$ of a stochastic model $X \sim P(\cdot \mid \theta), \theta \in \Theta$ the generalization of confidence sets is possible in the following way:
Let $\kappa(\cdot, \cdots, \cdot)$ be a classical confidence function for the parameter θ with coverage probability $1 - \alpha$. Based on the vector-characterizing function $\zeta(\cdot, \cdots, \cdot)$ of the combined fuzzy sample $\underline{x}^\star$, the generalized confidence region $\Theta_{1-\alpha}^\star$ of Θ is a fuzzy subset of Θ, whose membership function $\varphi(\cdot)$ is given by its values $\varphi(\theta) \quad \forall \quad \theta \in \Theta$, using the notation $\underline{x} = (x_1, \cdots, x_n) \in M_X^n$, where

$$\varphi(\theta) := \begin{Bmatrix} \sup\{\zeta(\underline{x}) : \underline{x} \in M_X^n, \theta \in \kappa(\underline{x})\} & \text{if } \exists\ \underline{x} : \theta \in \kappa(\underline{x}) \\ 0 & \text{if } \nexists\ \underline{x} : \theta \in \kappa(\underline{x}) \end{Bmatrix} \quad \forall \quad \theta \in \Theta.$$

Example 7.1. Lenght measurements. In order to obtain the best realistic information concerning the length of an object, n measurements $x_1^\star, \cdots, x_n^\star$ are taken. The individual measurement results $x_i^\star$ are fuzzy numbers. Assuming there are no systematic errors present, usually the mean value $\frac{1}{n}\sum_{i=1}^{n} x_i$ is taken as estimate for the length in the classical context.

For fuzzy measurement data $x_1^\star, \cdots, x_n^\star$ with corresponding characterizing functions $\xi_1(\cdot), \cdots, \xi_n(\cdot)$ the fuzzy estimate $\ell^\star$ of the length is obtained by application of theorem 1 to the function $g_1(\cdot, \cdots, \cdot)$ from section 6 and the combined fuzzy sample $\underline{x}^\star$ which is obtained from the fuzzy sample by application of the minimum-t-norm. Therefore the vector-characterizing function $\zeta(\cdot, \cdots, \cdot)$ of $\underline{x}^\star$ is given by its values

$$\zeta(x_1, \cdots, x_n) = \min\Big\{\xi_1(x_1), \cdots, \xi_n(x_n)\Big\} \quad \forall\ (x_1, \cdots, x_n) \in I\!R^n.$$

The characterizing function $\psi(\cdot)$ of $\ell^\star$ is given by its values $\psi(y)\ \forall \quad y \in I\!R$ in the following way:

$$\psi(y) = \sup\Big\{\min\{\xi_1(x_1), \cdots, \xi_n(x_n)\} : \text{for } \frac{1}{n}\sum_{i=1}^{n} x_i = y\Big\} \quad \forall \quad y \in I\!R$$

The δ-cuts $C_\delta\left[\psi(\cdot)\right]$ are finite unions of compact intervals obtained by

$$C_\delta\left[\psi(\cdot)\right] = \bigcup_{\sum\limits_{i=1}^{n} x_i = y} \left\{ \frac{1}{n}\sum_{i=1}^{n} x_i : x_i \in C_\delta\left[\xi_i(\cdot)\right] \right\}.$$

The proof of the equation above is given in the paper[6].

The characterizing function $\psi(\cdot)$ can be obtained also by application of the so-called *characterization lemma* for membership functions (compare[5]):

$$\psi(x) = \max\left\{\delta \cdot I_{C_\delta[\psi(\cdot)]}(x) \colon \delta \in [0,1]\right\} \quad \forall \quad x \in \mathbb{R}$$

In order to describe the variability of measurement results the generalization of the *sample dispersion*

$$s_n = \sqrt{\frac{1}{n-1}\sum_{i=1}^{n}(x_i - \overline{x}_n)^2} = f(x_1, \cdots, x_n)$$

is possible by application of theorem 1 to the combined fuzzy sample $\underline{x}^\star$. The resulting fuzzy number is denoted by $s_n^\star$.

The δ-cuts of $s_n^\star$ are given by

$$C_\delta(s_n^\star) = \left[\min_{(x_1,\cdots,x_n)\in C_\delta(x_1^\star)\times\cdots\times C_\delta(x_n^\star)} f(x_1,\cdots,x_n) \quad , \quad \max_{(x_1,\cdots,x_n)\in C_\delta(x_1^\star)\times\cdots\times C_\delta(x_n^\star)} f(x_1,\cdots,x_n)\right].$$

The chracterizing function $\varphi(\cdot)$ of $s_n^\star$ is determined by the characterization lemma.

A concrete example of fuzzy measurement data and the results of the generalized analysis method are displayed in figures 5 to 7.

Remark 7.1. The characterizing functions $\psi(\cdot)$ and $\varphi(\cdot)$ are the best information available concerning the length of the measurand.

Fig. 5. Characterizing function of 5 fuzzy measurement results

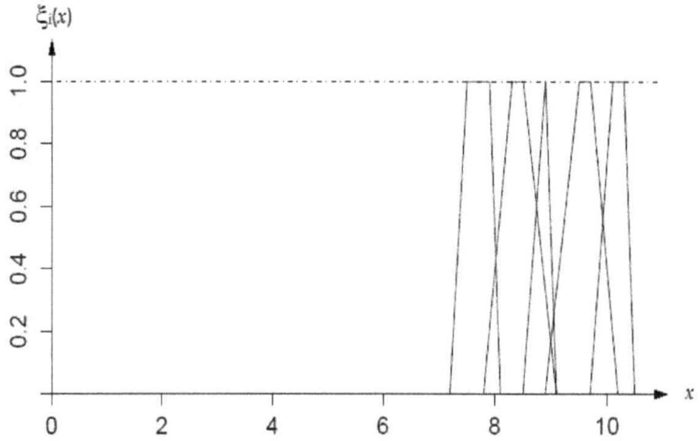

Fig. 6. Fuzzy sample mean

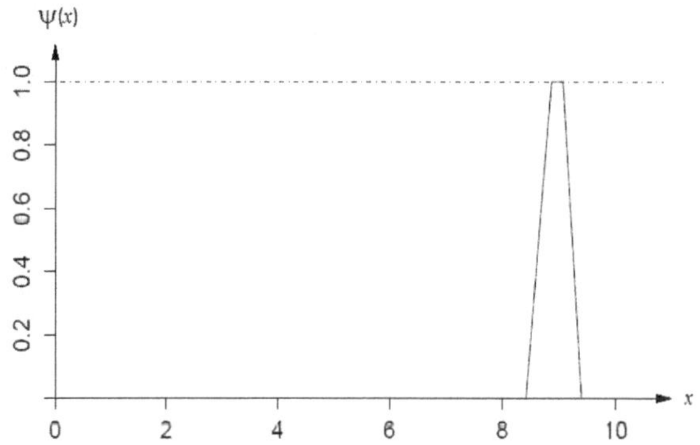

Fig. 7. Fuzzy sample dispersion $s_n^{\star}$

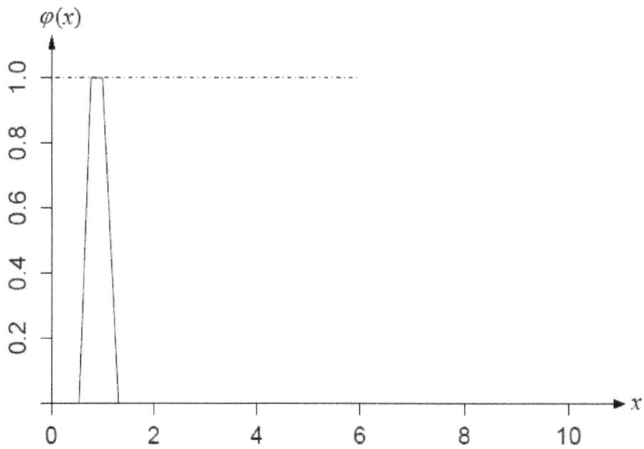

8. Vector-valued quantities

In case of continuous multivariate quantities, idealized modelled by $\underline{x} \in \mathbb{R}^k$, real measurement results are also more or less fuzzy. The best up-to-date mathematical model for this are fuzzy vectors $\underline{x}^\star$ which are defined in section 4.

For 2-dimensional quantities in section 4 a method for obtaining the vector-characterizing function is given.

In case if only fuzzy coordinates $x_1^\star, \cdots, x_k^\star$ of the fuzzy vector are given, it is not a trivial problem to obtain the vector-characterizing function.

For light points in the 2-dimensional Euclidean space $\mathbb{R}^2$ the situation is the following:
A reasonable model would be a fuzzy vector whose δ-cuts are ellipses. If only fuzzy coordinates $x_1^\star$ and $x_2^\star$ are given with corresponding characterizing functions $\xi_1(\cdot)$ and $\xi_2(\cdot)$, where $\xi_i(\cdot)$ are truncated exponential functions, i. e.

$$\xi_i(x) = \exp\left(-\left|\frac{x-\mu_i}{\lambda_i}\right|^p\right) \cdot I_{[\mu_i - c_i,\, \mu_i + c_i]}(x), \tag{8.1}$$

then the δ-cuts of $\xi_i(\cdot)$ are compact intervals, i. e.

$$C_\delta\Big[\xi_i(\cdot)\Big] = [\mu_i - c_i,\ \mu_i + c_i] \quad \forall\, \delta \in \left(0, \exp\left(-\left|\frac{c_i}{\lambda_i}\right|^p\right)\right],$$

and

$$C_\delta\Big[\xi_i(\cdot)\Big] = \Big[\mu_i - \lambda_i(-\ln\delta)^{1/p}, \mu_i + \lambda_i(-\ln\delta)^{1/p}\Big]$$

$$\forall\, \delta \in \left(\exp\left(-\left|\frac{c_i}{\lambda_i}\right|^p\right), 1\right].$$

In this situation the minimum-t-norm is not suitable to obtain the vector-characterizing function of the fuzzy vector $(x_1, x_2)^\star = \underline{x}^\star$ because the δ-cuts of $\underline{x}^\star$ would be rectangles.

If we take the *product-t-norm* for the combination of the fuzzy coordinates of type (8.1) with $p = 2$, the δ-cuts of $\underline{x}^\star$ would be ellipses. Therefore in this situation the vector-characterizing function of the fuzzy vector $\underline{x}^\star$ is given by

$$\zeta(x_1, x_2) = \xi_1(x_1) \cdot \xi_2(x_2) \qquad \forall \quad (x_1, x_2) \in \mathbb{R}^2.$$

For $p = 1$ the δ-cuts of the combined fuzzy vector would be rhombs.

For k-dimensional quantities $(x_1, \cdots, x_k)$ with $k > 2$, and fuzzy coordinates $x_i^\star$, $i = 1(1)k$ with characterizing functions $\xi_i(\cdot), i = 1(1)k$, the vector-characterizing function $\zeta(\cdot, \cdots, \cdot)$ of the fuzzy vector $\underline{x}^\star$ is obtained by application of the associativity of t-norms T by

$$\zeta(x_1, \cdots, x_k) = T\Big(\xi_1(x_1), T(\xi_2(x_2), T(\cdots T(\xi_{k-1}(x_{k-1}), \xi_k(x_k)) \cdots))\Big) \quad \forall \quad (x_1, \cdots, x_k) \in I\!R^k.$$

Again for $\xi_i(\cdot)$ of the form (8.1) the product-t-norm is suitable because the δ-cuts of the combined fuzzy vector, for $p = 2$, are hyperellipsoids. In case of $k = 3$ the δ-cuts are ellipsoids.

For two-dimensional quantities the best information concerning its value is obtained by taking repeated measurements $\underline{x}_1^\star, \cdots, \underline{x}_n^\star$ which are fuzzy vectors $\underline{x}_i^\star$ with vector-characterizing functions $\xi_i(\cdot, \cdot)$, $i = 1(1)n$. In order to estimate the quantity, generalized arithmetics for fuzzy vectors, based on the extension principle can be applied. This yields a fuzzy estimate $\hat{\underline{x}}^\star$ similar to one-dimensional quantities.

Remark 8.1. This is also possible for dimensions $k > 2$.

References

1. H. Bandemer: *Mathematics of Uncertainty - Ideas, Methods, Application Problems*, Springer, Berlin, 2006
2. P. Ciarlini, M. Cox, R. Monaco and F. Pavese (Eds.): *Advanced Mathematical Tools in Metrology*, World Scientific, Singapore, 1994
3. G. Klir, B. Yuan: *Fuzzy Sets and Fuzzy Logic*, Prentice Hall, Upper Saddle River, N.J., 1995
4. S. Salicone: *Measurement Uncertainty*, Springer, New York, 2007
5. R. Viertl: *Statistical Methods for Fuzzy Data*, Wiley, Chichester, 2011
6. R. Viertl: Measurement of Continuous Quantities and their Statistical Evaluation, *Austrian Journal of Statistics*, Vol. 44, pp. 25-32 (2015)

F Pavese, A B Forbes, N F Zhang, A G Chunovkina (Eds):
Advanced Mathematical and Computational Tools in Metrology and Testing XII
Series on Advances in Mathematics for Applied Sciences, Vol. 90

https://doi.org/10.1142/9789811242380_0007

Uncertainty calculation in nanoflow measurements using interferometry

E. Batista*, A. Furtado, J. Pereira, and I. Godinho
Instituto Português da Qualidade
Caparica, 2829-513, Portugal
**E-mail: ebatista@ipq.pt*
www.ipq.pt

R. F. Martins
UNIDEMI, Faculdade de Ciências e Tecnologia, Universidade Nova de Lisboa
Caparica, 2828-516, Portugal

The calibration of drug delivery devices and flow meters at microflow range is often performed by the gravimetric method, but for flow rates lower than 100 nL/min this method presents some limitations, mainly evaporation, flow instabilities and balance characteristics. In order to overcome these problems a new calibration method for microflow measurements down to 16 nL/min using interferometry was developed in a partnership between the Laboratory of Volume and Flow of the Portuguese Institute for Quality and the Department of Mechanical and Industrial Engineering of the New University of Lisbon under the EMPIR project MeDD II – Metrology for drug delivery. One of the main goals of this project is to develop a primary standard for measuring microflows down to 100 nL/min together with the appropriated uncertainty calculation. To validate the obtained results the Guide to the expression of Uncertainty in Measurement (GUM) and the Monte Carlo Method (MCM) were applied.

Keywords: Microflow, Uncertainty, Drug delivery devices, Interferometry.

1. Introduction

The precise measurement of micro and nanoflow, i.e. below 1 µL/h (16 nL/min), of incompressible liquids is a complex task due to several factors, mainly related with the calibration fluid of ultra-pure water, like its unknow air content, formation of air bubbles inside the lines, different adsorption rates depending on syringes and lines material types, and lastly its evaporation during weighing [1]. Nevertheless, the importance of its measurement is of undeniable importance in medical drug delivery devices such as insulin pumps and equipment used in newborns infusion therapeutics, to mention a few.

Thus, in 2019, the 3-years joint research project (JRP) "Metrology for Drug Delivery – MeDD II" [2] was funded by EMPIR programme, is coordinated by IPQ and involves 15 partners. This JRP aims to develop and improve the required metrology tools to assure the traceability of clinical data, allowing the metrological comparability of diagnostic and treatment information for flow rates within the interval from 100 nL/min down to 5 nL/h targeting a relative uncertainty of 1%. This project will develop calibration methods to enable traceable flow rate measurements with high accuracy.

2. Interferometry applied to microflow measurements

Laser interferometry technique is used to measure the intensity of a wave resulting from the overlapping of two or more waves that have travelled over different distances and are superimposed on a single point [3]. Therefore, this methodology can be applied to measure the distance of the pusher block of a flow generator connected to a glass syringe in order to determine the flow rate.

The optical interferometry for flow measurements includes the following components: a He-Ne laser unit with a detector incorporated (an optical arrangement composed by two retroreflector cubes - one of which with a beam splitter attached), a control unit, a pusher block and a Nexus® 3000 syringe pump with a glass removable syringe (Fig. 1). This experimental setup is now proven to be able to cover a flow measurement interval from 0,1 μL/h (1,6 nL/min) up to 10 mL/h (160 nL/min) with a relative uncertainty of 3 %.

Figure 1 – IPQ's microflow experimental setup using optical interferometry technique

3. Uncertainty calculation

The measurement's uncertainty of the interferometric method used for microflow determination is normally estimated according to the Guide to the expression of Uncertainty in Measurement (GUM) [4].

The measurement model is presented along with the standard uncertainties' components, the sensitivity coefficients' values, the combined standard uncertainty and the expanded uncertainty.

It was also considered to perform a validation process by a robust method, being used for that purpose, the Monte Carlo Method (MCM) simulation, as described in GUM Supplement 1 [5], using *MATLAB*® programming software.

The computational simulation was carried out using the same input information used by the Law of Propagation of Uncertainties (LPU) evaluation, namely, the same mathematical model (Eq.1), estimates and the assigned probability density functions (PDF) characterizing each input quantity. It was considered a number of Monte Carlo trials equal to $1{,}0\times10^6$.

3.1. *Measurement model*

The flow quantity in the presented setup is expressed by (Q) [1] and is given by equation (1), where v is the velocity of the fluid and A is the internal cross-section circular area of the syringe (πr^2).

Moreover, velocity is obtained by dividing the displacement d (in the x-axis between point 1 and 2) of the syringe piston through a period of time t; equation (2) is obtained from equation (1) considering the sum of the expansion correction factor γ of the syringe glass, for a reference temperature T of 20 ºC :

$$Q = v \times A = \frac{x_2 - x_1}{\Delta t} \times \pi r^2 = \frac{d\pi r^2}{t} \tag{1}$$

$$Q_{20} = \frac{d\pi r^2}{t} \times [1 - \gamma(T - 20)] \tag{2}$$

3.2. *Uncertainty evaluation*

The main uncertainties of components considered are: distance $u(d)$, time $u(t)$, radius of the syringe $u(r)$, water temperature (t), expansion coefficient of the glass syringe (γ), stability of the setup $u(\delta Q_{sta})$ and standard deviation of the mean $u(\delta Q_{rep})$.

Detailed information regarding the uncertainty components is described in Table 1.

Table 1. Resume of the main uncertainty components of the microflow measurements using IPQ's apparatus.

Uncertainty source	***Standard uncertainty***	***Uncertainty evaluation process***	***Uncertainty evaluation type***	***Uncertainty distribution***
Distance	$u(d)$	Calibration certificate	B	Normal
Radius	$u(r)$	Calibration certificate	B	Normal
Time	$u(t)$	Calibration certificate	B	Normal
Temperature	$u(T)$	Calibration certificate	B	Normal
Expansion coefficient	$u(\gamma)$	Literature	B	Rectangular
Stability	$u(\delta Q_{sta})$	Standard deviation of stability measurements	A	Normal
Repeatability	$u(\delta Q_{rep})$	Mean standard deviation of the flow measurements	A	Normal

Combined uncertainty of the flow measurement ($u(Q)$) by IPQ's interferometric method is given in equation (3), where ($\partial Q/\partial x_i$) represent the sensibility coefficients.

$$u(Q) = \sqrt{\begin{array}{l}\left(\frac{\partial Q}{\partial r}\right)^2 u^2(r) + \left(\frac{\partial Q}{\partial d}\right)^2 u^2(d) + \left(\frac{\partial Q}{\partial t}\right)^2 u^2(t) + \\ + \left(\frac{\partial Q}{\partial T}\right)^2 u^2(T) + \left(\frac{\partial Q}{\partial \gamma}\right)^2 u^2(\gamma) + u^2(rep) + u^2(stab)\end{array}} \qquad (3)$$

From the determined values of the coverage factor k and the combined standard uncertainty of the measurand, the expanded uncertainty is deduced by equation (4):

$$U(Q) = k \times u(Q) \qquad (4)$$

4. Results

A *Nexus*® 3000 pump (microflow generator) was calibrated using the interferometric method described in [6], for a programed flow rate of 1 μL/h. The measurements results were collected using a *Labview*® application and the average of 562 values were used to determine the measured flow, schematically presented in Fig. 2, with a mean value equal to $3,003\times10^{-4}$ μL/s (1,083 μL/h) and a standard deviation of the mean of $4,70\times10^{-6}$ μL/s .

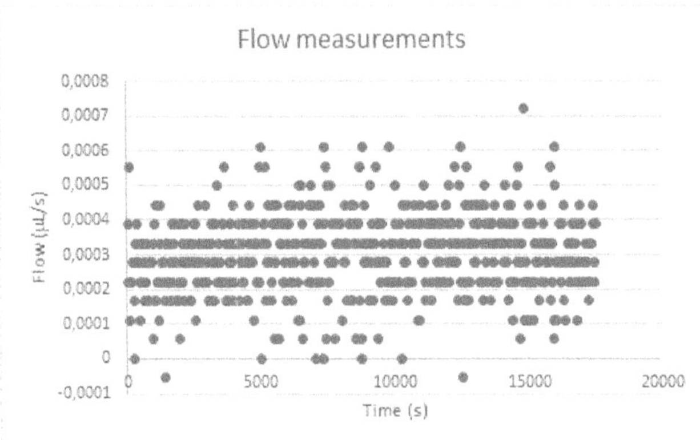

Figure 2. Flow measurement results

The uncertainty results using GUM approach are presented in Table 2.

A comparison between GUM and MCM approaches (considering coverage intervals of 68 %, 95 % and 99 %) is presented in Table 3, where is indicated the estimated values for the output quantity with the associated standard uncertainties and the limits determined for each measurement interval. A difference of $5,6\times10^{-9}$ μL/s was obtained for the output quantity, which is much smaller than the experimental system accuracy ($4,9\times10^{-6}$ μL/s).

Table 2. Uncertainty components in the calibration of a *Nexus*® 3000 pump - GUM

Uncertainty components	**Estimation**	$u(x_i)$	c_i	$(c_i \times x_i)^2$
Distance (mm)	0,316	$1,85\times10^{-4}$	$9,5\times10^{-4}$	$3,1\times10^{-14}$
Time (s)	17490	$7,0\times10^{-4}$	$-1,7\times10^{-8}$	$3,62\times10^{-21}$
Radius (mm)	2,3	5×10^{-3}	$2,6\times10^{-4}$	$1,71\times10^{-12}$
Temperature (ºC)	20,47	5×10^{-3}	$3,01\times10^{-9}$	$2,26\times10^{-22}$
Expansion coefficient (ºC)	$1,0\times10^{-5}$	$2,88\times10^{-8}$	$1,43\times10^{-4}$	$1,69\times10^{-23}$
Stability (μL/s)	0	$6,69\times10^{-7}$	1	$4,48\times10^{-13}$
Repeatability (μL/s)	0	$4,70\times10^{-6}$	1	$2,21\times10^{-11}$
Flow (μL/s)	$3,01\times10^{-4}$			
u_{comb} (μL/s)	$4,9\times10^{-6}$			
U_{exp} (μL/s)	$9,9\times10^{-6}$			

Table 3. Comparison between GUM and MCM approach

	MMC		GUM		
M	*y* (µL/s)	± *u* (µL/s)	*y* (µL/s)	± *u* (µL/s)	÷ y_{GUM} - y_{MCM}÷ (µL/s)
1,0E+06	3,0106E-04	4,931E-06	3,011E-04	4,93E-06	5,6E-09
Probability Density	$u_{probability}$	**Limits**	$u_{probability}$	**Limits**	(u_{GUM} - u_{MMC}) (µL/s)
68 % Û (*Y* ± *u*)	4,91E-06	2,9613E-04 3,0510E-04	4,9E-06	2,961E-04 3,060E-04	2,3E-08
95 % Û (*Y* ± 1,96 ′ *u*)	9,62E-06	2,914E-04 3,107E-04	9,7E-06	2,914E-04 3,107E-04	4,5E-08
99 % Û (*Y* ± 2,68 ′ u)	1,32E-05	2,878E-04 3,143E-04	1,3E-05	2,879E-04 3,143E-04	6,1E-08

From the Monte Carlo simulation performed it was obtained a normal probability density function of the output quantity as presented in Fig. 3.

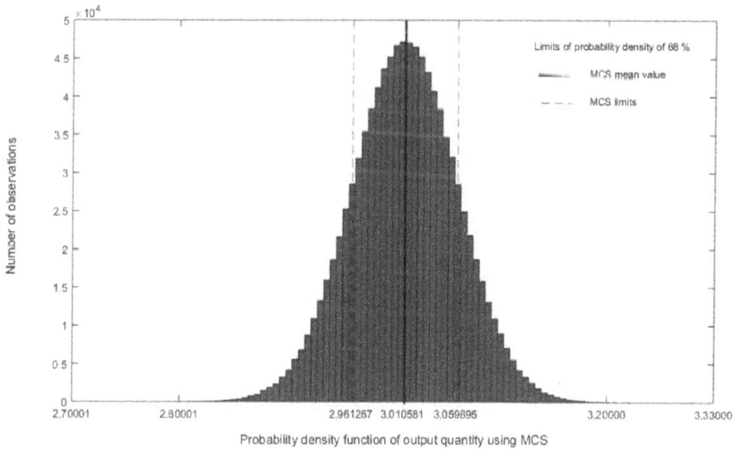

Figure 3. Probability density function of output quantity flow *Q* using the MCM.

5. Conclusions

In the measurement of nanoflow using the interferometric method, several influence factors were found that have a great contribution to the uncertainty calculation at 1 µL/h, namely the measurement of the radius of the syringe used and the calibration of the interferometer. In this method, one of the main uncertainty components is also the repeatability of the measurements.

Comparing the results and the corresponding uncertainties obtained by the two approaches, it can be concluded that the estimated output quantity values, considering both GUM and MCM approaches, show an excellent agreement ($\approx 10^{-9}$ μL / s), much smaller than the experimental system accuracy ($\approx 10^{-6}$ μL / s), which allows the validation of the methodology used for nanoflow measurements.

Acknowledgements

This work performed under MeDDII project has received funding from the EMPIR programme co-financed by the Participating States and from the European Union's Horizon 2020 research and innovation programme.

References

1. Batista E., et al., Assessment of drug delivery devices, Biomedizinische Technik (Biomed Tech). 2015, 60 (4); 347–57
2. E. Batista et al., New EMPIR project – Metrology for Drug Delivery, Flow Measurement and Instrumentation, Volume 72, April 2020, 101716.
3. A Donges, R Noll, Laser Measurement Technology – Fundamentals and applications, Springer, ISBN: 978-3-662-43633-2
4. JCGM 2008, Evaluation of measurement data - Guide to expression of uncertaity in measurement, 1ª ed., 2008.
5. BIPM, IEC, IFCC, ILAC, ISO, IUPAP and OIML, "Evaluation of measurement data – Supplement 1 to the Guide to the Expression of Uncertainty in Measurement – Propagation of distributions using a Monte Carlo method, Joint Committee for Guides in Metrology, JCGM 101, 2008.
6. Batista E., et al., Development of an experimental setup for microflow measurement using interferometry, Flow Measurement and Instrumentation, Volume 75 (2020) 101789.
7. F White (2007). Fluid Mechanics, 6th edition, McGraw-Hill, ISBN: 978-85-63308-21-4

F Pavese, A B Forbes, N F Zhang, A G Chunovkina (Eds.):
Advanced Mathematical and Computational Tools in Metrology and Testing XII
Series on Advances in Mathematics for Applied Sciences, Vol. 90

https://doi.org/10.1142/9789811242380_0008

What if we use almost-linear functions instead of linear ones as a first approximation in interval computations

M. Ceberio

Department of Computer Science,
University of Texas at El Paso,
500 W. University,
El Paso, TX 79968, USA
E-mail: mceberio@utep.edu

O. Kosheleva

Department of Teacher Education,
University of Texas at El Paso,
500 W. University,
El Paso, TX 79968, USA
E-mail: olgak@utep.edu

V. Kreinovich

Department of Computer Science,
University of Texas at El Paso,
500 W. University,
El Paso, TX 79968, USA
E-mail: vladik@utep.edu

In many practical situations, the only information that we have about measurement errors is the upper bound on their absolute values. In such situations, the only information that we have after the measurement about the actual (unknown) value of the corresponding quantity is that this value belongs to the corresponding interval: e.g., if the measurement result is 1.0, and the upper bound is 0.1, then this interval is $[1.0-0.1, 1.0+0.1] = [0.9, 1.1]$. An important practical question is what is the resulting interval uncertainty of indirect measurements, i.e., in other words, how interval uncertainty propagates through data processing. There exist feasible algorithms for solving this problem when data processing is linear, but for quadratic data processing techniques, the problem is, in general, NP-hard. This means that (unless P = NP) we cannot have a feasible algorithm that always computes the exact range, we can only find good approximations for the desired interval. In this paper, we propose two new metrologically motivated approaches (and algorithms) for computing such approximations.

Keywords: interval computations, measurement uncertainty, NP-hard problems, monotonicity, indirect measurements, uncertainty quantification

1. Why Interval Computations

Measurement uncertainty is ubiquitous. Measurements are never absolutely accurate: the measurement result $\widetilde{x}$ is, in general, different from the actual (unknown) unknown value x of the corresponding quantity; see, e.g.,[8].

Case of interval uncertainty. Traditional metrological techniques assumes that we know the probability distribution of the measurement error

$$\Delta x \stackrel{\text{def}}{=} \widetilde{x} - x.$$

However, in many real-life situations, the only information that we have about the measurement error is the upper bound Δ on its absolute value: $|\Delta x| \le \Delta$. In these situations, we do not have any information about the probability of different values, we do not even know which values are more probable and which are less probable. In principle, any probability distribution on the interval $[-\Delta, \Delta]$ is possible.

In this case, after we get the measurement result $\widetilde{x}$, the only information that we have about the actual value x is that x belongs to the interval $[\widetilde{x} - \Delta, \widetilde{x} + \Delta]$; see, e.g.,[8].

Why interval uncertainty. The usual way of determining the probability distribution of the measurement errors is to *calibrate* the measuring instrument, i.e., to compare, several times, the results of measuring the same value by this instrument and by a much more accurate ("standard") one. Since the measurement error of the standard instrument is much smaller that the measurement error of our instrument, we can safely ignore this standard measurement error and assume that the values $\widetilde{x}_{st}$ measured by the standard instrument represent the actual values of the corresponding quantity. Under this assumption, the difference $\widetilde{x} - \widetilde{x}_{st}$ is approximately equal to the measurement error $\Delta x = \widetilde{x} - x$. After we perform this comparison several times, we get a sample of values of measurement error – and from this sample, we can determine the desired probability distribution.

This procedure is reasonable, and it is often implemented, but there are two important classes of situations in which this calibration is not performed. First, this procedure is not done for cutting-edge measurements, when the measuring instrument that we use is among the most accurate, and thus, there is no much more accurate instrument that could be used as a standard. Second, this procedure is often not done in practice simply because calibration is an expensive procedure: e.g., when high school kids build robots, they can buy very cheap sensors, but calibrating each sensor would require the use of expensive high-accuracy measuring instrument and would thus cost much more than sensors themselves. As a result, in manufacturing, often, instead of calibrating a sensor, practitioners simply use the upper bound on the measurement error – bound provided by the manufacturer of the measuring instrument.

Need for indirect measurements. Some quantities we can measure directly. Other quantities y are difficult to measure directly. To estimate these quantities:

- we find (and measure) easier-to-measure quantities $x_1, \ldots, x_n$ which are related to the desired quantity y by a known dependence

$$y = f(x_1, \ldots, x_n),$$

- and then we plug in the measurement results $\widetilde{x}_i$ into this formula, producing an estimate $\widetilde{y} = f(\widetilde{x}_1, \ldots, \widetilde{x}_n)$.

This estimation process is known as *indirect measurement* or, alternatively, *data processing.*

Need for take measurement uncertainty into account in indirect measurements. The measurement results $\widetilde{x}_i$ are, in general, somewhat different from the actual (unknown) values x_i of the corresponding quantities. As a result, the estimate $\widetilde{y} = f(\widetilde{x}_1, \ldots, \widetilde{x}_n)$ is, in general, different from the actual value $y = f(x_1, \ldots, x_n)$ of the desired quantity. A natural metrological equation is: how big is the difference $\Delta y \stackrel{\text{def}}{=} \widetilde{y} - y$? What can we say about the measurement error Δy of the indirect measurement?

Why not use uniform distributions? At first glance, the situation can be covered by the traditional probabilistic methods. Indeed, since we do not know which values of a measurement error Δx_i are more probable and which are less probable, a reasonable idea is to assume that all these values are equally probable, i.e., that we have a uniform distribution on the corresponding interval $[-\Delta_i, \Delta_i]$. However, it is easy to show that this seemingly natural idea may lead to a drastic overestimation of the accuracy of the indirect measurement.

Indeed, let us consider a simple situation when $f(x_1, \ldots, x_n) = \sum\limits_{i=1}^{n} x_i$, all the measurement results are zeros, i.e., $\widetilde{x}_1 = \ldots = \widetilde{x}_n = 0$, and all the upper bounds on the measurement errors are equal to 1:

$$\Delta_1 = \ldots = \Delta_n = 1.$$

In this situation, the result of data processing is 0: $\widetilde{y} = \sum\limits_{i=1}^{n} \widetilde{x}_i = 0$, so $\Delta y = y$. The value $\Delta y = y$ attains its largest possible value when all the terms x_i attain their largest possible value $x_i = 1$. In this case, $y = n$, so the largest possible value of Δy is equal to n.

But what if we assume that all measurement errors Δx_i are uniformly distributed on the interval $[-\Delta_i, \Delta_i] = [-1, 1]$? In this case, for large n, the value y is the sum of a large number of independent identically distributed random variables. Due to the Central Limit Theorem, the distribution of the sum y is thus close to normal. The mean of the sum is equal to the sum of the means, i.e., to 0, and the variance σ^2 of the sum is equal to the sum of n variances, i.e., to $n/3$. Thus, the distribution of y is close to a normal distribution with mean 0 and standard deviation $\sqrt{n}/\sqrt{3}$. In practice, deviations larger than 6σ are so improbable that they are ignored. So, we conclude that all the values of y are bounded by $6\sigma = (6/\sqrt{3}) \cdot \sqrt{n}$.

For large n, this value $\text{const} \cdot \sqrt{n}$ is much smaller than the actual possible value n of the measurement error – a drastic underestimation of the error of indirect measurement (and thus, a drastic overestimation of accuracy). In many critical situations, an underestimation of the measurement error can lead to a disaster: e.g., when, based on the measurement results, we think that we are still within the safe zone, but in reality, we have already crossed the threshold to a danger zone.

Summarizing: from the metrological viewpoint, we cannot simply replace the interval uncertainty with a uniform distribution, we have to consider interval uncertainty – i.e., in effect, consider all possible probability distributions on each interval.

Need for interval computations. As we have mentioned earlier, often, we have the case of interval uncertainty, when the only information that we have about each value x_i is that this value belongs to the corresponding interval $[\underline{x}_i, \overline{x}_i]$. In this case, the only information that we have about the value $y = f(x_1, \ldots, x_n)$ is that this value belongs to the range $[y^-, y^+]$ of all possible values $f(x_1, \ldots, x_n)$ when each x_i is in the corresponding interval:

$$[y^-, y^+] = f([\underline{x}_1, \overline{x}_1], \ldots, [\underline{x}_n, \overline{x}_n]) \stackrel{\text{def}}{=}$$

$$\{f(x_1, \ldots, x_n) : x_1 \in [\underline{x}_1, \overline{x}_1], \ldots, x_n \in [\underline{x}_n, \overline{x}_n]\}.$$

Computation of this range is known as *interval computations*; see, e.g.,[2,4,5].

2. Interval Computations – Successes and Challenges: A Very Brief Overview

Measurement errors are usually small. Each interval of possible values of x_i has the form $[\underline{x}_i, \overline{x}_i] = [\widetilde{x}_i - \Delta_i, \widetilde{x}_i + \Delta_i]$. Each value x_i from this interval has the form $x_i = \widetilde{x}_i - \Delta x_i$, where $|\Delta x_i| \le \Delta_i$. Thus, the actual value $y = f(x_1, \ldots, x_n)$ has the form $y = f(\widetilde{x}_1 - \Delta x_1, \ldots, \widetilde{x}_n - \Delta x_n)$, and the measurement error Δy of the indirect measurement has the form

$$\Delta y = \widetilde{y} - y = f(\widetilde{x}_1, \ldots, \widetilde{x}_n) - f(\widetilde{x}_1 - \Delta x_1, \ldots, \widetilde{x}_n - \Delta x_n). \tag{1}$$

The dependence $f(x_1, \ldots, x_n)$ is usually analytical. Thus, the expression (1) can be expanded into power series in terms of the unknown values Δx_i, i.e., represented as the sum of terms which are linear in Δx_i, terms quadratic in terms of Δx_i, terms cubic in terms of Δx_i, etc. The measurement errors Δx_i are usually relatively small: usually, no more than 20% (and in most cases, much smaller than that). In this case, terms quadratic in Δx_i are of order $(20\%)^2 = 4\%$, terms cubic in Δx_i are of order $(20\%)^3 = 0.8\%$, etc. The higher the order, the smaller corresponding terms. Thus, from the practical viewpoint, we can safely ignore higher order terms in this expansion and only keep terms up to a certain power. For example, we can keep:

- only linear terms or
- only linear and quadratics terms.

Comment. If we want a more accurate estimate, then, instead of ignoring higher order terms, we can add one of the known bounds for the remaining terms.

What if we only keep linear terms. If we only keep linear terms, then we get a formula

$$\Delta y = \sum_{i=1}^{n} c_i \cdot \Delta x_i, \tag{2}$$

where $c_i \stackrel{\text{def}}{=} \dfrac{\partial f}{\partial x_i}$.

In this case, each term $c_i \cdot \Delta x_i$ in the sum is independent – in the sense that each terms depend only on its own variable Δx_i. So the sum attains its largest possible value Δ when each of the terms is the largest possible. Here:

- If $c_i > 0$, then the expression $c_i \cdot \Delta x_i$ is increasing in Δx_i, so its largest possible value is attained when the value Δx_i is the largest, i.e., when $\Delta x_i = \Delta_i$. In this case, the expression has the value $c_i \cdot \Delta_i$.
- If $c_i < 0$, then the expression $c_i \cdot \Delta x_i$ is decreasing in Δx_i, so its largest possible value is attained when the value Δx_i is the smallest, i.e., when $\Delta x_i = -\Delta_i$. In this case, the expression has the value $-c_i \cdot \Delta_i$.

In both case, the largest possible value of each term $c_i \cdot \Delta x_i$ is $|c_i| \cdot \Delta_i$. Thus, the largest possible value Δ of the sum Δy of these n terms is equal to the sum of these largest value:

$$\Delta \stackrel{\text{def}}{=} \sum_{i=1}^{n} |c_i| \cdot \Delta_i. \tag{3}$$

Similarly, the sum attains its smallest possible value Δ when each of the terms is the smallest possible. Here:

- If $c_i > 0$, then the expression $c_i \cdot \Delta x_i$ is increasing in Δx_i, so its smallest possible value is attained when the value Δx_i is the smallest, i.e., when $\Delta x_i = -\Delta_i$. In this case, the expression has the value $-c_i \cdot \Delta_i$.
- If $c_i < 0$, then the expression $c_i \cdot \Delta x_i$ is decreasing in Δx_i, so its smallest possible value is attained when the value Δx_i is the largest, i.e., when $\Delta x_i = \Delta_i$. In this case, the expression has the value $c_i \cdot \Delta_i$.

In both case, the smallest possible value of each term $c_i \cdot \Delta x_i$ is $-|c_i| \cdot \Delta_i$. Thus, the smallest possible value of Δy is equal to

$$-\sum_{i=1}^{n} |c_i| \cdot \Delta_i,$$

i.e., to $-\Delta$, where Δ is the expression (3). Thus, the range of Δy is the interval $[-\Delta, \Delta]$.

The expression (3) is easy to compute – it requires $O(n)$ steps, i.e., linear time.

What if we also keep quadratic terms: general case. In this case, we have a quadratic expression

$$\Delta y = \sum_{i=1}^{n} c_i \cdot \Delta x_i + \sum_{i=1}^{n} \sum_{j=1}^{n} c_{ij} \cdot \Delta x_i \cdot \Delta x_j, \tag{4}$$

where $c_{ij} = c_{ji} \stackrel{\text{def}}{=} \frac{1}{2} \cdot \frac{\partial^2 f}{\partial x_i \partial x_j}$. We want to find the minimum $\underline{y}$ and the maximum $\overline{y}$ of the expression (4) when $\Delta x_i \in [-\Delta_i, \Delta_i]$.

It turns out that this problem is already NP-hard; see, e.g.,[6]. This means that (unless P = NP) no feasible algorithm can always compute the bounds y^- and y^+; see, e.g.,[3,7].

Since we cannot compute the exact bounds, we need to provide estimates for these bounds – to be more precise, upper bounds, since one of the main purposes of metrology is to provide guaranteed upper bounds on the measurement errors.

Using monotonicity. One of the main ideas in interval computations is that if a function is monotonic with respect to one of the variables Δx_i, then to compute its range, it is sufficient to consider only the endpoints of the range $[-\Delta_i, \Delta_i]$.

Specifically, if the expression (4) is increasing in Δx_i, then:

- to find y^+, it is sufficient to consider the value $\Delta x_i = \Delta_i$, and
- to find y^-, it is sufficient to consider the value $\Delta x_i = -\Delta_i$.

Similarly, if the expression (4) is decreasing in Δx_i, then:

- to find y^+, it is sufficient to consider the value $\Delta x_i = -\Delta_i$, and
- to find y^-, it is sufficient to consider the value $\Delta x_i = \Delta_i$.

How can we check whether the expression (4) is increasing or decreasing with respect to x_i? According to calculus:

- an expression is increasing on some domain if and only if the partial derivative with respect to x_i is non-negative for all the points from this domain, and
- an expression is increasing on some domain if and only if the partial derivative with respect to x_i is non-negative for all the points from this domain.

For a quadratic expression, the partial derivative is linear, it is equal to

$$c_i + 2\sum_{j=1}^{n} c_{ij} \cdot \Delta x_j.$$

We already know how to compute the range of a linear function. So, the range of the partial derivative is equal to:

$$\left[c_i - 2\sum_{j=1}^{n} |c_{ij}| \cdot \Delta_j, c_i + 2\sum_{j=1}^{n} |c_{ij}| \cdot \Delta_j\right].$$

Here:

- If the lower endpoint of this interval is non-negative, this means that the derivative is always non-negative, so the expression is increasing in Δx_i.
- If the upper endpoint of this interval is non-positive, this means that the derivative is always non-positive, so the expression is decreasing in Δx_i.

Thus, for each i, we compute the value $s_i \stackrel{\text{def}}{=} 2 \sum\limits_{j=1}^{n} |c_{ij}| \cdot \Delta_j$. If $c_i - s_i \geq 0$, then:

- to compute y^+, we replace Δx_i in the expression (4) with Δ_i; and
- to compute y^-, we replace Δx_i in the expression (4) with $-\Delta_i$.

Similarly, if $c_i + s_i \leq 0$, then:

- to compute y^+, we replace Δx_i in the expression (4) with $-\Delta_i$; and
- to compute y^-, we replace Δx_i in the expression (4) with Δ_i.

After we do this for all variables, we get – for each of the two problems of computing y^- and of computing y^+ – an expression of the similar type (4), but with fewer variables – namely, only with variables x_i with respect to which the original expression was neither everywhere increasing not everywhere decreasing.

So, we end with the same problem of computing the range of the expression (4), but with fewer variables than originally. How can we compute it?

Straightforward approach and a natural question. If we find upper bounds for each term in the expression (4), then by adding them, we clearly get an upper bound for the expression (4). When $|\Delta x_i| \leq \Delta_i$ and $|\Delta x_j| \leq \Delta_j$, then we have $|\Delta x_i \cdot \Delta x_j| \leq \Delta_i \cdot \Delta_j$. Similarly, we conclude that the range of possible values of $(\Delta x_i)^2$ is the interval $\left[0, (\Delta_i)^2\right]$. Thus, for expression

(4), we get the following upper bound $\overline{Y}$ and lower bound $\underline{Y}$:

$$\overline{Y} = \sum_{i=1}^{n} |c_i| \cdot \Delta_i + \sum_{i:c_{ii}>0} c_{ii} \cdot (\Delta_i)^2 + \sum_{i \neq j} |c_{ij}| \cdot \Delta_i \cdot \Delta_j;$$

$$\underline{Y} = -\sum_{i=1}^{n} |c_i| \cdot \Delta_i - \sum_{i:c_{ii}<0} |c_{ii}| \cdot (\Delta_i)^2 - \sum_{i \neq j} |c_{ij}| \cdot \Delta_i \cdot \Delta_j.$$

These formulas are clearly feasible while the problem of computing the exact range is, as have mentioned, NP-hard. Thus, these formulas do not always produce exact ranges.

So, a natural question is: can we – and if yes, how – find more accurate bounds, i.e., bounds which are closer to the actual difficult-to-compute ranges?

Two natural simplifications. Computations can be somewhat less cumbersome if:

- instead of the original variables Δx_i – which take any values from Δ_i to Δ_i,
- we consider auxiliary variables $z_i \stackrel{\text{def}}{=} \text{sign}(c_i) \cdot \dfrac{\Delta x_i}{\Delta_i}$ for which $z_i \in [-1, 1]$ and $\Delta x_i = \Delta_i \cdot z_i$.

Substituting the expression for Δx_i in terms of z_i into the formula (4), we get an expression

$$\sum_{i=1}^{n} a_i \cdot z_i + \sum_{i=1}^{n} \sum_{j=1}^{n} a_{ij} \cdot z_i \cdot z_j, \tag{5}$$

where $a_i \stackrel{\text{ref}}{=} c_i \cdot \text{sign}(c_i) \cdot \Delta_i = |c_i| \cdot \Delta_i \geq 0$ and

$$a_{ij} \stackrel{\text{def}}{=} c_{ij} \cdot \text{digns}(c_i) \cdot \text{sign}(c_j) \cdot \Delta_i \cdot \Delta_j.$$

The task it then to find the largest and the smallest value of the expression (5) when $z_i \in [-1, 1]$.

This was the first simplification. The second simplification is that to compute the minimum of an expression, it is sufficient to consider the maximum of minus this expression. Since minus quadratic expression is still a quadratic expression, it is therefore sufficient to learn how to compute the maximum.

What we do in this paper. In this paper, we describe two approaches to compute such more accurate bounds.

3. First Approach: Taking Major Inputs into Account

Idea. In many cases, one input z_i is the most influential. In such cases, it is reasonable to assume that the effect of this input provides the largest contribution to the quadratic part of the expression (5). The corresponding quadratic terms is $a_{ii} \cdot z_i^2$.

We may have another input which is almost as influential, for which the corresponding major term is $a_{jj} \cdot z_j^2$.

Since we cannot estimate the exact range of the expression (5) that contains *all* the quadratic terms, a natural idea is:

- to estimate the expression that have all linear terms and the above-mentioned major quadratic terms, and then
- to use the straightforward estimation method to take all other terms into account – hoping that these other terms are smaller and thus, their overestimation will be smaller.

Resulting problem. We want to find the largest possible value $\overline{e}$ of the expression

$$e \stackrel{\text{def}}{=} \sum_{i=1} a_i \cdot z_i + \sum_{i=1}^{n} a_{ii} \cdot z_i^2 \tag{6}$$

when $z_i \in [-1, 1]$.

How to solve this problem: idea and resulting algorithm. The expression (6) is the sum of n independent expressions, each of which depends only on one of the variables z_i. Thus, the desired maximum $\overline{e}$ is simply equal to sum of the maxima of the corresponding n expressions.

For a quadratic expression $a_i \cdot z_i + a_{ii} \cdot z_i^2$ of one variable, its maximum on the interval $[-1, 1]$ is attained:

- either at one of the endpoints, i.e., for $z_i = -1$ or for $z_i = 1$,
- or at the point where the derivative of this expression is equal to 0, i.e., at the point $z_i = -\dfrac{a_i}{2a_{ii}}$ – provided that this point is located inside the interval $[-1, 1]$.

For each i, the values at these three (or two) points can be easily computed, their largest of these three (or two) points can also be easily computed. The sum of these n maxima is the desired bound $\overline{e}$.

For each of n values of the index i, we need a fixed number of computational steps, so overall, this algorithm requires linear time to compute.

4. Second Approach: Taking Major Combinations of Inputs into Account

Idea. In the previous section, we considered the case when one of the inputs makes a major contribution to the measurement error. In practice, however, we may have a situation in which the major contributor is not *one* of the inputs, but rather a *linear combination* of such inputs.

Let us clarify what we mean. In the linear approximation, the measurement error is equal to $\sum_{i=1}^{n} a_i \cdot z_i = \sum_{i=1}^{n} c_i \cdot \Delta x_i$. This value is equal to the difference Δx between the estimated and actual values of the linear combination $x \stackrel{\text{def}}{=} \sum_{i=1}^{n} c_i \cdot x_i$. This linear combination is responsible for all the linear terms. Since the quadratic terms are usually much smaller than the linear terms, this means that the linear combination x is the major contributor to the measurement error.

If we only take this linear combination into account, but allow quadratic dependence, then we get an expression of the type

$$\sum_{i=1}^{n} a_i \cdot z_i + C \cdot \left(\sum_{i=1}^{n} a_i \cdot z_i \right)^2,$$

for some constant C.

Of course, there may be other ("secondary") linear combinations

$$\sum_{i=1}^{n} b_{kj} \cdot x_j \quad (k = 1, 2, \ldots)$$

whose contribution is smaller than the contribution of x but which we would like to also take into account. We are considering quadratic expressions, so we may have quadratic terms of two types:

- terms proportional to the product of the major linear combination and one of the secondary linear combinations, and
- terms proportional to the square of a secondary combination or to a product of two secondary combinations.

Since the major combination has the largest effect on the measurement error, it is reasonable to assume that the joint effect of two combinations is larger when one of these combinations is the major one. Since we cannot take *all* the terms into account – this will make the problem NP-hard – it makes sense to only take such larger terms into account, i.e., to consider the expression of the type:

$$\sum_{i=1}^{n} a_i \cdot z_i + C \cdot \left(\sum_{i=1}^{n} a_i \cdot z_i\right)^2 + \sum_k C_k \cdot \left(\sum_{i=1}^{n} a_i \cdot z_i\right) \cdot \left(\sum_{j=1}^{n} b_{kj} \cdot z_j\right),$$

for some coefficients C_k. All non-linear terms in this formula have the same factor $\sum_{i=1}^{n} a_i \cdot z_i$. By combining these terms, we get the following expression:

$$e \stackrel{\text{def}}{=} \sum_{i=1}^{n} a_i \cdot z_i + \left(\sum_{i=1}^{n} a_i \cdot z_i\right) \cdot \left(\sum_{j=1}^{n} b_j \cdot z_j\right), \tag{7}$$

where $b_j \stackrel{\text{def}}{=} C \cdot a_j + \sum_k C_k \cdot b_{kj}$.

Resulting problem. We want to find the largest possible value $\overline{e}$ of the expression (7) when $z_i \in [-1, 1]$.

Natural simplification. If for some i and j, we have $\frac{b_i}{a_i} = \frac{b_j}{a_j}$, then

$$b_i \cdot z_i + b_j \cdot z_j = c \cdot (a_i \cdot z_i + a_j \cdot z_j),$$

where

$$c \stackrel{\text{def}}{=} \frac{b_i}{a_i} = \frac{b_j}{a_j}.$$

Thus:

- instead of two independent variables z_i and z_j,
- we have, in effect, a single variable $a_i \cdot z_i + a_j \cdot z_j$.

When $z_i \in [-1, 1]$ and $z_j \in [-1, 1]$, this variables takes all possible values from the interval $[-(a_i + a_j), a_i + a_j]$. We can therefore simplify the problem if follow the same idea that we used to introduce the variables a_i; namely:

- in the "a-term", we replace the expression $a_i \cdot z_i + a_j \cdot z_j$
- with the expression $(a_i + a_j) \cdot z_{ij}$ for a new variable $z_{ij} \in [-1, 1]$.

Correspondingly, in the "b-term":

- we replace the sum $b_i \cdot z_i + b_j \cdot z_j$
- with the expression $[c \cdot (a_i + a_j)] \cdot z_{ij}$.

By applying this simplification, we will end up with a problem with fewer variables, in which all the ratios $\frac{b_i}{a_i}$ are different.

How to solve our problem: idea. According to calculus, the maximum of a function $f(x_i)$ on an interval – in particular, on the interval $[-1, 1]$ is attained in one of the three cases:

- The maximum can be attained at the left endpoint – in our case, at the point $z_i = -1$. In this case, at this point, we must have $\frac{\partial f}{\partial z_i} \leq 0$. Indeed, if this partial derivative was positive, the value of the function would increase when we slightly increase z_i from -1, so we would not have the maximum at the point -1.
- The maximum can be attained at the right endpoint – in our case, at the point $z_i = 1$. In this case, at this point, we must have $\frac{\partial f}{\partial z_i} \geq 0$. Indeed, if this partial derivative was negative, the value of the function would increase when we slightly decrease z_i from 1, so we would not have the maximum at the point 1.
- The maximum can also be attained inside the interval. In this case, the partial derivative should equal to 0: $\frac{\partial f}{\partial z_i} \leq 0$.

So:

- If at the point where the maximum is attained, we have $\frac{\partial f}{\partial z_i} > 0$, this would mean that at this point, $z_i = 1$ – otherwise, the partial derivative would be non-positive.
- Similarly, if at the point where the maximum is attained, we have $\frac{\partial f}{\partial z_i} < 0$, this would mean that at this point, $z_i = -1$ – otherwise, the partial derivative would be non-negative.

For the expression (7), the partial derivative is equal to

$$\frac{\partial f}{\partial z_i} = a_i \cdot B + b_i \cdot A,$$

where

$$A \stackrel{\text{def}}{=} \sum_{j=1} a_j \cdot z_j \text{ and } B \stackrel{\text{def}}{=} 1 + \sum_{j=1}^{n} b_j \cdot z_j.$$

To decide for which values z_i the maximum is attained, we need to analyze the sign of this partial derivative.

Here, $a_i \geq 0$, so the inequality $a_i \cdot B + b_i \cdot A > 0$ is equivalent to $B + \frac{b_i}{a_i} \cdot A > 0$. Let us consider two possible cases: $A > 0$ and $A < 0$.

Case when $A > 0$. If $A > 0$, then the above inequality $B + \frac{b_i}{a_i} \cdot A > 0$ is, in its turn, equivalent to

$$\frac{b_i}{a_i} > -\frac{B}{A}.$$

We know that all the values $\frac{b_i}{a_i}$ are different. Let us order the indices i so that these ratios are increasing:

$$\frac{b_1}{a_1} < \frac{b_2}{a_2} < \ldots < \frac{b_n}{a_n}. \tag{8}$$

There exists a threshold $-\frac{B}{A}$ so that:

- for all indices for which the ratio $\frac{b_i}{a_i}$ is larger than this threshold, the partial derivative $\frac{\partial f}{\partial z_i}$ is positive and thus, $z_i = 1$;
- similarly, for all indices i for which the ratio $\frac{b_i}{a_i}$ is smaller than this threshold, the partial derivative $\frac{\partial f}{\partial z_i}$ is negative and thus, $z_i = -1$.

There can be no more than one value i for which the ratio $\frac{b_i}{a_i}$ is exactly equal to the threshold. For this index i, we cannot say anything about the value z_i. So, when $A > 0$, in the order of indices, the optimal sequence of values z_i should be

$$(-1, -1, \ldots, -1, z_i, 1, \ldots, 1) \tag{9}$$

Case when $A < 0$. If $A > 0$, then the inequality $B + \frac{b_i}{a_i} \cdot A > 0$ is, in its turn, equivalent to

$$\frac{b_i}{a_i} < -\frac{B}{A}.$$

So, in this case:

- for all indices for which the ratio $\frac{b_i}{a_i}$ is smaller than this threshold, the partial derivative $\frac{\partial f}{\partial z_i}$ is positive and thus, $z_i = 1$;
- similarly, for all indices i for which the ratio $\frac{b_i}{a_i}$ is larger than this threshold, the partial derivative $\frac{\partial f}{\partial z_i}$ is negative and thus, $z_i = -1$.

There can be no more than one value i for which the ratio $\frac{b_i}{a_i}$ is exactly equal to the threshold. For this index i, we cannot say anything about the value z_i. So, when $A < 0$, in the order of indices, the optimal sequence of values z_i should be

$$(1, 1, \ldots, 1, z_i, -1, \ldots, -1). \tag{10}$$

General comment. A priori, we do not know whether the maximum will be attained when $A > 0$ or when $A < 0$. So, a reasonable idea is to try all possible combinations (9) and (10).

Thus, we arrive at the following algorithm.

Resulting algorithm. First, we sort all the indices in the increasing order (8) of the ratios $\frac{b_i}{a_i}$. Then, for each i_0 from 1 to n, we substitute the following values into the expression (7):

- $z_i = -1$ for $i < i_0$, and
- $z_i = 1$ for $i > i_0$.

The value z_{i_0} is obtained from the condition that the resulting quadratic function of the only remaining variable z_{i_0} attains the largest possible value.

Similarly, for each i_0 from 1 to n, we substitute the following values into the expression (7):

- $z_i = 1$ for $i < i_0$, and
- $z_i = -1$ for $i > i_0$.

The value z_{i_0} is obtained from the condition that the resulting quadratic function of z_{i_0} attains the largest possible value.

This way, we get $2n$ different values of the expression (7). The largest of these values is the desired maximum $\overline{e}$ of the expression (7).

Comment. Sorting requires time $O(n \cdot \log(n))$; see, e.g.,[1]. The remaining part of the algorithm requires computing $1n$ values, each of which requires $O(n^2)$ operations, so the overall time is $O(n^3)$ – very feasible.

How to find the values b_j? The above algorithm assumes that we know the values b_j. But in general, all we know are the values a_i and a_{ij}. How can we find the values b_j based on this information?

Our objective is to minimize the remaining part of the quadratic form – i.e., the terms which are not covered by the algorithm. For each i and j:

- The original coefficient at $z_i \cdot z_j$ was a_{ii} for $i = j$ and $2a_{ij}$ for $i \neq j$.
- In the form (7), we have the coefficient $a_i \cdot b_i$ for $i = j$ and $a_i \cdot b_j + a_j \cdot b_i$ for $i \neq j$.
- Thus, in the remaining part, the coefficient at $z_i \cdot z_j$ is $a_{ii} - a_i \cdot b_i$ for $i = j$ and $2a_{ij} - (a_i \cdot b_j + a_j \cdot b_i)$ for $i \neq j$.

For this remainder, the above-mentioned straightforward estimate of the value of this term is (taking into account that the upper bound for each $|z_i|$ is 1):

$$\sum_{i=1}^{n} \sum_{j=1}^{n} \left| a_{ij} - \frac{a_i \cdot b_j + a_j \cdot b_i}{2} \right|. \tag{11}$$

We want to find the values b_j that minimize this estimate. The expression (11) is a convex function, and there exist feasible algorithms for minimizing convex functions. Thus, computing the values b_j is also feasible.

5. Conclusions

In many practical situations, the only information that we have about each measurement error Δx_i is the upper bound Δ_i on its absolute value: $|\Delta x_i| \leq \Delta_i$. In such situations, we do not know which values Δx_i are more probable and which are less probable. In this case, once we know the result $\widetilde{x}_i$ of measuring the corresponding quantity, the only conclusion that we can make about the actual (unknown) value x_i of this quantity is that this value is located in the interval $[\widetilde{x}_i - \Delta_i, \widetilde{x}_i + \Delta_i]$. Since the measurement results $\widetilde{x}_i$ are, in general, different from the actual values x_i, the result $\widetilde{y} = f(\widetilde{x}_1, \ldots, \widetilde{x}_n)$ of processing the measured values is different from the value $y = f(x_1, \ldots, x_n)$ that we would get if we could process the actual values of the corresponding quantities. A practically important question is to gauge the difference $\Delta y \stackrel{\text{def}}{=} \widetilde{y} - y$, i.e., the error of the corresponding indirect measurement.

In most real-life situations, the measurement errors are relatively small, so in the Taylor expansion of the formula for Δy, we can safely ignore terms which are of higher order in Δx_i, and keep only linear and quadratic terms. Sometimes, the measurement errors are so small that even the quadratic terms can be safely ignored. For such cases, there exist feasible algorithms for gauging Δy.

However, in general, when quadratic terms cannot be ignored, the problem of gauging Δy is NP-hard. This means that, in general, it is not possible to have a feasible algorithm that would always compute the exact upper bound on $|\Delta y|$; at best, we can find upper bounds which are not always exact.

In this paper, we describe two metrologically motivated methods for improving the resulting upper bounds.

Acknowledgments

This work was supported in part by the National Science Foundation grants 1623190 (A Model of Change for Preparing a New Generation for Professional Practice in Computer Science), and HRD-1834620 and HRD-2034030 (CAHSI Includes).

The authors are thankful to the anonymous referees for valuable suggestions.

References

1. T. H. Cormen, C. E. Leiserson, R. L. Rivest, and C. Stein, *Introduction to Algorithms*, MIT Press (Cambridge, Massachusetts, 2009).
2. L. Jaulin, M. Kiefer, O. Didrit, and E. Walter, *Applied Interval Analysis, with Examples in Parameter and State Estimation, Robust Control, and Robotics* (Springer, London, 2001).
3. V. Kreinovich, A. Lakeyev, J. Rohn, and P. Kahl, *Computational Complexity and Feasibility of Data Processing and Interval Computations* (Kluwer, Dordrecht, 1998).
4. G. Mayer, *Interval Analysis and Automatic Result Verification* (de Gruyter, Berlin, 2017).
5. R. E. Moore, R. B. Kearfott, and M. J. Cloud, *Introduction to Interval Analysis* (SIAM, Philadelphia, 2009).
6. H. T. Nguyen, V. Kreinovich, B. Wu, and G. Xiang, *Computing Statistics under Interval and Fuzzy Uncertainty* (Springer Verlag, Berlin, Heidelberg, 2012).

7. C. Papadimitriou, *Computational Complexity* (Addison-Wesley, Reading, Massachusetts, 1994).
8. S. G. Rabinovich, *Measurement Errors and Uncertainties: Theory and Practice* (Springer, New York, 2005).

F Pavese, A B Forbes, N F Zhang, A G Chunovkina (Eds.):
Advanced Mathematical and Computational Tools in Metrology and Testing XII
Series on Advances in Mathematics for Applied Sciences, Vol. 90
 (pp. 167–187)
https://doi.org/10.1142/9789811242380_0009

Data provenance, curation and quality in metrology

James Cheney

University of Edinburgh
The Alan Turing Institute, UK
E-mail: jcheney@inf.ed.ac.uk

Adriane Chapman

University of Southampton
The Alan Turing Institute, UK
E-mail: adriane.chapman@soton.ac.uk

Joy Davidson

Digital Curation Centre, UK
E-mail: joy.davidson@glasgow.ac.uk

Alistair B. Forbes

National Physical Laboratory, UK
E-mail: alistair.forbes@npl.co.uk

Data metrology – the assessment of the quality of data – particularly in scientific and industrial settings, has emerged as an important requirement for the UK National Physical Laboratory (NPL) and other national metrology institutes. Data provenance and data curation are key components for emerging understanding of data metrology. However, to date provenance research has had limited visibility to or uptake in metrology. In this work, we summarize a scoping study carried out with NPL staff and industrial participants to understand their current and future needs for provenance, curation and data quality. We then survey provenance technology and standards that are relevant to metrology. We analyse the gaps between requirements and the current state of the art.

Keywords: provenance, curation

1. Introduction

Metrology, the science of measurement, underpins commerce and manufacturing in the UK and worldwide. International System of Units (SI) provides a coherent foundation for the representation and exchange of measurement data, enabling interoperability and reproducibility in all scientific and technological domains. Concepts such as metrological traceability require a documented, unbroken chain of calibration experiments, each with a stated uncertainty. Confidence in measurement is provided through these traceability chains that establish the provenance of the measurement results. As science, engineering and industry increasingly rely on large-scale data analysis, it is increasingly critical to adapt the principles of metrology to digital settings (sometimes called *data metrology*), to ensure confidence in the use of data. Provenance, curation, and data quality need to be better understood in this context. Over the past 15 years, there has been a great deal of academic research and development in different scientific communities concerned with data provenance, data curation, and data quality; however, there seems to be relatively little interaction or mutual awareness between this academic community and the metrology community.

Provenance is information that explains "where" some data came from, or "why" it was produced in the result of a database query[1]. Provenance is "information that helps determine the derivation history of a data product. [It includes] the ancestral data product(s) from which this data product evolved, and the process of transformation of these ancestral data product(s)"[2]. Many overviews of provenance exist, from use in scientific computation and workflow systems[3–6] to types of provenance[7,8], and security over provenance[9]. Meanwhile, Digital curation (alias data curation) 'involves maintaining, preserving and adding value to digital research data throughout its lifecycle...As well as reducing duplication of effort in research data creation, curation enhances the long-term value of existing data by making it available for further high quality research'[10]. These definitions have been proposed originally with the scientific research data context in mind, but apply equally to data in industry; henceforth we just say 'data'.

In 2017–18, we conducted a study initiated by NPL (the UK's National Measurement Institute), interviewing 22 participants from NPL and industry partners in sectors such as energy & environment, life sciences and health, and advanced manufacturing, all with interests in this area, to analyze the needs of the metrology community. This paper summarizes the key findings of this study, and is intended as a guide to existing resources for readers who need *Context and Understanding* of where data came from, better information for *Curation and Reuse*, *Identify Good Practices*, to perform *Integrity* checks on the data, to provide *Interoperability*, to facilitate *Linking entities*, to understanding data *Quality*, to enable *Reproducibility* in research, and assess *Uncertainty*. We analyze a large number of use cases drawn from across the metrology domain, and identify their needs.

The contributions of this work include:

(1) A brief survey of provenance and available tools that allow the metrology community to immediately use concepts and tools within their domain, in Section 2.
(2) Identification of existing data curation and provenance standards that are appropriate for provenance within this domain in Section 3.
(3) An analysis of a wide range of use cases from NPL, Advanced Manufacturing, Life Sciences and Health and Energy and Environmental domains, in Section 4, concluding with suggested next steps.

2. Overview of provenance and digital curation

The World Wide Web Consortium (W3C) has developed standards for provenance under the auspices of the Provenance Incubator Group (2010) and Provenance Interchange Working Group (2011–2013). From the W3C Provenance Incubator group, "Provenance refers to the sources of information, such as entities and processes, involved in producing or delivering an artifact. The provenance of information is crucial in deciding whether information is to be trusted, how it should be integrated with other diverse information sources, and how to give credit to its originators when reusing it"[11]. Provenance, occasionally referred to as lineage, creates a "family tree" of how information, processes and people interacted to create a given artefact[a].

Several other related terms are used together with provenance and curation. These include metadata, context and data quality. In this report, we use the following definitions for each term:

- Metadata is extra information describing a data artefact. Based on definitions from the W3C Provenance Incubator Group[11], "Descriptive metadata of a resource only becomes part of its provenance when one also specifies its relationship to deriving the resource." Thus, metadata information that reflects information about the object, e.g. file size is not provenance information, while metadata that reflects information about creation or modification, e.g. creation date, is provenance information.
- Context is domain-relevant information about the state of the world at the time information is created or modified. Some context can be

[a]Artefact, in this context, denotes a digital object such as a data set.

collected with the provenance, e.g., the temperature in the laboratory at time of measurement; some can be added later by a user, e.g., annotations; some context is maintained separately, e.g., the accuracy of the measurement instrument used.

- Data quality assessments are estimates of accuracy, precision, etc., of the information. These assessments may be automatically or manually generated.
- Annotations can be made on either data or provenance, and are typically used to associate context and quality information with particular portions of the data and/or provenance.

2.1. *Provenance requirements and other considerations*

Groth et al.[12] identified three distinct dimensions of requirements for provenance information: content, management and use. In this work, we discuss content in Section 2.2, while management and use are expanded here along with Capture and Annotation. In addition to these concerns, Allen et al.[13] outlined practical considerations when designing a system to capture provenance that include capture, annotation, storage, and management. There are core implementation concerns for any system that contains or uses provenance. For instance, a mechanism must exist to gather or *capture* provenance information. This can either be a mechanism, e.g. an API, by which other systems report provenance information to the provenance system, or a set of tools that proactively gather provenance information from other applications. Additionally, it can be beneficial to associate data quality concerns and context information with specific portions of provenance information. Obviously, some thought must be given to how provenance information and annotations are stored for later retrieval and use. Finally, "Provenance management refers to the mechanisms that make provenance available and accessible in a system"[11], and include managing security and privacy concerns as well as tools for validation, visualisation, representation transformation, etc.[14].

2.2. *Provenance content*

Several dimensions related to "provenance content" were identified by the W3C community, including:

(1) Object: The artefact that a provenance statement is about.
(2) Attribution: The sources or entities that contributed to create the artefact in question.

(3) Process: The activities (or steps) that were carried out to generate or access the artefact at hand.
(4) Versioning: Records of changes to an artefact over time and what entities and processes were associated with those changes.
(5) Justification: Documentation recording why and how a particular decision is made.
(6) Entailment: Explanations showing how facts were derived from other facts.

However, there can be many possible representations of how processes interacted, or how data was derived, that are equally valid. It is possible to collect large amounts of provenance data, and still learn nothing from it. So what IS provenance?

We know no universal definition, and doubt one exists. Instead, the end usage of the information should be considered when deciding what information should be collected and stored as provenance. For instance, if an analysis over the provenance to determine process bottlenecks in a manufacturing setting is required, then it is imperative that the provenance contains information about those processes, as well as the starting and ending timestamps of activities. On the other hand, if the provenance is to be used to detect anomalous file-access by operating system processes in a security setting, then the processes and all of the file-read and file-write actions must be recorded. Of course, these uses may not always be fully anticipated at the time provenance needs to be recorded, which is why it is important to consider the possible uses and needs at an early stage.

Before any provenance information is recorded, the first consideration should be "what is it intended to be used for." Standardization may be required to identify agreed sets of provenance for particular applications, for example for measuring device metrology this could include uncertainty budget and calibration certificate information. If the provenance that is actually captured and stored, cannot be used for the desired purpose, the systems should be reconsidered. The Use Cases discussed in Section 4 involve different sectors/domains and each sector provides an immediate grounding of what should be in their provenance system. Additionally, how much provenance is actually required, or the *granularity* of the provenance, should be carefully considered. Granularity refers to the level of detail captured. For instance, in our first example above, the granularity is the set of processes. In the second example, the granularity is individual functions (read/writes) within a process. In terms of capture, finer granularity often

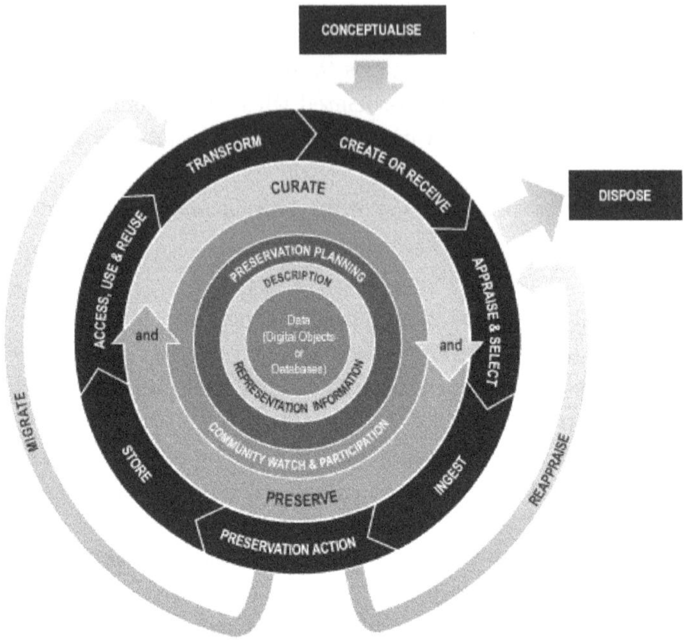

Fig. 1. Digital Curation Lifecycle Model[15]

requires the capture agents to be embedded deeper in the applications in order to get the appropriate detail. Additionally, on the storage side, the higher the granularity, the more provenance will actually be recorded, and an implementation of the storage component must be sufficiently scalable to deal with the resulting data size.

2.3. *Digital Curation Lifecycle Model*

Digital curation enhances the value of digital information through provision of provenance information, the production of metadata descriptions, and by creating links between outputs to provide context which ultimately supports intelligent reuse. The UK Digital Curation Centre (DCC)'s data-centric Curation Lifecycle Model (Figure 1) was developed to provide a graphical overview of the various stages required for successful curation. We briefly describe of each stage of the digital curation lifecycle; further information can be found on the DCC web site[15].

2.3.1. *Digital Curation Lifecycle Components*

During the **conceptualise** stage, active planning for the creation of outputs should be carried out. Planning should include an assessment of data capture equipment, quality assessment and storage methods as well as any relevant standards that will be adopted. Planning should also consider who needs access to the data during the active stage of data collection and how access will be facilitated.

During the **creation stage**, researchers should ensure that sufficient contextual information is captured to make their data FAIR (findable, accessible, interoperable, reusable)[b]. Researchers should consider what information may be necessary to validate any published findings. This may include information about hardware used, software or code that is produced and any models or algorithms used to analyse and visualise the data.

Researchers should **select and appraise** outputs necessary to support reproducibility or for validation purposes but should also be mindful of any additional outputs that may have longer term value. However, not all data generated during a given project can or should be retained.

Selected outputs should be deposited (**ingested**) to a suitable digital repository or data centre. Digital Object Identifiers (DOIs) should be assigned to all outputs selected for deposit. The use of identifiers for researchers (e.g., ORCiD[c], equipment, funding bodies (e.g., Funder Registry offered by Crossref[d]) and grant numbers should also be used to provide unambiguous context.

The chosen repository should undertake **preservation actions** to ensure that data remains authentic, reliable and usable while maintaining its integrity. Researchers may wish to make use of certified repositories (such as CoreTrustSeal[e]).

The key objective for digital curation is to support longer term **access and reuse**. Researchers should consider what context a future reuser may need to understand and make use the given output. FAIR data can be transformed into new research outputs to facilitate new investigations and help researchers to solve grand challenges.

The derived outputs feed back into the **conceptualise** stage of the lifecycle model and the cycle begins again.

---

[b] https://www.force11.org/group/fairgroup/fairprinciples
[c] https://orcid.org/
[d] https://www.crossref.org/services/funder-registry/
[e] https://www.coretrustseal.org/

3. Existing resources and standards

There are many resources, including standards and prototype tools, which may be useful for provenance applications in data metrology. In this section, we briefly survey these, with references to starting points in the computer science and scientific data management literature. The coverage we can provide here is necessarily shallow; further details are in the references.

3.1. *Standardization of provenance concepts*

Provenance information exists in many domain-specific standards[16,17]. There is also a World Wide Web Consortium (W3C) provenance interoperability standard called W3C PROV[18,19]. The World Wide Web Consortium (W3C) is the main standards body for the World Wide Web, and has created many standards (called Recommendations) including HTML (used for web pages), RDF (a data model for descriptive information using subject-predicate-object triples), and OWL, a language for defining ontologies for RDF data. In this context, "ontology" means a set of classes (defining kinds of things), properties (defining kinds of relationships between the things), and rules that impose constraints on the classes and properties, or define ways to infer new knowledge from existing facts. RDF and OWL are used to represent data in the Semantic Web / Linked Data settings, which have been proposed as ways to make data on the Web more machine-readable and more useful for automated analysis.

The Provenance Interchange Working Group was active from 2011–2013 and produced several W3C Recommendations as well as a number of supporting W3C Notes (documents that support and explain standards, but do not themselves specify standard behaviour). The four main Recommendations are:

- PROV-DM - which specifies the PROV Data Model, a collection of vocabulary terms used in the PROV standards and their intended meanings[f]
- PROV-N - which specifies the PROV Notation (also called PROV-N), a serialization format for PROV information that is intended to be both human and machine readable [g]
- PROV-O - which specifies a PROV Ontology (also called PROV-O), an OWL ontology that specifies basic constraints and usage for PROV

[f]https://www.w3.org/TR/prov-dm/
[g]https://www.w3.org/TR/prov-n/

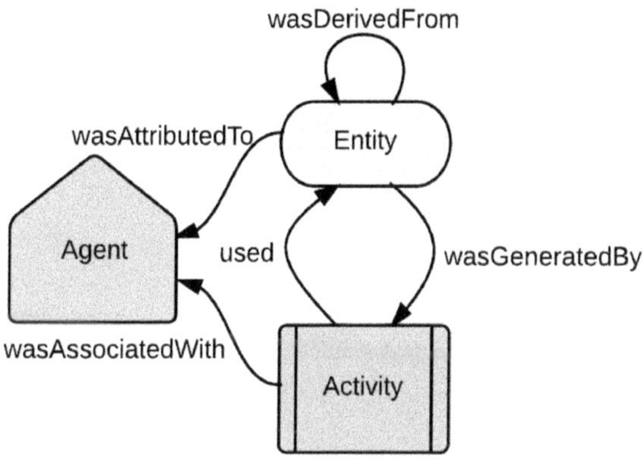

Fig. 2. PROV Data Model core concepts and relationships

represented in RDF format [h]

- PROV-CONSTRAINTS - which specifies more detailed constraints and inference rules for PROV, and defines a notion of validity and equivalence of PROV datasets [i]

Of these standards, PROV-DM and PROV-O are probably the most relevant for typical uses of provenance in the Semantic Web (that is, using RDF and OWL). Other formats for PROV based on XML and JSON have also been defined, but as supporting documents rather than formal Recommendations. The design rationale and lessons learned from the development of PROV are described by Moreau et al.[20]. Since PROV has now superseded earlier proposals, we do not go into further details about them.

In the rest of this section, we will summarize the key points of PROV with respect to its potential use in data metrology.

Figure 2 summarizes the core concepts and relationships of the PROV Data Model. The three main classes introduced in PROV are:

- Entities: representing some real-world or informational entity (e.g. a book, an article, a website, an instrument, an image file, a measurement)

[h]https://www.w3.org/TR/prov-o/
[i]https://www.w3.org/TR/prov-constraints/

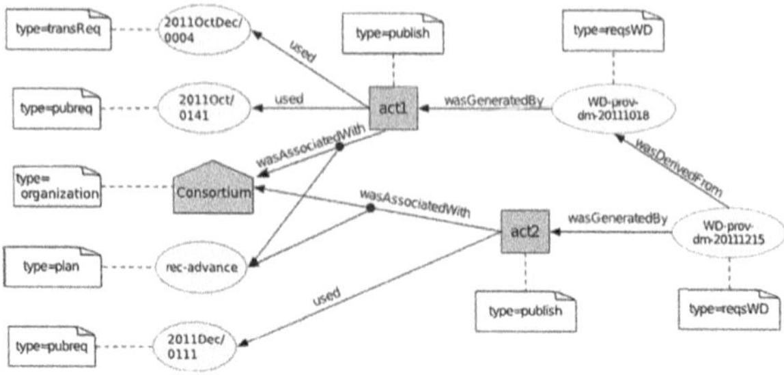

Fig. 3. Example provenance graph using W3C PROV

- Activities: representing processes that can be started or ended by agents, and can generate, use, or change entities
- Agents: representing people, organizations, or software systems that can control or bear responsibility for activities

The basic relationships between these concepts are:

- Usage: an activity may use an entity, meaning its behaviour or outcome is influenced in some way by it
- Generation: an activity may generate an entity, meaning the entity is created by the activity
- Association: an agent may be associated with an activity in different ways, such as controlling or participating
- Information: an activity may inform another activity (typically by some communication)
- Derivation: an entity may be derived from another entity by an activity. As a special case, derivation includes versioning where an old version of an entity is replaced by a new one derived from it
- Delegation: an agent may act on behalf of another agent

Using these concepts, it is possible to highlight relationships between artefacts and how they were manipulated. A simple example of how the W3C PROV standard might be used to represent provenance of some online documents is shown in Figure 3.

These concepts and relationships are intentionally open-ended and broad. Typically, a particular application of PROV will make use of additional concepts and relationships suitable to the domain of application. This is in keeping with the open-ended and distributed nature of the Semantic Web/Linked Data technologies; PROV intentionally does not try to cover all possible application areas, but instead it is encouraged that other vocabularies or ontologies can be used instead (or developed as needed).

For example, to represent executions of scientific computations, the ProvONE vocabulary[21] extends PROV with terminology for describing scientific workflow computations. As another example, to represent the recording and analysis of neuroimaging experiments, the Neuroimaging Data Model (NIDM)[22] extends PROV with concepts and relationships that standardize terminology relating to neuroimaging instruments and software packages.

3.2. *Available tools and techniques*

There is a great deal of research on provenance, extending from theoretical aspects[1,7], uses[23,24], tools[14] and implementation concerns[3,25,26]. While much of this research is still in the academic sphere, some has ripened into more mature tools that can be leveraged. In response to the US Federal Government's interest in implementing provenance, and based on the set of questions frequently encountered, the technical report[13] summarizes good practices and engineering design decisions. In the following subsections, we identify many of these more industrial-ready tools.

Capture tools and technology. There are no generic provenance capture tools that are suited to any kind of application or system. All provenance capture to date involves modifying the underlying system, tools and processes identified as being important to the organisation. In some cases, specific systems have been modified to provide built-in provenance tracking capabilities; the most relevant example of this approach is scientific workflow management systems, which have been proposed for high-level scientific computation in a number of areas[27–29]. Recently, some of the approaches developed for workflow systems have been adapted to scripting languages such as Python[30,31], but this still leaves the problem of how to track provenance when several different systems or languages are used together, each tracking provenance in its own way, or some not at all. Recent work considers a lazy-capture approach, which involves system replay and capture of provenance only when needed[32].

Annotation tools and technology. Annotation allows the enrichment of data by providing additional context or quality information about the underlying data. While it does not necessarily require provenance, it is a piece of metadata that can be attached either to the data, or to the provenance-representation of that underlying data. To date, there are many tools that assist with the creation of annotations and using them. However, most of these tools are very domain specific. For instance,[33] works with chemistry-specific annotations, by highlighting chemistry-specific text, like chemical equations, and providing easy mechanisms for chemistry-specific annotations. Many projects still provide annotation in a do-it-yourself, ad-hoc way. For instance, EMBL-EBI's UniProt[j] started by custom-building user annotation into their web application, and have since moved on to automatically suggest annotations. More general tools[k] [l] focus on ingesting text documents, extracting entities, topic models, social tags, etc., and providing the ability to use and organise this semantic information.

Storage. Provenance is data, and therefore the same storage options exist for provenance as for traditional data including: relational[25]; flat file[34]; graph-based[18]. These storage options are not mutually exclusive. For instance, the data and provenance may be permanently stored in a database, but may be instantiated in a file when passing between systems. Provenance storage introduces a number of challenges, since provenance data can often grow to dwarf the size of the actual data of interest, requiring compression or choices regarding what provenance to keep. Provenance can also be modelled in different ways, for example as annotations directly on the underlying data, or separately using identifiers for cross-referencing. The choice of which technology is best for provenance storage will depend upon how that provenance information is to be used, local technology expertise within the organisation, and network and trust architectures within the organisation.

Administration. Like ordinary data, provenance may need to be secured and protected; in particular, recording provenance in a secure system may introduce new risks to confidentiality (e.g. confidential data might be leaked via provenance) or privacy (e.g. recording user activity may inadvertently record protected personal information). At a minimum, we must apply classic access control techniques to the provenance information[35]. However, classic access controls may not be sufficient for provenance, because attackers may be able to leverage knowledge of how provenance is captured to infer missing information even if sanitisation is applied. As such, several protection mechanisms that work more gracefully with provenance

[j] https://www.ebi.ac.uk/uniprot
[k] Ontotext: https://ontotext.com/
[l] Open Calais: http://www.opencalais.com/

have been identified[36–39]. Cheney and Perera[9] surveyed and compared a number of techniques in this area.

Beyond security and protection, any provenance solution must have basic administration concerns addressed, including classic considerations for most systems such as facilitating user access, managing users, managing size, scalability, performance, etc.

Manipulation. A suite of tools exists to facilitate working with and manipulating W3C PROV[14]. These include:

(1) Validator: A RESTful web service that validates PROV descriptions against the PROV Constraints specification
(2) Translator: Translates between different representations of PROV. Supports PROV-N, PROV-XML, PROV-O and PROV-JSON
(3) Store: A provenance repository that allows storing, browsing, and managing provenance documents via a Web interface or a REST API
(4) ProvToolbox: a Java toolbox for handling PROV
(5) ProvPy: a Python implementation of the PROV data model
(6) ProvExtract: for dealing with PROV embedded in web pages
(7) ProvVis: visualisations of PROV
(8) PROV-N Editor: a text editor with PROV-N syntax highlighted

4. Summary of study: interviews and use cases

As part of our study, individuals from several organisations that relate to NPL's mission, as well as NPL staff, have been interviewed in order to identify the scope of possible interest in provenance, and identify specific use cases and requirements from each area. A brief summary of the interviewed groups and their activities is presented below. These have been organised into the following groups: Activities within NPL; Advanced Manufacturing sector; Life Science & Health sector; Energy & Environment sector.

The interviews with NPL staff helped us to understand the current capabilities and needs within NPL regarding "data science" aspects of data quality, provenance and curation. For the interviews with participants from other organisations, or initiatives where NPL staff are already collaborating with external organisations, we asked the following questions:

- What is the mission of the organisation/initiative and what role does it play in the UK economy?
- Why are data quality, provenance or curation important to that mission?

- What are the current practices regarding data quality, provenance and curation and what are the unmet challenges or future needs?
- What role does NPL currently play, and how could NPL provide better support for these needs?

For each interview summary, one or more use cases (scenarios illustrating a challenge or unmet need) relating to provenance, curation, and data quality have been identified. These are listed in Table 1. Space limits preclude discussing all of the use cases in detail; instead, we will briefly summarize three representative use cases for each domain.

NPL activities. We held discussions with several groups within NPL, including staff members representing NPLs core activities in metrology,such as the measurement of mass, force, pressure, and density, and a staff member in the early stages of developing a provenance-tracking prototype for stress-strain laboratory data. Based on these discussions we identified several common or significant needs. Use case 1 involves maintaining records of calibration and measurement methods for measurement equipment, including the measurement procedure, resulting artefacts, calibration methods, and times of (re)calibrations. Use case 2 considers the possibility that errors might initially go undetected and influence other results, leading to the need to understand how an error has propagated through a system so that users may be alerted and the errors corrected. Use case 4 considers the reverse scenario, when errors are noticed in derived results and the goal is to diagnose the root cause of the observed error. All of these use cases raise new challenges as measurements and calibration records move from largely analog, manually-maintained documents recording only a few steps of measurement and computation, to electronic records of possibly long *data supply chains.*

Advanced manufacturing. Advanced Manufacturing refers to the sophisticated use of cloud computing, cognitive computing, Internet of Things, and robotics in manufacturing. Because of the reliance on digital models and data analytics, ensuring the quality of the end products also depends upon understanding and extending techniques for data quality, such as tracking provenance through computations, and curation of digital artefacts. In this sector standards such as the Quality Information Framework (QIF) are used to record manufacturing processes, which is useful for understanding existing processes and where they might break down (Use case 6), and to provide traceability, optimisation, and quality control in manufacturing (Use case 7). Use case 8 likewise highlights the need to identify and calculate failure points in manufacturing, which may involve working backwards from observations of failed tests to identify root causes, or forward to identify other risks that may be correlated with the root causes identified.

Life science & health. The Life Science & Health sector covers medical, pharmaceutical and biological research, as well as healthcare and public health. Use of computation and data management in these areas is already widespread, particularly in biomedical science (bioinformatics). Some NPL researchers are involved in developing computational methodologies that have the potential to improve the quality of patients computerised medical records both prospectively by automating the correction of submitted data and retrospectively by reconstructing missing data from historical datasets. These methods enable incomplete or missing data to be made available for analysis and comparison with published results. Use case 9 highlights the potential benefits that AI can offer but introduces issues around transparency as it will be essential that researchers are able to easily distinguish between original data versus those that have been automatically enhanced or reconstructed using algorithms. Many life science researchers depend on having access to reliable physical samples to carry out their analyses. They find that the quality and consistency of the contextual information associated with these physical samples varies greatly which in turn impacts the quality of the analysis they can carry out. Use case 10 highlights that the ability to understand a samples quality is directly related to how the physical sample has been sourced and treated prior to its arrival in the lab for digitisation and analysis. NPL researchers are working to improve and standardise sample documentation practices by identifying points over the lifecycle where relevant contextual information is produced and assessing how it should be captured most efficiently and effectively. There is growing interest among researchers in all domains and industry to improve the availability and reuse of existing data. For the life sciences, this may include sharing pre-competitive data across organisations, data collected from hospitals, and subcontracting research to external organisations. Use case 17 introduces a need for improved communication about the context and meaning of the archived data to support informed reuse. As many research performing organisations (RPOs) transition to infrastructure based on cloud storage and object-based data repositories, there is an opportunity to rethink the data capture, curation and management processes to support future reuse.

Energy & environment. The Energy & Environment sector comprises oil & gas, nuclear, hydroelectric and other renewable sources of energy, as well as Earth observation, sustainability, climate change impact assessment, and environmental monitoring. Sensor networks or IoT devices are seeing increasing use for environmental monitoring or monitoring conditions of conventional or renewable energy technologies, but face challenges regarding calibration and uncertainty propagation through data analysis pipelines. Use case 23 identifies the need for transparency and inter-compatibility of standards used by different groups so that data can be exchanged, compared and reused. Environmental monitoring via Earth observation satellites plays an important role both in understanding the climate and making detailed information about the environment available to government or business, and requires the need to understand and put historical data in context of how it was captured and can be combined with current data, as described in Use Case 21. In a related vein, given the longevity of many environmental studies, it is essential that the provenance survives beyond the lifetime of the software used to create the data or capture the provenance, so that an understanding of this historical data can facilitate data usage later, as discovered in Use Case 22. In all of these cases, the increasing use of large amounts of data, gathered under varying conditions by devices with varying levels of accuracy, implies the need for provenance and data quality management techniques similar to those that have already been investigated for scientific data management and analysis.

4.1. *Recommendations*

We made several recommendations to NPL based on our study, which we have adapted for consideration by the broader metrology community as follows:

- **Metrologists should establish best practices for provenance standards** to provide greater clarity regarding what techniques are sufficient to meet current or future needs.
- **Flagship case studies should be developed** to consolidate expertise with developing and using provenance techniques.
- **Metrology institutes should raise awareness of provenance as part of data metrology** through augmenting training materials or other dissemination pathways.
- **The importance of data management planning should be highlighted** to promote data reuse and curation in industrial settings.

Table 1. Use cases identified during the study

		Context & Understanding	Curation & Reuse	Id Good Practice	Integrity	Interoperability	Linking Entities	Quality	Reproducibility	Uncertainty
	NPL Activities									
1	Understanding Measurements and Calibration Measures	•		•	•	•				
2	Taint Propagation	•	•	•	•	•				•
3	Analysis of Digital Twins	•					•		•	
4	Common Cause Analysis	•		•	•	•				•
5	Understanding data from stress-strain experiments	•	•	•		•			•	
	Advanced Manufacturing									
6	Modelling and improvement of manufacturing processes	•		•		•		•	•	
7	Traceability, optimisation, and quality control in manufacturing	•			•			•	•	
8	Calculation and Identification of Failure Points	•		•	•		•			•
	Life Science and Health									
9	Automating the reconstruction of missing data	•				•		•		•
10	Linking digital provenance with physical artefacts for quality analysis	•				•	•	•		
11	Interpretable confidence/quality assessment for machine learning	•						•	•	•
12	Electronic lab notebooks and reproducibility	•	•				•		•	
13	Identifying human gaps	•		•						
14	Standards and best practices for reuse	•	•							
15	Facilitating communication and understanding across organisations	•				•		•		
16	Provenance and Quality in trials	•		•	•	•		•		
17	Data Reuse for additional purposes	•	•	•	•					
18	Identifying duplication in rare genetic data	•	•				•	•		•
19	Diagnosing unexpected differences in results of analytical models	•		•			•	•		•
20	Understanding changes to data over time	•	•		•					
	Energy and Environment									
21	Reuse and understand historical data with recent data	•	•		•	•				•
22	Recording and using past configurations and performance	•	•			•				
23	Transparency and inter-compatibility of standards	•		•						
24	Reflecting good practice in new tools and services			•	•					

- **Academic research and industry need to be brought together** to facilitate transfer of solutions to practice, or communicate unmet needs to researchers.

Addressing these recommendations is the subject of current work by NPL and collaborators within the metrology and data science communities. For example, the EURAMET Technical Committee for Inter-disciplinary Metrology (TC-IM) has initiated a project on Research Data Management and the Consultative Committee for Units (CCU) of International Bureau for Weights and Measures (BIPM) has set up a task group (D-SI) to promote the digitisation of metrology data exchange, references, services and products and to transform the SI into the digital world. In these initiatives, it is seen as essential to address issues relating to data provenance, curation and the FAIR principles and that the metrology community joins with the wider scientific and technical communities to provide well-founded, coherent and inter-operable solutions.

5. Conclusion

Our study highlighted that there are a number of situations in metrology and industrial settings where provenance information is needed, and thus there are many settings where techniques developed by the provenance research community could be beneficial, but also showed that there is relatively little awareness or adoption of these techniques. This may in part be because of the initial focus of the provenance research community on settings such as scientific data and computation and on Web technologies that seem not to be in widespread use in industry settings. For example, only in the life science and health sector (where RDF is already widely used) did we find pre-existing awareness of RDF and PROV. Thus there remains a significant gap between what is technically possible in research prototypes and the day-to-day needs expressed in the use cases we encountered. We hope that this paper will serve to catalyze further discussion between the metrology and data provenance communities to address these challenges.

Acknowledgments This work was supported by EPSRC (grant number EP/S028366/1) and an ISCF Metrology Fellowship grant provided by the UK governments Department for Business, Energy and Industrial Strategy (BEIS).

References

1. P. Buneman, S. Khanna and W. C. Tan, Why and Where: A Characterization of Data Provenance, in *ICDT*, (Springer, 2001).
2. K. K. Muniswamy-Reddy, D. A. Holland, U. Braun and M. I. Seltzer, Provenance-Aware Storage Systems, in *USENIX Annual Technical Conference, General Track*, (Springer, 2006).
3. A. Barker and J. I. van Hemert, Scientific Workflow: A Survey and Research Directions, in *PPAM*, , LNCS Vol. 4967 (Springer, 2007).
4. R. Bose and J. Frew, Lineage retrieval for scientific data processing: a survey, *ACM Comput. Surv.* **37**, 1 (2005).
5. J. Freire, D. Koop, E. Santos and C. T. Silva, Provenance for Computational Tasks: A Survey, *Computing in Science and Engineering* **10**, 11 (2008).
6. Y. Simmhan, B. Plale and D. Gannon, A survey of data provenance in e-science., *SIGMOD Record* **34**, 31 (2005).
7. J. Cheney, L. Chiticariu and W.-C. Tan, Provenance in databases: Why, how, and where, *Found. Trends databases* **1**, 379 (April 2009).
8. B. Glavic and K. R. Dittrich, Data Provenance: A Categorization of Existing Approaches, in *BTW*, (GI, 2007).
9. J. Cheney and R. Perera, An analytical survey of provenance sanitization, in *IPAW*, (Springer, 2014).
10. Digital Curation Centre, What is digital curation? (2018), http://www.dcc.ac.uk/digital-curation/what-digital-curation/, accessed March 26, 2018.
11. W3C Incubator Group Report, *Provenance XG Final Report*, tech. rep., http://www.w3.org/2005/Incubator/prov/XGR-prov-20101214 (2010).
12. P. Groth, Y. Gil, J. Cheney and S. Miles, Requirements for provenance on the web, *IJDC* **7**, 39 (2012).
13. D. Allen, L. Seligman, B. Blaustein and A. Chapman, *Provenance Capture and Use: A Practical Guide*, tech. rep., The MITRE Corporation (2010).
14. ProvToolbox (2020), `http://lucmoreau.github.io/ProvToolbox/`.
15. Digital Curation Centre, DCC curation lifecycle model (2018), http://www.dcc.ac.uk/resources/curation-lifecycle-model, accessed March 26, 2018.
16. International Standards Organization, *ISO 19115*, tech. rep., https://www.iso.org/standard/53798.html (2014).
17. GWG Metadata Focus Group (MFG) Staff / NGA/NCGIS Metadata Team, *National System for Geospatial Intelligence Metadata Foundation (NMF)*, tech. rep., National System for Geospatial Intelligence (2010).
18. P. Missier, K. Belhajjame and J. Cheney, The W3C PROV family of specifications for modelling provenance metadata, in *EDBT*, (ACM, 2013).
19. W3C PROV Working Group, *An Overview of the PROV Family of Documents*, tech. rep., http://www.w3.org/TR/prov-overview/ (2013).
20. L. Moreau, P. Groth, J. Cheney, S. Miles and T. Lebo, The rationale of PROV, *Journal of Web Semantics* **35**, 235 (2015).
21. V. Cuevas-Vicenttín, B. Ludäscher, P. Missier, K. Belhajjame, F. Chirigati, Y. Wei, S. Dey, P. Kianmajd, D. Koop, S. Bowers, I. Altintas, C. Jones, M. B. Jones, L. Walker, P. Slaughter, B. Leinfelder and Y. Cao, ProvONE:

A PROV extension data model for scientific workflow provenance (May 2016), https://purl.dataone.org/provone-v1-dev.
22. C. Maumet, T. Auer, A. Bowring, G. Chen, S. Das, G. Flandin, S. Ghosh, T. Glatard, K. J. Gorgolewski, K. G. Helmer, M. Jenkinson, D. B. Keator, B. N. Nichols, J.-B. Poline, R. Reynolds, V. Sochat, J. Turner and T. E. Nichols, Sharing brain mapping statistical results with the neuroimaging data model, *Sci. Data* **3** (2016).
23. M. D. Allen, A. Chapman, L. Seligman and B. Blaustein, Provenance for collaboration: Detecting suspicious behaviors and assessing trust in information, in *CollaborateCom*, (IEEE, Oct 2011).
24. A. Chapman, B. T. Blaustein, L. Seligman and M. D. Allen, Plus: A provenance manager for integrated information, in *Information Reuse and Integration (IRI), 2011 IEEE International Conference on*, (IEEE, 2011).
25. M. Anand, S. Bowers, T. M. McPhillips and B. Ludaescher, Efficient Provenance Storage over Nested Data Collections, in *Procs. EDBT*, (ACM, March 2009).
26. A. Chapman, H. V. Jagadish and P. Ramanan, Efficient provenance storage, in *SIGMOD Conference*, (ACM, 2008).
27. I. Altintas, O. Barney and E. Jaeger-Frank, Provenance Collection Support in the Kepler Scientific Workflow System, in *IPAW*, (Springer, 2006).
28. J. Freire, S. P. Callahan, E. Santos, C. E. Scheidegger, C. T. Silva and H. T. Vo, VisTrails: visualization meets data management, in *SIGMOD Conference*, (ACM, 2006).
29. J. Kim, E. Deelman, Y. Gil, G. Mehta and V. Ratnakar, Provenance trails in the Wings/Pegasus workflow system, *Concurr. Comput.* **20** (2008).
30. T. McPhillips, T. Song, T. Kolisnik, S. Aulenbach, K. Belhajjame, R. K. Bocinsky, Y. Cao, J. Cheney, F. Chirigati, S. Dey, J. Freire, C. Jones, J. Hanken, K. W. Kintigh, T. A. Kohler, D. Koop, J. A. Macklin, P. Missier, M. Schildhauer, C. Schwalm, Y. Wei, M. Bieda and B. Ludäscher, Yesworkflow: A user-oriented, language-independent tool for recovering workflow information from scripts, *International Journal of Digital Curation* **10**, 298 (2015).
31. L. Murta, V. Braganholo, F. Chirigati, D. Koop and J. Freire, noworkflow: Capturing and analyzing provenance of scripts, in *IPAW*, eds. B. Ludäscher and B. Plale (Springer, 2015).
32. M. Stamatogiannakis, E. Athanasopoulos, H. Bos and P. Groth, Prov2r: Practical provenance analysis of unstructured processes, *ACM Trans. Internet Technol.* **17** (August 2017).
33. L. Hawizy et al., ChemicalTagger: A tool for semantic text-mining in chemistry, *J Cheminform* **3** (2011).
34. H. Park, R. Ikeda and J. Widom, Ramp: A system for capturing and tracing provenance in mapreduce workflows, in *VLDB*, (VLDB Foundation, August 2011).
35. A. Rosenthal, L. Seligman, A. Chapman and B. Blaustein, Scalable access controls for lineage, in *TAPP*, (USENIX, 2009).
36. R. Hasan, R. Sion and M. Winslett, The Case of the Fake Picasso: Preventing

History Forgery with Secure Provenance, in *FAST*, (USENIX, 2009).
37. J. Zhang, A. Chapman and K. Lefevre, Do you know where your data's been? — tamper-evident database provenance, in *SDM*, (Springer-Verlag, 2009).
38. J. Cheney, A formal framework for provenance security, in *CSF*, (IEEE, 2011).
39. B. Blaustein, A. Chapman, L. Seligman, M. D. Allen and A. Rosenthal, Surrogate parenthood: Protected and informative graphs, *Proc. VLDB Endow.* **4**, 518 (May 2011).

F Pavese, A B Forbes, N F Zhang, A G Chunovkina (Eds.):
Advanced Mathematical and Computational Tools in Metrology and Testing XII
Series on Advances in Mathematics for Applied Sciences, Vol. 90

https://doi.org/10.1142/9789811242380_0010

On estimation of linear regression confidence bands: Analytical solution and Monte Carlo simulation

A. G. Chunovkina and A. V. Stepanov+

D. I. Mendeleev Institute for Metrology, Russia
+E-mail: stepanov17@yandex.ru

G. Wimmer*

Mathematical Institute, Slovak Academy of Sciences, Slovakia
E-mail: wimmer@mat.savba.sk

The problem of calculating the uncertainty bands for a linear regression with correlated initial data is considered. The coverage factors for the uncertainty bands with different models of errors in the initial data are obtained analytically and by means of Monte Carlo method. The linear regression coefficients are estimated by the generalized method of least squares.

Keywords: linear regression, confidence bands, correlated noise

1. Introduction

Least Square Method (LSM) is widely used in metrology for data evaluation, especially in case of linear or nonlinear regression fitting. LSM provides optimal estimates of regression parameters for Gaussian noise. However, in practice the information about the distribution of a noise is often not available. Nevertheless LSM is formally applied, but the accuracy of parameters obtained is questionable.

The effect of correlated errors in regression considered in numerous work, see[1,2] and corresponding references provided in these articles. The correlated measurement errors of a very diverse nature may be described by a wide variety of models: an autoregressive (AR) model ([2,3,4] contains a chapter devoted to AR noise modelling with a practical example), a colored (flicker) noise model ([5,6] and corresponding references) etc. In this work a

*The work was partially supported by the Slovak Research and Development Agency, project APVV-15-0295, and by the Scientific Grant Agency of the Ministry of Education of the Slovak Republic and the Slovak Academy of Sciences, projects VEGA 2/0054/18 and VEGA 2/0081/19.

trivial AR model to be considered, for simplicity of analytical calculations and numerical modeling.

Consider a simple linear regression model with correlated errors. Based on N_0 measurements observed at equidistant time points $t_k = h(k-1)$ for $k = 1, \ldots, N_0$, the exact hyperbolic type confidence bands are constructed for the regression line over the whole interval (t_1, t_N), where $N \geq N_0$, which is used for prediction of the drift. In particular, we consider the regression model

$$y_k = \beta_0 + \beta_1 t_k + x_k. \tag{1}$$

Here, the time points t_k are assumed to be known, and the uncertainties associated with the measurement errors x_k are assumed to be equal, $u(x_k) = u(x(t_k)) = u$, with the covariances $cov(x_i, x_j) = u^2 \alpha^{|i-j|}$, where $|\alpha| < 1$ is a known correlation parameter. The correlation may occur, e.g., as a result of constant systematic errors.

In particular, an autoregressive AR(1) model for modeling the measurement errors is assumed:

$$x_k = \alpha x_{k-1} + \sqrt{1-\alpha^2}\, \varepsilon_k, \tag{2}$$

where ε_k represent the white noise, i.e. independent and identically distributed random quantities with zero mean and known variance σ^2, having a normal distribution or a symmetric two-side power distribution (TSP).[8]

As it was mentioned, the variance of x_k is supposed to be known and equal to $u^2 = \sigma^2$. The hyperbolic type confidence bands for the regression line in the finite interval are obtained using analytical solution (optimal for Gaussian noise), basing on the approach considered in,[7] and by Monte Carlo simulation in general case.

2. Analytical solution for normally distributed errors

Consider a set $C_t = [t_1,\ t_N]$ of possible values of the predictor variable; let $C = \{(1,t)' : t \subset C_t\}$. Then the hyperbolic confidence bands for all $c \in C$ are given by

$$c' \begin{pmatrix} \beta_0 \\ \beta_1 \end{pmatrix} \in c' \begin{pmatrix} \hat{\beta}_0 \\ \hat{\beta}_1 \end{pmatrix} \pm K\sigma \sqrt{c' \left(A'V^{-1}A\right)^{-1} c}, \tag{3}$$

when obtaining the LSM estimates for β_i:

$$\begin{pmatrix} \hat{\beta}_0 \\ \hat{\beta}_1 \end{pmatrix} = \left(A'V^{-1}A\right)^{-1} A'V^{-1} \begin{pmatrix} y_1 \\ \cdots \\ y_{N_0} \end{pmatrix}, \quad A' = \begin{pmatrix} 1 & 1 & \ldots & 1 \\ 0 & h & \ldots & (N_0-1)h \end{pmatrix},$$

$\sigma^2 V$ is a covariance matrix corresponding to the noise model (2):

$$V = \begin{pmatrix} 1 & \alpha & \dots \alpha^{N_0-2} & \alpha^{N_0-1} \\ \alpha & 1 & \dots \alpha^{N_0-3} & \alpha^{N_0-2} \\ & & \dots & \\ \alpha^{N_0-2} & \alpha^{N_0-3} & \dots \quad 1 & \alpha \\ \alpha^{N_0-1} & \alpha^{N_0-2} & \dots \quad \alpha & 1 \end{pmatrix},$$

and $K = K_{0.95}$ is a coverage factor to be calculated (corresponding to the probability level of 95%).

In the case of normally distributed errors LSM gives a BLUE (best linear unbiased estimator) if the covariance matrix V is given by the above expression. Moreover, the exact coverage factor $K_{0.95}$ in such a case is the 95% quantile of the distribution function specified by

$$F(x) = 1 - e^{-\frac{x^2}{2}} + \int_x^{\frac{\sqrt{2}x}{\sqrt{1+\rho}}} G\left(\frac{x}{r}\right) r e^{-\frac{r^2}{2}} dr, \quad x \geq 0, \tag{4}$$

where

$$G(u) = \begin{cases} \frac{\arccos(\rho) - 2\arccos(u)}{\pi} & \text{if } \frac{\sqrt{1+\rho}}{\sqrt{2}} \leq u < 1, \\ 1 & \text{if } \quad u = 1, \end{cases} \tag{5}$$

and the parameter ρ (which depends on the values N, N_0 and α) is given in[7]:

$$\rho = -\frac{(1,\,0)\,(A'V^{-1}A)^{-1}\begin{pmatrix}1\\(N-1)h\end{pmatrix}}{\sqrt{(1,\,0)\,(A'V^{-1}A)^{-1}\begin{pmatrix}1\\0\end{pmatrix}\cdot(1,\,(N-1)h)\,(A'V^{-1}A)^{-1}\begin{pmatrix}1\\(N-1)h\end{pmatrix}}}.$$

The outline of proof of the formula (4) is proposed in the Appendix.

Some values for $K_{0.95}$ for $N = 100, 1000$ are presented in Table 1. The

Table 1. Exact $K_{0.95}$ values, $N = 100, 1000$.

α	0.0	0.1	0.3	0.5	0.7	0.9
$N = 100,\ N_0 = 20$	2.4407	2.4405	2.4399	2.4389	2.4364	2.4235
$N = 100,\ N_0 = 35$	2.4389	2.4388	2.4384	2.4376	2.4358	2.4261
$N = 100,\ N_0 = 50$	2.4362	2.4360	2.4357	2.4350	2.4333	2.4244
$N = 1000,\ N_0 = 200$	2.4413	2.4413	2.4413	2.4412	2.4410	2.4399
$N = 1000,\ N_0 = 350$	2.4394	2.4394	2.4393	2.4393	2.4391	2.4383
$N = 1000,\ N_0 = 500$	2.4366	2.4366	2.4365	2.4365	2.4363	2.4355

other are given in[9] as well as the computation routine for $K_{0.95}$ for arbitrary values of N, N_0 and α.

Some plots for $K_{0.95}$ as a function of a ratio N_0/N are given in Figure 1 (for $\alpha = 0.7$ and some N values); some more are also available at[9] (as well as the polynomial approximation for them).

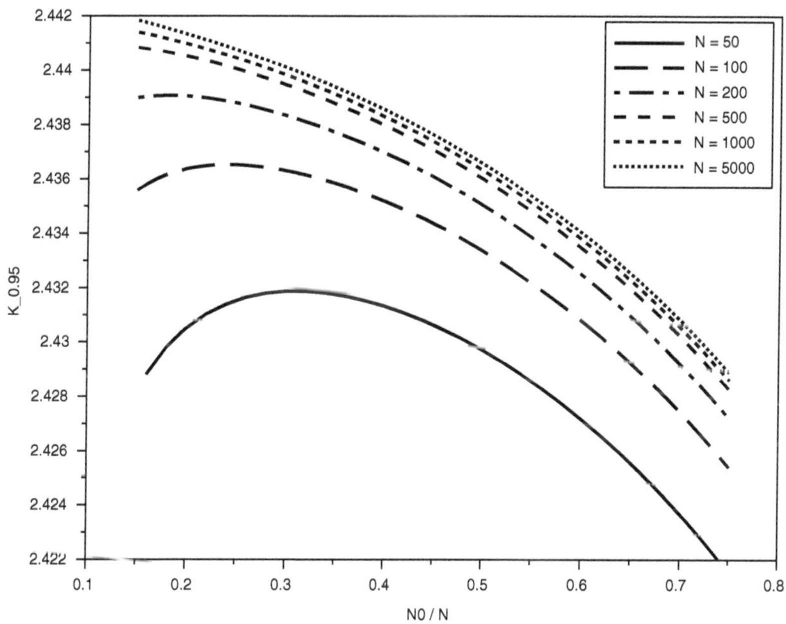

Fig. 1. $K_{0.95}(N_0/N)$, $\alpha = 0.7$.

Of course, the above approach could be extrapolated in an obvious way on a case of $AR(m)$ noise model for $m > 1$ (just need to rewrite the matrix V expression accurately).

3. Monte Carlo simulation

The section contains the numerical inference of the coverage factor $K_{0.95}$ for the regression bands by means of Monte Carlo techniques when ε_k from (2) has a distribution differing from the normal (so we can not simply apply the above analytial approach). Consider for ε_k a symmetric TSP distribution with a degree parameter d ($d = 1$ corresponds to a uniform distribution, $d = 2$ — to a triangular distribution etc.); u^2 hereafter is supposed to be equal to 1. A pdf for a symmetric TSP distribution, having a unity variance, is given by the following expression:

$$f(x) = \begin{cases} \frac{d}{2x_0}\left(1+\frac{x}{x_0}\right)^{d-1}, & -x_0 \le x < 0, \\ \frac{d}{2x_0}\left(1-\frac{x}{x_0}\right)^{d-1}, & 0 \le x \le x_0, \end{cases} \qquad x_0 = \sqrt{\frac{(d+1)(d+2)}{2}}.$$

It should be noted that the use of the exact $K_{0.95} = K_{0.95}^{normal}$ values obtained earlier analytically for the normally distributed ε_k, as well as the use of LSM, still give us quite acceptable values of coverage probability in some cases. E.g., for $N = 100$, the coverage probability values P (quite close to 0.95) of the band to contain the overall "true" (zero) trend line for the various values of TSP parameter d, are listed in Table 2. This example demonstrates that in terms of the coverage probability the character of ε_k distribution does not (at least sometimes) affect the results as significantly as it might be expected.

Table 2. Coverage probabilities when using LSM and $K_{0.95}^{normal}$ values, $N = 100$.

d	0.5	1.0	2.0	3.0	10.0	100.0
$N_0 = 20,\ \alpha = 0.1,\ K_{0.95}^{normal} = 2.4405$	0.954	0.953	0.952	0.950	0.947	0.944
$N_0 = 35,\ \alpha = 0.1,\ K_{0.95}^{normal} = 2.4388$	0.952	0.952	0.951	0.950	0.948	0.947
$N_0 = 50,\ \alpha = 0.1,\ K_{0.95}^{normal} = 2.4361$	0.951	0.951	0.951	0.950	0.949	0.948
$N_0 = 20,\ \alpha = 0.7,\ K_{0.95}^{normal} = 2.4364$	0.955	0.954	0.952	0.950	0.947	0.944
$N_0 = 35,\ \alpha = 0.7,\ K_{0.95}^{normal} = 2.4359$	0.953	0.952	0.951	0.950	0.948	0.947
$N_0 = 50,\ \alpha = 0.7,\ K_{0.95}^{normal} = 2.4334$	0.952	0.951	0.951	0.950	0.949	0.948
$N_0 = 20,\ \alpha = 0.9,\ K_{0.95}^{normal} = 2.4235$	0.953	0.952	0.951	0.950	0.948	0.946
$N_0 = 35,\ \alpha = 0.9,\ K_{0.95}^{normal} = 2.4261$	0.952	0.952	0.951	0.950	0.949	0.947
$N_0 = 50,\ \alpha = 0.9,\ K_{0.95}^{normal} = 2.4245$	0.951	0.951	0.951	0.950	0.949	0.948

Some $K_{0.95}^{TSP}$ values are given in Table 3. These values are obtained by means of Monte Carlo techniques using the LSM (3) for the regression estimation (a corresponding calculation routine is available at[10]).

Some plots for $K_{0.95}$ as a function of a ratio N_0/N are also presented in Figure 2 (for $\alpha = 0.7$, $d = 1, 3, 10$, and $N = 100, 500$).

Table 3. $K_{0.95}^{TSP}$ values, $N = 100$.

d	0.5	1.0	2.0	3.0	10.0	100.0
$\alpha = 0.1,\ N_0 = 20$	2.407	2.415	2.428	2.437	2.466	2.488
$\alpha = 0.1,\ N_0 = 35$	2.421	2.425	2.432	2.437	2.454	2.467
$\alpha = 0.1,\ N_0 = 50$	2.423	2.427	2.432	2.435	2.446	2.456
$\alpha = 0.3,\ N_0 = 20$	2.404	2.413	2.427	2.437	2.466	2.489
$\alpha = 0.3,\ N_0 = 35$	2.420	2.424	2.431	2.437	2.454	2.468
$\alpha = 0.3,\ N_0 = 50$	2.423	2.426	2.431	2.434	2.446	2.456
$\alpha = 0.5,\ N_0 = 20$	2.401	2.410	2.425	2.435	2.467	2.489
$\alpha = 0.5,\ N_0 = 35$	2.417	2.422	2.431	2.436	2.453	2.468
$\alpha = 0.5,\ N_0 = 50$	2.422	2.425	2.429	2.434	2.446	2.456
$\alpha = 0.7,\ N_0 = 20$	2.401	2.409	2.422	2.433	2.465	2.485
$\alpha = 0.7,\ N_0 = 35$	2.415	2.420	2.428	2.434	2.452	2.466
$\alpha = 0.7,\ N_0 = 50$	2.419	2.422	2.428	2.432	2.445	2.455
$\alpha = 0.9,\ N_0 = 20$	2.401	2.407	2.415	2.422	2.441	2.457
$\alpha = 0.9,\ N_0 = 35$	2.411	2.414	2.420	2.425	2.438	2.451
$\alpha = 0.9,\ N_0 = 50$	2.413	2.415	2.420	2.424	2.435	2.445

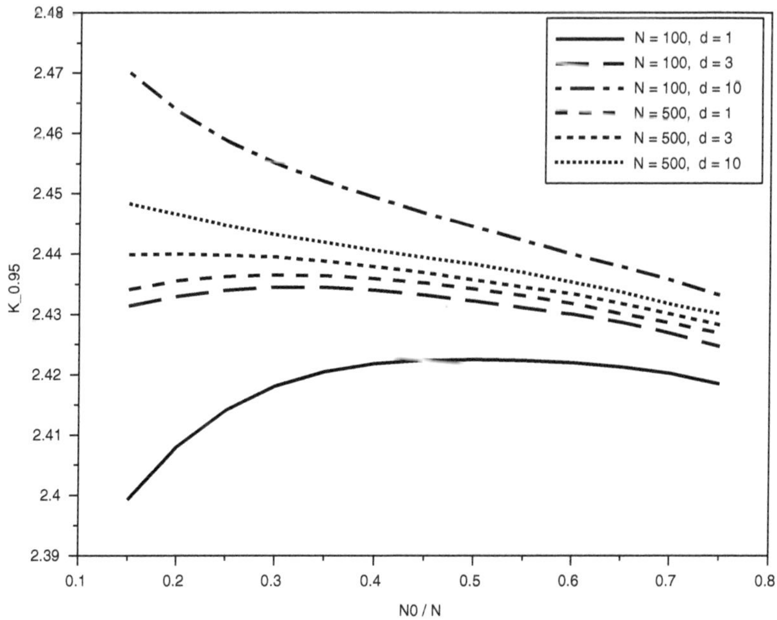

Fig. 2. $K_{0.95}^{TSP}(N_0/N)$, $\alpha = 0.7$.

Consider also an illustrative example of the effect of the noise correlation underestimation, i.e. the sensitivity of the coverage probability with respect to the correlation parameter α. Let someone made an incorrect assumption that $cov(x_i, x_j) = u^2\mu^{|i-j|}, |\mu| < |\alpha|$ and used the $K^{\mu}_{0.95}$ factors (as well as $V = V^{\mu}$ to calculate the bands corresponding to μ). Table 4 contains the real coverage probability values P for such a case. It could be noted that (again) a decrease of P for ε_k from TSP does not differ sufficiently from the case of normal ε_k distribution.

Table 4. Coverage probability P values, $N = 100$, $N_0 = 35$, $\mu = 0.7$.

α	0.75	0.80	0.85	0.90	0.95
P, $\varepsilon_k \in N(0,1)$	0.924	0.886	0.832	0.761	0.684
P, $\varepsilon_k \in$ TSP(0.5)	0.922	0.882	0.826	0.754	0.676
P, $\varepsilon_k \in$ TSP(1)	0.923	0.884	0.827	0.756	0.677
P, $\varepsilon_k \in$ TSP(10)	0.925	0.889	0.836	0.766	0.689

Of course, the Monte Carlo simulation techniques can be used not only for the noise model (2). It may be useful when the analytical inference is complicated (for example, for some combined noise models). E.g., some coverage factor values as well as the corresponding band estimates (LSM-based) for the $1/f^{\nu}$ colored noise are presented in[11].

4. Conclusion

So the conclusions may be summarized as follows. The coverage factor for the uncertainty bands was obtained analytically by analogy with[7] when the correlated errors are described by the trivial AR model. In more general case when analytical calculations are inapplicable (non-normal distribution for ε_k in (2) or the other noise models), the Monte Carlo method can be recommended (in case of normally distributed ε_k it will give results very close to the analytical). The proposed approaches admit obvious generalizations for more complicated noise models.

Appendix A. Outline of the proof of the formula (4)

The formula is proved by analogy with an Appendix of[7]. The exact confidence bands for all $c \in C$ are given by (3) where $K_{0.95}$ is a 95% quantile of the distribution function of a random variable

$$N_C = \sup_{c \in C} \frac{\left| c' \left(\begin{pmatrix} \hat{\beta}_0 \\ \hat{\beta}_1 \end{pmatrix} - \begin{pmatrix} \beta_0 \\ \beta_1 \end{pmatrix} \right) \right|}{\sigma \sqrt{c'(A'V^{-1}A)^{-1}c}}$$

The difference compared to[7] is that if B is any nonsingular matrix that $B'B = (A'V^{-1}A)^{-1}$, then

$$z = \frac{1}{\sigma} (B')^{-1} \left(\begin{pmatrix} \hat{\beta}_0 \\ \hat{\beta}_1 \end{pmatrix} - \begin{pmatrix} \beta_0 \\ \beta_1 \end{pmatrix} \right)$$

has normal $N_2(\mathbf{0}, I)$ distribution. If express $z = (z_1, z_2)'$ in polar coordinates,

$$z = \sqrt{z_1^2 + z_2^2} \begin{pmatrix} \cos\theta \\ \sin\theta \end{pmatrix} = Ru,$$

then R^2 has a χ_2^2 distribution and is statistically independent of θ, which has uniform distribution on $[0, 2\pi]$.

So u is independent of R^2 and has uniform distribution on the unit circle. The random variable N_C can now be rewritten in the form

$$N_C = \sup_{c \in C} \frac{|(Bc)'z|}{\sqrt{c'(A'V^{-1}A)^{-1}c}} = \sqrt{R^2} \sup_{c \in C} \left| \frac{c'B'}{\sqrt{c'(A'V^{-1}A)^{-1}c}} u \right| = RT,$$

where

$$T = \sup_{c \in C^*} |c'u|, \quad C^* = \left\{ \frac{Bc}{\sqrt{c'(A'V^{-1}A)^{-1}c}} : c \in C \right\}.$$

An expression for the distribution function F_N could also be obtained by analogy with the Appendix of[7]

$$F_N(w) = P\{RT < w\} = \int_0^\infty F_T(w/r) f_R(r) dr$$

$$= F_{R^2}(w^2) + 2 \int_w^{\frac{w\sqrt{2}}{\sqrt{1+\varrho}}} F_T(w/r) f_{R^2}(r^2) r \, dr.$$

The form of $F_T(t)$ is derived on the same way as in Appendix of[7] and so $K_{0.95}$ is the upper 0.05-point of the distribution function

$$F_N(w) = \frac{1}{2} \int_0^{w^2} e^{-\frac{x}{2}} dx + 2 \int_w^{\frac{w\sqrt{2}}{\sqrt{1+\varrho}}} G\left(\frac{w}{r}\right) \frac{r}{2} e^{-\frac{r^2}{2}} dr, \quad w \geq 0,$$

where G is given by (5). The latter expression gives (4).

References

1. Lee J., Lund R., Revisiting Simple Linear Regression with Autocorrelated Errors, *Biometrika*, Vol. 91, No. 1, 2004, pp. 240245.
2. Safi S., White A., The Efficiency Of OLS In The Presence Of Auto-Correlated Disturbances In Regression Models, *Journal of Modern Applied Statistical Methods*, 5(1), 2006, pp. 133-143.
3. Staudenmayer J., Buonaccorsi J. Measurement Error in Linear Autoregressive Models, *Journal of the American Statistical Association* Vol. 100, No. 471, 2005, pp. 841-852.
4. Cox M., Harris P. Statistical error modelling, NPL Report (2004) CMSC 45/04.
5. Allan D., Should the Classical Variance Be Used As a Basic Measure in Standards Metrology? *IEEE Transactions on Instrumentation and Measurement*, Vol. IM-36, No. 2, 1987, pp. 646-654.
6. Vernotte F., Lantz E., Metrology and $1/f$ noise: Linear regressions and confidence intervals in flicker noise context, *Metrologia*, Vol. 52, No. 2, pp. 222-237.
7. Uusipaikka E., Exact Confidence Bands for Linear Regression Over Intervals, *Journal of the American Statistical Association*, Vol. 78, No. 383, 1983, pp. 638-644.
8. van Dorp J.R., Kotz S., The Standard Two-Sided Power Distribution and its Properties, *The American Statistician*, 56(2), 2002, pp. 90-99.
9. https: //github.com/stepanov17/regressionConfidenceBands/tree/master/exactK
10. https://github.com/stepanov17/regressionConfidenceBands/tree/master/AR1Noise
11. https://github.com/stepanov17/regressionConfidenceBands/tree/master/coloredNoise

F Pavese, A B Forbes, N F Zhang, A G Chunovkina (Eds.):
Advanced Mathematical and Computational Tools in Metrology and Testing XII
Series on Advances in Mathematics for Applied Sciences, Vol. 90

https://doi.org/10.1142/9789811242380_0011

Repeatability, reproducibility and resampled effects models

Alistair B. Forbes* and Kavya Jagan

National Physical Laboratory, Teddington, Middlesex, UK
**E-mail: alistair.forbes@npl.co.uk*

One of the basic steps in characterising the behaviour of a system that is subject to random effects is to observe a number of outputs of the system and make inferences about the system on the basis of the observed outputs. The inferences are usually made in terms of a parametric model in which the samples are modelled as being drawn from a probability distribution that depends on a finite number of parameters. An important example of this approach is in determining the repeatability of a measurement system in the construction of its uncertainty budget. Repeatability and reproducibility studies in measurement system analysis and analysis of variance studies also follow this approach.

The Guide to the Expression of Uncertainty in Measurement considers the case of repeated sampling from a Gaussian distribution under the assumption that the samples are drawn independently. In practice, we cannot be sure that the samples are drawn independently (nor can we be sure that they are drawn from a Gaussian distribution). The objectives of this paper are, firstly, to examine the consequences of the assumption of independence in evaluating measurement uncertainty and, secondly, to provide alternative models for correlated effects that are both plausible and computationally tractable.

The motivation for the alternative models comes from the fact that in an uncertainty budget it is possible to associate an approximate time scale with each component of the budget. We show that such models can often be approximated by an autoregressive model that can also be expressed in terms of a Gaussian process model of temporal correlation. In this paper, we describe the models, analysis methods and algorithms and report on metrology applications.

Keywords: Repeatability, Reproducibility, Resampled effects, Bayesian inference

1. Introduction

One of basic steps in characterising the behaviour of a system that is subject to random effects is to observe a number of outputs of the system and make inferences about the system on the basis of the observed outputs. The inferences are usually made in terms of a parametric model in which the samples are modelled as being drawn from a probability distribution that depends on a finite number of parameters. An important example of this approach is in determining the repeatability of a measurement system in the construction of its uncertainty budget associated with a measurement result. Repeatability and reproducibility studies in measurement system analysis also follow this approach[4,9,15].

The Guide to the Expression of Uncertainty in Measurement[1] (GUM), considers the case of repeated sampling from a Gaussian distribution under the assumption that the samples are independently drawn:

$$x_i = \alpha + \epsilon_i, \quad \epsilon_i \in \mathrm{N}(0, \sigma^2).$$

In practice, we cannot be sure that the samples are drawn independently (nor can we be sure that they are drawn from a Gaussian distribution). The objectives of this paper are firstly to examine the consequences of the assumption of independence in evaluating measurement uncertainty and secondly to provide alternative models for correlated effects that are both plausible and computationally tractable.

The motivation for the alternative models comes from the fact that in an uncertainty budget it is possible to associate an approximate time scale with each component of the budget. Repeatability is associated with short time scales (seconds), environmental effects a range of times scales (hourly, daily, seasonally), bought-in components from calibration (yearly), and material parameters extracted from reference data may be regarded as constants (for all time) but still have associated uncertainties. This suggests a model[4] for the measurement x_t at time t of the form

$$x_t = \alpha + \sum_{j=1}^{p} \delta_{t,j}, \quad \delta_{t,j} \in \mathrm{N}(0, \sigma_j^2),$$

in which we associate to the jth effect δ_j, a variance σ_j^2 and a timescale τ_j , the latter representing the average rate at which the effect is resampled. The variances σ_j^2 defined by parameters $\boldsymbol{\sigma} = (\sigma_1, \ldots, \sigma_p)^\top$ and the timescale parameters $\boldsymbol{\tau} = (\tau_1, \ldots, \tau_p)^\top$ jointly specify a statistical model for the time series associated with the measurements x_t. We show that such models can often be approximated by an autoregressive model that can also be expressed in terms of a Gaussian process model[17] of temporal correlation with correlation kernel of the form

$$k(t, t') = \sigma^2 \exp\{-|t - t'|/\tau\}.$$

To take into account the temporal correlation, standard least-squares methods have to be modified but the modification can be implemented using bi-diagonal weighting matrices that introduce very little computational overhead.

The remainder of this paper is organised as follows. In section 2, we summarise approaches for analysing data associated with linear models and Gaussian random effects while in section 3 we look at the special case of correlated sampling governed by an autoregressive model of order one, AR(1). In section 4, we describe a resampled effects model that gives rise to correlated sampling and how these models can be approximated by an AR(1) model. In section 5, we give numerical examples relating to resampled effects and AR(1) models. Our concluding remarks are given in section 6.

2. Linear models, Gaussian random effects

2.1. *Independent sampling from a univariate Gaussian*

The GUM considers the case of sampling from a univariate Gaussian. Here we summarise this case from a Bayesian[7,14] point of view. Suppose observations $\boldsymbol{y} = (y_1, \dots, y_m)^\top$ arise according to the model $y|\alpha, \phi \sim \mathrm{N}(\alpha, \phi^{-1})$ where $\phi = 1/\sigma^2$ is unknown. Assuming a non-informative prior distribution $p(\alpha, \phi) \propto 1/\phi$, the posterior distribution $p(\alpha, \phi|\boldsymbol{y})$ is such that

$$p(\alpha, \phi|\boldsymbol{y}) \propto \phi^{m/2-1} \exp\left\{ -\frac{\phi}{2}(m-1)s^2 + m(a-\alpha)^2 \right\},$$

where

$$a = \bar{y} = \frac{1}{m}\sum_{i=1}^{m} y_i, \quad s^2 = \frac{1}{m-1}\sum_{i=1}^{m}(y_i - a)^2.$$

The quantity a is the least-squares estimate of α and s is a point estimate of σ. The marginalised distributions $p(\alpha|\boldsymbol{y})$ and $p(\phi|\boldsymbol{y})$ can be determined analytically and are associated with t and gamma distributions, respectively:

$$\alpha|\boldsymbol{y} \sim t_{m-1}(a, s^2/m), \quad \phi|\boldsymbol{y} \sim G((m-1)/2, (m-1)s^2/2). \qquad (1)$$

2.2. *Prior information about ϕ*

Often, the number of repeats associated with an experiment is small, in which case the information about both α and ϕ is weak. However, if repeatability of the measuring system has been previously assessed, for example, in measuring a similar artefact in a similar environment[12], then this information can be incorporated into the current analysis. If there is prior information about $\phi = 1/\sigma^2$ of the form

$$\phi \sim G(m_0/2, m_0\sigma_0^2/2), \tag{2}$$

then

$$\alpha|\boldsymbol{y} \sim t_\nu(a, \bar{s}^2/m), \quad \phi|\boldsymbol{y} \sim \mathrm{G}(\nu/2, \nu\bar{s}^2/2), \tag{3}$$

where

$$\bar{s}^2 = \frac{m_0\sigma_0^2 + (m-1)s^2}{m + m_0 - 1}, \qquad \nu = m + m_0 - 1,$$

The term $\bar{s}^2$ represents a point estimate of $\sigma^2 = 1/\phi$ and is seen to be weighted average σ_0^2 and s^2, the latter an estimate of σ^2 based on the sampled data. The prior information expression (2) corresponds to the information that would be derived about $\phi = 1/\sigma^2$ from observing m_0 draws z_i from $\mathrm{N}(0, \sigma^2)$ for which

$$\sigma_0^2 = \frac{1}{m_0}\sum_{i=1}^{m_0} z_i^2.$$

The distributions in (1) corresponds to those in (3) for the case $m_0 = 0$, i.e., no prior draws from $\mathrm{N}(0, \sigma^2)$.

2.3. *Linear regression associated with Gaussian random effects*

The case of repeated measurements is a special case of the more general model

$$\boldsymbol{y}|\boldsymbol{\alpha}, \phi, \boldsymbol{\tau} \sim \mathrm{N}(C\boldsymbol{\alpha}, \phi^{-1}V(\boldsymbol{\tau})),$$

where C is an $m \times n$ observation matrix, $m > n$, $\boldsymbol{\alpha}$ is a vector of n regression coefficients, $\phi = 1/\sigma^2$ as before, and $V(\boldsymbol{\tau})$ is an $m \times m$ full rank variance matrix. We allow for the possibility that V depends on additional parameters $\boldsymbol{\tau}$. We assume a prior distribution of the form

$$p(\boldsymbol{\alpha}, \phi, \boldsymbol{\tau}) = p(\phi)p(\boldsymbol{\tau})$$

where ϕ has a Gamma distribution as in (2). From Bayes' Theorem, the joint posterior distribution $p(\boldsymbol{\alpha}, \phi, \boldsymbol{\tau}|\boldsymbol{y})$ for the parameters is such that

$$p(\boldsymbol{\alpha}, \phi, \boldsymbol{\tau}|\boldsymbol{y}) \propto p(\boldsymbol{y}|\boldsymbol{\alpha}, \phi, \boldsymbol{\tau})p(\phi)p(\boldsymbol{\tau}).$$

The fact that the likelihood is a multivariate Gaussian and that the model depends linearly on the parameters $\boldsymbol{\alpha}$ means that marginalisation with respect to $\boldsymbol{\alpha}$ and ϕ can be performed analytically as follows. Given $\boldsymbol{\tau}$, let $\boldsymbol{a}$ the least squares estimate of $\boldsymbol{\alpha}$ given by

$$\boldsymbol{a} = \left(C^\top W(\boldsymbol{\tau})C\right)^{-1} W(\boldsymbol{\tau})\boldsymbol{y}, \quad W(\boldsymbol{\tau}) = V^{-1}(\boldsymbol{\tau}),$$

and set

$$s^2 = \frac{1}{m-n}(\boldsymbol{y} - C\boldsymbol{a})^\top W(\boldsymbol{\tau})(\boldsymbol{y} - C\boldsymbol{a}), \quad \bar{s}^2 = \frac{m_0\sigma_0^2 + (m-n)s^2}{m + m_0 - n}.$$

With $\boldsymbol{a}$ and $\bar{s}^2$ so defined and $\nu = m + m_0 - n$, we have

$$p(\boldsymbol{\alpha}, \boldsymbol{\tau}|\boldsymbol{y}) = p(\boldsymbol{\alpha}|\boldsymbol{\tau}, \boldsymbol{y})p(\boldsymbol{\tau}|\boldsymbol{y}),$$

where

$$\boldsymbol{\alpha}|\boldsymbol{\tau}, \boldsymbol{y} \sim t_\nu(\boldsymbol{a}, V_{\boldsymbol{a}}), \quad V_{\boldsymbol{a}} = \bar{s}^2\left(C^\top W(\boldsymbol{\tau})C\right)^{-1}, \tag{4}$$

a multivariate t-distribution with mean $\boldsymbol{a}$ and scale matrix $V_{\boldsymbol{a}}$, and

$$p(\boldsymbol{\tau}|\boldsymbol{y}) \propto \frac{p(\boldsymbol{\tau})}{\left[(\bar{s}^2)^\nu \left|V(\boldsymbol{\tau})\right| \left|C^\top W(\boldsymbol{\tau})C\right|\right]^{1/2}}. \tag{5}$$

Similarly,

$$p(\phi, \boldsymbol{\tau}|\boldsymbol{y}) = p(\phi|\boldsymbol{\tau}, \boldsymbol{y})p(\boldsymbol{\tau}|\boldsymbol{y})$$

where

$$\phi|\boldsymbol{\tau}, \boldsymbol{y} \sim G(\nu/2, \nu\bar{s}^2/2). \tag{6}$$

We note that in (4), (5) and (6) above, $\boldsymbol{a} = \boldsymbol{a}(\boldsymbol{\tau})$ and $\bar{s}^2(\boldsymbol{\tau})$ are functions of $\boldsymbol{\tau}$. Since the marginalised distributions $p(\boldsymbol{\alpha}|\boldsymbol{y})$ and $p(\phi|\boldsymbol{y})$ are given by

$$p(\boldsymbol{\alpha}|\boldsymbol{y}) = \int_T p(\boldsymbol{\alpha}, \boldsymbol{\tau}|\boldsymbol{y})\mathrm{d}\boldsymbol{\tau}, \quad p(\phi|\boldsymbol{y}) = \int_T p(\phi, \boldsymbol{\tau}|\boldsymbol{y})\mathrm{d}\boldsymbol{\tau},$$

the posterior distribution for $\boldsymbol{\alpha}$ is seen to be a weighted average of t-distributions while the posterior distribution for ϕ is a weighted average of gamma distributions, where the weighting function in both cases is given by $p(\boldsymbol{\tau}|\boldsymbol{y})$.

2.1. *Using the Cholesky factor of V to determine a and s^2*

If $V = V(\boldsymbol{\tau})$ has Cholesky[8] factor $L = L(\boldsymbol{\tau})$ with $V = LL^\top$, set $\tilde{C} = L^{-1}C$ and $\tilde{\boldsymbol{y}} = L^{-1}\boldsymbol{y}$, then $\boldsymbol{a}$ is the least squares solution of $\tilde{C}\boldsymbol{a} \approx \tilde{\boldsymbol{y}}$, and

$$s^2 = \frac{1}{m-n}(\tilde{\boldsymbol{y}} - \tilde{C}\boldsymbol{a})^\top(\tilde{\boldsymbol{y}} - \tilde{C}\boldsymbol{a}), \quad V_{\boldsymbol{a}} = \left(\tilde{C}^\top\tilde{C}\right)^{-1}.$$

The solution of linear least squares problems can be determined efficiently and numerically stably using the QR factorisation[8] of the observation matrix $\tilde{C}$ in the above.

2.5. *Making inferences about α and ϕ*

If the posterior distribution $p(\boldsymbol{\tau}|\boldsymbol{y})$ can be evaluated efficiently, i.e., its normalising constant can be calculated efficiently, then inferences about $\boldsymbol{\alpha}$ and ϕ can be made based on the fact that they are weighted averages of standard distributions, Gaussian and gamma distributions, respectively. If $\boldsymbol{\tau}$ involves only a small number of parameters, say three or less, the normalising constant can, in principle, be determined using numerical quadrature[16]. However each evaluation of the right hand side of (5) involves the solution of a least squares system and, likely, the factorisation of $V(\boldsymbol{\tau})$. Therefore the evaluation of $p(\boldsymbol{\tau}|\boldsymbol{y})$ could be computationally expensive for large m since both calculations could involve $O(m^3)$ operations. A more practical alternative is to find the mode $\hat{\boldsymbol{\tau}}$ of of $p(\boldsymbol{\tau}|\boldsymbol{y})$ using optimisation and then use $\hat{\boldsymbol{\tau}}$ as a representative value of $\boldsymbol{\tau}$. The posterior distributions for $\boldsymbol{\alpha}$ and ϕ can then be approximated by the distributions $p(\boldsymbol{\alpha}|\hat{\boldsymbol{\tau}}, \boldsymbol{y})$ and $p(\phi|\hat{\boldsymbol{\tau}}, \boldsymbol{y})$, t and gamma distributions, respectively. In practice $\hat{\boldsymbol{\tau}}$ is usually determined by minimising $E(\boldsymbol{\tau}) = -\log p(\boldsymbol{\tau}|\boldsymbol{y})$.

3. Autoregressive model of order 1, AR(1)

There is a rich and mature theory of autoregressive (AR) models with many applications across science and engineering[2]. In this section, we are concerned with a simple order one model, AR(1).

3.1. *Single parameter case*

Suppose $\boldsymbol{y}$ is generated according to the model

$$\begin{aligned}
y_1 &= \alpha + \delta_1, \quad \delta_1 = \epsilon_1, \\
y_2 &= \alpha + \delta_2, \quad \delta_2 = \sin(\theta)\delta_1 + \cos(\theta)\epsilon_2, \\
y_3 &= \alpha + \delta_3, \quad \delta_3 = \sin(\theta)\delta_2 + \cos(\theta)\epsilon_3, \\
&\vdots \quad \vdots \\
y_m &= \alpha + \delta_m, \quad \delta_m = \sin(\theta)\delta_{m-1} + \cos(\theta)\epsilon_m,
\end{aligned} \tag{7}$$

where $-\pi/2 \leq \theta \leq \pi/2$ and ϵ_i are independent draws from $\mathrm{N}(0, \phi^{-1})$. The model represents a sequence of random effects in which the random effect at the ith stage is positively ($\theta \geq 0$) or negatively ($\theta \leq 0$) correlated with the random effect at the $(i-1)$th stage. We write this model in matrix vector terms as

$$\boldsymbol{y} = C\alpha + \boldsymbol{\delta}, \quad \boldsymbol{\delta} = L(\tau)\boldsymbol{\epsilon},$$

where C is the $m \times 1$ matrix of ones and

$$L(\tau) = \begin{bmatrix} 1 & & & & \\ s & c & & & \\ s^2 & sc & c & & \\ s^3 & s^2c & sc\ c & & \\ \vdots & & & \ddots & \\ s^{m-1} & s^{m-2}c & \cdots & sc & c \end{bmatrix}, \tag{8}$$

with $s(\theta) = \sin\theta$, $c = c(\theta) = \cos\theta$ and $\boldsymbol{\epsilon} \in \mathrm{N}(\mathbf{0}, \phi^{-1}I)$. Thus, $\boldsymbol{y}$ is a sample according to the model of the form

$$\boldsymbol{y} \in \mathrm{N}(C\alpha, \phi^{-1}V(\tau)),$$

considered in section 2 above with

$$V(\tau) = L(\tau)L^\top(\tau) = \begin{bmatrix} 1 & s & s^2 & s^3 & \cdots & s^{m-1} \\ s & 1 & s & s^2 & \cdots & s^{m-2} \\ \vdots & & & \ddots & & \\ s^{m-1} & \cdots & s^2 & s & 1 & \end{bmatrix} \tag{9}$$

The inverse $L^{-1}(\tau)$ of $L(\tau)$ relates $\boldsymbol{\delta}$ and $\boldsymbol{\epsilon}$ through

$$\boldsymbol{\epsilon} = L^{-1}(\tau)\boldsymbol{\delta}$$

and from the model above it is seen that $\epsilon_1 = \delta_1$ and $\epsilon_i = (\delta_i - s\delta_{i-1})/c$ so that $L^{-1}(\tau)$ is the bi-diagonal matrix

$$L^{-1}(\tau) = \begin{bmatrix} 1 & & & & \\ -s/c & 1/c & & & \\ & -s/c & 1/c & & \\ & & & \ddots & \\ & & & -s/c & 1/c \end{bmatrix}.$$

For this model, $\tilde{C} = L^{-1}C$ and $\tilde{\boldsymbol{y}} = L^{-1}\boldsymbol{y}$ are given by

$$\tilde{C} = \begin{bmatrix} 1 \\ (1-s)/c \\ \vdots \\ (1-s)/c \end{bmatrix}, \quad \tilde{\boldsymbol{y}} = \begin{bmatrix} y_1 \\ (y_2 - sy_1)/c \\ \vdots \\ (y_m - sy_{m-1})/c \end{bmatrix}.$$

For non-informative prior $p(\alpha, \phi) \propto 1/\phi$, the least squares estimate $a = a(\theta)$ of α is given by

$$a = \frac{\tilde{C}^\top \tilde{\boldsymbol{y}}}{\tilde{C}^\top \tilde{C}} = \frac{1}{\tilde{m}} \tilde{C}^\top \tilde{\boldsymbol{y}}$$

where

$$\tilde{m} = \tilde{C}^\top \tilde{C} = 1 + (m-1)\left(\frac{1-s}{1+s}\right),$$

and

$$\tilde{C}^\top \tilde{\boldsymbol{y}} = \frac{1}{1+s}\left[(1-s)m\bar{y} + s(y_1 + y_m)\right], \quad \bar{y} = \frac{1}{m}\sum_{i=1}^{m} y_i.$$

The vector of residuals $\tilde{\boldsymbol{r}} = \tilde{\boldsymbol{y}} - \tilde{C}a$ provides an estimate $\tilde{s}$ of the variance $\sigma^2 = 1/\phi$:

$$\tilde{s}^2 = \frac{\tilde{r}^\top \tilde{r}}{m-1}.$$

Following the general theory in section 2, the posterior $p(\alpha|\theta, \boldsymbol{y})$ for α, given θ is given by the t-distribution $t_{m-1}(a, \tilde{s}^2/\tilde{m})$. The quantity $\tilde{m}$ represents the effective number of independent samples and governs how the uncertainty associated with the estimate a depends on the strength of the correlation. For $\theta = 0$, corresponding to complete independently sampling, we have $s = 0$, $a(0) = \bar{y}$ and $\tilde{m} = m$; for $\theta \approx \pi/2$, corresponding to near total positive correlation, $s \approx 1$, $a(\tau) \approx (y_1 + y_m)/2$, and $\tilde{m} \approx 1$. For intermediate values of θ, $a(\theta)$ is a weighted average of $\bar{y}$ and $(y_1 + y_m)/2$. The relationships above also holds for negatively correlated sampling in which case $\tilde{m} > m$ and tends to infinity as θ tends to $-\pi/2$. For perfectly negative correlation the sum of the first two samples gives an exact estimate of α.

3.2. *AR(1) model as a Gaussian process*

The auto-correlation model above can also be described in terms of a Gaussian process[17] (GP) defined by a particular correlation kernel. The GP model has the form

$$y = \alpha + \delta, \quad \delta \in N(0, \phi^{-1}),$$

where a measurement y is associated with an auxiliary variable t (often interpretable as time). The covariance of effects δ and δ' associated with (y, t) and (y', t') depends on t and t' according to

$$\mathrm{cov}(\delta, \delta') = \phi^{-1} k(t, t'), \quad k(t, t') = \exp\{-|t - t'|/\tau\},$$

where now τ is a timescale parameter controlling the strength of the correlation as a function of temporal separation. Suppose the measurements y_i are made at regularly spaced times t_i, $t_1 < t_2 < \cdots < t_m$, $t_i = t_{i-1} + \Delta$.

Then the covariance matrix associated with $\boldsymbol{y}$ is precisely that given in (9) but where now $s = s(\tau)$ is given by

$$s(\tau) = \exp\{-\Delta/\tau\}.$$

For irregularly spaced time series, the counterpart of (7) is

$$\begin{aligned}
y_1 &= \alpha + \delta_1, \quad \delta_1 = \epsilon_1, \\
y_2 &= \alpha + \delta_2, \quad \delta_2 = s_2\delta_1 + c_2\epsilon_2, \\
y_3 &= \alpha + \delta_3, \quad \delta_3 = s_3\delta_2 + c_3\epsilon_3, \\
&\vdots \quad \vdots \\
y_m &= \alpha + \delta_m, \quad \delta_m = s_m\delta_{m-1} + c_m\epsilon_m,
\end{aligned}$$

where

$$s_i = s_i(\tau) = \exp\{-(t_i - t_{i-1})/\tau\}, \; c_i = c_i(\tau) = (1 - s_i^2(\tau))^{1/2}, \; i = 2, \ldots, m.$$

We also set $s_1 = 0$ and $c_1 = 1$. The (ij)th element $l_{ij} = l_{ij}(\tau)$ of counterpart of $L(\tau)$ in (8) is given by

$$l_{ij}(\tau) = c_j \exp\{-(t_i - t_j)/\tau\}, \quad i \geq j$$

while the inverse of $L(\tau)$ is the bi-diagonal matrix

$$L^{-1}(\tau) = \begin{bmatrix} 1 & & & & \\ -s_2/c_2 & 1/c_2 & & & \\ & -s_3/c_3 & 1/c_3 & & \\ & & & \ddots & \\ & & & -s_m/c_m & 1/c_m \end{bmatrix}.$$

The fact that L^{-1} is bi-diagonal means that calculations associated with this model can be performed very efficiently. Simple quadrature schemes can be used to determine the normalising constants associated with the posterior distributions $p(\alpha|\boldsymbol{y})$, $p(\phi|\boldsymbol{y})$ and $p(\tau|\boldsymbol{y})$.

For the more general linear model $\boldsymbol{y}|\phi, \tau \in \mathrm{N}(C\boldsymbol{\alpha}, \phi^{-1}V(\tau))$, the fact that $L^{-1}(\tau)$ can be constructed explicitly and is bi-diagonal means that $\tilde{C} = L^{-1}C$ and $\tilde{\boldsymbol{y}} = L^{-1}\boldsymbol{y}$ can be constructed cheaply.

4. Resampled effects model

One of the key features of the GUM is the concept of the uncertainty budget, and how to construct the uncertainty budget for the measurement in terms of uncertainties associated with the listed influence factors. Similarly, in measurement systems analysis and repeatability and reproducibility studies (R&R)[9,11,15], the variance contribution of different factors such as tool, operator, environment, laboratory, etc., are assessed through analysis of variance experiments in which the different factors are varied in a systematic way. Influence factors in general can be associated with different timescales. Repeatability effects operate on short time scales, environmental effects are likely to have a hourly, daily and possibly seasonal timescales. The bought-in uncertainty associated with calibration is likely to be associated with the long timescales dictated by the period between re-calibration of standard artefacts, for example. Values of material properties specified in reference documents could be regarded as fixed for all time (but have an associated uncertainty).

In an R&R study, various influence factors such as tool or test method, operator, laboratory, etc., are deliberately varied, or resampled, according to the experimental design in order to assess the uncertainty contribution from each factor[10]. In general, any experiment, other that one undertaken in strict repeatability conditions, is subject to instrument drift and changing environmental conditions, etc. We may think of such an experiment as a random R&R study for which the underlying design is unknown but the effects of the resampled influence factors is reflected in the measured data.

Consider a model[4] of a random effect $\delta(t) \in \mathrm{N}(0, \sigma^2)$ composed of sub-effects $\epsilon_j(t)$ according to

$$\delta(t) = \sum_{j=1}^{p} \epsilon(t), \quad \epsilon_j(t) \in \mathrm{N}(0, \sigma_j^2), \quad \sigma^2 = \sum_j \sigma_j^2$$

Associated to the sub-effects are time scale parameters τ_j, that determine the rate of resampling that is proportional to $1/\tau_j$ governed by a Poisson distribution. The probability of no resampling in the time interval t to t' is given

$$\Pr(\epsilon_j(t') = \epsilon_j(t)) = e^{-(t'-t)/\tau_j}.$$

If the resampling occurs at regular time steps $i = 1, \ldots, m$, then the co-variance of δ_i with δ_{i+q} is given by

$$\mathrm{cov}(\delta_{i+q}, \delta_i) = \sum_{j=1}^{p} \sigma_j^2 p_j^q, \quad p_j = \Pr(\epsilon_j(t+1) = \epsilon_j(t))$$

defining the $m \times m$ variance matrix $V(\boldsymbol{\sigma}^2, \boldsymbol{\tau})$. If all the timescale parameters are the same, $\tau_j = \tau_j$, then the δ_i follow an AR(1) process exactly. For random samples of timescales $\boldsymbol{\tau}$ and variances $\boldsymbol{\sigma}^2$, the evolution of the δ_i

is likely to similar to a AR(1) process. Hence, if there is some doubt about the independence of a sequence of effects, an AR(1) model is a plausible model to use to test the degree of correlation, e.g., by determining the t that maximises the posterior distribution $p(\tau|\boldsymbol{y})$, where the single timescale parameter τ takes the role of a representative value of the p timescale parameters $\boldsymbol{\tau}$.

4.1. *Test of AR(1) correlation in fitted residuals*

Suppose $m \times 1$ data vector $\boldsymbol{y}$ is gathered according model of the form

$$\boldsymbol{y} \in N(C\boldsymbol{\alpha}, \phi^{-1}V)$$

involving $m \times n$ observation matrix C, where the variance matrix is V is usually taken to be the identity matrix $V = I$, corresponding to independently sampled random effects, but it is suspected that the random effects are in fact correlated due to instrument or environmental drift, for example. One approach to test for the presence of correlation is to determine the ordinary least squares fit to the data producing estimates $\boldsymbol{a}$ of $\boldsymbol{\alpha}$ and $\hat{\sigma}^2$ of $\sigma^2 = \phi^{-1}$ according to

$$\boldsymbol{a} = (C^\top C)^{-1}C^\top \boldsymbol{y}, \quad \hat{\sigma}^2 = \frac{\boldsymbol{r}^\top \boldsymbol{r}}{m-n}.$$

Suppose C has QR factorisation[8]

$$C = QR = [Q_1\ Q_2]\begin{bmatrix} R_1 \\ \boldsymbol{0} \end{bmatrix} = Q_1 R_1,$$

where Q is an $m \times m$ orthogonal matrix with submatrices Q_1 and Q_2 of dimensions $m \times n$ and $m \times (m-n)$, respectively, and R_1 is and $n \times n$ upper-triangular matrix. Then the variance matrix $V_{\boldsymbol{r}}$ associated with the residuals $\boldsymbol{r}$ is approximated by

$$V_{\boldsymbol{r}} \approx \hat{\sigma}^2 Q_2 Q_2^\top,$$

generally, a full matrix of rank $m-n$. However, the projected residuals $\tilde{\boldsymbol{r}} = Q_2^\top \boldsymbol{r}$ is associated with the $(m-n) \times (m-n)$ variance matrix $\hat{\sigma}^2 I$ so a test can be made on $\tilde{\boldsymbol{r}}$ for independence. In particular, we can assess the extent to which $\tilde{\boldsymbol{r}}$ can be explained by an $AR(1)$ process with parameters $\tilde{\alpha}$, $\tilde{\phi}$ and $\tilde{\tau}$. If the posterior distribution for $p(\tilde{\tau}|\tilde{\boldsymbol{r}})$ has low probability mass near $\tilde{\tau} = 0$, there is evidence that the assumption of independence is not justified. A representative value for τ can be the basis for better inferences about $\boldsymbol{\alpha}$ and σ^2.

5. Numerical examples

5.1. *Simulated linear response*

In this simulation $m = 100$ data points were generated according to the model

$$y_i = \alpha_1 + \alpha_2 t_i + \delta_i, \quad t_i = 1, \dots, m,$$

involving $p = 100$ sub-effects

$$\delta_i = \sum_{j=1}^{p} \epsilon_{i,j}, \quad \epsilon_{i,j} \in \mathrm{N}(0, \sigma_0^2), \quad \sigma^2 = p\sigma_0^2,$$

with associated timescale parameters τ_j drawn at random from a gamma distribution with mean 2 and variance 8. The simulated data are graphed in figure 1 and the time scale parameters graphed in figure 2. The distribution of timescales reflect the underlying gamma distribution. The vast majority of the timescale parameters are less the 2 but 8 of the 100 are greater that 5, with the largest equal to 26.1. For 40 of the timescale parameters, the probability of being re-drawn at each time step is at least 0.9. For the effect associated with the largest timescale parameter, the expected time between redraws is approximately 18 time steps. The data in figure 1 shows a reasonably random spread around the fitted linear response.

An ordinary least squares fit to the model was determined and then, following the procedure outlined in section 4.1, the projected residuals $\tilde{\boldsymbol{r}}$ associated with the fit were analysed on the assumption that they can be explained in terms of AR(1) process. Figure 3 graphs the posterior distribution $p(\tau|\tilde{\boldsymbol{r}})$ associated with the time scale parameter τ, along with the vague prior $p(\tau)$, a gamma distribution. The graph shows also shows the location of the mode $\tau = 1.1$ and the locations of the 5 % and 95 % quantiles given by $\tau = 0.73$ and $\tau = 2.1$, respectively. The quantiles bracket the mean 1.9 of the timescales τ_j used to generate the data. The distribution $p(\tau|\tilde{\boldsymbol{r}})$ is noticeably skewed with a longer tail supporting larger timescales τ. Figure 3 shows that, on the basis of the projected residuals, it is not plausible that the data in figure 1 arises from independently sampled random effects.

The effective number $\tilde{m}$ of independent samples (see section 3.1) for a value of τ can be calculated through the relationships $s(\tau) = e^{(-1/\tau)}$ (for unit time step $\Delta = 1$) and

$$\tilde{m}(\tau) = 1 + (m - 3)\left(\frac{1 - s(\tau)}{1 + s(\tau)}\right),$$

(see section 3). For the mode and 5 % and 95 % quantile values of τ, we have

$$\tilde{m}(\hat{\tau}) = 30.1, \quad \tilde{m}(\tau_5) = 59.0, \quad \tilde{m}(\tau_{95}) = 23.9,$$

respectively, showing that the observed correlation significantly reduces the strength of the inferences that can be derived from the data, compared with the case of independent sampling for which $\tilde{m} = m - 2 = 98$, the number of projected residuals $\tilde{\boldsymbol{r}}$.

The projected residuals $\tilde{\boldsymbol{r}}$ used to determine the distribution $p(\tau|\tilde{\boldsymbol{r}})$ arises from an ordinary least squares fit. The ordinary least squares fit is appropriate for independent sampling but there is strong evidence of correlated sampling. Therefore, $\tilde{\boldsymbol{r}}$ are not fully representative of the random effects associated with the model so inferences based on them, for example, the strength of correlation as represented by τ, might not be fully valid. To circumvent the slightly circular argument based on analysing $\tilde{\boldsymbol{r}}$, we look at the comprehensive model,

$$\boldsymbol{y} \in \mathrm{N}(C\boldsymbol{\alpha}, \phi^{-1}V(\tau)),$$

with gamma priors for ϕ and τ:

$$\phi \sim G(m_{0,\phi}/2 m_{0,\phi}\hat{\sigma}^2/2), \quad \tau \sim G(m_{0,\tau}/2 m_{0,\tau}\hat{\tau}_0^{-1}/2),$$

where $\hat{\sigma}$ is the estimate of the standard deviation of the random effects associated with the least squares fit and $\hat{\tau}_0$ is the estimate of τ given by the mode of the posterior distribution $p(\tau|\tilde{\boldsymbol{r}})$. Since both these estimates are derived from the data $\boldsymbol{y}$ and since both these estimates arise for models that are somewhat suspect, the degree of belief parameters $m_{0,\phi}$ and $m_{0,\tau}$ are assigned values of 1, so that the respective prior distributions are vague.

A univariate optimisation scheme was used to maximise the marginalised posterior distribution $p(\tau|\boldsymbol{y})$, given by (5) for the general case, to give modal estimate $\hat{\tau}$ of τ. From the optimisation scheme also is calculated $\bar{s}^2$, a point estimate of $\sigma^2 = 1/\phi$. Regarding $\hat{\tau}$ as an exact representative value, the posterior distribution for $\boldsymbol{\alpha}|\hat{\tau}, \boldsymbol{y}$ is a multi variate t-distribution specified as in (4) with

$$\nu = m - 2 + m_{0,\tau}, \quad W(\tau) = V(\hat{\tau})^{-1}.$$

The standard deviations $u(\boldsymbol{a})$ of the parameters $\boldsymbol{\alpha}$ of this distribution are such that

$$u^2(\boldsymbol{a}) = \frac{\nu}{\nu - 2} \mathrm{diag}\,(V_{\boldsymbol{a}})$$

calculated from the diagonal elements of $V_{\boldsymbol{a}}$.

Table 1 gives, in the first two numerical rows, point estimates $\boldsymbol{a}$ of the linear response parameters $\boldsymbol{\alpha}$ for a number of models. The first column, labelled $V(\boldsymbol{\sigma}, \boldsymbol{\tau})$, gives the estimate of $\boldsymbol{\alpha}$, given exact knowledge of $\boldsymbol{\sigma}$ and $\boldsymbol{\tau}$ and represents the best (maximum likelihood) estimate of $\boldsymbol{\alpha}$, given the data $\boldsymbol{y}$. The second column, labelled $V_{\tau=0}$, is the least squares estimate based on the assumption that the effects are independently distributed, i.e., $\tau = 0$. The next three columns, gives the estimates of $\boldsymbol{\alpha}$, given three point estimates of τ, treated as exact. In the fourth column is the estimate based on the mode $\hat{\tau}$ of $p(\tau|\boldsymbol{y})$, in this case, $\hat{\tau} = 1.1$. This estimate is equal to the modal estimate derived from $p(\tau|\tilde{\boldsymbol{r}})$ to two significant figures. The two estimates of the mode are derived from slightly different models and inferences based on $p(\tau|\tilde{\boldsymbol{r}})$ alone should be used with caution. To illustrate the dependence on the point estimate of τ, in the third and fifth columns, we provide estimates $\boldsymbol{a}$, given the quantile estimates $\tau_5 = 0.7$ and $\tau_{95} = 2.1$ derived from $p(\tau|\tilde{\boldsymbol{r}})$, respectfully. These quantile estimates may be somewhat different from the corresponding quantiles associated with $p(\tau|\boldsymbol{y})$.

Table 1 also gives, in the last two numerical rows, the standard uncertainties $u(\boldsymbol{a})$ associated with the corresponding estimates. In the first column, the uncertainties are derived from the multivariate Gaussian distribution $\mathrm{N}(\boldsymbol{a}, V_{\boldsymbol{a}})$ where $V_{\boldsymbol{a}} = \left(C^\top V^{-1}(\boldsymbol{\sigma}, \boldsymbol{\alpha})C\right)^{-1}$. These uncertainties represent a reference as they are derived according to the exact model used to generate the data. For the remaining four columns, the uncertainties are those derived from the corresponding multivariate t-distribution, arising from taking the four different point estimates of τ as exact. The assumption that $\tau = 0$, independent sampling, underestimates the uncertainties by a factor of 2, approximately, compared to the uncertainties in column 1. The uncertainties in the remaining three columns show how they depend on a plausible range of point estimates from τ_5 to τ_{95} and straddle the uncertainties in column 1. They show that using point estimates based on $p(\tau|\tilde{\boldsymbol{r}})$, derived from very straightforward calculations, can be used to quantify the uncertainties associated with $\boldsymbol{\alpha}$, at least for this example. Importantly, the table shows that the uncertainties associated with $\boldsymbol{\alpha}$ depend strongly on value of τ and demonstrate that correlated sampling can significantly reduce the information content represented by the data.

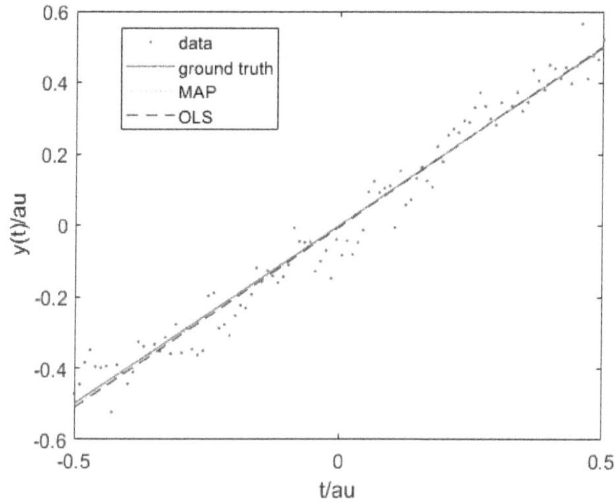

Fig. 1. Data simulating a measured linear response arising from a resampled effects model. Also show is the 'ground truth' response, $y = t$, along with the ordinary least-squares best with model, (OLS), and the AR(1) model determined using the point estimate $\hat{\tau}$ estimate of τ, (MAP).

Table 1. Estimation of linear response parameters $\boldsymbol{\alpha}$ the data in figure 1. The first two numerical rows give estimates $\boldsymbol{a}$ of $\boldsymbol{\alpha}$ for different estimation schemes (see main text). The last two numerical rows given the standard uncertainties $u(\boldsymbol{a})$ associated with the estimates.

$V(\boldsymbol{\sigma}, \tau)$	$V_{\tau=0}$	$V(\tau_5)$	$V(\hat{\tau})$	$V(\tau_{95})$
-0.0025	-0.0042	-0.0039	-0.0038	-0.0030
1.0061	1.0138	1.0133	1.0130	1.0115
0.0098	0.0054	0.0073	0.0084	0.0156
0.0311	0.0185	0.0249	0.0283	0.0513

5.2. *Drift in coordinate measurement*

In coordinate metrology, both instruments and artefacts are prone to environment effects, particularly temperature. The coordinate measuring machine (CMM) is used widely for artefact calibration. In a calibration of a high-density two-dimensional artefact using an error separation approach[3,5,6], substantial drift in the measurements was detected, possibly due to heat being generated by the CMM as it performed a large number

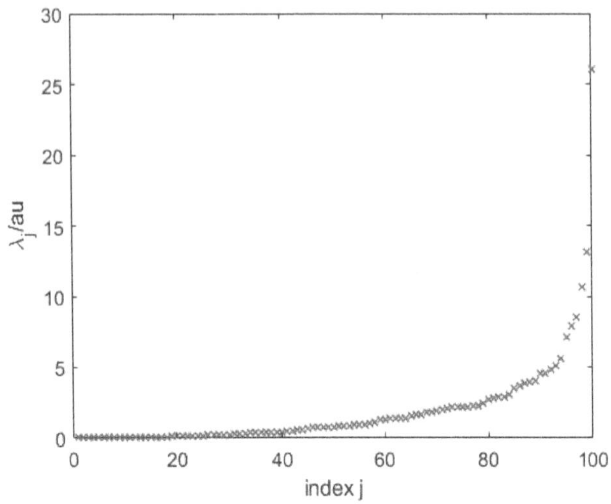

Fig. 2. Timescale parameters λ_j associated with the resampled effects model used to generate the data in figure 1.

of scans across the artefact. A drift correction approach was implemented based on repeated measurements of the same points on the artefact at regular time intervals. Figure 4 shows a typical set of measurements of drift in one coordinate over 53 scans. The drift can be corrected for by fitting a low degree polynomial to the data and then using the fitted curve as the basis of the correction. In order to assign an uncertainty to the correction, it is necessary to estimate the uncertainties associated with the fitted polynomial coefficient estimates. Under the assumption that the remaining effects not accounted for by the polynomial model are statistically independent, these uncertainty estimates can be derived using standard least squares techniques. However, is the assumption of independence justified?

To assess the validity of the independence assumption we can perform an analysis very similar to that described above in section 5.1, only this time with a quartic polynomial fit (degree four, order five) rather than a linear polynomial fit (degree one, order two). (For this example, there is no 'ground truth' with which to compare results.) Figure 5 graphs the posterior distribution $p(\tau|\tilde{\boldsymbol{r}})$ derived from the analysis of the projected residuals $\tilde{\boldsymbol{r}}$ associated with an ordinary least squares fit to the data. The posterior shows that the bulk of probability mass is centred around its mean at $\tau = 1.0$, between the 5 % and 95 % quantiles of $\tau = 0.48$ and $\tau = 1.83$,

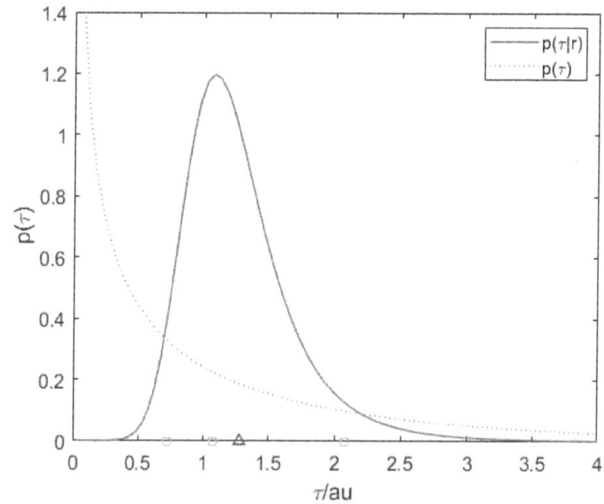

Fig. 3. Posterior distribution $p(\tau|\tilde{\boldsymbol{r}})$ determined from an analysis of the residuals $\tilde{\boldsymbol{r}}$ associated with an ordinary least squares fit of a linear model to the data in figure 1. The parameter τ is the parameter specifying the correlation kernel $\text{cov}(t,t') \propto \exp(-|t-t'|/\tau)$ of a Gaussian process equivalent to an AR(1) process. Also indicated on the axis by squares, the locations of the 5 % and 95 % quantiles and the mode of the distribution and, by triangle, its mean.

respectively. The posterior is non-zero at $\tau = 0$, indicating that an assumption of independence is somewhat plausible. The data provides little information to differentiate between values of τ near 0 and the likelihood $p(\tilde{\boldsymbol{r}}|\tau)$ is essentially flat near zero. For this reason, the posterior follows the shape of the prior $p(\tau)$ near zero which is why the posterior increases towards zero.

As in the example above, we can calculate the effective number of independent samples given the mode and 5 % and 95 % quantile values of τ:

$$\tilde{m}(\hat{\tau}) = 23.4, \quad \tilde{m}(\tau_5) = 42.9, \quad \tilde{m}(\tau_{95}) = 14.1,$$

respectively, and these numbers can compared with the case of independent sampling for which $\tilde{m} = m - 2 = 51$, the number of projected residuals $\tilde{\boldsymbol{r}}$.

From the comprehensive model,

$$\boldsymbol{y} \in \text{N}(C\boldsymbol{\alpha}, \phi^{-1}V(\tau))$$

with gamma priors for ϕ and τ we again use a univariate optimisation scheme to maximise the marginalised posterior distribution $p(\tau|\boldsymbol{y})$ with resulting point estimate $\hat{\tau} = 1.0$, essentially the same as the estimate derived from the mode of $p(\tau|\tilde{\boldsymbol{r}})$.

Table 2 gives, in the first five numerical rows, the standard uncertainties $u(\boldsymbol{a})$ associated estimates $\boldsymbol{\alpha}$ corresponding to four point estimates of τ, as in table 1. (The estimates themselves are very similar and are not shown.) The table shows that uncertainties associated with the case $\tau = 0$ are significant underestimates compared to the other, more likely, point estimates. Also shown are the point estimates of σ^2 associated with the four models. Larger estimates of σ^2 corresponds to larger values of τ. The reason for this is that for larger values of τ the time interval associated with the data represents a smaller section of the evolution of the time series and that the future evolution of the time series could show greater variation.

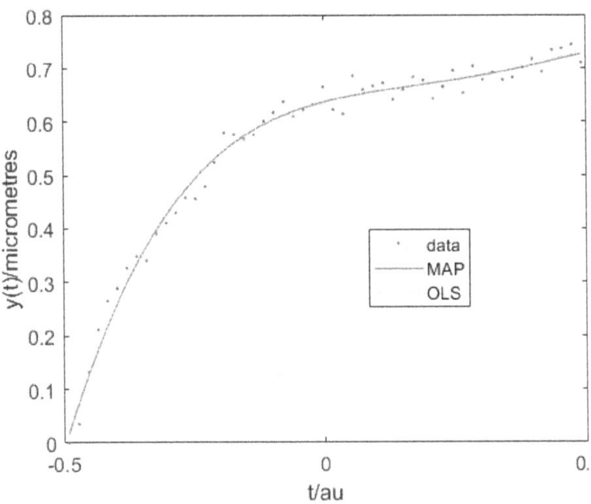

Fig. 4. Data measuring the drift in the measurement of a coordinate undertaken by a CMM in the calibration of a two dimension artefact. Also graphed are the fits of a quartic polynomial to the data, one determined using an AR(1) model (MAP), and one using an ordinary least squares fit (OLS).

5.3. *Computational requirements*

The calculations above have all been performed using standard numerical linear algebra tools and computational components that are straightforward to implement in software. An addition to standard least squares computa-

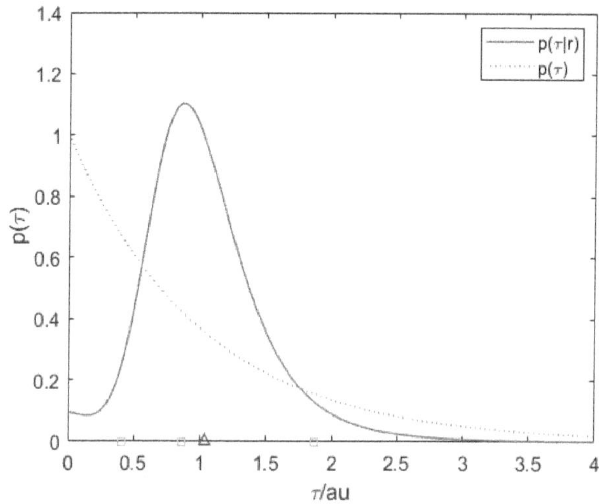

Fig. 5. Posterior distribution $p(\tau|\tilde{\boldsymbol{r}})$ determined from an analysis of the residuals $\tilde{\boldsymbol{r}}$ associated with an ordinary least squares fit (OLS) of a linear model to the data in figure 4, along with the prior distribution $p(\tau)$. Also indicated on the axis by squares, the locations of the 5 % and 95 % quantiles and the mode of the distribution, squares, and its mean, triangle.

Table 2. Estimation of parameters $\boldsymbol{\alpha}$ associated with a quartic polynomial fit to the data in figure 4. The first five numerical rows give the standard uncertainties $u(\boldsymbol{a})$ associated with the estimates. The final numerical row gives the estimate of the standard deviation of the random effects for the four models.The associated unit is micrometres.

$V_{\tau=0}$	$V(\tau_5)$	$V(\hat{\tau})$	$V(\tau_{95})$
0.0067	0.0074	0.0097	0.0147
0.0055	0.0060	0.0078	0.0115
0.0051	0.0056	0.0071	0.0102
0.0046	0.0050	0.0064	0.0093
0.0045	0.0049	0.0062	0.0086
0.0213	0.0215	0.0225	0.0264

tion components[8] such as the QR factorisation and Cholesky factorisation we have implemented the following calculations in software.

Evaluation of $p(\tau|y)$. Given a time series $\boldsymbol{y}$ arising according to the model

$$y_i = \alpha + \epsilon, \quad \boldsymbol{\epsilon} \in \mathrm{N}\left(, \mathbf{0}, \phi^{-1}V(\tau)\right),$$

$p(\tau|\boldsymbol{y})$ is calculated explicitly according to (5) with the normalising constant calculated using quadrature in one dimension.

Evaluation of the mode of $p(\tau|y)$ for an AR(1) model. Given a time series $\boldsymbol{y}$ arising according to the model

$$\boldsymbol{y} \in \mathrm{N}(C\boldsymbol{\alpha}, \phi^{-1}V(\tau)),$$

where $V(\tau)$ is the variance matrix arising from an AR(1) model with time scale parameter τ, the mode of $p(\tau|\boldsymbol{y})$ is calculated using a univariate optimisation algorithm to minimise $-\log(p(\tau|\boldsymbol{y})$ where $p(\tau|\boldsymbol{y})$ is given as in (5). The $m \times m$ bi-diagonal matrices $L^{-1}(\tau)$ are calculated and stored as two m vectors and matrix multiplication involving $L^{-1}(\tau)$ are implemented efficiently in terms in terms of these two vectors. The software also calculates the point estimates $\boldsymbol{a}$ and $\hat{\sigma}^2$ of $\boldsymbol{\alpha}$ and $\sigma^2 = 1/\phi$ along with degrees of freedom parameter ν as in (4).

If necessary, these calculations can be used as the basis for a Metropolis-Hastings algorithm[7,13] to sample from the joint posterior distribution $p(\boldsymbol{\alpha}, \phi, \tau|\boldsymbol{y})$.

6. Concluding remarks

This paper has been concerned with analysing data arising from measurements of a linear response subject to Gaussian random effects. Under the assumption that the random effects are independent, standard approaches based on least squares regression can be applied safely and successfully. However practical systems usually do not provide any guarantee that the effects are independently drawn and often there is evidence in the data that makes us question the assumption of independence. If a model based on the assumption of independence is suspect, what should we replace it with? In this paper, we have looked at a resampled effects model that uses the idea that individual effects are associated with timescales governing when these effects are resampled, similar to a repeatability & reproducibility study in which different influence factors (operator, tool, environment) are deliberately changed at different times in the study. The resampled effects model can be approximated by an AR(1) model and with this approximation the analysis of the data can be addressed using efficient computational methods.

Even a modest amount of correlation can reduce the effective number of independent samples and an assumption of independence can significantly overestimate the amount of information that the data provides about the underlying system. We have illustrated the approach on simulated data and data arising in coordinate metrology. We believe that the use of an AR(1) model to check for correlation is very worthwhile as it can quickly provide evidence that the assumption of independence is not valid and provides a simple but effective first step for accounting for possible correlation.

Acknowledgements

This work was funded by the UK's National Measurement System programme for Data Science. Examples in coordinate metrology were developed in collaboration with the National Physical Laboratory's Engineering Measurement Department.

References

1. BIPM, IEC, IFCC, ILAC, ISO, IUPAC, IUPAP, and OIML. Evaluation of measurement data — Guide to the expression of uncertainty in measurement. Joint Committee for Guides in Metrology, JCGM 100:2008.
2. G. E. P. Box and G. M. Jenkins. *Time Series Analysis.* Holden-Day, San Francisco, 1976.
3. C. J. Evans, R. J. Hocken, and W. T. Estler. Self-calibration: reversal, redundancy, error separation, and absolute testing. *Ann. CIRP*, 45(2):617–634, 1996.
4. A. B. Forbes. Measurement systems analysis and statistical process control. In P. Ciarlini, E. Felipe, A. B. Forbes, F. Pavese, C. Perruchet, and B. Siebert, editors, *Advanced Mathematical and Computational Tools in Metrology VII*, pages 161–170, Singapore, 2006. World Scientific.
5. A. B. Forbes and M. McCarthy. Calibration of grid plates bearing a high density of targets. In *euspen 15th International Conference & Exhibition, Leuven, Belgium, June, 2015*, 2015.
6. A. B. Forbes and I. M. Smith. Self-calibration and error separation techniques in metrology. In P. Ciarlini, M. G. Cox, E. Filipe, F. Pavese, and D. Richter, editors, *Advanced Mathematical and Computational Tools in Metrology V*, pages 149–163, Singapore, 2001. World Scientific.
7. A. Gelman, J. B. Carlin, H. S. Stern, D. B. Dunson, A. Vehtari, and

D. B. Rubin. *Bayesian Data Analysis.* Chapman & Hall/CRC, Boca Raton, Fl., third edition, 2014.
8. G.H. Golub and C.F. Van Loan. *Matrix Computations.* Johns Hopkins University Press, Baltimore, fourth edition, 2013.
9. International Organization for Standardization, Geneva. *ISO 5725-1: Accuracy (trueness and precision) of measurement methods and results – Part 1: General principles and definitions*, 1994.
10. International Organization for Standardization, Geneva. *ISO/TS 21748: Guidance for the use of repeatability, reproducibility and trueness estimates in measurement uncertainty estimation*, 2004.
11. International Organization for Standardization, Geneva. *ISO/TR 22971: Accuracy (trueness and precision) of measurement methods and results – Practical guidance for the use of ISO 5725-2:1994 in designing, implementing and statistically analysing interlaboratory repeatability and reproducibility results*, 2005.
12. K. Jagan and A. B. Forbes. Analysing data streams of doubtful provenance. In F. Pavese, A. Chunovkina, A. B. Forbes, and N.-F. Yang, editors, *Advanced Mathematical and Computational Tools for Metrology XII*, Singapore. World Scientific. Submitted.
13. K. Jagan and A. B. Forbes. User manual for NLLSMH: software implementating a Metropolis-Hastings algorithm for non-linear least squares problems. Technical Report MS 22, National Physical Laboratory, Teddington, April 2018.
14. H. S. Migon and D. Gamerman. *Statistical Inference: an Integrated Approach.* Arnold, London, 1999.
15. D. C. Montgomery and G. C. Runger. *Applied Statistics and Probability for Engineers.* John Wiley and Sons, 2003.
16. W. H. Press, S. A. Teukolsky, W. T. Vetterling, and B. P. Flannery. *Numerical Recipes: the Art of Scientific Computing.* Cambridge University Press, Cambridge, third edition, 2007.
17. C. E. Rasmussen and C. K. I. Williams. *Gaussian Processes for Machine Learning.* MIT Press, Cambridge, Mass., 2006.

F Pavese, A B Forbes, N F Zhang, A G Chunovkina (Eds):
Advanced Mathematical and Computational Tools in Metrology and Testing XII
Series on Advances in Mathematics for Applied Sciences, Vol. 90

https://doi.org/10.1142/9789811242380_0012

Uncertainty evaluation in rheology measurements

A. Furtado, J. Pereira and J. A. Sousa
Portuguese Institute for Quality,
Rua António Gião, 2/2829-513 Caparica, Portugal
E-mail: jasousa@ipq.pt

M. G. Cox
National Physical Laboratory,
Teddington/TW11 0LW, UK
E-mail: maurice.cox@npl.co.uk

A. S. Ribeiro
National Laboratory for Civil Engineering,
Avenida do Brail, 101/1700-066 Lisbon, Portugal
E-mail: asribeiro@lnec.pt

This work specifies general methods for the calibration of rheometers and evaluates the measurement uncertainty associated with the complex mathematical expression of viscosity derived for rotational rheometer equipment using an alternative method to that in the GUM. This method is a Monte Carlo implementation of the propagation of distributions (GUM-S1), which can treat measurement models having any degree of non-linearity and possibly large input uncertainties. It is therefore particularly suited to handling the complexity of the applicable measurement model, with the advantage of avoiding the need to calculate partial derivatives. In this work the difference between both approaches proved negligible for all practical purposes using baseline conditions. However further simulation showed how changing input conditions could alter this conclusion.

Keywords: uncertainty evaluation, model complexity, validation, rheology

1. Metrology traceability of rheological determinations

The establishment of standards is a practical and useful way of providing comparable results for a specified type of measurement. Metrology, the science of measurement, goes beyond by determining whether quantities are generally transferable between physical relations. Two key issues are traceability to common base units (the SI) and evaluation of the uncertainty of each quantity. The steps for evaluating the uncertainty of a measurement result are summarized in the Guide to the expression of uncertainty in measurement or GUM (JCGM

100:2008) [1]. The first step consists of expressing a mathematical function f that relates the measurand to the input quantities on which the measurand depends, i.e. "function f should contain every quantity, including all corrections and correction factors, which can contribute to a significant component of uncertainty in the result of the measurement" [1].

This paper aims to specify general methods for the calibration of rheometers and additionally covers the uncertainty budget of viscosity measurements performed by rotational methods, since the complexity of the mathematical model raises doubts on the validity of the GUM approach.

According to the International Vocabulary of Metrology (VIM) [2], a metrological traceability chain is described as a "sequence of measurement standards and calibrations that is used to relate a measurement result to a reference". In order to relate the measurement results obtained by viscosity sensors, specifically rotational rheometers, to the SI base units, the following sequence of steps needs to be accomplished: (1) calibration of the standard rotational rheometer traceable to the SI units; (2) use of Certified Reference Materials (CRMs) for viscosity (by tracing flow and viscosity curves); (3) calibration of a laboratory rheometer (each of the following parameters need to be calibrated independently: viscosity by using a CRM; shear rate (indirectly through rotational velocity) and shear stress (indirectly through torque); temperature; dimensions of the measuring geometries, including gap geometry; (4) production of a reference material (RM) with specified viscosities by a standardized method and subsequent determination of its viscosities at specified shear rates by using the laboratory rheometer calibrated in step 3; (5) calibration of the viscosity sensors by using the RM from step 4. Uncertainty necessarily increases with every step in this sequence. Steps 1 and 2 would be carried out by an NMI and steps 3–5 by the laboratory itself.

Input quantities are specific to a given measurement principle. The main quantities considered significant when determining the viscosity of non-Newtonian liquids by using a rotational rheometer are torque; rotational speed; temperature (due to the increase in uncertainty at high shear rates generated by frictional heating); dimensions (radii, lengths, angles, measuring gap); end-effect correction [3][4][5][6]; mathematical model (choice of representative location, "narrow-gap" approximation); measurement time (affects temperature via frictional heating and laminar flow field establishment within the gap) [7]; repeatability; reproducibility; sample shear history. These quantities influence the evaluation of the uncertainty of the viscosity of a CRM and represent parameters that must be controlled carefully when applying the CRM to the subsequent calibration of rheometers.

2. Rheometer calibration methods

As the purpose of use of a rotational rheometer is not always the same, the calibration method applied should fit the purpose. For this reason, two calibration methods are considered. A direct method, where all the measured input quantities, i.e. torque, angular velocity, temperature and geometric dimensions are compared with reference values, and an indirect method where the CRM viscosity is used to determine the shear rate and shear stress measurement errors of a set composed of the measuring instrument (rotational viscometer or rheometer) and a measuring geometry (together with the temperature sensors). In order to obtain reliable viscosity results, steady state flow should be achieved in each measurement.

Direct calibration. Taking as an example the determination of the viscosity of a Newtonian sample by a rotational rheometer with a concentric cylinder system, the defining equation of dynamic viscosity η is given by,

$$\eta = \frac{\tau}{\dot{\gamma}} \tag{1}$$

where τ is the shear stress and $\dot{\gamma}$ is the shear rate. According to the standards ISO 3219 (1993) [8] and DIN 53019-2 (2001) [9], the dynamic viscosity η at reference temperature T_{ref}, is determined by,

$$\eta(T_{\text{ref}}) = \frac{M\left[\left(\frac{R_{\text{a}}}{R_{\text{i}}}\right)^2 - 1\right]}{8\pi^2 nLR_{\text{i}}^2 c_L \left(\frac{R_{\text{a}}}{R_{\text{i}}}\right)^2}\left[1 - \beta_T(T - T_{\text{ref}})\right] \tag{2}$$

where M is torque, R_{a} is the outer radius of the gap, R_{i} is the inner radius of the gap, n is the number of rotations per time unit, L is the length of the cylinder mantle, c_L is the end-effect correction factor, β_T is the linear coefficient of change of viscosity with temperature, T is measurement temperature and T_{ref} is reference temperature.

Thus, the direct method for calibration of a rotational viscometer or a rheometer requires a separate calibration of the torque and the angular velocity of the measuring geometry. Additionally, the dimensions (diameters and angles) of the measuring geometry and the temperature should also be calibrated by adequate means. The direct method is described in standard documents such as DIN 53019-2 (2001) [9]; ASTM E2510-07 (2013) [10] (for torque calibration) and ASTM E2509-14 (2014) [11] (for temperature calibration). Expression (2) will be the mathematical model used in this study for the evaluation of measurement uncertainty together with the data of Table 1.

Indirect calibration. The calibration of a rheometer by an indirect method is described in ASTM E2975 (2016) [12]. The calibration procedure consists of determining the correction factor F for the selected measuring geometry in its whole viscosity range, by changing its rotational speed or its torque, thus in controlled rate (CR) and controlled stress (CS) modes by using a Newtonian CRM with a traceable viscosity value F defined as the ratio of the viscosity value of the standard liquid (η_{std}) and the average of the viscometer readings (η_{rdg}) at the test temperature of 20 °C.

3. Evaluation of measurement uncertainty in rheometer calibration

Measurement uncertainty evaluation includes components arising from systematic effects, such as components associated with corrections and the assigned quantity values of measurement standards, as well as definitional uncertainty.

The aim of this work is to compare the evaluation of measurement uncertainty followed the methods described in GUM [2], which entails a set of assumptions to be applicable, and a more general and reliable approach suggested in the GUM Supplement 1 [13] based on the propagation of distributions by a Monte Carlo method.

GUM methodology. The GUM method applied to a univariate model relates a single output quantity Y to input quantities $\boldsymbol{X} = (X_1, \dots, X_N)^{\top}$ by a functional relationship f in the form of expression (1). The estimate of the output quantity is taken as $y = f(\boldsymbol{X})$, evaluated at $\boldsymbol{X} = \boldsymbol{x}$, the estimate of $\boldsymbol{X}$. The associated standard uncertainty $u(y)$ is evaluated from the *law of propagation of uncertainty* (LPU).

$$u^2(y) = \sum_{i=1}^{N} \sum_{j=1}^{N} c_i u(x_i, x_j) c_j \tag{3}$$

given in the GUM, where c_i is the partial derivative of f with respect to $\boldsymbol{X}$, evaluated at $\boldsymbol{X} = \boldsymbol{x}$, and is known as the ith sensitivity coefficient. By defining the covariance matrix.

$$\boldsymbol{U}_x = \begin{bmatrix} u(x_1, x_1) & \cdots & u(x_1, x_N) \\ \vdots & \ddots & \vdots \\ u(x_N, x_1) & \cdots & u(x_N, x_N) \end{bmatrix}$$

of dimension $N \times N$, containing the covariances $u(x_i, x_j)$, and the (row) vector $\mathbf{c}^{\top} = [c_1, \dots, c_N]$ of dimension $1 \times N$, containing the sensitivity coefficients, a compact way of writing expression (3) is

$$u^2(y) = \boldsymbol{c}^{\mathrm{T}}\boldsymbol{U}_x\boldsymbol{c} \tag{4}$$

For independent input quantities, a simpler formulation is often used, where the variance $u^2(y)$ can be viewed as a sum of terms $u^2(x_i)$, so that each input quantity contributes to the final value of the squared standard uncertainty (variance). In this case, expression (3) can be written as

$$u^2(y) = \sum_{i=1}^{N}[c_i u(x_i)]^2 = \sum_{i=1}^{N} u_i^2(y)$$

with the various $u_i(y)$ in the uncertainty budget often being used to give a measure of the importance of each input quantity.

Given a coverage probability p, a $100p$ % coverage interval for Y is defined as $y \pm U_p$ with expanded uncertainty $U_p = k_p u(y)$ and coverage factor k_p. In Metrology it is common to take $p = 0.95$, yielding a value of $k_p = 2$ for a normal distribution (a scaled and shifted t-distribution is used in some cases to determine kp [2]).

Propagation of distributions. A Monte Carlo method is an implementation of the propagation of distributions and is generally considered a very reliable approach for uncertainty propagation. The PDFs (probability density functions) for the input quantities are propagated through the measurement model to provide the PDF for the output quantity, with no restriction in relation to shape or symmetry. The expectation (mean) of the resulting PDF is used as the best estimate of the quantity and the standard deviation of the PDF is used as the standard uncertainty associated with that estimate.

Appropriate use of Monte Carlo methods will provide valid results when the applicability of the GUM uncertainty framework is questionable. Aspects to be carefully considered include the number of samples taken of the model input quantities, and the related numerical accuracy of the results obtained. Guidance is provided in the GUM Supplement 1 [13].

From the PDF for the output quantity Y, a coverage interval for Y for any coverage probability p (usually taken as 0.95) can be obtained. Such an interval contains the value of Y with probability p. A coverage interval is not unique: a shortest 95 % coverage interval will provide values with the highest density, whereas for a symmetrical PDF the percentiles 0.025 and 0.975 should be used.

An advantage of this approach is that it provides a PDF for Y that is consistent with the input PDFs. The assumption in the GUM that the PDF for Y is Gaussian (or a scaled and shifted t-distribution) is not required.

4. Uncertainty assessment of the direct calibration method

GUM Methodology. Consider the analysis of a measurement of a Newtonian oil, at 20 °C, using a rheometer with a concentric cylinder system, according to DIN 53019 [11], at a shear rate of 1 s^{-1}, using the input quantities indicated in Table 1 and the mathematical model (1). The result is a viscosity of 1.091 Pa·s with an associated expanded uncertainty of 0.018 Pa·s. The construction of the uncertainty budget poses an immediate difficulty in the evaluation of the sensitivity coefficients (partial derivatives). This task is far from trivial when applied to such a complicated model: the usual approach employs a numerical method to avoid the need to provide analytic expressions for the partial derivatives. The suggested approach is the complex-step method [14] based on the Taylor expansion of a function of a complex variable: it is accurate, avoiding subtractive cancellation.

The results in Table 1 show that 31 % of the uncertainty originates from the end-effect correction factor, 30 % from torque, 20 % from temperature stability and 13 % from R_a. These numbers can change drastically by varying measurement conditions: any conclusions drawn are only valid for a specific condition. At lower shear rates, for example, the measured torque will be smaller, but its absolute uncertainty is constant, which increases its share in the viscosity uncertainty. Conversely, at higher shear rates, because the temperature stability becomes questionable due to frictional heating, the viscosity uncertainty will increase drastically and be dominated by temperature uncertainty. Also, as it will be shown in this study there are input quantities which when having their uncertainties increased may change dramatically the output PDF: the Monte Carlo method reveals such situations.

This example refers only to Newtonian fluids, and the mathematical model is unlikely to cover all input quantities but is a first approach to give an impression of the complexity associated with the treatment of uncertainty in rheology. In the more complex cases of non-Newtonian fluids cases exist where there is not a mathematical model, uncertainty evaluation relying on interlaboratory comparisons or other experimental means. Additionally, it demonstrates that uncertainty evaluation needs to be backed by real measurements.

Monte Carlo method. Using the same input values of Table 1, the results given by the Monte Carlo method are displayed in Table 2 and the output PDF in the form of a scaled histogram is shown in Figure 1. Although there is considerable similarity to a Gaussian PDF, the GUM overstates the uncertainty by some 4 %, which is hinted at by the graph and is not surprising because of the

Table 1. Example of an uncertainty budget of the measurement of a Newtonian oil using a rheometer with a concentric cylinder system according to DIN 53019 standard [9] (and t = 20 °C).

Quantity/Symbol	Estimate x_i	Dist. PDF	$u(x_i)$	Sensivity coefficient	$u_i(y)$ Relative contribution
Torque / M	$5.7938 \cdot 10^{-5}$ Nm	R	$2.89 \cdot 10^{-7}$ Nm	1.88248	29.5 %
Rotations per time unit / n	0.0128985 s^{-1}	G	$5.00 \cdot 10^{-7}$ s^{-1}	−0.00846	0.2 %
Outer gap radius / R_a	0.014458 m	R	$2.89 \cdot 10^{-6}$ m	0.08549	13.4 %
Inner gap radius / R_i	0.0133295 m	R	$2.89 \cdot 10^{-7}$ m	−0.10909	1.7 %
Cylinder mantle length / L	0.040009 m	R	$2.89 \cdot 10^{-7}$ m	−0.00272	–
End-effect correct factor / c_L	1.1	R	$5.77 \cdot 10^{-3}$	−0.00010	31.0 %
Measurement temperature / T	20 °C	G	$5 \cdot 10^{-2}$ C	−0.00001	20.2 %
Reference temperature / T_{ref}	19.99 °C	G	$1 \cdot 10^{-2}$ C	0.00001	4.0 %
Linear coefficient of change of viscosity with temperature / β_T	0.0683 C^{-1}	R	$5.77 \cdot 10^{-5}$ C^{-1}	−0.00000	–
Viscosity η	1.091		$u(y) = 0.009$		$U_{95\%} = 0.018$

model non-linearity and the key input quantities with because of the model non-linearity and the key input quantities with rectangular PDFs. The GUM assumes the approximate applicability of the central limit theorem, that is, the PDF for the output quantity will be approximately Gaussian when the input quantities are independent and the output standard uncertainty is much larger than that of any single input quantity with a non-Gaussian PDF. In a case such as the current example, validation should be mandatory, even though the difference may be acceptable for many applications. It should be pointed out, nevertheless, that in decision-making situations this difference may prove significant. Moreover, the difficulty in applying GUM because of the above assumption and entailed by the determination of the sensitivity coefficients can be avoided by using a Monte Carlo method.

Table 2. Comparison between GUM and Monte Carlo method with data from Table 1.

Viscosity	Estimate y/Pa.s	$u(y)$/Pa.s	$U_{95\%}(y)$/Pa.s
GUM	1.091	0.0091	0.018
Monte Carlo	1.091	0.0088	0.017

For a very different set of conditions, a completely different result would have been obtained. Table 3 and Figure 2 show the results for a hypothetical situation where the uncertainty associated with the input quantity R_i is increased by a factor of 10^3 (all other input quantities are unchanged). In this case, the use of the GUM approach would more visibly be totally inadequate for the evaluation of measurement uncertainty, in this instance overstating the expanded uncertainty by about 22 %. Similar results would be obtained if the uncertainties associated with the Gaussian distributions would have been reduced.

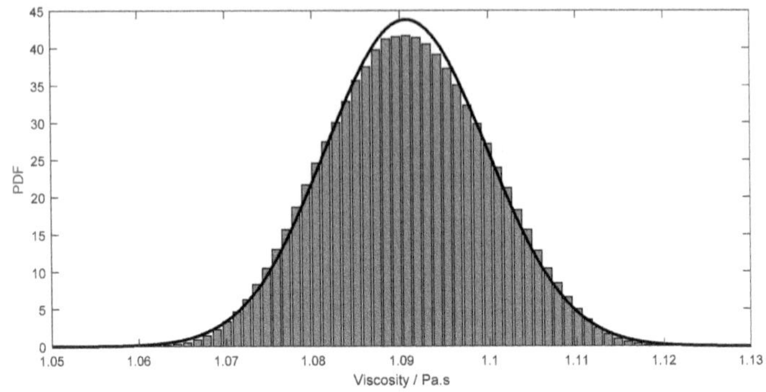

Fig. 1. Output quantity PDF for the model (2) using the GUM (continuous curve) and a Monte Carlo approach (scaled histogram).

Table 3. Comparison between GUM and MC method increasing uncertainty associated with R_i.

Viscosity	Estimate y/Pa.s	$u(y)$/Pa.s	$U_{95\%}(y)$/Pa.s
GUM	1.091	0.315	0.63
Monte Carlo	1.101	0.260	0.52

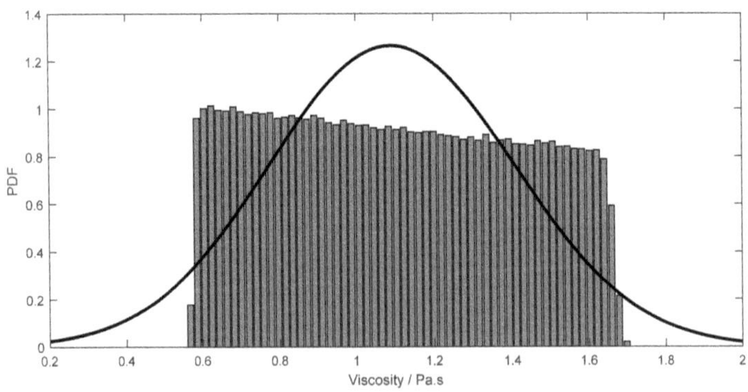

Fig. 2. Output quantity PDF for the model (2) using the GUM (continuous curve) and a Monte Carlo approach (scaled histogram), with a greatly increased uncertainty assigned to R_i.

The purpose of this simulation is to point out that for such complex, nonlinear models, conclusions are very much dependent on the values and distribution of input quantities, in the same way as presented elsewhere [13, clause 9.3], and that no blind recipes are to be applied in metrology

5. Conclusions

This study showed the complexity associated with measurement in the field of rheology. Even for the least complicated case, as is the case presented here for Newtonian liquids, there may be influencing quantities not reflected in the existing measurement models. Moreover, for the more complicated cases encountered for non-Newtonian liquids, the conclusions drawn from this study are at least as important, since the complexity of the models are expected to be enhanced. The case studies presented have shown it is important to apply adequate statistical tools to evaluate measurement uncertainty, as illustrated in several examples in [13]. The use of conventional recipes might bring erroneous results which, depending on the application, can have serious implications, e.g., in the oil and mine industry or in health applications.

Acknowledgments

The EMPIR projects 17RPT02-rhoLiq and 17NRm05-EMUE that supported this work have received funding from the EMPIR programme co-financed by the Participating States and from the European Union's Horizon 2020 research and innovation programme.

References

1. JCGM 200:2008. International vocabulary of metrology – Basic and general concepts and associated terms (VIM).
2. JCGM 100:2008. Evaluation of Measurement Data – Guide to the expression of uncertainty in measurement (GUM 1995 with minor corrections).
3. Highgate DJ and Whorlow RW 1969 *Rheologica Acta* **8**(2) 142-151.
4. Bauer H and Boese N 1990 In *Third European Rheology Conf. and Golden Jubilee Meeting of the British Society of Rheology* (pp. 37-40). Springer, Dordrecht.
5. Kelessidis VC *et al* 2010 *J. Petroleum Science and Engineering* **71**(1) 37-46.
6. Wein O *et al* 2015 *Rheologica Acta* **54**(11-12) 903-914.
7. Giesekus H and Langer G 1977 *Rheologica Acta* **16**(1) 1-22.

8. ISO 3219:1993, Plastics - Polymers/resins in the liquid state or as emulsions or dispersions — Determination of viscosity using a rotational viscometer with defined shear rate.
9. DIN 53019-2:2001, Viscosimetry - Measurement of viscosities and flow curves by means of rotation viscosimeters – Part 2: Viscosimeter calibration and determination of the uncertainty of measurement.
10. ASTM E2510-07(2013), Standard Test Method for Torque Calibration or Conformance of Rheometers.
11. ASTM E2509–14, Standard Test Method for Temperature Calibration of Rheometers in Isothermal Mode.
12. ASTM E2975–16e1, Standard Test Method for Calibration or Calibration Verification of Concentric Cylinder Rotational Viscometers.
13. JCGM 101:2008. Evaluation of Measurement Data - Supplement 1 to the 'Guide to the Expression of Uncertainty in Measurement' – Propagation of distributions using a Monte Carlo method.
14. R. Boudjemaa, MG Cox, AB Forbes, and PM Harris. Automatic differentiation and its applications to metrology. In Patrizia Ciarlini, MG Cox, F Pavese, and GB Rossi, editors, Advanced Mathematical and Computational Tools in Metrology VI, pages 170–179, Singapore, 2004. World Scientific.

F Pavese, A B Forbes, N F Zhang, A G Chunovkina (Eds):
Advanced Mathematical and Computational Tools in Metrology and Testing XII
Series on Advances in Mathematics for Applied Sciences, Vol. 90

https://doi.org/10.1142/9789811242380_0013

Measuring instruments comparison for calibration and check: Four typical tasks of data processing

V. A. Granovskii
Institute of Fundamental Training and Technological Innovations,
State University of Aerospace Instrumentation,
Saint Petersburg, 190000, Russia
E-mail: int@guap.ru, valgr39@gmail.com
http://suai.edu.ru

The problem of measuring instrument calibration and check by its comparison with other instrument used as a reference standard is discussed. The typical tasks of data processing are formulated and solved. The results being published earlier and new ones are presented.

Keywords: Measuring instrument; Reference standard; Comparison; Data processing.

1. Introduction

Comparisons in practical metrology are aimed differently and related to measuring instruments (MI) of different metrological status. Firstly, a comparison may confirm the measurand estimate obtained with one of two MI [1, 2]. MIs under comparison can be of the same accuracy level and even the same type. Secondly, two MIs readings comparisons can be aimed at determination (and exclusion) of a constant systematic error inherent in one of the MI [3]. It is important at that the other MI is of another type (the better case is if the second measurement procedure differs from the one). Thirdly, reference standards (RSs) are compared for establishing their accuracy relation in course of international round-robin comparisons of national standards [4-6].

But the most typical case is a MI comparison with RS in course of periodic verification, calibration, and tests for a MI's type approval [7]. The comparison principle is as follows. When a quantity Q is measured directly readings of MI φ and of RS φ_r are represented with the symbolic equations

$$\varphi = Q + \delta, \tag{1}$$
$$\varphi_r = Q + \delta_r, \tag{2}$$

where δ, δ_r – errors of MI and RS accordingly. The MI error can be determined from (1), (2) as

$$\delta = \varphi - \varphi_r + \delta_r. \tag{3}$$

The problem consists in checking whether the MI error meets an established norm δ_N:

$$\delta = \varphi - \varphi_r + \delta_r \leq \delta_N. \tag{4}$$

Under MI calibration the problem is determining δ in accordance with (3). It is to note that the problem (4) generally does not include structurally the problem (3) what illustrated below.

RS in problems of MI error control and determination is interpreted usually as MI which is more accurate than MI under verification (calibration):

$$\delta_r << \delta, \tag{5}$$

so (3) and (4) can be simplified:

$$\delta \cong \varphi - \varphi_r, \tag{6}$$

$$\delta \cong \varphi - \varphi_r \leq \delta_N. \tag{7}$$

Quantitative relations according to (5) differs among themselves but

$$\delta_r \leq 0{,}1\delta, \tag{8}$$

is considered practically sufficient whence it follows the known relation of standard deviations (SD) σ, σ_R of random errors δ, δ_R:

$$\sigma \geq 3\sigma_r. \tag{9}$$

The condition (5) is hardly ensured in some technical branches because of necessary access to the RS of required accuracy level. So, for instance, in course of marine test, it is necessary to compare heading channels of navigation system under test and reference system provided that the reference system is slightly more precisely than the test subject [8]. MIs comparison problem if the condition (5) cannot be ensured is valid for all metrological practice taking into account economical price (including an operation price) of access to RS of different accuracy.

So the general metrological comparison problem can be formulated of MI and RS with arbitrary accuracy relation [9]. Experimental (but technological) part of the problem is principally simple – it has to ensure readings synchronization of devices compared, and to form a readings difference. It permits to exclude possible measurand Q changes out of consideration. Provided that the problem comes to data processing. Attempts are made earlier for solving partial tasks of the problem [10-12] and solving it for some application [13]. This article is aimed at solving the problem in general statement.

The problem under solution fall to four typical tasks:
1 test of comparison data flow stationarity as a necessary condition for using a stochastic data model;
2 test of data generation process ergodicity as a requirement for statistical processing a dynamic series;
3 analysis of comparison data structure, that is systematic and random part relation in general stochastic model, and random part probabilistic modeling;
4 comparison data processing aimed at check or determination of MI precision characteristics.

Two first parts became practically as part and parcel of the problem, and generate together the data quality determination (or check) task.

The article contain the earlier derived results [8-13] and new ones as well.

2. Data general character analysis

As it is supposed a priori that comparison data processing rest upon MI and RS statistical properties so initial data (general population) quality and sample representativity are governing factors for solving the task. To make single valued conclusions concerning calibrated MI properties it is desirable to have the data flow being stationary. To this end measurement experiment conditions are made stable and/or measurement time is established so that they remain stable during the time. But foremost they try to ensure MI properties stability in the course of MI construction. The combination of MI stability and its use conditions permits to hope at comparison data flow stationarity.

Per se data stationarity problem is not principally complicated. Stationarity hypothesis check criteria are known but they need to get samples for fixed points of time that is samples out of general populations for random values corresponding to temporal cut sets of random process studied. But practically we have only one time implementation of the process. Its use as a carrier of the process statistical properties is possible if it is only ergodic. To inspect this condition it has to establish firstly that the process is stationary. Therefore a closed disk arises. The only way out is the temporal implementation uniformity checking [15].

So practically we confront foremost with the first two above-mentioned tasks concerning data stationarity and ergodicity closed with each other in actual situation of comparison data processing.

Because of restriction, most often, than not with first two probabilistic moments at real data processing then stationarity is treated in wide sense, and ergodicity is considered below in respect of averages and dispersions.

If the inspection gives beneficial effect (more precisely – not negative result) known ergodicity criteria may be used.

As the ergodicity criterion by averages it is expedient to use the Slutsky condition [16]:

$$\lim_{T\to\infty}\frac{1}{T}\int_0^T \psi(\tau)d\tau = 0 \tag{10}$$

where $\psi(\tau)$– correlation function of the stationary random process $\{x_i\}$.

The ergodicity criterion on dispersion is generalization of (10) and have the form [17]:

$$\lim_{T\to\infty}\frac{1}{T}\int_0^T \psi^2(\tau)d\tau = 0 \tag{11}$$

Since the correlation function $\psi(\tau)$ is evaluated by a discrete temporal series $\{x_i\}$, it is expedient to use the analogues of formulas (10), (11) for discrete stationary random processes:

$$\lim_{n\to\infty}\frac{1}{n}\sum_0^n \psi[m] = 0 \tag{10a}$$

$$\lim_{n\to\infty}\frac{1}{n}\sum_0^n \psi^2[m] = 0 \tag{11a}$$

Bearing in mind both properties of stationarity and ergodicity it is possible to accept the processing order of compared devices temporal series suggested in [18]. Stationarity inspection can be carried out by calculation of average and mean square for every part of data series. Collating averages among themselves they establish that there is no bias and deviations are into well-founded limits. If the result is positive, data can be treated as stationary. Of course, stationarity checking can be intensified by using different statistical criteria, firstly non-parametric (for unknown probably distributions). As it was underlined fairly in [18], lack of harmonic parts into analysable data merits special check. The latter is carried out by filtering or correlation analysis.

3. Analysis of data structure and probabilistic nature

Analysis of data structure is study of random and systematic parts relation of either of the compared device error. This task is topical because its solution determines processing methods used for characteristics test or determination of MI under verification or calibration.

If the first two tasks are solved positively, the sample systematic part can be picked out without difficulty. If the systematic part is constant, data number for its determination as average (mathematical expectation) practically is quite enough. The situation is more complicated if it is supposed that the systematic part changes periodically; this way data must be enough for revealing a period of change. When periodicity has the strong theoretical foundation, data including one period is enough. Otherwise it has to make sure that change is periodical what requires several periods to be revealed. If the systematic part changes regularly but more complicated, it must be evaluated at the interval sufficient for change modelling. Nevertheless difficulties, if they are, have technological character, and relevant methods are well known.

Probabilistic model determination of a sample random part proves to be more complicated. The point is not that there are no methods for hypothesis test of random part distribution – there are enough and more [19]. As a rule, necessary sample volume turns out a stumbling point. In most practical cases sample volume is insufficient. Moreover plenty of hypothesis test methods permit this fact. If sample volume were sufficient for reliable evaluation of high order moments [20], then plotting bar chart of the distribution can be approximated by one of the standard distributions [21]. Presence of many statistical hypothesis test methods can be explained by striving to make up for the shortage of a priory information about distribution.

4. Formulation and solution of data processing task

Fourth task is formally a statistical estimation of error probabilistic characteristics. It comes apart at variants differing firstly by systematic nature of compared devices errors. Systematic error of every device is supposed being equal zero, constant or limited by given interval. Data processing task, herewith, comes apart at nine variants (see Table 1).

As to error random parts, they approximated by means of two most popular probability distribution laws – normal and uniform. These laws meet the kind of a priory information concerning random value dispersion: quadratic measure or possible limits. The laws are the worst one-modal in corresponding distribution class. In the case when a priory data presented by the limits into which can lie random part realisation, uniform law is considered as the worst among possible one-modal laws. If dispersion characteristic (variance) is known a priory, then normal law is recognised as the worst. Thus the following distribution combinations of MI and RS random errors are examined:

$$\begin{cases} \varphi_r(t) = \varphi_0(t) + \xi_r(t),\ \xi_r(t) \sim N(0, \sigma_r^2) \text{ with any } t = t_i; \\ \varphi(t) = \varphi_0(t) + \xi(t),\ \xi(t) \sim N(0, \sigma^2) \text{ with any } t = t_i, \end{cases} \quad (12)$$

$$\begin{cases} \varphi_r(t) = \varphi_0(t) + \xi_r(t),\ \xi_r(t) \sim N(0, \sigma_r^2) \text{ with any } t = t_i; \\ \varphi(t) = \varphi_0(t) + \xi(t),\ \xi(t) \sim U(-\Delta, \Delta) \text{ with any } t = t_i, \end{cases} \tag{12a}$$

$$\begin{cases} \varphi_r(t) = \varphi_0(t) + \xi_r(t),\ \xi_r(t) \sim U(-\Delta_r, \Delta_r) \text{ with any } t = t_i; \\ \varphi(t) = \varphi_0(t) + \xi(t),\ \xi(t) \sim N(0, \sigma^2) \text{ with any } t = t_i; \end{cases} \tag{12б}$$

$$\begin{cases} \varphi_r(t) = \varphi_0(t) + \xi_r(t),\ \xi_r(t) \sim U(-\Delta_r, \Delta_r) \text{ with any } t = t_i; \\ \varphi(t) = \varphi_0(t) + \xi(t),\ \xi(t) \sim U(-\Delta, \Delta) \text{ with any } t = t_i. \end{cases} \tag{12в}$$

Table 1. Data processing formal tasks.

№ п/п	RS sample parameter	MI sample parameter	MI parameter estimated	Note
1	$a_r = 0$	$a = 0$	σ	–
2	$a_r = 0$	$a = const$	a, σ	–
3	$a_r = 0$	$a \in [a_l, a_u]$	$a(j), \sigma$	$j = \overline{1, m}$ – sample number
4	$a_r = const$	$a = 0$	σ	–
5	$a_r = const$	$a = const$	a, σ	–
6	$a_r = const$	$a \in [a_l, a_u]$	a, σ	see № 3
7	$a_r \in [a_{r,l}, a_{r,u}]$	$a = 0$	σ	see № 3
8	$a_r \in [a_{r,l}, a_{r,u}$	$a = const$	a, σ	see № 3
9	$a_r \in [a_{r,l}, a_{r,u}$	$a \in [a_l, a_u]$	a, σ	see № 3

To get analytical solution taking into account MI study experience, and data nature and volume of trials, and the fact that any sample moment is asymptotically normal, it is expedient to use as the basic the stochastic model (12) with unbiased normal (Gaussian) distributions of both device errors, where σ_R – root mean square (RMS) of RS random error (known with definite accuracy); σ – RMS of verified MI random error (under check); $\varphi_0(t)$ – unknown measurand at the time moment t.

In general case, data processing remains the same independently of that comparison is carried out aiming at MI verification or calibration because the aims are achieved using the same processing results. The results that is actual accuracy characteristics estimates are used as such in the course of calibration, or are compared with norms for establishing whether MI is up (out of order) in the course of verification. But verification may be organised by check without estimating actual characteristics of MI under verification what is illustrated in following section.

Data processing in the course of verification by check

a) Let us consider the check task when RS and MI has no noticeable systematic errors: $a_r = a = 0$.

It is supposed that MI and RS readings are not correlated: $cov[\xi(t), \xi_r(t)] = 0$ for any $t = t_i$.

In this case the variance $D_x = D[x(t)]$ of readings difference $x(t) = \varphi(t) - \varphi_r(t)$ of two compared devices

$$D_x = D[\xi(t) - \xi_r(t)] = \sigma^2 + \sigma_r^2. \quad (13)$$

Expression (13) serves as a base of a simple criterion for examination whether actual (remained unknown) RMS of verified MI meets presented requirements. First of all, the case has to be excluded when the actual RMS of RS is known very precisely that is the actual RMS σ_r, its estimate $\tilde{\sigma}_r$, and the norm $\sigma_{r,max}$ tie practically and there is opportunity to examine directly RMS of verified MI on the base of (13). The stated case serves as a limit for any processing method: the closer to trivial case a measurement situation approaches the better a method is.

In general case, determining the sample variance $\widetilde{D}_x$ of differences $x_i = \varphi_i - \varphi_{r,i}$ of compared devices discrete readings:

$$\widetilde{D}_x = \Sigma_i (x_i - \bar{x})^2 / (n-1), \quad (14)$$

where $\bar{x} = n^{-1} \cdot \Sigma x_i$, and taking into account its closeness to the general population variance (13), we compare it with the sum of norms for the devices variances, which is taken as permissible upper limit of the estimate (14). If the specified normative limit is overrun:

$$\widetilde{D}_x > \sigma_{max}^2 + \sigma_{r,max}^2, \quad (15')$$

the tested MI is surely useless, because then it is uniquely $\sigma > \sigma_{max}$. But to confirm that MI is fit, the mentioned limit should be made tougher controlling whether the inequality (15″) holds:

$$\widetilde{D}_x \le \sigma_{max}^2 + M^2 \cdot \sigma_{r,max}^2, \quad (15'')$$

where the coefficient $M + \inf\{\frac{\sigma_r}{\sigma_{r,max}}\} < 1$ determines "dead band" of the criterion. The coefficient *M* must be introduced because generally the RS error variance σ_r^2 is less than its normative limit $\sigma_{r,max}^2$ on an unknown value. The more the value is the less may be *M*.

It is important that the criterion (15) in the form above is invariant to the kind of a MI error distribution, and in this regard it oversteps far the limits of the probabilistic model, for instance, (12). It ignores a possible random error of the estimate of the total variance (14), and it has the deterministic nature. As to an unaccounted random error of the estimate $\widetilde{D}_x$, then this error can be practically neglected provided that the sample is of sufficient volume $n = 50$-100.

b) An examination of a verified MI having a remarkable systematic error differs from the above in that at first a sample mathematical expectation of the difference series processed is compared with the norm for a MI systematic error. Then the procedures presented above (section a)) are carried out.

c) When a systematic part is only in the RS error then this part must be excluded from the difference series what brings the case to the section a).

If both compared devices have significant systematic errors then the above-mentioned method of examination cannot be realised. In that case a RS properties must be studied additionally for its systematic error evaluation what permits to come back to the section c).

Data processing in the course of a MI test

The more detailed criterion is based on a RMS determination as actual accuracy characteristic of the calibrated MI. To find a sample RMS estimate of tested MI and its variance the distribution moments in (13) have to be changed, and the following formula for calculation of an estimate σ can be determined from the converted expression:

$$\tilde{\sigma} = \sqrt{\widetilde{D}_x - \tilde{\sigma}_r^2}. \tag{15b}$$

By means of sample method a desired actual RMS can be represented only by an aggregate of statistical estimate and its accuracy characteristic.

Finding RMS of the calibrated MI is represented below as the sequential solution of the nine tasks specified in Table 1. Subject to above-mentioned and asymptotic normality of sample characteristics, a solution of the tasks comes to finding dispersion characteristics of the estimate (15b).

Task 1

Here MI and RS errors are supposed centered random values. In accordance with section 4, the task is separated into four variants which differ by MI and RS dispersion laws.

As only sample estimate of RMS can be found experimentally an accuracy of the estimate becomes of crucial importance. Really, to determine whether actual RMS meets the established norm it becomes necessary to determine the practically reliable upper limit of actual RMS with a foundation of the derived estimate. The confidence limit for a sample RMS by a confidence probability close to one may serve as the above-mentioned upper limit. To determine the confidence limit we need to estimate the RMS sample variance.

For finding the sample estimate of the tested MI we replace in (13) the distribution moments by their estimates, and get from the transformed expression the following formula for calculating the estimate σ:

$$\tilde{\sigma} = \sqrt{\tilde{D}_x - \tilde{\sigma}_r^2}. \tag{16}$$

If in the course of the parameter $\tilde{\sigma}_R^2$ estimation there is significant random error presented by its sample variance $D(\tilde{\sigma}_R^2)$ then taking into account the expression [15]

$$D(\tilde{\sigma}_r^2) = 4\sigma_r^2 D(\tilde{\sigma}_r), \tag{17}$$

we can get the variance estimate of the estimate (16) through*:

$$D(\tilde{\sigma}) = \frac{D(\tilde{D}_x)}{4\sigma^2} + \frac{D(\tilde{\sigma}_r^2)}{4\sigma^2} = \frac{D(\tilde{D}_x)}{4\sigma^2} + \frac{\sigma_r^2}{\sigma^2} D(\tilde{\sigma}_r). \tag{18}$$

The parameters $\sigma^2, \sigma_r^2, D_x$ are determined through their estimates $\tilde{\sigma}^2$, $\tilde{\sigma}_r^2$, $\tilde{D}_x$ on the base of data from quite different experiments what permits, similarly to (13), do not account "correlation item" by derivation of formulas (18).

Task 1.1.

When the compared devices readings have the normal distribution (model (12)) taking into account the well-known [15] estimate of the variance $D(\tilde{D}_x)$ of the sample variance (14)

$$D(\tilde{D}_x) = 2n^{-1}(D_x)^2, \tag{19}$$

the expression (18) is transformed to the form:

$$D(\tilde{\sigma}) = \frac{D_x^2}{2n\sigma^2} + \frac{1}{4\sigma^2} D(\tilde{\sigma}_r^2) = \frac{D_x^2}{2n\sigma^2} + \frac{\sigma_r^2}{\sigma^2} D(\tilde{\sigma}_r). \tag{20}$$

Coming back to verification and beginning with (16), (20), we arrive at the following verification criterion of the tested MI:

$$\bar{\sigma} \pm 3\sqrt{D(\tilde{\sigma})} \lesseqgtr \sigma_{max}, \tag{15'''}$$

where upper signs correspond to a suitable MI, and lower signs – unsuitable one. As to the confidence interval centre $\bar{\sigma}$, it is determined in the following way. If the sample $\{x_i\}$ can be divided into parts of sufficient volume then we can determine the average $\bar{\sigma} = m^{-1} \sum_j \tilde{\sigma}_j$, where j- sample part number ($j = \overline{1,m}$), at that $m \geq 10$. Otherwise $\bar{\sigma} = \tilde{\sigma}$ has to be taken.

---

*The formula (18) can be proved by different methods, particularly on the base of the estimate (16) expansion into bivariate Taylor series and general properties of variance. The general expression for such kind of estimates is given for instance in [22].

The criterion (15‴) is based on the fact that, on the assumption with the normal distribution law of the estimate $\tilde{\sigma}_R^2$, the estimate (16) is asymptotically normal and asymptotically unbiased accurate within an item $O(n^{-1})$.

The formula (20) analysis shows that the estimate (16) accuracy is restricted by increasing *n*:

$$D(\tilde{\sigma}) \geq \frac{\sigma_r^2}{\sigma^2} D(\tilde{\sigma}_r) \approx \frac{const}{\left(\frac{\sigma}{\sigma_r}\right)^2}. \tag{21}$$

This level is achieved practically (if take 1:10 the items relation in (20)) for the readings number

$$n_{max} = \frac{20(D_x)^2}{D(\tilde{\sigma}_r^2)} = \frac{5(D_x)^2}{\sigma_R^2 D(\tilde{\sigma}_r)}. \tag{22}$$

Therefore when the relation σ/σ_r is decreased, by fixed σ_r and $D(\tilde{\sigma}_r^2)$, the random errors level of the estimate (16) increase inevitably.

It is interesting to analyse the formula (20) by different relation its parameters. Thus with increase of the variance $D(\tilde{\sigma}_r^2)$ the variance $D(\tilde{\sigma})$ decreases at that roughly linearly in line with (21). At the same time if two compared devices RMS relation σ/σ_r increases then relative variance $D(\tilde{\sigma})/\sigma^2$ decreases as well remaining bounded from below by the value $(2n)^{-1}$.

As a variance estimate by expression (18) do not use directly data for the accuracy relation $k = \sigma/\sigma_r$ of compared devices we introduce normalisation:

$$\begin{cases} \varepsilon_r = \sqrt{D(\tilde{\sigma}_r)}/\sigma_r \\ \varepsilon = \sqrt{D(\tilde{\sigma})}/\sigma, \end{cases} \tag{23}$$

and transform on this base the formula (20) to the shape:

$$\varepsilon = \varepsilon(k, \varepsilon_r) = \frac{1}{k^2}\sqrt{\frac{1}{2n}(1+k^2)^2 + \varepsilon_r^2}. \tag{20'}$$

The expression (20′) represents a relative accuracy of the obtained RMS estimate (16) of a tested MI depending on the MI and RS accuracy relation *k* what permits to determine permissible limits for this relation.

The functions shown in Figures 1a, b, c give an equal introduction of the estimate ε behavior. On diagrams and below, ε_r and ε are denoted accordingly by ε_1 and ε_2.

It is useful to note that introducing relative accuracy figures $\varepsilon_r, \varepsilon$ $(\varepsilon_1, \varepsilon_2)$ simplifies the criterion (15‴): if the estimate (16) accuracy provided on the level $\varepsilon \leq \left|\frac{\sigma_{max}}{\sigma} - 1/3\right|$ then the tested MI is accepted as suitable / unsuitable depending on execution of the conditions $\tilde{\sigma} < \sigma_{max}$ or $\tilde{\sigma} \geq \sigma_{max}$. But if ε exceeds the mentioned level then the criterion is non-applicable because then $\tilde{\sigma}$ falls into its “dead zone” $\sigma_{max} \pm 3\sqrt{D(\tilde{\sigma})}$.

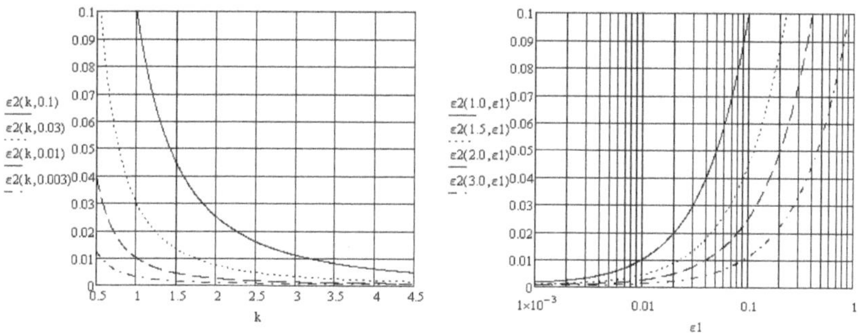

Figure 1a. The functions $\varepsilon_2(k)$ and $\varepsilon_2(\varepsilon_1)$ by $n=10^6$.

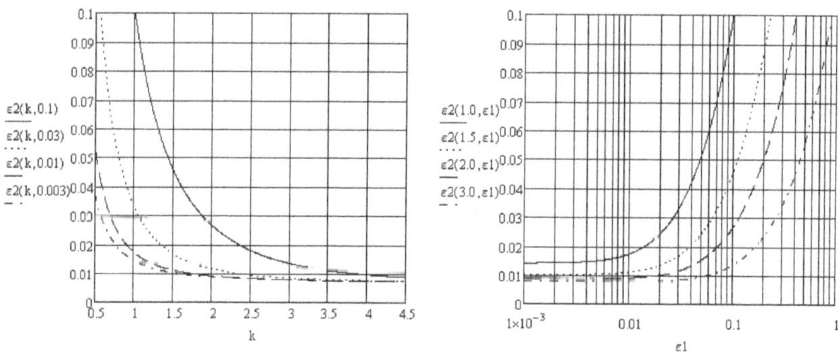

Figure 1b. The functions $\varepsilon_2(k)$ and $\varepsilon_2(\varepsilon_1)$ by $n=10^4$.

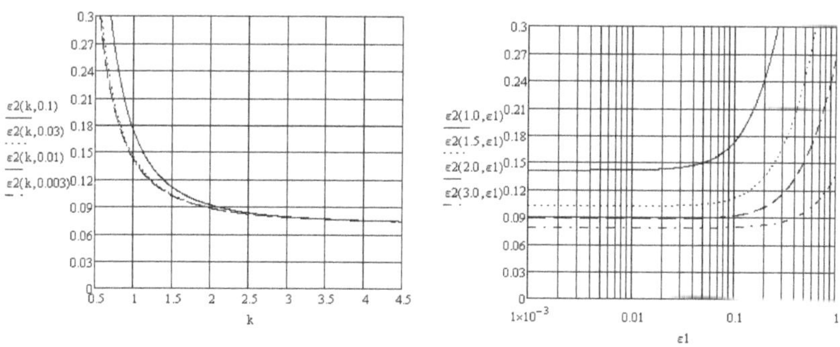

Figure 1c. The functions $\varepsilon_2(k)$ and $\varepsilon_2(\varepsilon_1)$ by $n=10^2$.

As to confidence limits of the tested MI full error, they are equal $\pm 3[\tilde{\sigma} + 3\sqrt{D(\tilde{\sigma})}]$, where $\tilde{\sigma}$ – the estimate (16), $D(\tilde{\sigma})$ – the estimate (20).

Therefore, for the above-mentioned a priori information (task 1 in Table 1), the task of data processing when data have got by MI comparison is solving by means of formulas (13)–(20) on the assumption with that Gaussian model (12) is true.

The obtained results particularly the functions shown in Figure 1 point at necessity of the RS treatment expansion. It emerges that quality of the measuring device as the RS is determined not only by its accuracy characteristic but by the characteristic estimate accuracy as well.

Task 1.2.

If it is no possibility to establish that the tested MI error has the Gaussian distribution and there is only some interval $[-\Delta, \Delta]$ which the error realization are in, then instead of model (12) it has to use model (12a), where Δ – the limit of the MI error module which is under evaluation and must not exceed the given normative error Δ_{max}. At that, for the used statistical methods uniformity, instead of the parameter Δ the equivalent RMS $\sigma = \Delta/\sqrt{3}$ of the uniform distribution is considering below where the norm is $\sigma_{max} = \Delta_{max}/\sqrt{3}$.

In this case expressions (13)–(20) hold but expressions (17)–(20) must be revised on the base of the mixture of normal and uniform distributions (12a). At that it is expedient, as a rule, to choose the limit σ_{max} as the RMS reliable limit of the tested MI error. However in the case when the actual limit of the RMS estimate $\tilde{\sigma}$ proved to be much less than the norm then its use instead of σ_{max}, as discussed above, is more reasonable.

The formula (17) is refined on the base of the general expression for the estimate (14) variance in the case of an arbitrary distribution:

$$D(\tilde{D}_x) = \frac{\mu_4 - \mu_2^2}{n} \tag{24}$$

where μ_k – k-th central moment ($k = 2, 4$) of distribution of the compared devices readings differences x_i. Under conditions (12a) the density of mentioned distribution is composition of normal and uniform distributions [15].

Calculating the density moments of the mentioned composition

$$\begin{cases} \mu_2 = \mu_2^{(r)} + \mu_2 = \sigma_r^2 + \sigma^2 \\ \mu_4 = \mu_4^{(r)} + \mu_4 + 6\mu_2^{(r)}\mu_2 = 3\sigma_r^4 + \frac{9}{5}\sigma^4 + 6\sigma_r^2\sigma^2 \end{cases} \tag{25}$$

and substituting them into (24), we will get the following formula (19) modification:

$$D(\widetilde{D}_x) = \frac{2}{n}[(D_x)^2 - \frac{3}{5}\sigma^4]. \qquad (19a)$$

Analysing this formula we have to take into account that within the framework of the model (12a) the actual estimate $\widetilde{D}_x$ variance can be significant less than the value obtained on the base of the model (12) (if accepted), especially by $\sigma > \sigma_r$. So by $\sigma/\sigma_r \approx 3$ the estimate (19a) is half of the estimate (19). Thus, a not well-founded choice of MI error models can lead to a faulty estimate of calibration results reliability.

It must be added to this that the RS choice is made before the comparison when instead of the actual RMS σ_r, σ we get their norms $\sigma_{r,max}, \sigma_{max}$ which we have to compare. It is thought that, due to the fact that the RS is studied carefully, its norm $\sigma_{r,max}$ is close to the actual characteristic σ_r, so $\frac{\sigma}{\sigma_r} \leq \frac{\sigma_{max}}{\sigma_{r,max}}$ what softens a little the note about estimates relation by incorrect model choice.

Finally, the following estimate of the estimate (16) variance runs from (19a): (Note: From Eq. (20a) to Eq. (22c), the numbering is not ordered as it follows the reasoning):

$$D(\tilde{\sigma}) = \frac{1}{2n\sigma^2}[(D_x)^2 - \frac{3}{5}\sigma^4 + \sigma_r^4], \qquad (20a)$$

from which we get the expression

$$D(\tilde{\sigma}) > \frac{\sigma_r^2}{\sigma^2} D(\tilde{\sigma}_r) - \frac{3}{10n}\sigma^2, \qquad (21a)$$

and

$$n_{max} = \frac{20(\widetilde{D}_x)^2}{D(\tilde{\sigma}_r^2)}[1 + \frac{3}{50}\frac{\sigma^4}{(\widetilde{D}_x)^2}]. \qquad (22a)$$

Introducing normalisation (23), we get similarly (20′)

$$\varepsilon = \varepsilon(k, \varepsilon_r) = \frac{1}{k^2}\sqrt{\frac{1}{2n}\left[(1 + k^2)^2 - \frac{3}{5}k^4\right] + \varepsilon_r^2}, \qquad (20a')$$

or, taking into account the RS normal distribution,

$$\varepsilon = \frac{1}{k^2}\sqrt{\frac{1}{2n}(1 + k^2)^2 - \frac{3}{5}k^4 + 1}. \qquad (20a'')$$

The confidence limits of the tested MI full error, as in Task 1.2, are equal $\pm\sqrt{3}[\tilde{\sigma} + \sqrt{D(\tilde{\sigma})}]$, where $\tilde{\sigma}, D(\tilde{\sigma})$ – estimates (16), (20a).

Task 1.3

In the case, when the MI error distribution is normal, and the RS error is specified by its limits, we get (12b), where $\sigma_r = \Delta_r/\sqrt{3}$ – is known; σ – is under estimate, instead of (12a).

As in 1.2, a revision of expressions (17)–(20) is based on the normal and uniform distributions (12b) composition.

Conducting transformations like above for model (12a), we get:

$$D(\widetilde{D}_x) = \frac{2}{n}[(\widetilde{D}_x)^2 - \frac{3}{5}\tilde{\sigma}_r^4] \quad \text{(19b)}$$

and accordingly the estimate of the estimate (16) variance

$$D(\tilde{\sigma}) = \frac{1}{2n\sigma^2}\left[(D_x)^2 - \frac{3}{5}\sigma_r^4\right] + \frac{\sigma_r^2}{\sigma^2}D(\tilde{\sigma}_r), \quad \text{(20b)}$$

from which

$$n_{max} = \frac{20(D_x)^2}{D(\tilde{\sigma}_r^2)}[1 + \frac{3}{50}\frac{\sigma_r^4}{(D_x)^2}].$$

The analogue of the expression (20a′) in this case is

$$\varepsilon = \frac{1}{k^2}\sqrt{\frac{1}{2n}\left[(1+k^2)^2 - \frac{3}{5}\right] + \varepsilon_r^2}. \quad \text{(20b′)}$$

The confidence limits of the tested MI full error are equal $\pm 3[\tilde{\sigma} + D(\tilde{\sigma})]$, where $\tilde{\sigma}, D(\tilde{\sigma})$ – estimates (16), (20б).

Task 1.4

If the both devices errors are specified by their limits, the model (12c) has to be used, where $\sigma_r = \Delta_r/\sqrt{3}$ – known; $\sigma = \Delta/\sqrt{3}$ – under estimation. Calculating the moments of the uniform distributions composition (triangular distribution) we get estimates:

$$D(\widetilde{D}_x) = \frac{4}{5n}[(D_x)^2 + 3\sigma^2\sigma_r^2], \quad \text{(19c)}$$

$$D(\tilde{\sigma}) = \frac{1}{5n\sigma^2}[(D_x)^2 + 3\sigma^2\sigma_r^2] + \frac{\sigma_r^2}{\sigma^2}D(\tilde{\sigma}_r), \quad \text{(20c)}$$

$$\varepsilon = \frac{1}{k^2}\sqrt{\frac{1}{5n}[(1+k^2)^2 + 3k^2] + \varepsilon_r^2}, \quad \text{(20c′)}$$

on the base of which we get the expression (21в) matching with (21), and

$$n_{max} = 8\frac{(D_x)^2}{D(\tilde{\sigma}_r^2)}[1 - \frac{3}{10}\frac{\sigma^2\sigma_r^2}{(D_x)^2}]. \quad \text{(22c)}$$

Confidence limits of the tested MI full error in this case are equal $\pm 3[\tilde{\sigma} + 3D(\tilde{\sigma})]$, where $\tilde{\sigma}, D(\tilde{\sigma})$ – estimates (16), (20c).

Task 1 is basic for all other tasks given in Table 1 because parameters evaluation methods of random error of MI calibrated by comparison with RS are developed in its framework. General approach to a solution of the other tasks lies in finding estimation methods of systematic parts of RS and MI errors after that found estimates are excluded from the sample. By “corrected” sample the dispersion parameters of error random parts are determined as it is made in Task 1.

Task 2

Here the RS is supposed having only random error ($a_r = 0$), and the calibrated MI – both random and systematic errors ($a = const$). It means the following Gaussian stochastic model is accepted:

$$\begin{cases} \varphi_r(t) = \varphi_0(t) + \xi_r(t),\ \xi_r(t) \sim N(0, \sigma_r^2) \text{ for any } t = t_i; \\ \varphi(t) = \varphi_0(t) + \xi(t),\ \xi(t) \sim N(a, \sigma^2) \text{ for any } t = t_i, \end{cases} \quad (12)$$

where σ_r – is known; a, σ – are evaluated on the assumption that the compared devices readings are uncorrelated: $cov[\xi_r(t), \xi(t)] = 0$ for any $t = t_i$. At that the mathematical expectation a is interesting for us, as a rule, not in itself but in respect of correct estimating the RMS σ and the confidence interval limits $[a \pm k(P) \cdot \sigma; P]$ of the calibrated MI full error alike.

The estimate $\tilde{a}$ of the constant mathematical expectation a can be found with necessary precision by the sample $\{x_i\}$ after that excluding $\tilde{a}$ from the sample we will come to Task 1. In this case the task solution comes to an end by determining limits of the above-mentioned "shifted" confidence interval where the coefficient $k(P)$ is under specification. The coefficient remains permanent in the tasks 1.1, 1.3 (normal distribution of ξ). In the tasks 1.2 and 1.4, $k(P) = \sqrt{3}P$.

Task 3

It is remarkable for the presence of only random error of the RS ($a_r = 0$), and both random and systematic errors of the calibrated MI: $a \in [a_l, a_{up}]$ (point 3, Table 1).

The task solution depends on the systematic part nature of the calibrated MI error. It is expedient to consider the following three variants.

Task 3.1: the permanent systematic error $a = const \in [a_l, a_{up}]$.
This case do not differ principally from the previous one unless consider the centre of the confidence interval as the random value distributed uniformly. The latter is hardly sensible in practice. It is more expedient to determine two intervals with the centres a_l and a_{up} or to confine oneself to the second as the largest.

Task 3.2: periodic systematic error $a = f(t), f(t_0) = f(t_0 \pm nT), T > 0$.
First of all, the error change period T is under determination. It can be done, for instance, by means of moving average. If, as it is usual in practice, T lies out (mostly - below) of the function $x(t)$ spectrum, after its determination the periodic part can be picked out from $x(t)$ whereupon Task 1 is under solution. In this case the confidence limits change periodically, and they have to be fixed by special rule. For instance, the widest limits can be accepted as final ones.

***Task 3.3**: so called *progressive* systematic error which increases in time, as a first approximation linearly $a = At$.

As in the previous case, this part can be determined and excluded from the sample $\{x_i\}$ whereupon Task 1 is solved. Here as above the rule for forming final confidence limits is under introduction. It is to take into account that the limits generally depend on the observation time.

Tasks 2, 3 contains in their formulations all elements which combinations lead to other tasks. Because of that data processing methods in Tasks 4-9 reduce them to formulation of Tasks 2, 3 and therefore to Task1.

Task 4

Here the RS error is centered random value, and the calibrated MI error is not centered. So, the conditions are symmetric to Task 2 concerning RS and MI. Because of that this task is solved as the latter.

Task 5

In that task errors of both compared devices are modelled by centered ($a_r \neq 0, a \neq 0$) random values. If the calibrated MI error variance (RMS) estimate is in mind then its determination comes down to solving Task 2 where firstly the difference $a - a_r$ estimate is found and extracted from the sample $\{x_i\}$ whereupon Task 1 is solved. To determine the MI error confidence interval, the MI has to be studied separately for finding a estimate. The alternative is the RS study (finding a_r estimate) and then determining $\tilde{a}$ from the mentioned difference.

Task 6.

This task comes apart on the variant similarly to Task 3 which algorithm can be used. It is possible to determine and extract sums: $a_r + a = const$; $const + f(t), f(t_0) = f(t_0 \pm nT)$; $const + At$. The RS preliminary study maybe is more practicable because it gives the estimate $\tilde{a}_r$ whereupon we get into Task 3 conditions.

Task 7

This task is symmetric to Task 3 concerning RS and calibrated MI: $a_r \in [a_{r,l}, a_{r,up}]$; $a = 0$. Accordingly, it falls into some sub-variants depending on the a_r behavior. After a_r estimate determination and extraction from the sample it remains to solve Task 1 where the confidence interval is not shifted.

Task 8

This task is symmetric to Task 6 concerning RS and calibrated MI: $a_r \in [a_{r,l}, a_{r,up}]$; $a = const$. Data processing can be done using algorithm analogous to Task 6, and the $\tilde{a}$ estimate can be found, if possible, directly from the calibrated MI readings.

Task 9

This task is most bulky among the tasks indicated in Table1 and so it has to be solved in the framework of the following nine variants.

Task 9.1: $a_r = const \in [a_{r,l}, a_{r,up}]; a = const \in [a_l, a_{up}]$.

These variant conditions accord with Task 5 which algorithm is to follow.

Task 9.2: $a_r = f(t), f(t_0) = f(t_0 \pm nT); a = const$.

The periodic systematic part a_r is picked out from the sample $\{x_i\}$ with relative ease. If the part has zero average so after its extraction from the sample, we fall into Task 2 conditions. But if the periodic part has not zero average then it has to be analysed additionally readings $\varphi(t)$ or $\varphi_r(t)$ to determine a.

Task 9.3: $a_r = At; a = const$

The task is similar to one of the Task 6 variants. Accordingly, the same data processing algorithm must be used.

Task 9.4: $a_r = const; a = f(t), f(t_0) = f(t_0 \pm nT)$.

The task is symmetric to Task 9.2 concerning RS and calibrated MI. But the same problem arises as in Tasks 3.2, 3.3 concerning the confidence interval construction for the MI error.

Task 9.5: $a_r = g(t), g(t_0) = g(t_0 \pm nS); a = f(t), f(t_0) = f(t_0 \pm nT)$.

The total systematic part $a_r + a$ is of course deterministic signal having beating habit. However, if there is no data on frequencies, amplitude and phase relations of signals $a_r = a_r(t), a = a(t)$ then picking out beats from their composition with random parts can become non-trivial problem. Because of that it is more practicable to study readings $\varphi(t)$ and $\varphi_r(t)$ aimed at determining at least periods S, T what simplify the problem cardinally.

Task 9.6: $a_r = At; a = f(t), f(t_0) = f(t_0 \pm nT)$.

Difficulty of such a kind as in Task 9.5, but of less level, attend solving this task. Moving average method permits to determine the incline A of the straight line $a_r(t)$. After that, extracting a_r from the sample, we can find and exclude the periodic part a. As a result, we get to conditions of Task 1 (periodic signal with zero average) or Task 2 (non-zero average of periodic part).

Task 9.7: $a_r = const; a = At$.

The task is symmetric to Task 9.3, and to one of variants of Task 8 or Task 6. So above-mentioned method can be used.

Task 9.8: $a_r = f(t), f(t_0) = f(t_0 \pm nT); a = At.$

The task is symmetric to Task 9.6 thus the above-mentioned statements on data processing methods stand.

Task 9.9: $a_r = Bt; a = At.$

Moving average method permits to determine and extract the total straight line $(A + B)t$ from the sample $\{x_i\}$ whereupon we get to Task 1 conditions. If the lines $a_r(t), a(t)$ are phase-shifted then data processing becomes a little complicated because of that polyline section with initial slopes A, B must be determined.

Solution of the tasks presented in Table 1 is ended by Task 9.3.

5. Conclusions

The formulation and solutions of the four basic problems of data processing by measuring devices comparison aimed at device verification and calibration are presented in the article. Necessity of experimental data quality check as necessary condition of using traditional data processing methods is shown. The data quality is treated as their ergodicity and stationarity.

The data processing problem is solved for nine specific tasks characterized by the systematic parts nature and random error distribution laws of the reference standard and calibrated measuring instrument.

It is shown that the compared devices errors relation can be decreased principally if assume the new, extended, interpretation of the reference standard . The latter has to be regarded as not only a device more precise than calibrated measuring instrument but the device with precisely known metrological properties.

References

1 Josef Bukac, Comparison of measuring devices / AStA 92008) 92: 217-227.

2 S. Dikstein, K. Katz, Y. Broun, D. Wilson, H. Maibach, Comparison of different instruments for measuring stratum corneum moisture content / Int. J. Cosmet. Sci., 1986, Dec; 8(6), 289-92

3 A. Momeni-Boroujeni, M.R. Picus, Systematic Error Detection in Laboratory Medicine / https://www.intechopen.com/books/quality-control-in-laboratory/systematic-error-detection-in-laboratory-medicine

4 L. Marais, S. Yang, B. Lam, and oth., Comparison of standards for the calibration of voltage, current and resistance meters / Metrologia, v. 55, Technical Supp., 2018

5 R.E. Beehler, R.C. Mockler, C.S. Snider, A Comparison of Atomic Beam Frequency Standards / Nature 187, 681-682 (1960)
6 Bureau International des Poids et Mesures. Key and supplementary comparisons / https://kcdb.bipm.org/AppendixB/KCDB_ApB_search.asp
7 https://en.wikipedia.org/wiki/Calibration
8 V.A. Granovskii, M.D. Kudryavtsev, Comparison of Inertial Heading Sensors: Reference Sensor with Zero Systematic Error / The First Intern/ Conf/ on Advances in Sensors, Actuators, Metering and Sensing (ALLSENSORS 2016), Apr 24-28, Venice, Italy
9 V.A. Granovskii, M.D. Kudryavtsev, Error ratio of a measuring instrument under calibration and the reference standard: conditions and possibilities of decrease / Advances in Intelligent Systems and Computing, vol. 543 "Recent Advances in Systems, Control and Information" / ed. Roman Szevczyk, Małgorzata Kaliczyńska // Springer, Warsaw, 2016
10 V.A. Granovskii, M.D. Kudryavtsev, Accuracy relation decreasing of a reference standard and a measuring instrument having considerable random and constant systematic errors (in Russian) / 6th Int. Conf. "Measurement and test in shipbuilding and adjacent branches" (SUDOMETRIKA-2016)» / Concern CSRI Elektropribor, St. Petersburg, 2016, 232-241
11 V.A. Granovskii, K.S. Gorokhovsky, Measuring instrument accuracy check: reference standard requirements and control risks / 1st Int. Conf. ICONMET 2017, Oct. 2017, Jakarta, Indonesia
12 V.A. Granovskii, K.S. Gorokhovsky, Accuracy proportion of a measuring instrument under calibration and a reference standard having both predominant random error / XXII IMEKO World Congress, Belfast, UK, Sept. 2018
13 V.A. Granovskii, M.D. Kudryavtsev, A.A. Untilov, Testing the INS Heading Channel by Comparing it to the Reference Channel with a Zero Systematic Error / GYROSCOPY AND NAVIGATION, Vol. 9 No. 1, 2018
15 H. Cramér, "Mathematical methods of statistics". Uppsala: Almqvist and Wiksells, 1945.
16 Yaglom A. M. Introduction to the theory of stationary random functions / UMN vol. 7, no. 5, 1952
17 Kosarev E. L. Methods of experimental data processing / Fizmatlit, Moscow, 2008.
18 Bendat J. S., Piersol A. G. Measurement and analysis of random data. John Wiley & Sons, Inc., 1967.
19 Lemeshko B. Yu. at al. Statistical analysis of data, modelling and study of probabilistic laws. Computer approach / NETI Publishing, Novosibirsk, 2011. (In Russian).

20 Feller W. An Introduction to Probability Theory and its Applications / John Wiley & Sons, Inc. NJ, 3d Ed., 1968.
21 Wilks S. S. Mathematical Statistics / Princeton Univ. Press, Princeton New Jersey, 1947.
22 Granovskii V. A., Siraya T. N. Experimental Data Processing Methods of Measurements / Energoatomizdat, Leningrad, 1990. (In Russian).

F Pavese, A B Forbes, N F Zhang, A G Chunovkina (Eds.):
Advanced Mathematical and Computational Tools in Metrology and Testing XII
Series on Advances in Mathematics for Applied Sciences, Vol. 90

https://doi.org/10.1142/9789811242380_0014

Discrete wavelet transform on uncertain data: Efficient online implementation for practical applications

M. Gruber[1], T. Dorst[2], A. Schütze,[2,3] S. Eichstädt[1] and C. Elster[1 1]

[1]*Physikalisch-Technische Bundesanstalt, Braunschweig/Berlin, Germany*
[2]*ZeMA - Center for Mechatronics and Automation Technology gGmbH, Saarbruecken, Germany*
[3]*Lab for Measurement Technology, Department of Mechatronics, Saarland University, Saarbruecken, Germany*

The discrete wavelet transform has a huge number of applications in science and engineering. It allows to extract features and obtain information often hidden in raw signals. For applications in metrology the propagation of uncertainties for this transform is relevant. By using and extending existing techniques for propagating uncertainties in the context of discrete filtering, we propose an efficient implementation of the discrete wavelet transform that enables the online propagation of uncertainties. The proposed approach is illustrated by its application to simulated data and to data from an experimental testbed.

Keywords: discrete wavelet transform; uncertainty propagation; online filter methods

1. Introduction

The discrete wavelet transform (DWT) is a well established tool within the digital signal processing community[1,2]. Typical application scenarios include signal filtering and smoothing[3], dimensionality reduction[4], audio analysis[5], image compression[6], feature extraction for machine learning (ML)[7] and condition monitoring[8].

State of the art implementations are based on the lifting scheme[9] or a filterbank structure[10]. The latter one is based on finite impulse response

(FIR) filters and therefore suitable for efficient online implementation. This enables feature extraction and data reduction with low latency.

Within the field of metrology, the evaluation of uncertainties is of major importance[11]. Uncertainty propagation for the DWT between input signals with time-varying input covariance and its output have been analyzed in depth[12,13], and algorithms for uncertainty propagation in digital filters have been proposed[14,15]. However, to the best of our knowledge, approaches for the online-capable propagation of the uncertainty associated with the input signal through a DWT are not available yet.

In this paper we use and extend discrete filter techniques to develop an efficient implementation of the discrete wavelet transform that enables the online propagation of uncertainties. We illustrate the developed technique first by using simulated data. Then the proposed approach is applied to data of a testbed for electromechanical cylinders[16]. Within this application, a subset of calculated DWT coefficients forms the input to an ML classification method which estimates the remaining useful lifetime (RUL) of a component in a production line[7]. It is of interest to evaluate the influence of sensor uncertainties in the result of a linear discriminant analysis (LDA) classification.

The paper is organized as follows. In section 2 the online capable uncertainty propagation for the DWT is developed. The approach is then applied to simulated data and to data from a testbed in section 3. Finally, we present some conclusions and an outlook.

2. Methods

In general, a wavelet transform (WT) consists of a wavelet analysis that denotes the transformation from time to time-frequency domain and the wavelet synthesis that denotes the back transformation. DWT uses discretely sampled wavelets to operate on time-discrete signals.

In this paper, we aim at calculating a multi-level DWT of a scalar signal with given time-dependent standard uncertainty. The signal is known at equidistant points in time and becomes available in consecutive blocks of variable length, which is often referred to as a data stream. The uncertainty associated with the DWT result should be propagated following the

Guide to the expression of uncertainty in measurement (GUM)[17], to indicate the quality of a (processed) measurement consistent with international metrological guidelines.

The DWT can be implemented very efficiently using a filter bank structure[10]. The proposed multi-level DWT algorithm builds on a single-level DWT method, which itself is based on a discrete filter method. Online uncertainty propagation will be enabled at the filter level by largely following an existing approach[15].

2.1. *Online Discrete Filter*

At the lowest implementation level, a discrete filter is applied to an equidistant sequence of the input signal $\underline{\chi_i} \in \mathbb{R}^m$ with associated standard uncertainty values $\underline{u}_{\chi_i} \in \mathbb{R}^m$. We assume that no correlations are present in the input signal. The elements in $\underline{\chi}_i$ and $\underline{u}_{\chi_i}$ are given by equations (1) and (2).

$$\underline{\chi}_i = [\chi[i-m+1], \ldots, \chi[i-1], \chi[i]]^T \tag{1}$$

$$\underline{u}_{\chi_i} = [u_\chi[i-m+1], \ldots, u_\chi[i-1], u_\chi[i]]^T \tag{2}$$

The algorithm uses incremental update schemes from discrete filter theory[14,15] to minimize necessary computing operations. The latter citation already provides equations that are suitable for online implementation, but additional efforts are made to use their full potential.

To build on previous work, every DWT filter $\underline{\phi} \in \mathbb{R}^n$ with compact support (an FIR) can be rewritten as an infinite impulse response (IIR) filter characterized by its coefficients $\underline{b} = \underline{\phi}$ and $\underline{a} = [1.0]$. This filter can then be transformed into the state-space system equation (3) with matrices $A \in \mathbb{R}^{n \times n}$, $B \in \mathbb{R}^{n \times 1}$, $C \in \mathbb{R}^{1 \times n}$ and $D \in \mathbb{R}^{1 \times 1}$, see[18]. The output $\underline{\psi} \in \mathbb{R}^m$ of the filter is given by evaluating equation (4) iteratively.

$$\underline{z}[j+1] = A\underline{z}[j] + B\chi[j] \tag{3}$$

$$\psi[j] = C\underline{z}[j] + D\chi[j] \tag{4}$$

The output uncertainty $\underline{u}_\psi \in \mathbb{R}^m$ is evaluated using equation 12 in[15]. This requires updating the state $\underline{z}$, the state's sensitivity $\frac{\partial z}{\partial t}$ and the covariance of the state P_z using matrix-vector-expressions. These terms form the internal state $\hat{s} = \left\{\underline{z}, \frac{\partial z}{\partial t}, P_z\right\}$ of the filter method.

Algorithm 2.1 Online IIR-Filter

procedure IIR_FILTER($\underline{\chi}_i, \underline{u}_{\chi_i}, \underline{b}, \underline{a}, \hat{s}$)
  if $\hat{s}$ not given **then**
    calculate $\underline{z}_0$ from equation (5)
    calculate $\frac{\partial \underline{z}_0}{\partial t}$ from equation (6)
    calculate P_0 from equation (7)
    $\hat{s} \leftarrow \left\{\underline{z}_0, \frac{\partial \underline{z}_0}{\partial t}, P_0\right\}$
  for j in $i, \ldots, i+m-1$ **do**
    update $\hat{s}$
    calculate $\psi[j], u_\psi[j]$ from equation (4)
  return $\underline{\psi}, \underline{u}_\psi, \hat{s}$

The internal state $\hat{s}$ represents an abstract history of the processed input signal, allows to calculate the filter response to a given signal history and enables online capabilities by efficient incremental updates. The state can either be set to the previous last value or initialized by solving the linear equations (5) and (6) and the discrete Lyapunov equation (7). The equations can be derived by assuming stationarity in equations 6, 17 and 18 in[15] (i.e. set $\underline{z}[n] = \underline{z}[n+1] = \underline{z}_0$ and solve for a steady state $\underline{z}_0$). The implementation is presented in algorithm 2.1.

$$\underline{z}_0 = (I_A - A)^{-1} * B * \chi[0] \tag{5}$$

$$\frac{\partial \underline{z}_0}{\partial t} = (I_A - A)^{-1} * \frac{\partial A}{\partial \underline{\theta}} * \underline{z}_0 \tag{6}$$

$$P_0 = AP_0A^T + BB^T * u_\chi[0]^2 \tag{7}$$

2.2. *Single-level DWT*

In the intermediate implementation step a single level of the DWT is performed by using the previously described online discrete filter method. This is done by applying a wavelet high-pass $\underline{h} \in \mathbb{R}^n$ and a corresponding wavelet low-pass $\underline{g} \in \mathbb{R}^n$ to the incoming signal $\underline{\chi}_i, \underline{u}_{\chi_i} \in \mathbb{R}^m$. The vectors $\underline{h}$ and $\underline{g}$ define the coefficients of FIR filters. The internal states of both filter operations $\tilde{s}_{low}$ and $\tilde{s}_{high}$ are used and returned as $\tilde{s}$:

$$\tilde{s} = \{\tilde{s}_{low}, \tilde{s}_{high}\} \tag{8}$$

Algorithm 2.2 Single-Level DWT

procedure DWT($\underline{\chi}_i, \underline{u}_{\chi_i}, \underline{g}, \underline{h}, \tilde{s}, l_{off}$)
    $\tilde{s}_{low}, \tilde{s}_{high} \leftarrow \tilde{s}$
    $\underline{a}, \underline{u}_a, \tilde{s}_{low} \leftarrow$ IIR_FILTER($\underline{\chi}_i, \underline{u}_{\chi_i}, \underline{g}, 1.0, \tilde{s}_{low}$)
    $\underline{d}, \underline{u}_d, \tilde{s}_{high} \leftarrow$ IIR_FILTER($\underline{\chi}_i, \underline{u}_{\chi_i}, \underline{h}, 1.0, \tilde{s}_{high}$)
    subsample $\underline{a}, \underline{u}_a, \underline{d}, \underline{u}_d$ by 2 with offset l_{off}
    $\tilde{s} \leftarrow \{\tilde{s}_{low}, \tilde{s}_{high}\}$
    return $\underline{a}, \underline{u}_a, \underline{d}, \underline{u}_d, \tilde{s}$

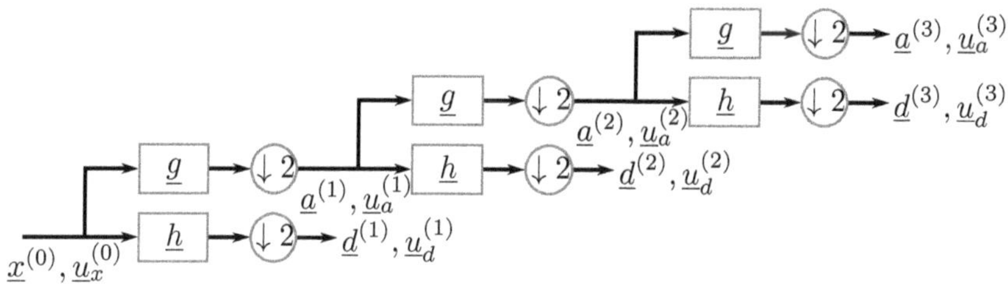

Figure 1. Three level analysis filter bank structure for the DWT implementation of an input signal $x^{(0)}$.

The output of both filter operations is then subsampled by a factor of two, yielding the detail coefficients $\underline{d}, \underline{u}_d \in \mathbb{R}^{\lfloor \frac{m}{2} \rfloor}$ and approximation coefficients $\underline{a}, \underline{u}_a \in \mathbb{R}^{\lfloor \frac{m}{2} \rfloor}$. The implementation is presented in algorithm 2.2.

2.3. *Multi-level DWT*

A multi-level DWT of a signal is performed by connecting multiple single-level DWTs, known as the filterbank method[10]. As a consequence of the data stream assumption, there is no theoretical limit on the maximum reachable DWT level n_{max}, but a user defined limit n.

The input of the first DWT level is the currently available block $\underline{x}^{(0)}, \underline{u}_x^{(0)} \in \mathbb{R}^m$ of the time-discrete signal $x^{(0)}[i]$ with uncertainty $u_x^{(0)}[i]$. The sequence of blocks is established by passing through the approximation coefficients $\underline{a}^{(l)}$, $\underline{u}_a^{(l)}$ of the l-th level to the input of the $(l+1)$-th DWT level. This results in the filter bank structure shown in figure 1.

The result c, u_c of the multi-level DWT is then formed by the detail coefficients of all levels and the approximation coefficient of the highest level:

$$c = \left\{\underline{a}^{(n)}, \underline{d}^{(n)}, \ldots, \underline{d}^{(2)}, \underline{d}^{(1)}\right\} \tag{9}$$

$$u_c = \left\{\underline{u}_a^{(n)}, \underline{u}_d^{(n)}, \ldots, \underline{u}_d^{(2)}, \underline{u}_d^{(1)}\right\} \tag{10}$$

The implemented algorithm can be applied to incoming blocks of arbitrary length m. A basic book-keeping scheme is used to track which coefficients at which level of the DWT can be calculated. The book-keeping is realized using the counter i_0 at the zeroth level (updated after every block using equation (11)) and the current wavelet level l. The counter i_l for the output coefficients of the l-th level is then calculated using equation (12). The subsampling at the l-th level returns only values, for which i_{l-1} is uneven (implemented using an offset l_{off}). If some level returns no coefficients, no filter application is triggered in the higher levels.

$$i_{0,updated} = (i_0 + m) \mod 2^n \tag{11}$$

$$i_l = \left\lfloor \frac{i_0}{2^l} \right\rfloor \tag{12}$$

The internal states $s^{(l)}$ with $l \in \{1, \ldots, n\}$ of every single-level DWT and the global index i_0 are cached in s:

$$s = \left\{s^{(1)}, \ldots, s^{(n)}, i_0\right\} \tag{13}$$

The index scheme is visualized for $n = 3$ in figure 2. With the exemplary input block lengths given in this figure the second incoming block (length 3) allows to calculate only one coefficient at the first level, whereas the fourth block (length 2) allows to calculate one coefficient at the first level and one coefficient at the second level. The implementation is presented in algorithm 2.3.

3. Results

The implemented algorithm is applied in two different scenarios. In both cases a Daubechies-4 (D4) wavelet is considered for the DWT as an example. Application of the proposed method for other wavelets is possible in the same way.

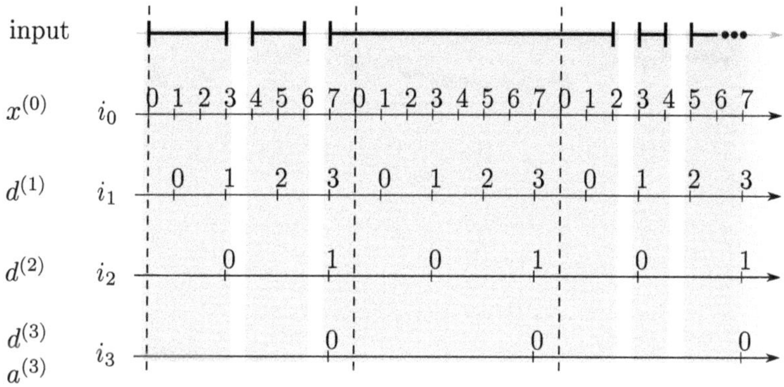

Figure 2. Index scheme for multi-level DWT

Algorithm 2.3 Multi-Level DWT

procedure DWT_MULTI($\underline{x}^{(0)}, \underline{u}_x^{(0)}, \underline{g}, \underline{h}, n, s_{prev}$)
  if s_{prev} is given **then**
    $s \leftarrow s_{prev}$
  $\underline{a}^{(0)}, \underline{u}_a^{(0)} \leftarrow \underline{x}^{(0)}, \underline{u}_x^{(0)}$
  for l in $1, \ldots, n$ **do**
    set i_n from equation (12)
    set l_{off}
    $\underline{a}^{(l)}, \underline{u}_a^{(l)}, \underline{d}^{(l)}, \underline{u}_d^{(l)}, s^{(l)} \leftarrow$ DWT($\underline{a}^{(l-1)}, \underline{u}_a^{(l-1)}, \underline{g}, \underline{h}, s^{(l)}, l_{off}$)
    update $s^{(l)}$ in s
  update i_0 in s using equation (11)
  construct c, u_c based on equations (9) and (10)
  return c, u_c, s

The D4 wavelet consists of a low-pass filter $\underline{g} \in \mathbb{R}^4$ and a corresponding high-pass filter $\underline{h} \in \mathbb{R}^4$. The filter coefficients can be found in the initial publication by Daubechies[19] and are available within the Python package PyWavelets[20].

3.1. *Application to Simulated Data*

Figure 3 shows a single frame of a raw data signal $x^{(0)}$ with its corresponding time-dependent uncertainties $u_x^{(0)}$. The signal is used as input to an

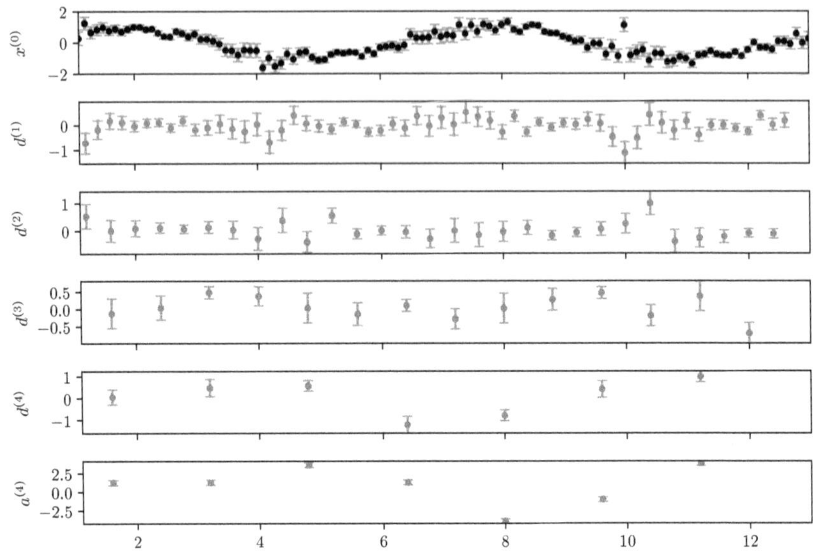

Figure 3. DWT output visualization of the implemented online method with uncertainties

online DWT with D4 wavelet of length four. The result of the DWT is visualized by the detail coefficients $d^{(l)}$ for corresponding l-th level and the approximation coefficients $a^{(4)}$ of the highest wavelet level which is chosen as $n = 4$. The uncertainties of the coefficients are shown by error bars.

3.2. *Application to Experimental Data*

Predictive maintenance, i.e. for example estimating the RUL of critical components in production lines, can help to reduce repair costs and machine downtime. The RUL estimation of an electromechanical cylinder (EMC) with a spindle drive is considered here. To this end, long-term speed driving and high load tests were carried out with a testbed until a position error of the EMC occurs due to increased friction. During the test, the EMC was loaded axially by a pneumatic cylinder that is constantly pulling on the EMC with 7 kN. A typical working cycle lasts 2.8 s and consists of a forward stroke, a waiting time (150 ms) and a return stroke.

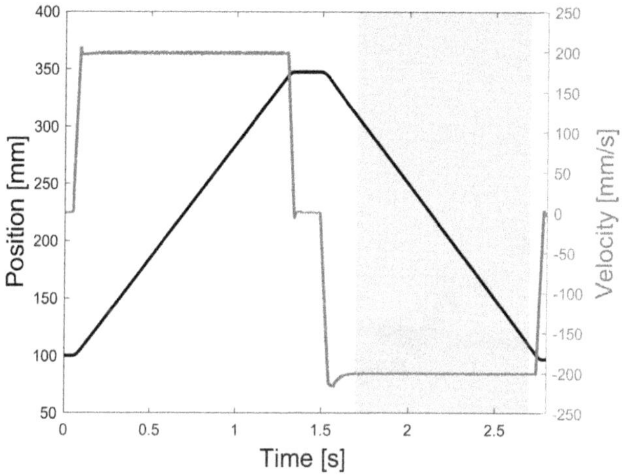

Figure 4. Position (black) and velocity (gray) of the cylinder during one working cycle, only the last second (gray area) of each cycle is considered for estimation of RUL.

The eleven included sensors in the testbed are:

- one microphone with a sampling rate of 100 kHz,
- three accelerometers with 100 kHz sampling rate, attached at the ball bearing, at the plain bearing, and at the piston rod,
- four process sensors (active current, axial force, pneumatic pressure, and velocity) with 10 kHz sampling rate each,
- three electrical motor current sensors with 1 MHz sampling rate each.

The collected data represents the functionality of the EMC and its degradation over time. A typical lifetime of an EMC under these test conditions is approx. 629000 cycles (20.4 days) and generates 12 TB of raw data. The dataset[16] of "axis 3" used in this contribution is resampled to 2 kHz for all sensors and only one second at constant velocity recorded during the return stroke of every 100th cycle is included (cf. figure 4). The reduced data set used to test the proposed algorithm therefore consists of 6292 cycles with a total of 1.03 GB.

An LDA with Pearson feature selection implemented using Scikit-learn[21] was calculated to discriminate ten classes representing the RUL percentages

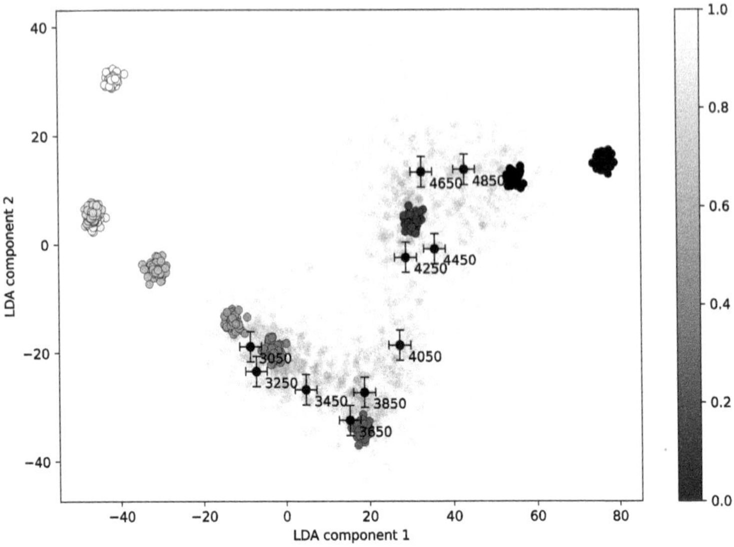

Figure 5. Projection of the online DWT into the LDA components (light gray markers in the background), selected projections show the propagated uncertainty (black), RUL of training data shown by clusters of descending color

1 %, 11 %, . . . , 91 % of the total lifetime. The LDA input is formed by the wavelet decomposition of the incoming sensor data followed by Pearson selection. The data used for determining the LDA model are displayed in figure 5 using a color scale to indicate the RUL. The trained model was then applied to the continuously calculated DWT after each cycle. Figure 5 shows a projection of the result of an online DWT for cycles 3000 to 5000 (corresponding to 52.3 % to 20.5 % RUL) projected into the first and second components of the model. The uncertainty of every 200th projection as determined by the online algorithm is indicated additionally starting at cycle 3050. The propagation of the calculated uncertainties associated with the DWT coefficients through the model was performed via matrix multiplication with uncertainty propagation according to GUM supplement 2 [22].

The propagated uncertainties in figure 5 are relatively small in comparison to the general prediction capabilities of the LDA model, as indicated by the scatter of the projections for individual cycles. Therefore, the inherent uncertainty coming from the sensors has only small influence on the classification outcome in this case.

4. Conclusion and Outlook

The discrete wavelet transform is becoming increasingly important for applications in metrology. Examples comprise the Industrial Internet of Things where it is applied as a preprocessing tool for subsequent machine learning approaches. It is crucial that these methods - including the preprocessing - can be applied online, run on low-powered hardware and enable metrological data processing by the propagation of uncertainties.

In this paper a calculation of the discrete wavelet transform has been developed that includes the online-capable propagation of uncertainties. The approach can help extending applications of the discrete wavelet transform within metrology to cases where online-capability is relevant. The proposed algorithm is implemented in Python 3 and will be published within the package PyDynamic[23] starting from release 2.0.0.

5. Acknowledgments

Part of this work has received funding within the project 17IND12 Met4FoF from the EMPIR program co-financed by the participating states and from the European Union's Horizon 2020 research and innovation program.

Part of this work has received funding within the project FAMOUS by the Federal Ministry of Education and Research (BMBF).

References

1. I. Daubechies, Orthonormal bases of compactly supported wavelets, *Communications on Pure and Applied Mathematics* **41**, 909 (1988).
2. F. Truchetet and O. Laligant, Wavelets in industrial applications: A review, in *Proceedings of SPIE - The International Society for Optical Engineering*, 2004.
3. A. Abbate, J. Koay, J. Frankel, S. Schroeder and P. Das, Signal detection and noise suppression using a wavelet transform signal processor: Application to ultrasonic flaw detection, *IEEE Transactions on Ultrasonics, Ferroelectrics, and Frequency Control* **44**, 14 (1997).

4. Y. Qu, B.-l. Adam, M. Thornquist, J. D. Potter, M. L. Thompson, Y. Yasui, J. Davis, P. F. Schellhammer, L. Cazares, M. Clements, G. L. Wright and Z. Feng, Data Reduction Using a Discrete Wavelet Transform in Discriminant Analysis of Very High Dimensionality Data, *Biometrics* **59**, 143 (2003).
5. R. Kronland-Martinet, The Wavelet Transform for Analysis, Synthesis, and Processing of Speech and Music Sounds, *Computer Music Journal* **12**, 11 (1988).
6. S. Mallat, A theory for multiresolution signal decomposition: The wavelet representation, *IEEE Transactions on Pattern Analysis and Machine Intelligence* **11**, 674 (1989).
7. T. Schneider, N. Helwig and A. Schütze, Industrial condition monitoring with smart sensors using automated feature extraction and selection, *Meas. Sci. Technol.* **29** (2018).
8. Z. K. Peng and F. L. Chu, Application of the wavelet transform in machine condition monitoring and fault diagnostics: A review with bibliography, *Mechanical Systems and Signal Processing* **18**, 199 (2004).
9. W. Sweldens, The lifting scheme: A construction of second generation wavelets (1995).
10. M. Vetterli and C. Herley, Wavelets and Filter Banks: Theory and Design, *Signal Processing, IEEE Transactions on* **40**, 2207 (1992).
11. BIPM, IEC, IFCC, ILAC, ISO, IUPAC, IUPAP and OIML, International vocabulary of metrology - Basic and general concepts and associated terms (VIM) (2012).
12. L. Peretto, R. Sasdelli and R. Tinarelli, Uncertainty propagation in the discrete-time wavelet transform, *IEEE Transactions on Instrumentation and Measurement* **54**, 2474 (2005).
13. S. Cambanis and C. Houdre, On the continuous wavelet transform of second-order random processes, *IEEE Transactions on Information Theory* **41**, 628 (1995).
14. C. Elster and A. Link, Uncertainty evaluation for dynamic measurements modelled by a linear time-invariant system, *Metrologia* **45**, 464 (2008).
15. A. Link and C. Elster, Uncertainty evaluation for IIR (infinite impulse response) filtering using a state-space approach, *Meas. Sci. Technol.* **20** (2009).
16. T. Dorst, Sensor data set of 3 electromechanical cylinder at ZeMA testbed (2kHz) (2019).

17. BIPM, IEC, IFCC, ILAC, ISO, IUPAC, IUPAP and OIML, Guide to the expression of uncertainty in measurement (2008).
18. J. D. Aplevich, *The Essentials of Linear State-Space Systems* (Wiley, New York, 2000).
19. I. Daubechies, *Ten Lectures on Wavelets* (Society for Industrial and Applied Mathematics, 1992).
20. G. R. Lee, R. Gommers, F. Waselewski, K. Wohlfahrt and A. O'Leary, PyWavelets: A Python package for wavelet analysis, *Journal of Open Source Software* **4**, p. 1237 (2019).
21. F. Pedregosa, G. Varoquaux, A. Gramfort, V. Michel, B. Thirion, O. Grisel, M. Blondel, P. Prettenhofer, R. Weiss, V. Dubourg, J. Vanderplas, A. Passos, D. Cournapeau, M. Brucher, M. Perrot and É. Duchesnay, Scikit-learn: Machine Learning in Python, *Journal of Machine Learning Research* **12**, 2825 (2011).
22. BIPM, IEC, IFCC, ILAC, ISO, IUPAC, IUPAP and OIML, *Supplement 2 to the "Guide to the Expression of Uncertainty in Measurement" – Extension to Any Number of Output Quantities* 2011.
23. S. Eichstädt, C. Elster, I. M. Smith and T. J. Esward, Evaluation of dynamic measurement uncertainty – an open-source software package to bridge theory and practice, *Journal of Sensors and Sensor Systems* **6**, 97 (2017).

F Pavese, A B Forbes, N F Zhang, A G Chunovkina (Eds.):
Advanced Mathematical and Computational Tools in Metrology and Testing XII
Series on Advances in Mathematics for Applied Sciences, Vol. 90

https://doi.org/10.1142/9789811242380_0015

Digital representation of measurement uncertainty for metrological traceability

B. D. Hall* and D. R. White

Measurement Standards Laboratory of New Zealand
Lower Hutt, 5010, New Zealand
**E-mail: blair.hall@measurement.govt.nz*

Widespread use of digital systems provides an opportunity to improve the quality of information reported for traceable measurements. By using digital technology to pass information along a traceability chain, the effects of common influences can be rigorously accounted for, so the provenance of any influence becomes traceable. This can be achieved by using a simple form of chain rule for differentiation to evaluate and propagate components of uncertainty. The method can achieve the goal of internal consistency identified in the *Guide to the expression of uncertainty in measurement*, thereby improving on current common practice.

1. Introduction

While modern digital systems strive for autonomy, national quality infrastructures still rely on experts with sheets of paper to deliver metrologically traceable measurements. Trends in technology suggest that digital systems will soon be capable of managing traceable-measurement chains, from the realisation of SI units through to application-specific measurements in the field. However, to match—or preferably to exceed—the performance of current processes, adequate digital representations of metrological concepts must be found. A growing number of initiatives would benefit from this. There are plans to produce an entirely digital, machine-readable, SI brochure;[a] machine-readable formats for calibration reports are in development [1]; and a secure cloud-based platform for the legal metrology infrastructure in Europe is proposed [2]. These projects intend to replace paper-based, expert-oriented processes with automated digital ones. In this article we consider the representation of measurement uncertainty, which is essential for traceable measurement.

[a]CIPM/108-28 in: https://www.bipm.org/en/committees/cipm/meeting/108(II).html

The information about uncertainty passed along a traceability chain is intended to satisfy principles enunciated in the *Guide to the expression of uncertainty in measurement* (GUM), but the GUM was written long before the modern scenario of a 'digital world' was envisaged. So, while the principles remain sound, some of the implementation details should be reviewed. In particular, the GUM explains that an ideal method of evaluating uncertainty will be both *internally consistent* and *transferable*[3] §0.4. However, these goals are not always achieved today.

The GUM describes transferability as the possibility to

> ... use directly the uncertainty evaluated for one result as a component in evaluating the uncertainty of another measurement in which the first result is used.

This goal is satisfied by the current practice of propagating a single summary value of the uncertainty at reporting stages along a traceability chain. However, the goal of internal consistency cannot be satisfied this way, because information about the provenance of different contributions to uncertainty is not propagated. The GUM describes internal consistency with reference to components that contribute to uncertainty and, in particular, by requiring that the combined uncertainty, at the end of a measurement, should be[3] §0.4

> ... independent of how these components are grouped and of the decomposition of the components into subcomponents.

We look more closely at this requirement in the next section. It is clearly intended that, between the realisation of SI units and the final measurement stage, the occurrence of intermediate reporting stages should not affect the uncertainty of the final result. However, more information about the factors influencing a measurement must be passed along traceability chains to consistently deliver both internal consistency and transferability.

The next section looks at how the GUM methodology should be applied when a traceable measurement involves a number of intermediate reporting stages. If the final results are to be independent of the specific choice of decomposition into stages, then a simple chain rule must be used to propagate components of uncertainty through the whole calculation. Digital systems can implement a method that is both internally consistent and transferable by using this chain rule. In section 3, we discuss this finding and consider its implications. In section 4 we present our conclusions.

Throughout the text, upper-case letters are used in expressions for quantities that are not precisely known while lower-case letters are used for definite values. So, for example, a measured value y is a definite value which provides an estimate of some quantity Y and $u(y)$ is the standard uncertainty of y as an estimate of Y.

2. Two views of a measurement scenario

The methodology of uncertainty analysis described in the GUM requires a measurement model, which is a mathematical description of the measurement that identifies the physical quantities and processes in play when numerical results are obtained; the model accounts for the variability of outcomes from a measurement procedure under reproducibility conditions.

In this section we consider two alternative model-based descriptions of a measurement: one with intermediate stages and one without. For the results of a GUM analysis to produce the same final uncertainty in each case, it will become apparent that information about uncertainty budgets must be passed from one stage to the next; the common practice of propagating a single summary value of uncertainty from stage to stage is not sufficient.

In the measurement scenario we consider, the intention is to use a calibrated instrument to measure a quantity of interest, Y. Two reference standards, for quantities Y_1 and Y_2, are used to calibrate the instrument. The characterisation of these references can be modelled individually:[b]

$$Y_1 = f_1(\cdots) \tag{1}$$

$$Y_2 = f_2(\cdots)\,. \tag{2}$$

Then, in the next stage, these reference standards are used to determine an instrument calibration factor

$$Y_3 = f_3(Y_1, Y_2, \cdots)\,. \tag{3}$$

Finally, the calibrated instrument is used to measure Y, which involves the instrument calibration factor and other unspecified arguments,

$$Y = f_4(Y_3, \cdots)\,. \tag{4}$$

In this way, the whole measurement procedure is partitioned into stages, as shown schematically in figure 1. However, the procedure can also be

[b]The unspecified function arguments '$\cdots$' are intended to imply quantities that influence the respective measurements; they may be external influences or intermediate results. The letter X will be used specifically for external influences and the letter Y will be used to indicate intermediate and final results.

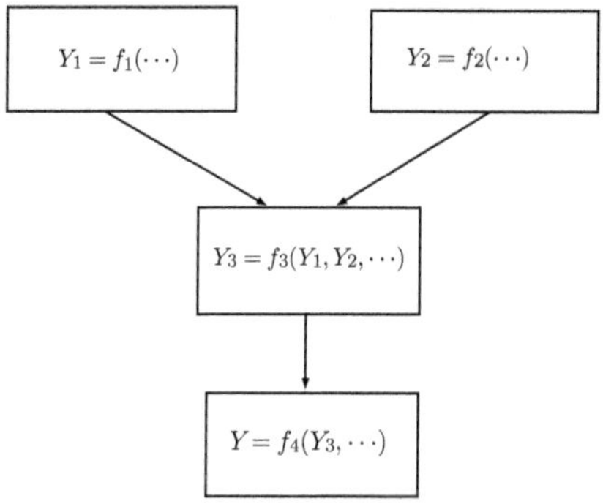

Fig. 1. Measurement of Y in four stages. Arrows indicate information transfer about results. The unspecified function arguments '$\cdots$' represent external quantities that influence the procedures.

described in a single model without stages

$$Y = f(X_1, X_2, \cdots) \,. \tag{5}$$

The intermediate results (Y_1, Y_2 and Y_3) generated by equations (1)—(4) are not among the arguments to (5); the $X_1, X_2, \cdots$ are all external influences that affect the realisation of the SI units, the determination of reference standard values, the determination of the instrument calibration factor, and the determination of a measured value for Y using the instrument.

These two model descriptions represent the same measurement procedure, so they are equivalent; i.e.,

$$f(X_1, X_2, \cdots) \equiv f_4(f_3(f_1(\cdots), f_2(\cdots), \cdots), \cdots) \,. \tag{6}$$

Data processing applied to either will produce the same measured value, y. However, the final measurement uncertainty $u(y)$ can depend on the model structure. If single summary values of uncertainty are passed between intermediate stages, then the value obtained for $u(y)$ can depend on the partitioning of the model into stages. This will be the case, for instance, if some stages have influences in common, which will cause intermediate results to be correlated. Before looking at this in more detail, we briefly review the GUM methodology for uncertainty analysis.

2.1. *GUM analysis of measurement uncertainty*

For a measurement model of the form

$$Y = f(X_1, X_2, \cdots) \,, \tag{7}$$

and with a measured value of y for Y, the component of uncertainty in y due to uncertainty in x_i, a measured value or estimate of X_i, is

$$u_i(y) = \frac{\partial f}{\partial X_i} u(x_i) , \tag{8}$$

where $u(x_i)$ is the standard uncertainty of x_i. The combined standard uncertainty in y is then

$$u(y) = \sqrt{\sum_i \sum_j u_i(y)\, r(x_i, x_j)\, u_j(y)} , \tag{9}$$

where $r(x_i, x_j)$ is the correlation coefficient between x_i and x_j.[c]

A number of degrees of freedom, ν_i, is associated with each $u(x_i)$. When input estimates are uncorrelated (independent), the degrees of freedom for the combined standard uncertainty is[d]

$$\nu_y = \frac{u^4(y)}{\sum_i \frac{u_i^4(y)}{\nu_i}} . \tag{10}$$

2.2. *Components of uncertainty for composed models*

When a measurement model is composed of stages, as is equation (6), components of uncertainty in equation (8) can be evaluated by the chain rule for partial differentiation. Suppose the model function at stage h is

$$y_h = f_h(\Lambda_h) , \tag{11}$$

where Λ_h denotes a set of direct function arguments, which may include prior intermediate results. The component of uncertainty in y_h due to uncertainty in x_i is

$$u_i(y_h) = \sum_{Y_k \in \Lambda_h} \frac{\partial f_h}{\partial Y_k} u_i(y_k) , \tag{12}$$

where $u_i(y_k)$ is the component of uncertainty in y_k, due to uncertainty in x_i. The term y_k is a direct input to f_h.

To propagate uncertainty information through the model, equation (12) can be applied sequentially to the stages. At each stage, the components

[c]Often, inputs are assumed to be independent, in which case (9) becomes the root-sum-squared of components.

[d]The GUM does not offer guidance on the evaluation of ν_y when any inputs are correlated.

of uncertainty of direct inputs are weighted by the sensitivity of the stage-function to those inputs. For example, the component of uncertainty of the intermediate result $y_3 = f_3(y_1, y_2, \cdots)$, due to uncertainty in x_i, is

$$u_i(y_3) = \frac{\partial f_3}{\partial Y_1} u_i(y_1) + \frac{\partial f_3}{\partial Y_2} u_i(y_2) + \cdots , \tag{13}$$

where $u_i(y_1)$ is the component of uncertainty in y_1 due to uncertainty in x_i, etc.

So, if the equivalence expressed in equation (6) is to hold, results produced at intermediate stages must include the components of uncertainty due to the uncertainty in external influence factors (the X_i). However, this is not current practice. Instead of using a model of the complete traceability chain, each reporting stage is usually taken to be the measurement model. For example, if this were done to our model with intermediate reporting stages, equations (9) and (10) would be applied to f_1, f_2, f_3 and f_4, producing a value of combined standard uncertainty and a number of degrees of freedom at each stage. The direct inputs to the function at each stage would be used to determine the components of uncertainty in (9) and (10), rather than external influences corresponding to $X_1, X_2, \cdots$ in (5).[e] The final calculation of combined standard uncertainty may then depend on how the measurement is partitioned into stages.

Current practice may be thought of as making the simplifying assumption that all external influence factors and intermediate results are independent. When that assumption is correct, expressions like (13) for the components of uncertainty reduce to single terms, but in more general situations the components of uncertainty account for common influences.[f] For instance, if X_i is a common influence on Y_1 and Y_2 in our example, then both $u_i(y_1)$ and $u_i(y_2)$ contribute to the value of $u_i(y_3)$ in (13). A conventional analysis would only handle this correctly if a correlation coefficient $r(y_1, y_2)$ were also reported at the third stage.

[e]The notation we have adopted helps to make the distinction clearer. We have explicitly stated that the $X_1, X_2, \cdots$ are external influences. However, if following current practice, the model of a measurement stage would be used in (7) and in that case the function arguments are a mixture of Xs and Ys in our notation.

[f]Readers will find more details relating to uncertainty calculations involving common influences in the examples given in reference[7].

2.3. *Maintaining internal consistency*

To illustrate how common influences easily arise when we look at the whole traceability chain as a measurement, we can extend our scenario to consider a batch of instruments calibrated with a single pair of reference standards (figure 2). If reference values, y_1 and y_2, are determined during the initial stages of the process, then the instrument calibration constants, $y_3[1], y_3[2], \cdots$, and any subsequent measurements using the calibrated instruments, are affected by those reference values: the residual errors in those numbers have a systematic effect on the overall measurement process.

Measurement procedures sometimes compensate for systematic measurement errors[g] by taking differences (or ratios) of intermediate results. When this is done, a model of the measured difference (or ratio) has to combine two instances of the stages shown in figure 2—one for each measurement—and correctly identify the common and the independent influences. Such a model will depend on whether the same instrument was used to make both measurements or whether two different instruments were used. Although the scenario is now more complicated, the stage functions

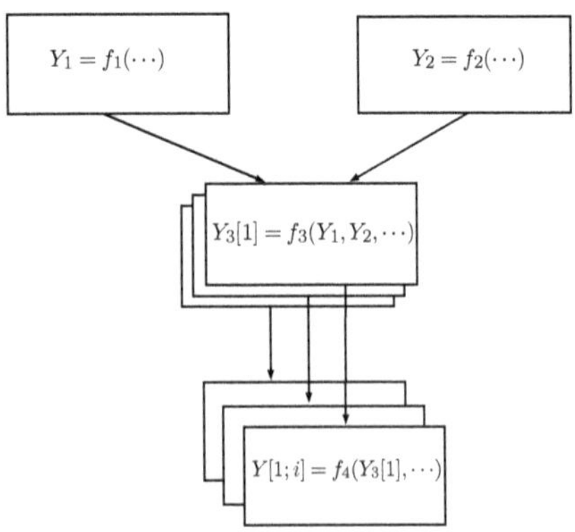

Fig. 2. Several instruments are calibrated with one pair of reference standards. The instrument calibration constants are indicated by $Y_3[j]$, where j identifies an instrument. $Y[j;i]$ represents the ith measurement with instrument j. Repeat measurements using the same instrument will be correlated, because the calibration constant is fixed; measurements with different instruments will be correlated too, because the reference standard values used during their calibration were fixed.

have not changed; the complexity lies in identifying the relevant terms and

[g]Understood to be a residual error of unknown value after corrections have been applied.

the sequencing of stages in the process. The procedure described in §2.2 to evaluate components of uncertainty would still apply. Digital systems can handle this type of computational complexity without difficulty.

3. Discussion

We take the view that a measurement starts with realisations of SI units and proceeds, through any number of stages, to a final determination of the measurand. This is surely not an unreasonable interpretation of metrological traceability, but it is not common to look at the evaluation of measurement uncertainty from this perspective. To do so highlights a significant difference between results obtained at intermediate stages of the process and the final result. Intermediate results relate to quantities and processes that make up the complete measurement model and they must be handled in a way that maintains the integrity of that model. In contrast, the information ultimately provided by a measurement will be used to make an inference about the measurand, usually to inform a decision-making process. This step is not actually part of the measurement model; it is a statistical problem that may be tailored to meet the end-user's requirements.

The distinction between intermediate and final results is important. A recent study identified a problem of transferability when a Monte Carlo (MC) method was used to propagate intermediate results through a staged measurement process[4]. At each stage, the model incorporated new influences, characterised by type-A evaluations of uncertainty. The final results were found to be unsatisfactory, and the problem was attributed to an improper prior distribution used in the Bayesian formulation of the MC calculation at intermediate stages. However, in our opinion, the application of a Bayesian inference method to the propagation of uncertainty at the intermediate stages was inappropriate, because it altered the internal structure of the measurement model. As a result, the model used for computation no longer represented the behaviour of measurement processes, and this led to the unsatisfactory outcomes in the study.

Internal consistency becomes important when common influences and systematic effects play a role in a measurement process. Experienced metrologists may deal with such situations in *ad hoc* ways, such as by differential and ratiometric measurements, but more detailed and rigorous reporting along a traceability chain could extend those benefits along multiple stages in a measurement process and be handled automatically. The approach has already shown its merit in the field of radio and microwave frequency metrology, where systematic effects must be considered: best-practice now mandates methods that can account for correlations due to systematic effects and special software is available to do this[5]. In situations involving calibration standards, systematic effects often arise. For example, the calibration of a radiation thermometer using a blackbody as a transfer medium will be influenced by the emissivity, but if two radiation thermometers have nearly the same spectral responsivity, then errors due to blackbody emissivity cancel. In photometry, recent work has shown that goniometric measurement systems require careful attention to shared influences[6]. In environmental and industrial settings, large networks of low-cost sensors need to account for systematic effects arising from batch manufacturing and perhaps batch calibrations.[h]

The main reporting requirements for a system that implements our proposal are summarised in §3.1. The additional effort is comparable to that of reporting conventional uncertainty budgets, which are currently prepared during uncertainty analysis but not reported. Digital systems could easily handle this. Indeed, there is software available which is based on the notion of an *uncertain number*—an entity that encapsulates the information being communicated between stages[8–10]. The twin goals of transferability and internal consistency can be achieved by reporting uncertain numbers at each stage. The uncertain-number approach is effective, and convenient,[i] but a conventional uncertainty analysis could also be adapted to provide the required components of uncertainty with little difficulty, because the sensitivity coefficients identified in (12) are already required at each stage.

Uncertainty calculation when common factors influence intermediate measurement stages has also been considered by Willink[11]. He proposes a solution with, as explained in[11] §2.2, an emphasis *"... on the representation of uncertainty information in a form that is particularly suitable for written communication"*. Willink was concerned about the amount of data a person tasked with calculating uncertainty might need to handle (large sparsely-populated correlation coefficient matrices) and proposed a method to reduce that burden. Digital systems have effective ways of storing and manipulating data and may use more direct methods[9].

---

[h]Reference[7] discusses the details of calculations involved for some simple sensor network scenarios.

[i]One advantage of uncertain-number software is that the algorithms automate the evaluation of components of uncertainty, so explicit routines for the various partial derivatives do not need to be developed.

3.1. *Recommendations*

The following are minimum reporting requirements at each stage to satisfy the requirements for transferability and internal consistency.[j]:

- the result (y_h, in §2.2),
- the components of uncertainty ($u_i(y_h)$, in §2.2),
- any non-trivial correlation coefficients ($r(x_i, x_j)$, in §2.1),
- the degrees of freedom (ν_i, in §2.1).

Also, the rather severe restriction imposed by the GUM on the Welch-Satterthwaite formula (our equation (10)) may be eased by identifying groups of correlated influence quantity estimates with finite degrees of freedom[9]. The degrees of freedom can then be calculated for other important cases, such as the type-A evaluation of a complex quantity or the parameters obtained by a least-squares regression.

Digital systems can use globally unique identifiers (UID) to play the role of the subscripts i and j in equations of §2.1 and §2.2, which identify the external influences in terms for the degrees of freedom, the components of uncertainty and the correlation coefficients. These UIDs must uniquely identify instances over time and space, because the different stages of a traceable measurement generally occur in different laboratories[9].

4. Conclusions

The ubiquity and data handling capability of digital systems provides an opportunity to improve the quality of information reported as the results of traceable measurements. The goals of transferability and internal consistency, identified in the GUM, but not compatible with current practice, can be realised with little extra effort. Information obtained at intermediate stages can be propagated to provide more detail about the factors influencing a measurement at the end of a traceability chain. Ultimately, this more detailed and accurate information can be made available to inform end-user decisions. The approach allows common influences in measurement chains to be rigorously accounted for and, because tracking of influences can be continued during any subsequent data analysis, the traceability can

[j]This information will be reported between stages along the traceability chain, as occurs already with paper-based reporting: the information is needed for the next stage of data processing, or by the client to make an inference about the result. However, it is not necessary to make this information more widely available.

extend to post-processing, where measurement results may be compared or combined.

Acknowldgement

This work was funded by the New Zealand government. The authors thank P Saunders for a careful review of the manuscript and the anonymous reviewer for insightful and helpful comments.

References

1. S. Hackel, F. Härtig, J. Hornig and T. Wiedenhöfer, *PTB-Mitteilungen* **127**, 75 (2017).
2. F. Thiel, *OIML Bulletin* **LIX**, 10 (2018).
3. BIPM, IEC, IFCC, ISO, IUPAC, IUPAP and OIML, *Evaluation of measurement data Guide to the expression of uncertainty in measurement JCGM 100:2008 (GUM 1995 with minor corrections)*, 1st edn. (BIPM Joint Committee for Guides in Metrology, Paris, Sèvres, 2008).
4. G. Wübbeler and C. Elster, *Metrologia* **57**, p. 015005 (2020).
5. M. Zeier, J. Hoffmann, P. Hürlimann, J. Rüfenacht, D. Stalder and M. Wollensack, *Metrologia* **55**, p. S23 (2018).
6. E. Molloy, P. Saunders and A. Koo, Effects of rotation errors on goniometric measurements, *In preparation.*
7. B. D. Hall, An opportunity to enhance the value of metrological traceability in digital systems, in *2019 II Workshop on Metrology for Industry 4.0 and IoT (MetroInd4.0&IoT)*, 2019.
8. B. D. Hall, *Metrologia* **43**, p. L56 (2006).
9. B. D. Hall, The GUM Tree Calculator: a Python package for measurement modelling and evaluating measurement uncertainty, *In preparation.*
10. M. Zeier, J. Hoffmann and M. Wollensack, *Metrologia* **49**, p. 809 (2012).
11. R. Willink, *Metrologia* **46**, p. 145 (2009).

F Pavese, A B Forbes, N F Zhang, A G Chunovkina (Eds.):
Advanced Mathematical and Computational Tools in Metrology and Testing XII
Series on Advances in Mathematics for Applied Sciences, Vol. 90

https://doi.org/10.1142/9789811242380_0016

Software representation of measured physical quantities

B. D. Hall

Measurement Standards Laboratory of New Zealand,
Lower Hutt, 5010, New Zealand
E-mail: blair.hall@measurement.govt.nz

Relationships between physical quantities are often expressed as mathematical equations, but evaluation of these equations is subject to restrictions related to the quantities involved and the scales of measurement. Calculations will be prone to error if quantity values are assigned to standard numeric data types in digital systems. To address this, the digital representation of quantities must capture the semantics of quantity calculations. The paper shows that simple dimensional analysis must be used to identify all quantities of interest in specific problems. This provides sufficient information for a framework to monitor computations and detect illegal operations. The approach can monitor calculations with so-called 'dimensionless' quantities and could also allow systems to resolve issues related to incompatible measurement units.

1. Introduction

When a relationship between physical quantities is expressed in a mathematical equation, the semantics go beyond mere computation. P W Bridgman, in *The Nature of Physical Theory*[1] p 72, put it this way:

> "The equations always have to be accompanied by a 'text' telling what the significance of the equations is and how to use them"

and he went on to warn that

> "The necessity for a text is almost always overlooked, but I think it must be recognised as essential."

For example, readers will recognise $F = m \cdot a$ and $I = V/R$ as Newton's second law and Ohm's law, respectively. The mathematical simplicity of these equations does not reflect the skill required to apply the laws correctly in situations of practical importance; to know exactly what F, m and a correspond to in a mechanical problem, or V, I and R in the analysis of an electrical network. Tacit knowledge, beyond computational competency, is needed to interpret information about a problem—this is the 'text' referred to by Bridgeman, and the ability to understand and apply it.

Calculations performed on digital systems are prone to error when values of physical quantities are assigned to standard numeric data types. Systems designed to evaluate numerical expressions are ill-equipped to identify incorrect operations for quantities. There is no provision, for example, to verify that a numeric substituted for 'm' in Newton's law corresponds to a measure of mass on an appropriate scale (unit). The knowledge that a skilled person would use to validate a calculation is not available to a computer, so contextual information is lost.

A growing number of initiatives would benefit from clarity in the requirements for representations of physical quantities and increased support for detecting errors during calculations in digital systems. There are plans to publish an entirely digital, machine-readable, SI brochure;[a] to develop a machine-readable format for calibration reports[2]; and to build a secure cloud-based platform for the legal metrology infrastructure in Europe[3]. Such projects intend to transform paper-based, person-oriented processes into digital forms. However, the ubiquity of skilful people working in today's quality systems complicates this task. If these digital transformations are to be successful, every effort should be made to identify situations where tacit knowledge is used and find satisfactory ways of encoding it. The representation of physical quantities is one aspect of this problem and its importance is almost certainly underestimated: Bridgman's warning should be heeded.

One useful step appears to be standardising the representations of measurement units. With support for unit systems, software can monitor the dimensional homogeneity of equations and detect some types of error[4–7]. However, important cases still slip through: not all quantities can be uniquely identified (e.g., torque and work are indistinguishable in the International System of units (SI)); ratios of quantities of the same kind are indistinguishable ("dimensionless" quantities); and some distinct quantities in a particular context will be indistinguishable if they are not recognised within the system of units (e.g., lengths measured along perpendicular directions).

So, unit-checking is not a panacea. Part of the problem is that unit-checking algorithms are constrained to use the conventional base dimensions for a unit system. In fact, an appropriate choice of dimensions is needed.

[a]CIPM/108-28 in: https://www.bipm.org/en/committees/cipm/meeting/108(II).html

Some of the difficulties identified here can be resolved if base dimensions more closely related to a problem, and the laws pertaining to it, are used instead. That is the approach proposed in this paper. The next section will briefly review aspects of dimensional analysis and quantity calculus, which are required in section 3, where we propose an encoding of dimensional information that allows digital systems to check the correctness of computations. We discuss the proposal in more detail in section 4 before presenting conclusions in section 5.

2. Background

2.1. *Dimensions*

The distinction between a quantity and a dimension is not always clear[8]. There is a tendency to use the term 'quantity' loosely, when sometimes 'dimension' would be more appropriate. In this work the notion of 'dimension' is important.

Following Ellis[9], we take a dimension to be *a class of similar measurement scales* for a particular quantity.[b] For example, the metre, foot and furlong all belong to the same dimension, because they are length scales related to each other by simple scale factors. On the other hand, measurements of the relative radiant flux transmitted through a material on a linear scale (transmittance) and on a logarithmic scale (optical density) belong to different dimensions. Note, we consider that a ratio like transmittance can also have a dimension. This seems compatible with the views expressed by Ellis[9], and by Johansson—who coined the term 'unit-less quantity' for such cases[10]; but in the SI, quantity ratios, such as transmittance, are referred to as 'dimensionless' quantities or as quantities of 'dimension unity'[11].

2.2. *Dimensional analysis*

In dimensional analysis, it is assumed that the relationship between measurements of a quantity P and measurements of other quantities $Q, R, S, \cdots$ can be expressed in the form

$$P = c \cdot Q^{\alpha} \cdot R^{\beta} \cdot S^{\gamma} \cdot \ldots \,, \tag{1}$$

[b]The word 'similar' here relates to a similarity transform: a multiplicative factor can be used to convert a value on one scale to a corresponding value on another scale.

where c depends on the scales chosen for $Q, R, S, \cdots$.[c] The dimensional exponents, α, β, γ, etc, are typically integers but sometimes assume rational values.[d] They do not depend on the scales used.

As Ellis has explained[9], when physical laws are expressed as unit-invariant equations, the terms for quantities in the expressions correspond to classes of similar scales (dimensions). This is why physical laws hold true when units change and indeed dimensional analysis would not be possible without this convention.

It is important in this work that when relation (1) holds, the dimension of P can be identified by the pattern of dimensional exponents for $Q, R, S, \cdots$ in an ordered sequence. For instance, if exponents are associated with dimensions for: mass, length and time, then $(0, 1, -1)$ identifies the dimension for velocity ($v = \mathrm{d}x/\mathrm{d}t$) and $(1, 1, -2)$ identifies the dimension for force ($F = m \cdot \mathrm{d}^2x/\mathrm{d}t^2$).

2.3. *Quantity calculus*

Although all calculations involving physical quantities do ultimately depend on the scales used to produce numerics for computation, quantity equations express relationships between terms corresponding to dimensions, not scales (units). The rules that govern such computations necessarily depend on both dimensions and scales but the rules for dimensions take precedence.

Addition and subtraction is permitted only when terms belong to the same dimension and then only when the numerics are represented on the same scale. Multiplication and division is permissible among terms with different dimensions, but the constant c, which provides equality between the left and right hand sides of equation (1), will depend on the associated scales.

In multiplication, the exponents associated with common dimensions in the arguments are added, and in division the exponents are subtracted. For

[c] Our earlier expression of Newton's and Ohm's laws takes this form but omits the dimensional constant c. In fact, $c = 1$ for coherent systems of units, like the SI. Computations often need to account for units by introducing a non-trivial value for this factor.

[d] Fourier introduced dimensional exponents in his *La Théorie Analytique de la Chaleur*[12] as a means of determining the scale factor needed to maintain equality when measurement scales were changed in an expression such as (1). For example, if a length measurement appears on the right raised to the power of two, then a change of measurement scale from centimetres to metres, say, requires the scale factor to be multiplied by 10^4.

instance, if quantities X and Y depend on lengths, L_X and L_Y, such that

$$X = c_X \cdot Q^\alpha \cdot L_X^{\beta_X}$$
$$Y = c_Y \cdot L_Y^{\beta_Y} \cdot S^\gamma ,$$

then the exponents for length in the product are added ($\beta_{X \cdot Y} = \beta_X + \beta_Y$)

$$X \cdot Y = c_X \cdot c_Y \cdot Q^\alpha \cdot L^{\beta_{X \cdot Y}} \cdot S^\gamma .$$

To evaluate a value for $L^{\beta_{X \cdot Y}}$, the units of length in L_X and L_Y should be the same.

It is important to note that individual contributions from L_X and L_Y become indistinguishable in the product. If the physical characters of those lengths contribute to the problem in fundamentally different ways, we might consider the alternative expression

$$X \cdot Y = c_X \cdot c_Y \cdot Q_X^{\alpha_X} \cdot L_X^{\beta_X} \cdot L_Y^{\beta_Y} \cdot S_Y^{\gamma_Y} ,$$

which retains the distinction between L_X and L_Y. This would also facilitate the use of different units for measures of L_X and L_Y. For example, if L_X and L_Y are measured along perpendicular directions, it might be sensible to retain separate dimensions for each.[e]

3. Dimensional signatures

In the context of a particular problem, we assume that base dimensions can be chosen so that every quantity of interest is uniquely identified by a 'dimensional signature'. This signature is the pattern of dimensional exponents in a pair of sequences, Θ^N and Θ^D.

To form a set of signatures for a context, base dimensions are chosen and then other quantities of interest can be declared as products and quotients of those base quantities. Signatures for derived dimensions are obtained by operating on the base-quantity signatures. The usual calculus for dimensional exponents is applied independently in Θ^N and Θ^D and a special form of division operation can be used to create signatures for ratio quantities (see §3.1).

When checking computations, the dimensional signatures of results must match those declared in context. We can illustrate this in a simple kinematical context.[f] With base dimensions: mass, length, and time interval, we

[e]The separation (gauge) between railway tracks and a measure of distance along the track is often cited as an example.

[f]Only the exponents in Θ^N will be shown if those in Θ^D are all zero, otherwise the sequences will be separated by ':' to denote a ratio.

can declare momentum, with signature $(1,1,-1)$ and force, with signature $(1,1,-2)$. Then, the product of a force with a time interval produces a result with the signature $(1,1,-1)$, which matches momentum. However, if a force were divided by a time interval, the signature $(1,1,-3)$ would be obtained, which cannot be matched. We expect intermediate computational states will occur during some computations that do not match dimensions declared in context. However, a final result must always match a declared dimension. So, if $(1,1,-3)$ was indeed a final result, there must have been a mistake in the computation.

In the same context, it is possible to declare a dimension for mass ratio, with signature $(1,0,0) : (1,0,0)$. During a calculation, the product of a mass and a mass-ratio will produce a signature $(2,0,0) : (1,0,0)$, because the calculus for dimensional exponents is applied independently to the sequences Θ^N and Θ^D. This signature cannot be matched. However, in appropriate circumstances, ratio signatures can be explicitly simplified. Doing so treats the exponents in Θ^N and Θ^D as the numerator and denominator, respectively, of a ratio of dimensions. So, simplifying $(2,0,0) : (1,0,0)$ produces $(1,0,0)$, which matches mass.

3.1. *Summary of signature operations*

The rules for manipulating dimensional signatures are summarised below. Each signature $\Theta = \Theta^N : \Theta^D$ consists of a pair of ordered sequences, Θ^N and Θ^D, in which the position of elements relates to base dimensions. By writing $\Theta_{\mathrm{l}}^N + \Theta_{\mathrm{r}}^N$, it is intended that the result is a new ordered sequence obtained by element-wise addition of the exponents, and similarly for subtraction.

The extended division operation, denoted $//$, mixes the exponents in the sequences Θ^N and Θ^D from the two arguments, which allows signatures for quantity ratios to be created. Simplification allows a signature with exponents in Θ^N and Θ^D to be reduced to a sequence with exponents in Θ^N only.

Multiplication: $\Theta = \Theta_{\mathrm{l}} \times \Theta_{\mathrm{r}}$

$$\Theta^N = \Theta_{\mathrm{l}}^N + \Theta_{\mathrm{r}}^N$$
$$\Theta^D = \Theta_{\mathrm{l}}^D + \Theta_{\mathrm{r}}^D$$

Division: $\Theta = \Theta_l / \Theta_r$

$$\Theta^N = \Theta_l^N - \Theta_r^N$$
$$\Theta^D = \Theta_l^D - \Theta_r^D$$

Extended division: $\Theta = \Theta_l // \Theta_r$

$$\Theta^N = \Theta_l^N + \Theta_r^D$$
$$\Theta^D = \Theta_l^D + \Theta_r^N$$

Simplification: $\Theta = \text{simplify}(\Theta_r)$

$$\Theta^N = \Theta_r^N - \Theta_r^D$$
$$\Theta^D = (0, \cdots)$$

3.2. *Other examples*

This section briefly presents three more contexts. The intent is to show how the proposal can be used in simple cases. More detail about these examples is available in a recent paper [13] and in documentation that accompanies a software implementation of the proposed method written in Python [14].

3.2.1. *Electrical dimensions*

With base dimensions for current (I), potential difference (V), and time interval (T), the quantities shown in table 1 can be declared with unique signatures. This context is suitable for quite a wide range of electrical measurements, and it avoids the use of dimensions for mechanical quantities, which are only a distraction.[g]

Quantity ratios arise in a number of ways in electrical networks. One example is a potential divider, created by a simple resistor network; another is the voltage gain of an amplifier. The last two rows in the table show how dimensions could be declared for these ratios and the corresponding signatures. Software can check dimensions to verify the nature of a ratio before using it in a calculation [13].

[g] For example, the dimension of capacitance is declared here by the relation $I \cdot V^{-1} \cdot T$. This retains an intuitive interpretation, which is hardly the case in the SI, where the dimensions of capacitance are $\mathsf{M}^{-1}\mathsf{L}^{-2}\mathsf{T}^4\mathsf{I}^2$.

Table 1. Some electrical quantities, with corresponding quantity expressions used to declare dimensional signatures. The base quantities are: current, voltage and time interval: I,V and T.

Dimension	Expression	Signature
impedance	$I^{-1} \cdot V$	$(-1, 1, 0)$
capacitance	$I \cdot V^{-1} \cdot T$	$(1, -1, 1)$
inductance	$I^{-1} \cdot V \cdot T$	$(-1, 1, 1)$
angular frequency	T^{-1}	$(0, 0, -1)$
power	$I \cdot V$	$(1, 1, 0)$
energy	$I \cdot V \cdot T$	$(1, 1, 1)$
voltage gain	$V//V$	$(0, 1, 0) : (0, 1, 0)$
potential divider	$(I^{-1} \cdot V)//(I^{-1} \cdot V)$	$(-1, 1, 0) : (-1, 1, 0)$

3.2.2. *Angle*

Adding a base dimension for angle to the SI would address some problems, such as the indistinguishably of torque and energy from their dimensions[15,16]. However, such a fundamental change would be widely disruptive and is not recommended. It has been suggested, however, that digital systems could use a dimension for angle without exposing that implementation detail to users[15,16].

For example, if there was a dimension for angle, the length of arc s on a circle of radius r subtended by an angle θ could be expressed as

$$s = \eta \cdot r \cdot \theta \, , \tag{2}$$

where η has the dimension of inverse angle. This constant depends on the units of θ such that $\eta \cdot \theta$ is the magnitude of θ when expressed in radians. Equation (2) does not depend on the units of r and θ; it is in the form of equation (1), with η in the role of c.[h]

To support this context, length and angle can be used as base quantities, in which case the signature for reciprocal angle will be $(0, -1)$. Each term on the right in (2) then has a unique signature and, on the left, the arc length s has a signature $(1, 0)$.

[h]The more familiar expression for arc length is $s = r \cdot \theta$, where θ must be in radians—the coherent SI unit for angle—so angles measured in other units cannot be accommodated. A skilled person could convert any measure of θ into radians, but that extra step implicitly changes $s = r \cdot \theta$ into equation (2).

3.2.3. *Fuel consumption*

Motor vehicle fuel consumption is usually reported as a ratio involving a volume of fuel and a distance (e.g., litres-per-100 km and miles-per-gallon). SI dimensions are not able to express this clearly, because the dimensions of distance and volume are both represented in terms of the base dimension for length.

In this context, the volume of a fluid is a quantity of an entirely different character to a distance, so it makes no sense for their dimensions to be related. We can define independent base dimensions for volume and distance. Then, any measure expressed as a ratio of volume to distance has the dimensional signature $(1, -1)$ and any measure expressed as a ratio of distance to volume has the signature $(-1, 1)$. In this context, the common units of litres-per-100 km and miles-per-gallon can be supported, but others would also be acceptable, such as cl/km.

4. Discussion

Although the difficulty in providing support for computations with physical quantities has been recognised for decades[17,18], the important role of physical dimensions in unit-invariant expressions of physical laws has been overlooked. A relationship between quantities, expressed as an equation, is related to the dimensions, not units, and so a suitable representation for dimensions must be used to validate computations. Of course, units of measurement are also important, but the requirements on units are of secondary importance to the dimensions. Units relate to numeric values substituted for terms in quantity expressions. Indeed, if there are inconsistencies among units, but an expression is dimensionally correct, then those inconsistencies could, in principle, be resolved if there is a policy to choose the units in which a result is to be expressed.

The initial step of choosing base dimensions for a problem is a normal and necessary part of any dimensional analysis. Although this is a new and unfamiliar step to take in data processing, it should not present too much difficulty. For the purpose of checking calculations, the only requirement on the base dimensions is that each quantity of interest should have a unique dimensional signature. That would be easy to verify when quantities of interest in a context are first declared.

The flexibility afforded by choosing arbitrary base dimensions is used in formal analysis to handle distinctly different features that are associated with the same kind of quantity. The case of lengths measured along perpendicular directions is discussed in textbooks by Huntley[19], and Focken[20], and was described as early as 1892 by Williams[21].

Dimensionless ratios are often used in scientific and engineering formulae and should be catered for by checking systems. The SI brochure acknowledges the importance of ratios, stating *"[it is] especially important to have a clear description of any quantity [] expressed as a ratio of quantities of the same kind"*[11] p.137. It also states that *"... ratios of quantities of the same kind [may be expressed] with units (m/m, etc) to aid the understanding of the quantity"*[11] p.151. Our proposal aligns well with these statements.

This study is concerned with maintaining the integrity of information about measured quantities during computational processes. It is part of a search for representations of metrological concepts that are suitable for digital systems. In that context, the management of data is important too[22]. Static representations—metadata—and techniques of data processing will have to be compatible, because information associated with measured values, such as units and the nature of quantities, needs to be propagated in computations and correctly interpreted in the final results.

In practice, computations are not always specified in terms of unit-invariant expressions; unit-dependent formulae are also used. A recent review, by Grozier, found that the distinction is not always clear[23]. He observes that *"... unit-invariance does not appear to have yet entered the scientific mainstream; instead, formulae which are unit-invariant are often not recognised as such, and formulae that are unit-dependent are often wrongly taken to be of universal application"*. This is somewhat alarming, because it suggests that the metrological notions of dimensions and scales are not widely understood. So, care will be needed when digital technologies are designed to take on important tasks. The proposal in this work should improve the quality of computational processes by explicitly recognising the constraints imposed by the dimensions and scales associated with terms; it will impose a certain discipline on the expression of calculations that might also prove helpful to data management and stewardship.

5. Conclusion

Support for checking the correctness of calculations with physical quantities must take into account physical dimensions in the problem context. Base dimensions should be selected to provide unique signatures for every quantity of interest, using dimensional exponents. In this way, a digital representation of the context can be constructed for all quantities of interest and the relationships between them. This makes it possible to verify the correctness of computational operations applied to terms representing quantities.

The proposal is based on a recognition that terms representing quantities in unit-invariant equations, such as physical laws, correspond to classes of similar scales (dimensions). Dimensions predominate measurement scales (units), so a framework for checking quantity correctness must be based on dimensional representations. Such a framework can then accommodate the further requirements imposed by units; indeed, it could be designed to resolve problems of unit inconsistency automatically. The proposal includes the ability to attribute dimensions to quantity ratios, which are usually categorised as dimensionless. Doing so strengthens the ability of the method to track the semantics of calculations and thereby to detect anomalies.

Acknowldgement

This work was funded by the New Zealand government. The author thanks J S Borbely for a careful review of the manuscript and D R White for helpful comments.

References

1. P. W. Bridgman, *The Nature of Physical Theory* (Dover Publications New York, 1936).
2. S. Hackel, F. Härtig, J. Hornig and T. Wiedenhöfer, *PTB-Mitteilungen* **127**, 75 (2017).
3. F. Thiel, *OIML Bulletin* **LIX**, 10 (2018).
4. A. Kennedy, Types for Units-of-Measure: Theory and Practice, in *Central European Functional Programming School*, eds. Z. Horvth, R. Plasmeijer and V. Zsk (Springer, 2010) pp. 268–305.
5. M. Contrastin, A. Rice, M. Danish and D. Orchard, *Computing in Science & Engineering* **18**, 102(January 2016).
6. H. E. Grecco, Pint [Online] Available: https://github.com/hgrecco/pint.
7. M. Schabel and S. Watanabe, Units: Zero-overhead dimensional analysis and unit/quantity manipulation and conversion (2008), [Online] Available: https://www.boost.org/doc.
8. W. H. Emerson, *Metrologia* **42**, L21 (2005).
9. B. Ellis, *Philosophy of Science* **31**, 357 (1964).
10. I. Johansson, *Metrologia* **47**, 219 (2010).

11. BIPM, *The International System of Units*, 9th edn. (BIPM, Sèvres, France, 2019).
12. J. B. J. Fourier, *The Analytical Theory of Heat* (Cambridge University Press, London, 1878).
13. B. D. Hall, Software for calculation with physical quantities, in *2020 IEEE International Workshop on Metrology for Industry 4.0 & IoT*, 2020.
14. B. D. Hall, Quantity-Value [Online] https://github.com/MSLNZ/Quantity-Value.
15. K. Brownstein, *American Journal of Physics* **65**, 605 (1997).
16. P. Quincey and R. J. C. Brown, *Metrologia* **54**, 454 (2017).
17. J. C. Cleaveland, Pouches, a programming language construct encouraging redundancy, msc, University of California, Los Angeles–Computer Science (1975).
18. N. Gehani, *Computer Languages* **2**, 93 (1977).
19. H. E. Huntley, *Dimensional analysis*, 2 edn. (MacDonald, 1952).
20. C. M. Focken, *Dimensional methods and their applications* (Edward Arnold & Co., 1955).
21. W. Williams, *Philosophical Magazine* **34**, 234 (1892).
22. R. J. Hanisch, I. S. Gilmore and A. L. Plant, *J. RES. NATL. INST. STAN.* **124**, p. 124024 (2019).
23. J. Grozier, *Studies in History and Philosophy of Science Part A* **80**, 9 (April 2020).

F Pavese, A B Forbes, N F Zhang, A G Chunovkina (Eds.):
Advanced Mathematical and Computational Tools in Metrology and Testing XII
Series on Advances in Mathematics for Applied Sciences, Vol. 90

https://doi.org/10.1142/9789811242380_0017

Combining data streams of doubtful provenance

Kavya Jagan* and Alistair B. Forbes

Data Science Department, National Physical Laboratory,
Teddington - TW110LW, UK
**E-mail: kavya.jagan@npl.co.uk*
www.npl.co.uk

With the advent of the Internet of Things, sensors have become almost ubiquitous. From monitoring air quality to optimising manufacturing processes, networks of sensors collect vast quantities of data every day, creating multiple data streams. While some applications may warrant the use of very high quality sensors that are regularly calibrated, this may not always be the case. In most applications, sensors may have been calibrated prior to installation but their performance in the field may be compromised, or a combination of low and high quality sensors may be used. Thus, the uncertainty of the data from a sensor operating in the field may be unknown, or the quality of an estimate of the uncertainty from a prior calibration of the sensor may not be known. Taking into account the unknown or doubtful provenance of the data is critical in its subsequent analysis in order to derive meaningful insights from it.

Many data analysis problems arising in metrology involve fitting a model to observed or measured data. This paper addresses the issue of how to make inferences about model parameters based on data from multiple streams, each of doubtful or unknown provenance. It is assumed the data arises in a system modelled as a linear response subject to Gaussian noise. In order to account for the lack of perfect knowledge of data quality, Bayesian hierarchical models are used. Hyper-parameters of a gamma distribution encode the degree of belief in uncertainty estimates associated with each data stream. In this way, information on the uncertainties can be learnt or updated based on the data. The methodology can be extended to data streams for which uncertainty statements are missing. The Bayesian posterior distribution for such hierarchical models cannot be expressed analytically in closed form but approximate inferences can be made in terms of the Laplace approximation to the posterior distribution determined by finding its mode i.e., the maximum a posteriori (MAP) estimate. The Laplace approximation can in turn be used in a Metropolis-Hastings algorithm to sample from the posterior distribution. The methodology is illustrated on metrological examples, including the case of inter-laboratory comparisons.

Keywords: data streams, doubtful provenance, Bayesian hierarchical models, Metropolis-Hastings, uncertainty evaluation

1. Introduction

Many data analysis problems arising in metrology involve fitting a model to observed or measured data. If the data arises from a measurements using a single, well characterised instrument, then standard methods such as least squares regression can be applied successfully[1,12]. The fit of the model of the data can also be used to provide a posterior estimate of the standard deviation of random effects associated with the measuring system. However, in the case analysing data streams from multiple instruments in Internet of Things and sensor network environments, the various instruments involved may be only partially characterised and their performance in changing environments will also change so that an *a priori* characterisation of the instruments may be only partially valid. For these systems, standard approaches can only be applied under the assumption that each instrument has been characterised sufficiently well to assign a weighting scheme for the various data streams.

In this paper, we are concerned with extending standard approaches to account for data streams of doubtful or unknown provenance. Our target applications are ones in which there are potentially many sensors measuring a small number of quantities related to a physical system, e.g, in environmental monitoring or in a Factory of the Future. The extended approaches use Bayesian hierarchical models that introduce hyper-parameters to model our lack of perfect knowledge about the measurement systems with prior distributions used to model our initial characterisations of the systems. The data is then used to provide estimates of parameters of interest but also update our knowledge about the hyper-parameters characterising the measuring systems behaviour. For the case of linear models and Gaussian random effects, much of the methodology associated with least squares methods can be exploited so that the computational overhead incurred by these extended models is manageable. In section 2, we describe the Bayesian hierarchical model while in sections 3–5, we describe computational approaches that account for different levels of knowledge about the data streams. Numerical examples are given in section 6 while our concluding remarks are given in section 7.

2. Basic model for multi-sensor systems

We are concerned in this paper with determining the parameters $\boldsymbol{\alpha} = (\alpha_1, \ldots, \alpha_n)^\top$ associated with a linear model from data $\boldsymbol{y} = (y_1, \ldots, y_m)^\top$, $m > n$. We assume that there are n_K instruments or sensors resulting in data vectors $\boldsymbol{y}_k$ of length m_k with $m = \sum_{k=1}^{n_K} m_k$. We assume that the measured data arises according to the model

$$\boldsymbol{y}_k|\boldsymbol{\alpha}, \phi_k \sim \mathrm{N}(C_k\boldsymbol{\alpha}, \phi_k^{-1}I_k), \quad \phi_k = 1/\sigma_k^2, \tag{1}$$

where C_k is an $m_k \times n$ observation matrix, and ϕ_k are precision parameters, the inverse of variances $1/\sigma_k^2$. Letting $\boldsymbol{\phi} = (\phi_1, \ldots, \phi_{n_K})^\top$,

$$V_{\boldsymbol{\phi}} = \begin{bmatrix} \phi_1^{-1}I_{m_1} & & \\ & \ddots & \\ & & \phi_{n_K}^{-1}I_{m_{n_K}} \end{bmatrix},$$

$$C = \begin{bmatrix} C_1 \\ \vdots \\ C_{n_K} \end{bmatrix}, \quad \boldsymbol{y} = \begin{bmatrix} \boldsymbol{y}_1 \\ \vdots \\ \boldsymbol{y}_{n_K} \end{bmatrix},$$

then the model is summarised by $\boldsymbol{y}|\boldsymbol{\alpha}, \boldsymbol{\phi} \sim \mathrm{N}(C\boldsymbol{\alpha}, V_{\boldsymbol{\phi}})$. $V_{\boldsymbol{\phi}}$ could be replaced by a full-rank variance matrix, V, that accounts for the correlation between sensor streams and treated as a Gauss-Markov problem as described in section 3.2.

We assume that, for each k, an estimate $\sigma_{k,0}^2$ of σ_k^2 is available, from a previous calibration exercise for instance. Our concern is that we may have little information about the quality of the estimate $\sigma_{k,0}^2$ - is it the output of a thorough characterisation of the measurement system or is it based on more meagre information? It may also be the case that the measurement system is affected by environmental factors; while the instrument may have been calibrated under certain controlled conditions the operating conditions may be quite different.

We consider a Bayesian hierarchical model that specifies different levels of information about σ_k^2, $k = 1, \ldots, n_K$. The levels are as follows.

L1 The parameters σ_k^2 are known exactly and equal to $\sigma_{k,0}^2$.

L2 Prior information about the σ_k^2 is available in the form

$$m_{k,0}\sigma_{k,0}^2\phi_k \sim \chi_{m_{k,0}}^2, \quad \phi_k = 1/\sigma_k^2,$$

where χ_ν^2 denotes a chi-squared distribution with ν degrees for freedom[17]. Equivalently, $\phi_k \sim \mathrm{G}(m_{k,0}/2, m_{k,0}\sigma_{k,0}^2/2)$, where $G(A, B)$ denotes a gamma distribution[17] with shape parameter α and rate parameter β. For this model, $m_{k,0} > 0$ is known and specifies the degree of belief in the estimate $\sigma_{k,0}^2$. The larger $m_{0,k}$ is, the more confidence we have in the estimate $\sigma_{k,0}^2$.

L3 Prior information about σ_k^2 is available in that $\phi_k|m_{k,0}$ is distributed according to

$$\phi_k|m_{k,0} \sim \mathrm{G}(m_{k,0}/2, m_{k,0}\sigma_{k,0}^2/2)$$

as above but now the degrees of freedom $m_{k,0}$ is regarded as an additional hyper-parameter for which there is prior information encoded by a prior distribution $p(m_{k,0})$, $k = 1, \ldots, n_K$.

2.1. *Motivation for the hierarchical model for lack of provenance*

This last situation, L3, corresponds most closely with the problem of analysing data of unknown provenance: we have an estimate $\sigma_{k,0}$ that characterises the uncertainty associated with the measurements taken by the kth instrument but we are not sure how good that estimate is. The model is appropriate for situations typified by the following.

Measurement instrument in a new environment. Suppose an instrument has been used in repeatability studies involving m repeated measurements y_i of a single parameter α modelled according to $y_i|\alpha, \phi \in N(\alpha, \phi^{-1})$, $\phi = 1/\sigma^2$ where very little is known about ϕ, prior to the measurements. With non-informative priors $p(\alpha) \propto 1$ and $p(\phi) \propto 1/\phi$. Then the posterior distribution for ϕ

$$\phi|\boldsymbol{y} \sim G((m-1)/2, (m-1)\sigma_0^2/2), \quad \sigma_0^2 = \sum_{i=1}^{m}(y_i - a)^2, \quad a = \frac{1}{m}\sum_{i=1}^{m} y_i. \quad (2)$$

This posterior distribution gives a state of knowledge about the repeatability of the instrument. Note that the more repeat measurements that are taken, the stronger is our belief in the estimate σ_0^2. If the instrument was to perform an additional set of repeat measurements of a parameter β in an equivalent experiment in the same environment gathering data z_j, then it would be appropriate to model the new experiment according to

$$z_i \in \mathrm{N}(\beta, \phi^{-1}), \quad \phi \sim G\left((m-1)/2, (m-1)\sigma_0^2/2\right).$$

The results from the previous experiment can be used to provide an informative prior for ϕ and subsequently benefit the estimation of β. However, if the instrument is used on a slightly different experiment or the environmental conditions have changed, then the performance of the instrument might not be the same as in the previous experiment and the prior (2) may be only partially valid but likely still relevant. A solution is to use the same estimate σ_0^2 but regard the degree of belief m_0, corresponding to $m-1$, as an additional hyper-parameter so that values of $m_0 < m-1$ are possible and can account for the changed circumstances.

Uncertainty statement with no supporting evidence. Suppose now that a statement of the repeatability of the instrument in terms of an estimate σ_0 is given but with no further information about how this estimate was arrived at. How can we use this estimate in a principled way? An approach would be to say that the estimate was derived in an experiment that gave information equivalent to a repeatability experiment involving m_0 repeated measurements. We do not know what m_0 is so we regard m_0 as a hyper-parameter to be determined. We choose a prior distribution for m_0 with an expectation ν_0 that can be pessimistic, say, $\nu_0 = 3$, or more optimistic, say $\nu_0 = 20$, about the degree of belief in σ_0. Importantly, using a disperse prior for ν we are not committed to either a pessimistic or optimistic point of view and the data arising from the physical system itself has an opportunity revise our characterisation.

Uncertainty statement missing. Finally, suppose now that an instrument has no uncertainty characterisation associated with it. As an analogy, consider the case of measurement information about α of the form $y \in \mathrm{N}(\alpha, \sigma^2)$ but the information about σ^2 is missing. What can be done? One approach is as follows. There may be some other metadata available around (historical data, typical performance of instruments commonly used in the domain) to allow some guess σ_0^2 about σ^2 to be made. The question remains as to how much weight is to be placed on this guess. A simple solution is to extend the model above according to $y \in \mathrm{N}(\alpha, \phi^{-1}\sigma_0^2)$ and $\phi \sim G(m_0/2, m_0/2)$, where the hyper-parameter $m_0 > 0$ specifies the degree of belief in the estimate σ_0^2. Assigning a non-informative prior for α, $p(\alpha) \propto 1$, the posterior distribution for α, ϕ, and m_0 is such that

$$p(\alpha, \phi, m_0|y) \propto p_{\mathrm{N}}(a, \phi^{-1}\sigma_0^2) p_{\mathrm{G}}(\phi|m_0/2, m_0/2) p(m_0),$$

with $a = y$. For fixed, known m_0, $\alpha|y$ is associated with the t-distribution $t_{m_0}(a, \sigma_0^2)$. The mean of this distribution is defined if $m_0 > 1$ and the variance of this distribution is defined if $m_0 > 2$. If we accept that the estimate a is sufficiently informative that the distribution associated with α should have a mean, then we should assign the distribution for the hyper-parameter m_0 to have zero probability mass for $0 < m_0 \leq 1$. Since the uncertainty statement associated with a is missing we do not want the probability mass associated with the region $m_0 > 2$ to be very large. This suggests assigning to m_0 a vague prior distribution with its mean between 1 and 2. For example, we can set

$$m_0 = 1 + \nu, \quad \nu \sim G(A, B), \quad B < A < 2B, \quad B > 0.$$

This model allows the information that the estimate exists to be used but does not make an assumption strong enough that would allow an uncertainty associated the estimate to exist. With this model, the information is used appropriately under the general scheme described by the hierarchical model.

These situations show how the model L3 can be used to analyse data where the provenance of the data is in doubt through one reason or another.

We assume that our primary interest is in making inferences about $\boldsymbol{\alpha}$ but that updating our information about $\boldsymbol{\phi}$ on the basis of the information provided by the data $\boldsymbol{y}$ could also be of interest. In the following sections, we describe computational approaches that enable us to make such inferences for the different levels of information available, starting with the most concrete. Section 3 deals with the case where all uncertainties are known exactly, i.e., L1 above. The case where there is partial information about the uncertainty, L2, is described in section 4. The doubtful provenance case, L3, is discussed in section 5. These methods are applied to laser tracker data and inter-laboratory comparison data and results are presented in section 6.

3. All uncertainties known exactly

For this case the variance matrix $V_{\boldsymbol{\phi}}$ is a fixed diagonal matrix constructed with $\phi_k = 1/\sigma_{k,0}^2$.

3.1. *Weighted least squares*

Assuming a non-informative prior for $\boldsymbol{\alpha}$, $p(\boldsymbol{\alpha}) \propto 1$, then

$$\boldsymbol{\alpha}|\boldsymbol{y} \sim \mathrm{N}(\boldsymbol{a}, V_{\boldsymbol{a}}), \tag{3}$$

where

$$\boldsymbol{a} = \left(C^\top V_{\boldsymbol{\phi}}^{-1} C\right)^{-1} C^\top V_{\boldsymbol{\phi}}^{-1} \boldsymbol{y}, \quad V_{\boldsymbol{a}} = \left(C^\top V_{\boldsymbol{\phi}}^{-1} C\right)^{-1}. \tag{4}$$

Let W be the diagonal weighting matrix where $W = V_{\boldsymbol{\phi}}^{-1/2}$ and $\tilde{C} = WC$ and $\tilde{\boldsymbol{y}} = W\boldsymbol{y}$, then $\tilde{\boldsymbol{y}} \in \mathrm{N}(\tilde{C}\boldsymbol{\alpha}, I)$, and

$$\boldsymbol{a} = \left(\tilde{C}^\top \tilde{C}\right)^{-1} \tilde{C}^\top \tilde{\boldsymbol{y}}, \quad V_{\boldsymbol{a}} = \left(\tilde{C}^\top \tilde{C}\right)^{-1}.$$

3.2. *Gauss-Markov problem*

If V is a more general full-rank variance matrix that has Cholesky factorisation[19] $V = LL^\top$ (or any such factorisation where L is a square matrix), then the model $\boldsymbol{y} \in \mathrm{N}(C\boldsymbol{\alpha}, V)$ is equivalent to $\tilde{\boldsymbol{y}} \in \mathrm{N}(\tilde{C}\boldsymbol{\alpha}, I)$, where $\tilde{\boldsymbol{y}}$ and $\tilde{C}$ are the transformed data vector and observation matrix, respectively, given by

$$\tilde{\boldsymbol{y}} = L^{-1}\boldsymbol{y}, \quad \tilde{C} = L^{-1}C. \tag{5}$$

The matrix L^{-1} plays the same role as W in the weighted least squares case.

Because of these transformations it is sufficient to consider the case $\boldsymbol{y}|\boldsymbol{\alpha} \in \mathrm{N}(C\boldsymbol{\alpha}, I)$, where C is an $m \times n$ observation matrix. If C has QR factorisation Q_1R_1 where Q_1 is an $m \times n$ orthogonal matrix and R_1 is an $n \times n$ upper-triangular matrix, the estimate $\boldsymbol{a}$ is found by solving $R_1\boldsymbol{a} = Q_1^\top \boldsymbol{y}$ by back substitution. The associated variance matrix is given by $V_{\boldsymbol{a}} = (R_1^\top R_1)^{-1}$. Thus, the case where all the uncertainties are known exactly can be addressed using standard computational approaches based on least-squares methods[13,19].

4. Partial information about σ_k

In this section we consider the model, L2, above where there is partial information about the parameters σ_k.

4.1. *Posterior distribution for α and ϕ*

We assume prior information about σ_k^2 is available in the form

$$\phi_k \sim \mathrm{G}(m_{k,0}/2, m_{k,0}\sigma_{k,0}^2/2)$$

where $m_{k,0}$ is known and specifies the degree of belief in the estimate $\sigma_{k,0}^2$. Assuming prior independence of $p(\phi_k)$ for $k = 1, \ldots, n_K$, the joint prior for $\boldsymbol{\phi}$ is given by

$$p(\boldsymbol{\phi}) \propto \left(\prod_{k=1}^{n_K} \phi_k^{m_{k,0}/2-1}\right) \exp\left\{-\frac{1}{2}\sum_{k=1}^{n_K} m_{k,0}\sigma_{k,0}^2\phi_k\right\}.$$

If $p(\boldsymbol{\alpha}) \propto 1$, then by Bayes' theorem, the posterior distribution for $\boldsymbol{\alpha}$ and $\boldsymbol{\phi}$ is such that

$$p(\boldsymbol{\alpha}, \boldsymbol{\phi}|\boldsymbol{y}) \propto p(\boldsymbol{\phi})|V_{\boldsymbol{\phi}}|^{-1/2} \exp\left\{-\frac{1}{2}(\boldsymbol{y} - C\boldsymbol{\alpha})^\top V_{\boldsymbol{\phi}}^{-1}(\boldsymbol{y} - C\boldsymbol{\alpha})\right\}. \quad (6)$$

This expression can also be rewritten making the partitioning of the data more explicit,

$$p(\boldsymbol{\alpha}, \boldsymbol{\phi}|\boldsymbol{y}) \propto \prod_{k=1}^{n_K} \phi_k^{(m_{k,0}+m_k)/2-1} \quad (7)$$
$$\exp\left\{-\frac{\phi_k}{2}\left[m_{k,0}\sigma_{k,0}^2 + (\boldsymbol{y}_k - C_k\boldsymbol{\alpha})^\top(\boldsymbol{y}_k - C_k\boldsymbol{\alpha})\right]\right\}.$$

4.2. *Marginalising with respect to ϕ_k*

A standard integration rule can be applied to each ϕ_k in (7), leading to

$$p(\boldsymbol{\alpha}|\boldsymbol{y}) \propto \prod_{k=1}^{n_K} \left[1 + \frac{1}{m_{k,0}}(\boldsymbol{y}_k - C_k\boldsymbol{\alpha})^\top \left(\frac{1}{\sigma_{k,0}^2} I\right)(\boldsymbol{y}_k - C_k\boldsymbol{\alpha})\right]^{-(m_{k,0}+m_k)/2}.$$

Comparing this expression with the probability density function (PDF) associated with a multivariate t-distribution[17], we see that

$$p(\boldsymbol{\alpha}|\boldsymbol{y}) \propto \prod_{k=1}^{n_K} p_t(\boldsymbol{y}_k|C_k\boldsymbol{\alpha}, \sigma_{k,0}^2 I, m_{k,0}), \quad (8)$$

the product of likelihoods given by multivariate t-distributions whose means are linear functions of $\boldsymbol{\alpha}$, variances are $\sigma_{k,0}^2 I$ and degrees of freedom are $m_{k,0}$. The posterior distribution in (8) will not in general correspond to a standard distribution and so inferences about $\boldsymbol{\alpha}$ have to be determined numerically.

4.3. *Posterior point estimate for $\boldsymbol{\phi}$*

The marginalised distribution $p(\boldsymbol{\phi}|\boldsymbol{y})$ can also be calculated analytically. If $\boldsymbol{a}$ is the least squares solution that minimises

$$(\boldsymbol{y} - C\boldsymbol{\alpha})^\top V_{\boldsymbol{\phi}}^{-1}(\boldsymbol{y} - C\boldsymbol{\alpha}),$$

and

$$F(\boldsymbol{\phi}) = (\boldsymbol{y} - C\boldsymbol{a})^\top V_{\boldsymbol{\phi}}^{-1}(\boldsymbol{y} - C\boldsymbol{a}),$$

the residual sum of squares weighted by V_{ϕ}^{-1}, then

$$p(\phi|y) \propto p(\phi)|V_{\phi}|^{-1/2}|C^{\top}V_{\phi}^{-1}C|^{-1/2}\exp\left\{-\frac{1}{2}F(\phi)\right\}.$$

The ϕ that maximises $p(\phi|y)$, i.e., gives the most probable value of ϕ, also known as the MAP estimate (maximum a posteriori estimate) can be found be minimising $-\log p(\phi|y)$ or, equivalently, minimising

$$E(\phi|y) = \frac{1}{2}\left(F(\phi) + \log|V_{\phi}| + \log|C^{\top}V_{\phi}^{-1}C|\right) - \log p(\phi). \qquad (9)$$

Minimising $E(\phi|y)$ is an unconstrained optimisation problem involving a smooth objective function and can be solved using standard optimisation techniques, based on Newton's method, for example[11,18]. The MAP estimates $\hat{\phi}$ can be used as representative values and the problem can be treated as a case where the uncertainties are known exactly, as described in section 3 using the diagonal variance matrix $V_{\hat{\phi}}$.

This approach is an extension of that in section 3 in that, rather than using the prior estimates $\sigma_{k,0}^2$ to estimate ϕ_k, the data has been used to provide updated estimates $\hat{\phi}_k$ that, hopefully, reflect better that actual physical system. The estimates $\hat{\phi}$ and, subsequently, a can be determined using standard numerical algorithms for nonlinear unconstrained optimisation and linear least squares methods.

4.4. *Approximate sampling from the joint posterior distribution $p(\alpha, \phi|y)$*

Uncertainties associated with the parameters a calculated as in section 3, treating $\hat{\phi}$ as exact, are likely to be underestimates as they ignore the uncertainty contribution associated with the hyper-parameters ϕ. Approximate samples from the joint posterior distribution $p(\alpha, \phi|y)$ can be drawn using the following approach.

The joint posterior distribution $p(\alpha, \phi|y)$ can be factored as

$$p(\alpha, \phi|y) = p(\alpha|\phi, y)p(\phi|y) \qquad (10)$$

Given ϕ_q sampled from $p(\phi|y)$, $q = 1, \ldots, M$, let a_q be the least-squares solution associated with C, y and variance matrix V_{ϕ_q} determined as described in section 3. Then a_q represents a sample from $p(\alpha|y)$ and can be used to construct an estimate of the mean and variance of $p(\alpha|y)$ as for standard Monte Carlo approaches[2,4,5]. There remains the issue of how to sample from $p(\boldsymbol{\phi}|\boldsymbol{y})$. A Markov chain Monte Carlo approach can be implemented. However, we can use the Hessian matrix $H_{\hat{\boldsymbol{\phi}}}$ of second partial derivatives of $E(\boldsymbol{\phi}|\boldsymbol{y})$ in (9) evaluated at $\hat{\boldsymbol{\phi}}$ to determine the Laplace approximation $p_0(\boldsymbol{\phi}|\boldsymbol{y})$ to $p(\boldsymbol{\phi}|\boldsymbol{y})$ with $p_0(\boldsymbol{\phi}|\boldsymbol{y})$ given by the PDF associated with the multivariate Gaussian distribution $\mathrm{N}(\hat{\boldsymbol{\phi}}, V_{\hat{\boldsymbol{\phi}}})$ with $V_{\hat{\boldsymbol{\phi}}} = -H_{\hat{\boldsymbol{\phi}}}^{-1}$. Samples $\boldsymbol{a}_q$ from a multivariate Gaussian $\mathrm{N}(\boldsymbol{\alpha}, V)$ with variance matrix V and Cholesky factorisation $V = LL^\top$ can be constructed as $\boldsymbol{a}_q = \boldsymbol{\alpha} + L\boldsymbol{e}_q$, where $\boldsymbol{e}_q = (e_{1,q}, \ldots, e_{n,q})^\top$ constructed from samples $e_{j,q} \in \mathrm{N}(0,1)$ from a standard univariate normal distribution. Approximate sampling from $p(\boldsymbol{\alpha}|\boldsymbol{y})$ can be constructed using samples from $p_0(\boldsymbol{\phi}|\boldsymbol{y})$. (In fact, it is usually better to work with parameters $\tau_k = \log \phi_k$ since the posterior distribution $p(\boldsymbol{\tau}|\boldsymbol{y})$ is more likely to be approximated well by a Gaussian.)

Sampling from $p_0(\boldsymbol{\phi}|\boldsymbol{y})$ (or $p_0(\boldsymbol{\tau}|\boldsymbol{y})$) can be used in a Metropolis-Hastings algorithm to construct samples from the true posterior $p(\boldsymbol{\phi}|\boldsymbol{y})$ (or $p(\boldsymbol{\tau}|\boldsymbol{y})$). Such sampling schemes exploiting the factorisation in (10) are not computationally efficient as constructing each sample $\boldsymbol{a}_q$ involves solving a least squares problem. However, they are straightforward to implement and can be used to assess whether the use of point estimate of $\boldsymbol{\phi}$, treated as exact, underestimates significantly the variance associated with the parameters $\boldsymbol{\alpha}$. If necessary, a more efficient MCMC sampling scheme can then be implemented to sample from $p(\boldsymbol{\alpha}, \boldsymbol{\phi}|\boldsymbol{y})$.

5. Partial information about σ_k^2 of doubtful provenance

In this section we consider the case where $m_{k,0}$ is considered as an additional hyper-parameter, model L3 above, and discussed in section 2.1. This enables us to model our doubt about the degree of belief in the quality of the estimate $\sigma_{k,0}$.

5.1. *Posterior distribution for $\boldsymbol{\alpha}$, $\boldsymbol{\phi}$, ν*

We assume that we can write $m_{k,0} = m_{k,0}(\nu) = m_{k,00} + \nu$, where $m_{k,00}$ is a initial pessimistic degree of belief in the estimate $\sigma_{k,0}$ for data stream k, $k = 1, \ldots, n_K$, and $\nu > 0$ is a parameter representing the additional degree of belief. Given a prior $p(\nu)$ on ν with expectation ν_0, $m_{k,0} + \nu_0$ represents our prior degree of belief in the estimate $\sigma_{k,0}$. For instruments that have been well characterised, we can choose $m_{k,00}$ to be quite large say, 50, while for instruments where the characterisation is suspect, we can choose a small value, say, 3. In the model, a single parameter ν has been considered for simplicity; the model can be extended to $m_{k,0}(\nu_k) = m_{k,00} + \nu_k$, $k = 1, \ldots, n_K$.

Given ν, the model assigns the prior

$$\phi_k|\nu \sim \mathrm{G}(m_{k,0}(\nu)/2, m_{k,0}(\nu)\sigma_{k,0}^2/2),$$

$k = 1, \ldots, n_K$, and the joint prior $p(\boldsymbol{\phi}, \nu)$ is given by

$$p(\boldsymbol{\phi}, \nu) = p(\boldsymbol{\phi}|\nu)p(\nu) = p(\nu)\prod_{k=1}^{n_K} p(\phi_k|\nu).$$

Assuming a noninformative prior for $\boldsymbol{\alpha}$, $p(\boldsymbol{\alpha}) \propto 1$, the posterior distribution $p(\boldsymbol{\alpha}, \boldsymbol{\phi}, \nu|\boldsymbol{y})$ is such that

$$p(\boldsymbol{\alpha}, \boldsymbol{\phi}, \nu|\boldsymbol{y}) \propto p(\boldsymbol{\phi}, \nu)|V_{\boldsymbol{\phi}}|^{-1/2} \exp\left\{-\frac{1}{2}(\boldsymbol{y} - C\boldsymbol{\alpha})^\top V_{\boldsymbol{\phi}}^{-1}(\boldsymbol{y} - C\boldsymbol{\alpha})\right\}. \quad (11)$$

This posterior can also be expanded as

$$p(\boldsymbol{\alpha}, \boldsymbol{\phi}, \nu|\boldsymbol{y}) \propto p(\nu)\prod_{k=1}^{n_K} K(A_k(\nu), B_k(\nu))\phi_k^{(m_{k,0}(\nu)+m_k)/2-1} \quad (12)$$
$$\exp\left\{-\frac{\phi_k}{2}\left[m_{k,0}(\nu)\sigma_{k,0}^2 + (\boldsymbol{y}_k - C_k\boldsymbol{\alpha})^\top(\boldsymbol{y}_k - C_k\boldsymbol{\alpha})\right]\right\},$$

where $A_k(\nu) = m_{k,0}(\nu)/2$, $B_k(\nu) = m_{k,0}(\nu)\sigma_{k,0}^2/2$ and $K(A,B) = B^A/\Gamma(A)$, where $\Gamma(A)$ is the gamma function. The term $K(A,B)$ is the normalising constant associated with the corresponding gamma distribution $G(A,B)$.

In the numerical examples discussed below in section 6, we have used a gamma prior for ν, regarding $\nu > 0$ as a kind of scale parameter.

5.2. *Approximate inferences based on point estimates of ϕ and ν*

From (11), the marginalised distribution $p(\boldsymbol{\phi}, \nu|\boldsymbol{y})$ can be calculated analytically and point estimates $\hat{\boldsymbol{\phi}}$ and $\hat{\nu}$ can be determined by minimising $E(\boldsymbol{\phi}, \nu|\boldsymbol{y}) = -\log p(\boldsymbol{\phi}, \nu|\boldsymbol{y})$. These point estimates can then be used to determine an approximate posterior distribution for $\boldsymbol{\alpha}$ using the approach in section 3. The fact that ν is an additional hyper-parameter means that there is more flexibility on how much weight to place on the prior estimates $\sigma_{k,0}$ in deriving the point estimate $\hat{\boldsymbol{\phi}}$; the data has a somewhat greater say on the determining the point estimate.

5.3. *Sampling from* $p(\boldsymbol{\alpha}, \boldsymbol{\phi}, \nu|\boldsymbol{y})$

The posterior distribution in (11) will not in general have a closed form. In order to make inferences about $\boldsymbol{\alpha}$, $\boldsymbol{\phi}$ and ν we instead use a Markov chain Monte Carlo approach based on a Metropolis-Hastings algorithm[17] to sample from $p(\boldsymbol{\alpha}, \boldsymbol{\phi}, \nu|\boldsymbol{y})$.

The Metropolis-Hastings (MH) algorithm is an iterative algorithm for sampling $\{\boldsymbol{a}_q\}$ from a *target distribution* $p(\boldsymbol{\alpha})$. Given a draw $\boldsymbol{a}_{q-1}$, a proposed draw $\boldsymbol{a}^*$ for the next member of the sequence is generated at random from a *jumping distribution* $p_0(\boldsymbol{a}^*|\boldsymbol{a}_{q-1})$. Then $\boldsymbol{a}_q$ is set to $\boldsymbol{a}^*$ with acceptance probability

$$P_q = \min\{1, r_q\}, \quad r_q = \frac{p(\boldsymbol{a}^*)p_0(\boldsymbol{a}_{q-1}|\boldsymbol{a}^*)}{p(\boldsymbol{a}_{q-1})p_0(\boldsymbol{a}^*|\boldsymbol{a}_{q-1})}. \tag{13}$$

The simplest way to implement the acceptance step is to draw u_q from the uniform distribution $R(0,1)$ and if $u_q < P_q$, set $\boldsymbol{a}_q = \boldsymbol{a}^*$, otherwise set $\boldsymbol{a}_q = \boldsymbol{a}_{q-1}$.

If $p_0(\boldsymbol{\alpha})$ is a distribution that approximates $p(\boldsymbol{\alpha})$ then the jumping distribution can be taken to be $p_0(\boldsymbol{\alpha})$ itself, with $p_0(\boldsymbol{a}^*|\boldsymbol{a}_{q-1}) = p_0(\boldsymbol{a}^*)$. Thus we can implement an MH algorithm for sampling from $p(\boldsymbol{\alpha}, \boldsymbol{\phi}, \nu|\boldsymbol{y})$ given a suitable approximating distribution $p_0(\boldsymbol{\alpha}, \boldsymbol{\phi}, \nu|\boldsymbol{y})$. We do so using the Laplace approximation, a multivariate Gaussian approximation determined by finding a quadratic approximation to

$$E(\boldsymbol{\alpha}, \boldsymbol{\phi}, \nu|\boldsymbol{y}) = -\log p(\boldsymbol{\alpha}, \boldsymbol{\phi}, \nu|\boldsymbol{y})$$

about its minimum as discussed in section 4.4, using the Hessian matrix of second partial derivatives of $E(\boldsymbol{\alpha}, \boldsymbol{\phi}, \nu|\boldsymbol{y})$ at the minimum to define the variance matrix of the Gaussian. As mentioned in section 4.4, it may be advantageous to work with transformed parameters $\boldsymbol{\tau} = \log \boldsymbol{\phi}$ (and $\log \nu$).

6. Numerical examples

6.1. *Laser tracker data*

A laser tracker determines the location of a target in space from a fixed measuring station by measuring the distance to the target and the azimuth and elevation angles of the target, relative to the measuring station coordinate system.The measurements of distance, azimuth and elevation angles can be considered as made by three sensors each providing a different data stream. The refractive index effects influence the distance measurements differently to the angle measurements and it is difficult to predict how the tracker will behave for a particular set of conditions. While the system characterisation can give a guide to the accuracy of the measurements, the actual measurements depend strongly on environmental effects and it is prudent to regard statements of performance or characterisations determined from previous experiments as being of doubtful provenance.

The data was collected from an experiment involving the calibration of a laser tracker[20]. Prior estimates $\sigma_{k,0}^2$ associated with the sensors, determined from the instrument manufacturer's specification sheet were used to weight data in a non-linear least squares (NLLS) fit and the problem was then linearised around NLLS estimates. In the linearised problem, the parameters $\boldsymbol{\alpha}$ are all nominally zero. The linearised problem can be expressed in the form

$$\boldsymbol{y}_k \in \mathrm{N}(C_k\boldsymbol{\alpha}, \phi_k^{-1}I), \quad \sigma_{k,0}^2 = 1.$$

A modest degree of belief $m_{00,k} = 1$ in $\sigma_{k,0}^2 = 1$ for each of distance, azimuth angle and elevation angle was assigned. As described in section 5, a vague gamma prior for ν was used.

A histogram of samples obtained from the marginal distribution of ν along with its prior distribution (solid line) is shown in the first panel in Figure 1. The prior distribution gives more weight to large values of ν than the posterior distribution. The mode of the distribution is around 2.5 which suggests that the initial degree of belief could be increased by 2.5.

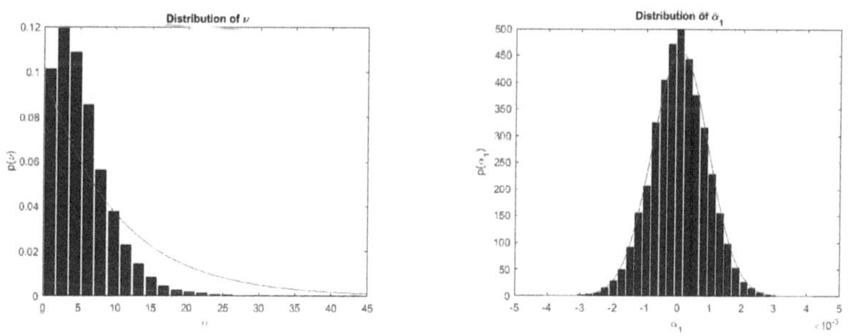

Fig. 1: Histogram of samples from the posterior distributions of ν and $\boldsymbol{\alpha}_1$ and plot of the prior distribution for ν and $\boldsymbol{\alpha}_1$ respectively.

A histogram of samples from the posterior distributions of α_1 and a superimposed Gaussian density with the same mean and variance as the samples is shown in the second panel in Figure 1. The marginal posterior distribution of the samples is more peaked than a Gaussian distribution with the same mean and variance. This suggests that the posterior distribution of α has fatter tails as it accounts for uncertainties in prior estimates $\sigma_{k,0}$.

The prior (solid line) and posterior distributions (histogram) for ϕ are shown in Figure 2. ϕ_1 represents the precision parameter of the distance measurements, ϕ_2 the azimuth measurements and ϕ_3 the elevation measurements. Since the data was re-weighted taking into account the prior uncertainty, the prior estimates of $\sigma_{k,0} = 1$ for each data stream. The posterior distribution of ϕ_1 has shifted to the right, which suggests that the distance measurements are more precise than what we expected *a priori*. The posterior distributions of the precision of both angle measurements, however, have been shifted to the left, indicating that the angle measurements are less precise than what we expected *a priori*.

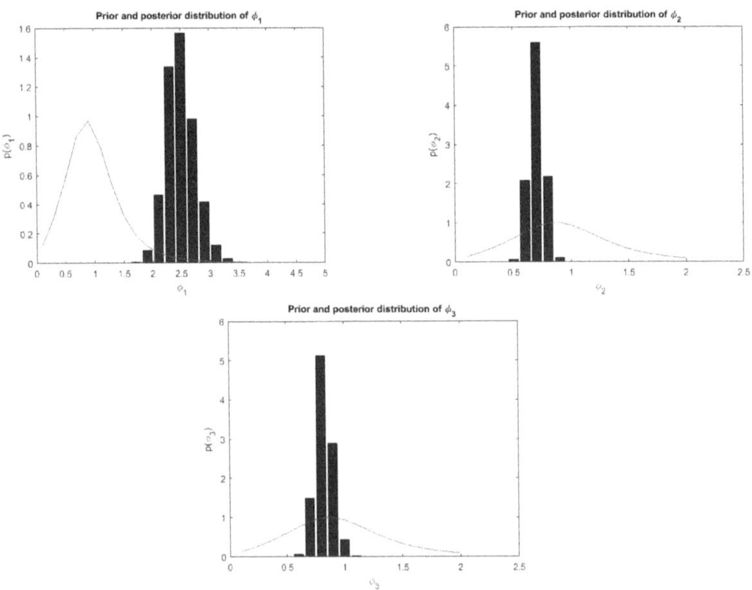

Fig. 2: Prior and Posterior distributions for ϕ for the laser tracker data.

The plots in Figure 3 show estimates of (x, y, z) coordinates obtained from the spherical coordinates measured using the laser tracker along with their 95% credible intervals using the 3 methods corresponding to varying degrees of information about $\sigma_{k,0}$. The parameter estimates based on a weighted least squares fit (LS) are all nearly zero and the uncertainties are the largest of the three methods. Using a representative value $\hat{\phi}$ for ϕ by maximising $p(\phi|y)$, parameter estimates a were derived by re-weighted data[14] using $\hat{\phi}$, as discussed in section 4.3. The Hierarchical Bayes approach (hierarchical model - HM) goes a step further in that it accounts for the uncertainties in the prior estimates $\sigma_{k,0}$, i.e., it accounts for the unknown provenance of the data. In this instance, the prior uncertainty estimates are not given much weight and the posterior distribution is dominated by the data; hence there are no pronounced differences between the MAP and HM methods.

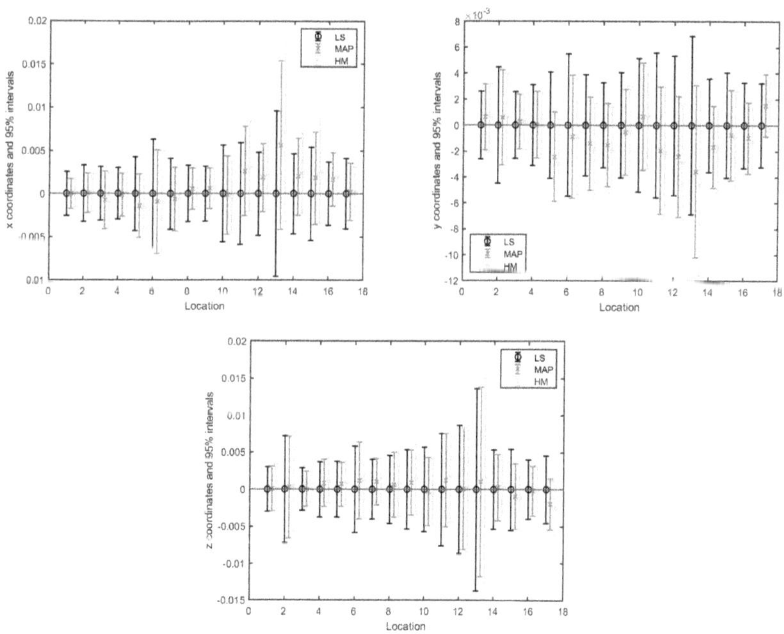

Fig. 3: A comparison of the 3 levels of information about uncertainty estimates.

6.2. *Inter-laboratory comparison data*

Inter-laboratory comparisons (ILCs) are one of the key activities in metrology. The exercise involves a number of laboratories that evaluate estimates of a quantity (along with their associated uncertainties) relating to the same artefact. A comparison of these estimates may be performed in order to i) determine a consensus value for the quantity being measured by combining experimental results from a number of laboratories, ii) assess the degree of equivalence of different laboratories or measurement methodologies. When the measurement technique is novel, data from an ILC may be treated as that of doubtful provenance. As there is little prior experience of using the measurement technique, there may be sources of uncertainty that have not been accounted for by some or all of the participating laboratories and one may not know how good the uncertainty estimates are.

The starting point for the analysis of ILC data is usually a model[7] of the form

$$y_k \in \mathrm{N}(\alpha, \sigma_k^2), \tag{14}$$

where α is the value of the measurand, y_k is the estimate of α produced by the kth laboratory and σ_k its associated standard uncertainty. It may be of particular relevance to apply the Bayesian methods described in this paper to ILC data when the estimates of the quantity, y_k, are considered to be inconsistent given their associated uncertainties, σ_k. In practice, adjustment procedures are sometimes applied to determine uncertainties associated with estimates that represent better consistency[6,8,9,16,23,25].

Treating each laboratory as a data stream that measures an estimate y_k of α, with an estimated uncertainty equal to $\sigma_{k,0}$, the Bayesian approach adjusts the uncertainty estimates based on the data in a way that results in better consistency[15].

Figure 4 illustrates this approach for different levels of confidence in the estimates $\sigma_{k,0}$ for Planck's constant data[10,26]. Assuming a vague gamma prior for ν as described in section 5, it can be observed that when the initial degree of belief $m_{k,00} = 0$, i.e., a non-informative prior on $\phi_{k,0} = 1/\sigma_{k,0}^2$, the uncertainties associated with most of the laboratories is increased slightly. This increase is larger for estimates that are further away from the bulk of the data relative to their uncertainties (laboratories 1 and 8). When there is modest initial degree of belief of $m_{k,00} = 5$ in the prior uncertainties, then only the uncertainties associated with laboratories 1 and 8 are adjusted. In the case where $m_{k,00} = 100$, i.e., near certainty in the uncertainty estimates, the problem reduces to the case where all uncertainties are known exactly, as described in Section 3 and none of the estimates are adjusted.

The posterior modes of ν are 7.2, 4.6 and 2.1 when $m_{k,00} = 0, 5, 100$ respectively. This suggests that the initial degree of belief could be increased by 7.2, 4.6 and 2.1 respectively. For the case $m_{k,00} = 100$, the data provides little information about ν and the posterior point estimate of ν is

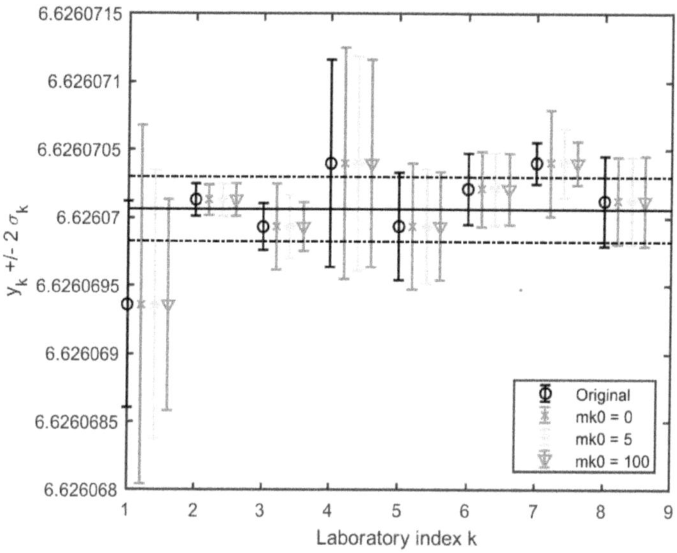

Fig. 4: Estimates of Planck's constant and 2 times the associated posterior uncertainty for each laboratory for different levels of prior belief in the uncertainty estimates.

determined largely by the prior $p(\nu)$.

6.3. *Discussion: relationship to the GUM and measurement system analysis*

The approach to uncertainty evaluation described in the *Guide to the Expression of Uncertainty in Measurement* (GUM[3]) can be regarded as a bottom-up approach in which the uncertainty contributions associated with various influence factors are assigned and then propagated through a known, fixed model in order to assign an uncertainty associated with the measurand. The validity of this approach depends on the validity of the model and the validity of the assigned distributions associated with the influence factors. The GUM approach is particularly appropriate for measurements in a metrology laboratory in which instrument responses and all major influence factors are well characterised.

In a measurement systems analysis (MSA) approach to uncertainty evaluation[21,22,24], the variance contribution of different factors such as tool, operator, environment, laboratory, etc., are assessed through experiments in which the different factors are varied in a systematic way and their effect on the output observed. The approach is very much data driven, using the redundancy in the measurements to separate out the uncertainty contributions from the various effects. In the MSA approach, the uncertainty contribution of effects that do not make themselves apparent in the observed data cannot be accounted for. By contrast, the GUM's model-based approach is able to account for such effects through their inclusion in the model.

The hierarchical model can be viewed as a blend of the GUM and MSA approaches. Rather than regarding the distributions associated with the influence factors as fixed, they are regarded as members of classes of distributions, weighted according to prior distributions. On the basis of redundancy in the measurement data, the weighting is updated. If the data adds no new information about a hyper-parameter (in particular, if there is no effective redundancy), then its posterior distribution is the same as its prior distribution. If the data includes new information, then the posterior distributions reflect the gain in information. In the laser-tracker example, the model includes a scale offset that accounts for the difference between the estimated refractive index and the actual refractive index of the air. For an experiment involving a single laser tracker, the data will supply no new information about this offset and the uncertainty contribution is that specified by the prior information. For an experiment involving two or more trackers measuring the same set of targets, each with their own scale offset, the data can provide information about the validity of the prior uncertainties assigned to the scale offsets, similarly to the way that a χ^2-test is applied.

The hierarchical model allows uncertainties associated with the parameters of interest to be quantified, taking into account the actual behaviour of the measurement systems. The traceability of estimates, determined by combining multiple data streams or determined using a GUM approach for a single instrument, depends on valid uncertainty evaluation. Averaging a large number of data streams is no guarantee of traceability if the data streams are subject to the same systematic effect(s). An analysis method based purely on observed data, as in MSA, will not in general deliver traceability. The hierarchical model, implemented correctly, is designed to retain those elements of the GUM approach that enables traceability to be maintained. However, we are reminded of the aphorism, well known in the statistical community (often attributed to the statistician George Box), that "all models are wrong, but some are useful". One of the roles of metrology is to provide ground truth references that help us assess whether our models are useful or not.

7. Concluding remarks

This paper is concerned with analysing data from multiple data streams arising in sensor networks or Internet of Things environments. It is often the case that measurement instruments is such applications are calibrated in a certain environments but used in similar but not identical conditions. While the characterisation of the instruments will provide some information about their accuracy, we may not know exactly how the instruments perform in the field. This paper has been concerned with how to use the information that is available in a principled way, extending standard data analysis methods. The paper has described methods of analysing such data using Bayesian Hierarchical models. Hyper-parameters are assigned to account for the lack of perfect knowledge of the instruments' performance, in particular, the lack of perfect knowledge of the uncertainty of the data obtained. The computational approaches described on the paper using standard least squares estimate and unconstrained numerical optimisation. While the Bayesian hierarchical model leads to nonstandard posterior distributions, the hierarchical model allows a straightforward and effective Metropolis-Hastings Markov chain Monte Carlo sampling procedure to sample from the posterior distributions. The methods have been applied to laser tracker data and inter-laboratory comparison data.

Acknowledgements

This work was supported by the UK's National Measurement System programme for Data Science and the European Metrology Programme for Innovation and Research (EMPIR) project 17IND12, Metrology for the Factory of the Future. The EMPIR programme is co-financed by the Participating States and from the European Unions Horizon 2020 research and innovation programme.

References

1. R. M. Barker, M. G. Cox, A. B. Forbes, and P. M. Harris. Software Support for Metrology Best Practice Guide no. 4: Modelling Discrete Data and Experimental Data Analysis. Technical Report DEM-ES 018, National Physical Laboratory, Teddington, 2007.

2. R. M. Barker and A. B. Forbes. Software Support for Metrology Best Practice Guide No. 10: Discrete Model Validation. Technical report, National Physical Laboratory, Teddington, March 2001.
3. BIPM, IEC, IFCC, ILAC, ISO, IUPAC, IUPAP, and OIML. Evaluation of measurement data — Guide to the expression of uncertainty in measurement. Joint Committee for Guides in Metrology, JCGM 100:2008.
4. BIPM, IEC, IFCC, ILAC, ISO, IUPAC, IUPAP, and OIML. Evaluation of measurement data — Supplement 1 to the "Guide to the expression of uncertainty in measurement" — Propagation of distributions using a Monte Carlo method. Joint Committee for Guides in Metrology, JCGM 101:2008.
5. BIPM, IEC, IFCC, ILAC, ISO, IUPAC, IUPAP, and OIML. Evaluation of measurement data — Supplement 2 to the "Guide to the expression of uncertainty in measurement" — extension to any number of output quantities. Joint Committee for Guides in Metrology, JCGM 102:2011.
6. R. T. Birge. Probable values of the general physical constants. *Rev. Mod. Phys*, 1:1–73, 1929.
7. M. G. Cox. The evaluation of key comparison data. *Metrologia*, 39:589–595, 2002.
8. M. G. Cox, A. B. Forbes, J. Flowers, and P. M. Harris. Least squares adjustment in the presence of discrepant data. In P. Ciarlini, M. G. Cox, F. Pavese, and G. B. Rossi, editors, *Advanced Mathematical and Computational Tools in Metrology VI*, pages 37–51, Singapore, 2004. World Scientific.
9. M.G. Cox. The evaluation of key comparison data: determining the largest consistent subset. *Metrologia*, 44(3):187–200, 2007.
10. S. Davidson, J. Berry, and P. Williams. Ensuring the stability of primary mass standards during the implementation of the new definition of the kilogram. *Journal of Physics: Conference Series*, 1065, 2018.
11. R. Fletcher. *Practical Methods of Optimization*. John Wiley and Sons, Chichester, second edition, 1987.
12. A. B. Forbes. Nonlinear least squares and Bayesian inference. In F. Pavese, M. Bär, A. B. Forbes, J.-M. Linares, C. Perruchet, and N.-F. Zhang, editors, *Advanced Mathematical and Computational Tools in Metrology VIII*, pages 103–111, Singapore, 2009. World Scientific.
13. A. B. Forbes. Parameter estimation based on least squares methods. In F. Pavese and A. B. Forbes, editors, *Data modeling for metrology and testing in measurement science*, pages 147–176, New York, 2009. Birkhäuser-Boston.

14. A. B. Forbes. Weighting observations from multi-sensor coordinate measuring systems. *Measurement Science and Technology*, 23(online:025004), 2012.
15. A. B. Forbes. A hierarchical model for the analysis of inter-laboratory comparison data. *Metrologia*, 53:1295–1305, 2016.
16. A. B. Forbes and C. Perruchet. Measurement systems analysis: concepts and computational approaches. In *IMEKO World Congress, September 18–22, 2006, Rio de Janeiro*, 2006.
17. A. Gelman, J. B. Carlin, H. S. Stern, D. B. Dunson, A. Vehtari, and D. B. Rubin. *Bayesian Data Analysis*. Chapman & Hall/CRC, Boca Raton, Fl., third edition, 2014.
18. P. E. Gill, W. Murray, and M. H. Wright. *Practical Optimization*. Academic Press, London, 1981.
19. G.H. Golub and C.F. Van Loan. *Matrix Computations*. Johns Hopkins University Press, Baltimore, fourth edition, 2013.
20. E. B. Hughes, A. B. Forbes, A. J. Lewis, W. Sun, D. Veal, and K. Nasr. Laser tracker error determination using a network measurement. *Meas. Sci. Tech.*, 22, 2011. doi:10.1088/0957-0233/22/4/045103.
21. International Organization for Standardization, Geneva. *ISO 5725 1: Accuracy (trueness and precision) of measurement methods and results – Part 1: General principles and definitions*, 1994.
22. International Organization for Standardization, Geneva. *ISO/TR 22971: Accuracy (trueness and precision) of measurement methods and results – Practical guidance for the use of ISO 5725-2:1994 in designing, implementing and statistically analysing interlaboratory repeatability and reproducibility results*, 2005.
23. K. Jagan and A. B. Forbes. Assessing interlaboratory comparison data adjustment procedures. *International Journal of Metrology and Quality Engineering*, January 2019.
24. D. C. Montgomery and G. C. Runger. *Applied Statistics and Probability for Engineers*. John Wiley and Sons, 2003.
25. R. C. Paule and J. Mandel. Consensus values and weighting factors. *J Res. Natl. Bur. Stand.*, 87:377–385, 1982.
26. M. Stock, S. Davidson, H. Fang, M. Milton, E. de Mirandes, P. Richard, and C. Sutton. Maintaining and disseminating the kilogram following its redefinition. *Metrologia*, 54(6):S99, 2017.

F Pavese, A B Forbes, N F Zhang, A G Chunovkina (Eds.):
Advanced Mathematical and Computational Tools in Metrology and Testing XII
Series on Advances in Mathematics for Applied Sciences, Vol. 90

https://doi.org/10.1142/9789811242380_0018

Using measurement uncertainty in a risk-based decision-making framework for clinical diagnosis

Kavya Jagan[1,*], Peter M. Harris[1], Nadia A.S. Smith[1], Jarmo Teuho[2], Reetta Siekkinen[2], Jussi Schultz[2] and Antti Saraste[2]

[1] *National Physical Laboratory, Hampton Road, Teddington - TW110LW, UK*
[2] *Turku PET Centre, Turku University Central Hospital and University of Turku, Turku, Finland*
**E-mail: kavya.jagan@npl.co.uk*

Clinical diagnosis can be approached as a problem of conformity assessment in which the patient takes the role of the item to be assessed and the classification of the patient as unhealthy or healthy is expressed as a requirement on the true value of a measured characteristic of the patient. The classification of patients into treatment and non-treatment groups is an important task in medicine with significant societal and personal implications. In this paper, a framework for clinical diagnosis that accounts for measurement uncertainty, based on the principles of conformity assessment, is described.

The framework is illustrated for the problem of deciding the extent of myocardial blood flow abnormalities in patients based on a predefined clinical guideline using clinical data from patients participating in a cardiac Positron Emission Tomography (PET) perfusion study at the Turku PET Centre.

By combining patient data with expert insights, the framework can help less experienced clinicians make better decisions regarding patient health, serve as a starting point for further clinical investigation, and be used as a screening to categorise patients so that the most severe cases can be prioritised on the clinical list. Furthermore, the framework could be used within a machine learning classification pipeline both to label unlabelled data and to assess the quality of labelled data.

Keywords: Conformity assessment, Measurement uncertainty, Cardiac perfusion, Positron emission tomography

1. Introduction and conformity assessment

Conformity assessment is an activity undertaken to decide whether an item fulfills a specified requirement. The item may be a product or process or system or patient, and the requirement typically takes the form of an interval of permissible values of a property or characteristic of the item. The interval may be defined by a lower or upper limit, giving a one-sided interval, or by both lower and upper limits, giving a two-sided interval. If the true value of the property lies within the interval, the item is said to be conforming and, otherwise, it is non-conforming.

The true value of the property is generally unknown and available knowledge of the property is obtained through measurement. The most complete expression of that knowledge is in the form of a probability distribution, although that distribution is often summarised by a measured value and an associated standard uncertainty corresponding to, respectively, the expectation and standard deviation of the distribution. Assessing conformity depends on using that probability distribution to calculate a conformance probability, which is the probability that the item fulfills the specified requirement. Furthermore, making a decision about whether the item conforms to the specified requirement depends on interpreting that calculated probability in a way that controls the risks of incorrect decisions about the item. The result is a documented decision rule that describes how measurement uncertainty is accounted for with regard to accepting or rejecting an item, given a specified requirement and the result of a measurement.

The evaluation of measurement uncertainty is treated in the 'Guide to the expression of uncertainty in measurement' (GUM)[1] and in its Supplement[2]. A further guidance document[3] considers the role of measurement uncertainty in conformity assessment, including how conformance probabilities and risks of incorrect decisions are calculated and decision rules for conformity assessment are constructed. Accounting for uncertainty in the context of conformity assessment and risk-based decision-making is also addressed in[4–9], and the topic has come even more into focus with the latest revision of ISO/IEC 17025:2017[10], which obliges calibration and testing laboratories to "document the decision rule employed, taking into account the level of risk (such as false accept and false reject and statistical assumptions) associated with the decision rule employed, and apply the decision rule".

In this study we apply the principles of these documents to a clinical application to investigate decision rules to support the medical diagnosis of a patient as being unhealthy (and classified into a treatment group) or healthy (and classified into a non-treatment group). Medical diagnosis can be regarded as a problem of conformity assessment with the patient taking the role of the item and the classification of the patient as unhealthy or healthy expressed as a requirement on the true value of a characteristic of the patient, in this case the patient's myocardial blood flow (MBF): see Section 2. Here, the MBF for a patient is not directly measurable, and knowledge of MBF is obtained indirectly from a measurement of the patient made using an imaging system that delivers information about each pixel in an image. The study is made complicated (and interesting) because the probability distribution for MBF is determined numerically using a Monte Carlo method and available information about each pixel in the image, the distribution can vary appreciably from patient to patient according to the measured property value, and it generally does not take a standard form, such as a Gaussian or other symmetric distribution, that is typically assumed in guidance documents on conformity assessment.

The paper is organised as follows. Section 2 provides the background to the clinical application and defines a clinical guideline that clinicians currently use to classify patients as healthy or not healthy. In Section 3 a risk-based decision-making framework for the degree of severity of myocardial perfusion abnormalities in patients is presented based on a clinical guideline that accounts for measurement uncertainty. Section 4 presents a summary and concluding remarks.

2. Background to the clinical application

Cardiovascular disease is a leading cause of mortality in Europe and is estimated to cost the European economy approximately €200 billion each year. Perfusion, the flow of blood, and hence oxygen, is essential to the functioning of the heart and reduced perfusion (or ischemia) is an early marker of chronic heart disease[11]. Myocardial perfusion imaging is a non-invasive technique used in the diagnosis, management and prognosis of cardiovascular disease as regions of underperfused myocardium are a reliable indicator of the presence of significant obstructive coronary artery disease. Information about the uncertainties associated with perfusion measurements are usually not taken into account when patients are diagnosed, which can be detrimental to the decision-making process. However, the use of such uncertainty information can help to quantify the risk of misdiagnosis of patients, especially in borderline cases. If the risk is found to be high, a clinician can order further tests rather than relying on a single measured value of blood perfusion whose quality is unquantified.

There are two ways in which a patient can be misdiagnosed: a patient who is ischemic is diagnosed as not ischemic and the patient does not receive necessary treatment, or a patient who is not ischemic is diagnosed as ischemic and the patient receives unnecessary treatment. Both risks are quantified by probabilities, and can be evaluated for a specific patient (known as specific risk) or for a patient drawn at random from the population of patients (known as global risk). Guidance documents on conformity assessment generally refer to risks to the 'consumer' (here, the patient) and to the 'producer' (here, the clinician). However, in medical applications the patient and clinician are not equal parties in the diagnosis and both risks are really risks to the patient, albeit realised in different ways and having different consequences.

The framework for classifying patients is illustrated using Positron Emission Tomography (PET) images of quantified stress MBF. PET imaging allows quantitative measurements of MBF by modelling mathematically the tracer distribution in the myocardium[12,13]. The quantification of MBF using PET imaging provides important physiological information, which clinicians can use to inform their diagnosis, and may also be used to individualise patient therapy. MBF evaluated over a given volume of tissue gives an estimate of perfusion and, in this paper, the terms MBF and perfusion are used interchangeably.

To decide if a patient is ischemic or not, clinicians use the perfusion measurements alongside other factors such as the patient's medical history and risk contributors such as age, body mass index, etc. Regarding the perfusion measurements, clinical guidelines define a myocardial perfusion scan as moderately-severely abnormal when the extent of ischemic myocardium exceeds 10 % of the left ventricle myocardium[14]. In this study, ischemia is defined as MBF during adenosine stress less than 1.5 ml/(g·min) in order to focus on the impact of the extent of low stress MBF in the analysis.

This guideline does not account for uncertainties in the values of perfusion, and the outcome of applying the guideline is that the patient is classified into one or other of the two states 'ischemic' or 'not ischemic' based on the estimate of the extent of severe myocardial ischemia. Figure 1 shows an example of a perfusion map based on the processing of an image obtained using PET. In the figure it can be seen that there are several pixels where the value of perfusion is close to the threshold value of 1.5 ml/(g·min). Errors in the values of perfusion for these pixels could alter the classification of the patient. Uncertainty factors affecting PET quantification can originate from several sources, beginning with patient preparation and ending with data analysis and interpretation of the results. When taken into account, the uncertainty can be used to provide a *probability* that the patient is ischemic or not ischemic. Then, the risk to the patient or to the clinician resulting from an incorrect classification can be quantified.

Clinical data was obtained from a prospectively recruited cohort of 138

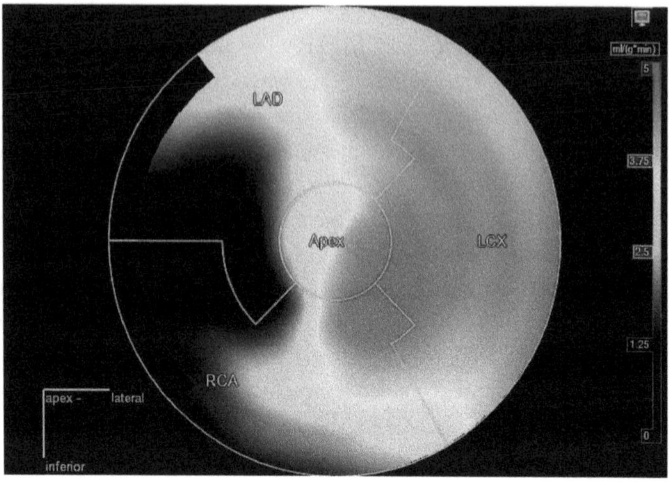

Figure 1: Perfusion map as generated by the software package Carimas 2.10.

patients who underwent PET myocardial ^{15}O-water perfusion imaging at the Turku PET Centre and the Turku University Hospital during 2007–2010[15]. The PET data was processed using Carimas 2.10[16] by an expert at the Turku PET Centre to obtain perfusion maps. The inputs to the methods developed within Sections 3 are such stress perfusion maps. The Carimas software also delineates the regions of the heart into the left coronary artery (LAD), the left circumflex (LCX), the right coronary artery (RCA) and the left coronary artery excluding the distal apical segment (LADwa). The results in this paper focus on the LCX region as the results from the other regions were observed to be similar.

3. Risk-based clinical guideline

In this section a risk-based decision-making framework for classifying a patient as ischemic based on perfusion measurements is presented, and illustrations of its application are given using the perfusion maps derived from the PET clinical data. A clinical guideline or decision rule that accounts for the uncertainties in measured values of perfusion is described and a quantification of the risks to a patient of a misdiagnosis based on that guideline is given. Two types of risk are considered[3,17]: *specific risk*, which is the risk that a specific patient is misdiagnosed, and *global risk*, which is the risk that a new (unknown) patient will be misdiagnosed.

3.1. *Conformance probability*

Let the property of interest be the proportion of the area of the myocardium of a patient for which the perfusion is less than 1.5 ml/(g·min), denoted by η, and suppose η_c denotes a predefined critical value for η, such as the value of 10 % used in the clinical guideline given in Section 2. PET provides a measured value η_m for η with associated standard uncertainty u_m. Let $g(\eta|\eta_m, u_m)$ denote the probability density function for the percentage of affected myocardium conditional on the measured value and its uncertainty. The density function reflects the available knowledge about the health of a patient given information obtained from a measurement of η made using a PET scan. It allows the calculation of the probability p, defined by

$$p = \int_{\eta_c}^{\infty} g(\eta|\eta_m, u_m)\, d\eta, \tag{1}$$

that the proportion η exceeds η_c based on the information from the scan. The clinical guideline (CG) described in Section 2 is expressed as

CG1 A patient is classified as ischemic if $\eta_m > \eta_c$,

which depends only on the measured value η_m. In contrast, the clinical guideline

CG2 A patient is classified as ischemic if $p > p_c$,

for a predefined critical probability p_c, accounts for both the measured value and its uncertainty. In the special case that η_m coincides with the median of the probability density function g, such as when g is symmetric, then CG1 and CG2 are equivalent when $p_c = 0.5$ in the sense that they will classify a patient in the same way irrespective of the uncertainty u_m. However, when $p_c < 0.5$, a patient for which $\eta_m < \eta_c$ can be classified as ischemic using CG2 and, in this sense, the guideline is less demanding than CG1. On the other hand, when $p_c > 0.5$, only patients for which $\eta_m > \eta_c$ will be classified as ischemic using CG2 so that generally fewer patients will be diagnosed as having ischemia compared with the use of CG1.

In practice, the density function $g(\eta|\eta_m, u_m)$ seldom takes a known form. However, the probability defined in expression (1) can be approximated using a Monte Carlo calculation[18,19]. Let f_k denote the perfusion for the

kth pixel in the perfusion map, so that

$$\eta = \frac{\#(\{f_k : f_k < 1.5\,\mathrm{ml/(g \cdot min})\})}{K} \times 100\,\%, \tag{2}$$

where K is the total number of pixels in the map and $\#(S)$ is the cardinality of the set S, i.e., the number of elements in S. There are several potential sources of uncertainty associated with the measurement of the perfusion f_k, from patient preparation to generating perfusion maps, that give rise to uncertainty in the measured value of η. Effects can be broadly categorised as random and systematic. A random effect yields errors in the values of perfusion for the pixels that are different and independent, whereas a systematic effect yields errors that are common to the values of perfusion for different pixels. An example of a systematic effect is an erroneous data correction (e.g., attenuation) which affects all of the pixels in the perfusion map. The statistical model

$$f_k = f_{\mathrm{m},k} + E_\mathrm{r} + E_\mathrm{s}, \tag{3}$$

is assumed for the measurement of perfusion, where $f_{\mathrm{m},k}$ is the measured value of the perfusion for the kth pixel, and E_r and E_s describe random and systematic effects, respectively. The quantities E_r and E_s are assumed to be characterised by Gaussian distributions having expectations of zero and known standard deviations of σ_r and σ_s, respectively. This error model is used to provide an illustration of the methods and a more realistic model is being developed in an extension to this work. Then, f_k has expectation equal to $f_{\mathrm{m},k}$, standard deviation σ, where

$$\sigma^2 = \sigma_\mathrm{r}^2 + \sigma_\mathrm{s}^2, \tag{4}$$

and the correlation coefficient for f_k and f_ℓ, $k \neq \ell$, is ρ, where

$$\rho = \frac{\sigma_\mathrm{s}^2}{\sigma^2}. \tag{5}$$

In the following, σ is sometimes reported as a relative standard deviation σ_rel in percent by expressing its value as a fraction of the average of the measured values of perfusion for all pixels in the perfusion map.

The jth Monte Carlo trial in the calculation involves making K random draws $e_{\mathrm{r},k}$ independently from the probability distribution for E_r and a single random draw e_s from the probability distribution for E_s. These draws are used to evaluate values $\widehat{f_k}$ of the perfusion from

$$\widehat{f_k} = f_{\mathrm{m},k} + e_{\mathrm{r},k} + e_\mathrm{s}, \quad k = 1, \ldots, K, \tag{6}$$

cf. expression (3), and then a value $\widehat{\eta}_j$ for η is evaluated from

$$\widehat{\eta}_j = \frac{\#(\left\{\widehat{f}_k : \widehat{f}_k < 1.5\,\text{ml}/(\text{g} \cdot \text{min}\right\})}{K} \times 100\,\%, \tag{7}$$

cf. expression (2). Finally, an approximation to p is

$$\widehat{p} = \frac{\#(\{\widehat{\eta}_j : \widehat{\eta}_j > \eta_\text{c}\})}{M}, \tag{8}$$

where M is the number of Monte Carlo trials. Illustrations of the approximations to the distributions $g(\eta|\eta_\text{m}, u_\text{m})$ based on measurements made in the LCX region for two different patients are shown in Figure 2, where $M = 10000$, $\sigma_\text{rel} = 12\,\%$ and $\rho = 0.7$. The graph on the left relates to a patient for whom $\eta_\text{m} > \eta_\text{c}$ and it is found that $\widehat{p} = 0.91$, and the graph on the right relates to a patient for whom $\eta_\text{m} < \eta_\text{c}$ and $\widehat{p} = 0.10$, where $\eta_\text{c} = 10\,\%$ in both cases. These probabilities express the "degree of belief" that the patients are ishemic. The first patient is almost certainly ischemic, while the second is almost certainly not ischemic.

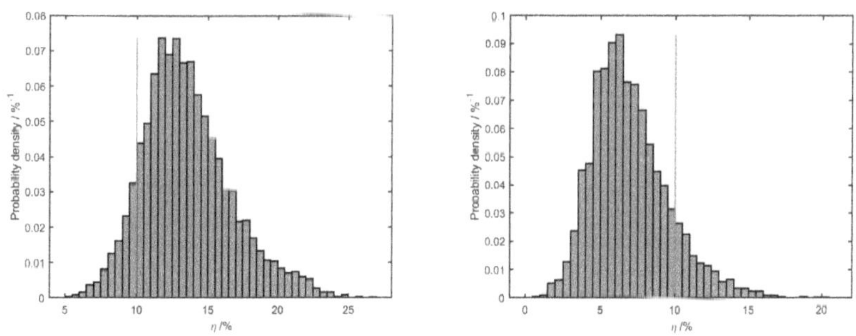

Figure 2: Approximations to the distribution $g(\eta|\eta_\text{m}, u_\text{m})$ based on measurements made in the LCX region for (left) a patient for whom $\eta_\text{m} > \eta_\text{c}$ and (right) a patient for whom $\eta_\text{m} < \eta_\text{c}$.

The graph on the left in Figure 3 shows the calculated values p for all 138 patients when typical levels of uncertainty and correlation, of $\sigma_\text{rel} = 12\,\%$ and $\rho = 0.7$, respectively, were chosen. There are 106 patients for which $p < 0.025$, 20 patients for which $p > 0.975$, and 12 patients for which p lies between 0.025 and 0.975. It follows that, even allowing for the uncertainties in the measured perfusion values, the results suggest strongly that 106 patients are not ischemic, 20 patients are ischemic, but there is some doubt about how to classify the remaining 12 patients. Applying CG2 with $p_c = 0.2$, 0.5 and 0.8, $p > p_c$ for 8, 4 and 3 of those 12 patients, respectively, so that the total number of patients classified as ischemic is 28, 24 and 23, respectively. The number classified as ischemic using CG2 is the same as for CG1 when $p = 0.5$ (even though the distributions g are not necessarily symmetric), and the number varies according to the choice of p_c. Understanding the risks to patients and clinicians of misdiagnosis using the clinical guideline CG2 is the topic of the following subsections.

Finally, the graph on the right in Figure 3 shows the calculated values p based on measurements made in the LCX region for all 138 patients when $\sigma_{\text{rel}} = 20\,\%$ and $\rho = 0.7$. In this case, there are 103 patients for which $p < 0.025$, 18 patients for which $p > 0.975$, and 17 patients for which p lies between 0.025 and 0.975. With larger uncertainties in the measured perfusion values, the number of patients for which there is some doubt about their classification as ischemic or not ischemic has increased correspondingly. In general as the uncertainty in the perfusion values increases, the probability p increases for patients whose measured value is less than the critical value ($\eta_m < \eta_c$) and decreases for those whose measured value is greater than the critical value ($\eta_m > \eta_c$). The decision about a patient becomes less clear as the uncertainty increases, and that is reflected in the two graphs in Figure 3 in which the values of p for patients at the extremes (near to zero or one) move towards the centre when σ_{rel} is increased from 12 % to 20 %.

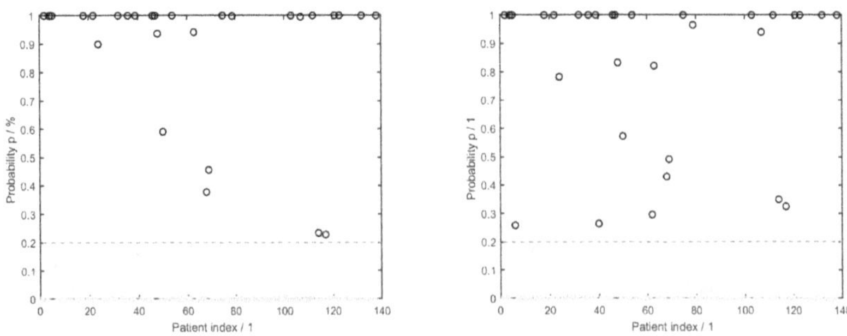

Figure 3: Values of p based on measurements made in the LCX region for all 138 patients when (left) $\sigma_{\text{rel}} = 12\,\%$ and $\rho = 0.7$ and (right) $\sigma_{\text{rel}} = 20\,\%$ and $\rho = 0.7$. The thick black circles indicate values of p that are greater than the critical value $p_c = 0.2$ (indicated by the dashed horizontal line), while the gray asterisks indicate values that are less than the critical value.

3.2. *Specific risk*

Specific risk is a property of a clinical guideline and is calculated for a particular (or 'specific') patient. Considering the clinical guideline CG2, if $p < p_\mathrm{c}$ so that the patient is classified as not ischemic, then the *specific patient's risk* R_p^* is defined by

$$R_\mathrm{p}^* = \int_{\eta_\mathrm{c}}^{\infty} g(\eta|\eta_\mathrm{m}, u_\mathrm{m})\,\mathrm{d}\eta = p. \tag{9}$$

It gives the probability that the patient is ischemic when the patient has been classified as not ischemic, and quantifies the risk, for example, of deciding incorrectly not to treat the patient when treatment is necessary. Conversely, if $p > p_\mathrm{c}$ so that the patient is classified as ischemic, then the *specific clinician's risk* R_c^* is defined by

$$R_\mathrm{c}^* = \int_{-\infty}^{\eta_\mathrm{c}} g(\eta|\eta_\mathrm{m}, u_\mathrm{m})\,\mathrm{d}\eta = 1 - p. \tag{10}$$

It gives the probability that the patient is not ischemic when the patient has been classified as ischemic, and quantifies the risk, for example, of deciding incorrectly to treat the patient when that treatment is unnecessary.

In the same way as in Section 3.1, the risks R_p^* and R_c^* can be approximated using a Monte Carlo calculation. When $p < p_\mathrm{c}$,

$$R_\mathrm{p}^* \approx \frac{\#(\{\widehat{\eta}_j : \widehat{\eta}_j > \eta_\mathrm{c}\})}{M}, \tag{11}$$

and, when $p > p_\mathrm{c}$,

$$R_\mathrm{c}^* \approx \frac{\#(\{\widehat{\eta}_j : \widehat{\eta}_j < \eta_\mathrm{c}\})}{M}. \tag{12}$$

Consider applying the clinical guideline CG2 with $p_\mathrm{c} = 0.5$ to a particular patient who is not diagnosed as ischemic based on measurements of perfusion in that part of the perfusion map defined by the region LCX. The graph on the left of Figure 4 shows the values of specific patient's risk calculated for different choices of σ_rel and ρ. For example, if the measurements of perfusion are described by a statistical model with $\sigma_\mathrm{rel} = 12\,\%$ and $\rho = 0.7$, then the risk that this patient is ischemic is about $10\,\%$. In general, as σ_rel and ρ are increased, so the risk that the patient might be ischemic also increases.

3.3. *Global risk*

Global risk is another property of a clinical guideline but is calculated for the population of possible patients. The *global patient's risk* is the probability that a future patient who is ischemic will be diagnosed as not ischemic based on a measurement for that patient. Conversely, the *global clinician's risk* is the probability that a future patient who is not ischemic will be diagnosed as ischemic based on a measurement for that patient. The evaluation of global risks depends on knowing the joint distribution $g(\eta, p)$ that, using the product rule of probability, is given by

$$g(\eta, p) = g(\eta|p)g(p). \tag{13}$$

There are different options for specifying the distribution $g(p)$, which captures information about possible values across the population of patients for the probability p used in the clinical guideline for diagnosing a patient. Here, the values p_n, $n = 1, \ldots, N$, obtained from a sample of N patients taken from the population of patients, have been used as a discrete approximation to the distribution. Similarly, the distribution $g(\eta|p)$ is approximated using the values $\widehat{\eta}_j|n$, $j = 1, \ldots, M$, calculated for the nth patient using a Monte Carlo procedure in the manner described in Section 3.1.

Using the discrete approximations to $g(p)$ and $g(\eta|p)$ described above, a discrete approximation $(\widehat{\eta}^j, \widehat{p}^j)$, $j = 1, \ldots, M$, to the joint density $g(\eta, p)$ is constructed as follows. For each j:

(1) Draw a patient at random with replacement, i.e., make a random draw n with replacement from the set $\{1, \ldots, N\}$, and store the calculated value of p for that patient as $\widehat{p}^j$;
(2) For the chosen patient, draw a value for the proportion of affected area at random from the Monte-Carlo sample generated in Section 3.1, i.e., make a random draw i with replacement from the set $\{1, \ldots, M\}$ and set $\widehat{\eta}^j = \widehat{\eta}_i|n$.

Then, the calculation of the global patient's risk R_{p} is given, approximately, by

$$R_{\mathrm{p}} \approx \frac{\#(\{(\widehat{\eta}^j, \widehat{p}^j) : \widehat{\eta}^j > \eta_{\mathrm{c}} \text{ and } \widehat{p}^j < p_{\mathrm{c}}\})}{M}, \tag{14}$$

and the calculation of the global clinician's risk is given, approximately, by

$$R_{\mathrm{c}} \approx \frac{\#(\{(\widehat{\eta}^j, \widehat{p}^j) : \widehat{\eta}^j < \eta_{\mathrm{c}} \text{ and } \widehat{p}^j > p_{\mathrm{c}}\})}{M}. \tag{15}$$

Again, consider applying the clinical guideline CG2 with $p_c = 0.5$ to a future patient based on measurements of perfusion in that part of the perfusion map defined by the region LCX. The graph on the right of Figure 4 shows the values of global patient's risk calculated for different choices of σ_{rel} and ρ. Note that the graph is not as smooth as that for specific risk shown on the left of the figure because the calculations are carried out in terms of a number $N = 138$ of patients used to describe the distribution $g(p)$, and that number is small and fixed. It is seen that the risk is quite low for this sample of patients, but increases as σ_{rel} and ρ are increased.

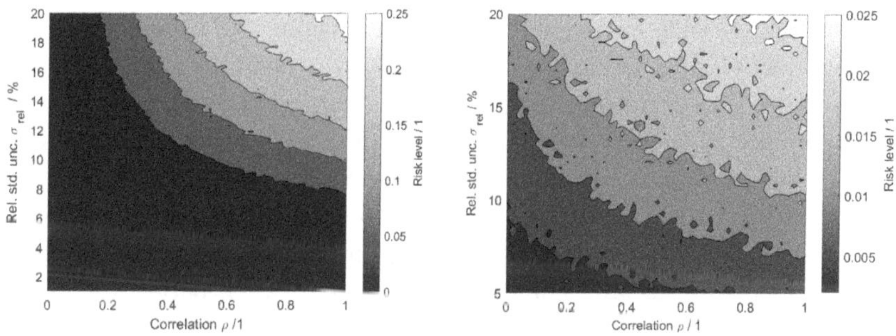

Figure 4: For the clinical guideline CG2 with $p_c = 0.5$, (left) specific risks for a particular patient, and (right) global risks for a future patient, based on measurements made in the LCX region and different choices of σ_{rel} and ρ.

4. Summary and concluding remarks

A risk-based decision-making framework for the extent of myocardial perfusion abnormalities in patients has been presented. It is based on regarding the diagnosis of a patient using information obtained from a measurement of perfusion as a problem of conformity assessment with the measurement delivering a measured value of perfusion together with an associated uncertainty. The framework is implemented by applying established principles and guidance widely used in metrology for evaluating measurement uncertainty and for taking that uncertainty into account when assessing conformity of an item against a specified requirement. The framework allows for the calculation of a probability of conformance, as well as the calculation of the risks to the patient and clinician of a misdiagnosis. The framework can be used to support the design or selection of an imaging system, in terms of its uncertainty characteristics, to achieve required levels of those risks.

The risk-based decision-making approach is based on a predefined clinical guideline and approaches are being investigated to decide whether the predefined clinical guideline is optimal in terms of reproducing 'ground-truth' values obtained independently using the measurement method of invasive coronary angiography. This analysis will be reported in a separate paper.

The risk-based decision-making framework has been illustrated using myocardial PET perfusion images collected at the Turku PET Centre. It is expected that the decision-making framework can be applied to other imaging modalities, such as magnetic resonance imaging (MRI) and computed tomography (CT). Classifying patients into treatment and non-treatment groups based on perfusion images is highly subjective, and it takes years of experience to be able to diagnose a patient with a high level of accuracy. There are a number of steps involved in deriving a measurement of perfusion from a PET image, with each step subject to different sources of uncertainty. Information about the uncertainties associated with perfusion measurements are usually not taken into account when patients are diagnosed, which is detrimental to the decision-making process. However, the use of such information can help to quantify the risk of misdiagnosis of patients, especially in borderline cases, which is a key difficulty in healthcare planning. If the risk is found to be high, a clinician can order further tests rather than relying on a single measured value of blood perfusion whose quality is unquantified. Thus, by combining patient data with expert insights, this framework can help make more informed decisions about patient health.

Acknowledgements

This project has received funding from the European Metrology Research Programme (EMPIR) co-financed by the Participating States and from the European Union's Horizon 2020 research and innovation programme, under project number 15HLT05. This report reflects only the authors' view and EURAMET is not responsible for any use that may be made of the information it contains.

The cardiac PET perfusion study was conducted within the Finnish Center of Excellence in Molecular Imaging in Cardiovascular and Metabolic Research supported by the Academy of Finland, University of Turku, Turku

University Hospital and Åbo Akademi University.

References

1. BIPM, IEC, IFCC, ILAC, ISO, IUPAC, IUPAP and OIML 2008 Evaluation of measurement data – Guide to the expression of uncertainty in measurement Joint Committee for Guides in Metrology, JCGM 100:2008
2. BIPM, IEC, IFCC, ILAC, ISO, IUPAC, IUPAP and OIML 2008 Evaluation of measurement data – Supplement 1 to the "Guide to the expression of uncertainty in measurement" – Propagation of distributions using a Monte Carlo method Joint Committee for Guides in Metrology, JCGM 101:2008
3. BIPM, IEC, IFCC, ILAC, ISO, IUPAC, IUPAP and OIML 2012 Evaluation of measurement data – the role of measurement uncertainty in conformity assessment Joint Committee for Guides in Metrology, JCGM 106:2012
4. Forbes AB 2006 Measurement uncertainty and optimized conformance assessment *Measurement* **39**, 808–14
5. EURACHEM/CITAC Guide 2007 Use of Uncertainty Information in Compliance
Assessment 1st edn http://www.eurachem.org/images/stories/Guides/pdf/Interpretation_with_expanded_uncertainty_2007_v1.pdf
6. Pendrill LR 2007 Optimised Measurement Uncertainty and Decision-Making in Conformity Assessment *NCSLi Measure*, **2**(2), 76–86
7. Pendrill LR 2014 Using measurement uncertainty in decision-making and conformity assessment *Metrologia*, **51**(S206), doi:10.1088/0026-1394/51/4/S206
8. Pendrill LR, Allard A, Fischer N, Harris P and Smith I 2019 Risk calculations for conformity assessment in practice *ENBIS 19th annual conference of the European Network for Business and Industrial Statistics*, 2–4 September
9. Allard A, Fischer N, Smith IM, Harris PM, Nguyen J and Pendrill LR 2019 Risk calculations for conformity assessment in practice *CFM 2019*, 25 September, doi:10.1051/metrology/201916001
10. International Organization for Standardization 2017 ISO/IEC 17025:2017 General requirements for the competence of testing and calibration laboratories
11. Shaw LJ, Berman DB, Maron DJ, Mancini GBJ, Hayes SW, Hartigan

PM, Weintraub WS, O'Rourke RA, Dada M, Spertus JA, Chaitman BR, Friedman J, Slomka P, Heller GV, Germano G, Gosselin G, Berger P, Kostuk WJ, Schwartz RG, Knudtson M, Veledar E, Bates ER, McCallister B, Teo KK and Boden WE 2008 Optimal Medical Therapy With or Without Percutaneous Coronary Intervention to Reduce Ischemic Burden *Circulation*, **117**(10),1283–91

12. Moody JB, Lee BC, Corbett JR, Ficaro EP and Murthy VL 2015 Precision and accuracy of clinical quantification of myocardial blood flow by dynamic PET: a technical perspective *Journal of Nuclear Cardiology*, **22**(5), 935–951
13. Schindler TH, Schelbert HR, Quercioli A and Dilsizian V 2010 Cardiac PET Imaging for the Detection and Monitoring of Coronary Artery Disease and Microvascular Health *JACC: Cardiovascular Imaging*, **3**(6), 623–640
14. Knuuti J, Wijns W, Saraste A, Capodanno D, Barbato E, Funck-Brentano C, Prescott E, Storey RF, Deaton C, Cuisset T, Agewall S. 2019 ESC Guidelines for the diagnosis and management of chronic coronary syndromes *European Heart Journal*
15. Stenström I, Maaniitty T, Uusitalo V, Pietilä M, Ukkonen H, Kajander S, Mäki M, Bax JJ, Knuuti J and Saraste A 2017 Frequency and angiographic characteristics of coronary microvascular dysfunction in stable angina: a hybrid imaging study *European Heart Journal: Cardiovascular Imaging*, **18**(11), 1206–13
16. Nesterov SV, Han C, Mäki M, Kajander S, Naum AG and Helenius H 2009 Myocardial perfusion quantitation with ^{15}O-labelled water PET: high reproducibility of the new cardiac analysis software (Carimas) *Eur. J. Nucl. Med. Mol. Imaging*, **36**(10),1594—602
17. Kuselman I, Pennecchi FR, da Silva RJNB and Hibbert D B 2017 Risk of false decision on conformity of a multicomponent material when test results of the components' content are correlated *Talanta*, **174**, 789–96
18. Metropolis N and Ulam S 1949 The Monte Carlo method *Journal of the American Statistical Association*, **44**(247), 335–41
19. Robert C and Casella G 2013 *Monte Carlo statistical methods* Springer Science & Business Media
20. Berman DS, Kang X, Slomka PJ, Gerlach J, de Yang L, Hayes SW, Friedman JD, Thomson LEJ and Germano G 2007 Underestimation of extent of ischemia by gated SPECT myocardial perfusion imaging in patients with left main coronary artery disease *Journal of Nuclear Cardiology*, **14**(4), 521–528

F Pavese, A B Forbes, N F Zhang, A G Chunovkina (Eds):
Advanced Mathematical and Computational Tools in Metrology and Testing XII
Series on Advances in Mathematics for Applied Sciences, Vol. 90

https://doi.org/10.1142/9789811242380_0019

Evaluation of measurement uncertainty in SBI – Single Burning Item reaction to fire test

L. L. Martins[1,*], A. S. Ribeiro[1], M. G. Cox[2] and J. A. Sousa[3]

[1] *LNEC, National Laboratory for Civil Engineering, Av. do Brasil, 101, 1700-066 Lisbon, Portugal*
[2] *NPL, National Physical Laboratory, Hampton Road, Teddington TW11 0LW, UK*
[3] *IPQ, Portuguese Institute for Quality, Rua António Gião, 2, 2829-513 Caparica, Portugal*
[*] *E-mail: lfmartins@lnec.pt*

This paper applies the Monte Carlo method (MCM) for measurement uncertainty propagation in the Single Burning Item (SBI) test, within the European normative framework of reaction to fire tests for building products: EN standard 13823:2010+A1. The use of an MCM is necessitated by (i) the multistage nature of the calculation and (ii) the multivariate, non-linear and complicated form of the functional relations, involving many input, intermediate and output quantities. The approach provided a validation of the GUM approach specified for use in the SBI uncertainty document CEN/TR 16988.

Keywords: Single Burning Item test; Measurement uncertainty; Monte Carlo method.

1. Introduction

The Single Burning Item (SBI) standard test [1] is intended to carry out the evaluation and classification of construction materials (excluding floorings) by measuring a set of quantities in order to characterize the contribution to the deflagration and propagation of fires in buildings, when exposed to adverse thermal conditions, by means of a combustion item.

In this test, the specimen retrieved from the tested material comprises two plates vertically positioned at right angles, being exposed to a burner located in the lower region of the junction of the plates. The performance of the specimen is evaluated for a period of 20 minutes, based on the indirect measurement of quantities related to heat release and smoke production. Complementary observations are performed regarding lateral flame propagation and the production of drops or particles from the combustion process.

This paper describes a numerical approach for evaluating the uncertainties for two main measurands of the SBI test given that the mathematical models involved are complicated and contain many quantities. Section 2 specifies the measurands and the testing process. Section 3describes how uncertainties are propagated through the models, compares the approach with that of the GUM [2] specified in the CEN document [3] for the uncertainty propagation of the SBI test. Section 4 gives specimen results. Section 5 gives our conclusions.

2. Main measurands of the SBI test

Two main measurands are defined in the SBI test: the *heat release rate*, HRR, which corresponds to the thermal power (in kW) released in a given time instant and, the *smoke production rate*, SPR (in $m^2 \cdot s^{-1}$), both being related to the combustion of the specimen.

In the SBI test, HRR is measured in two stages: (1) during the preliminary periodic combustion of propane and heptane and (2) during the combustion of the specimen. The only significant difference relates to the heat release per unit of consumed oxygen mass, which assumes different estimates and measurement uncertainties in each test case (propane, heptane or specimen combustion).

HRR is a key intermediate quantity in the determination of two main output quantities of the SBI test – THR, the total heat release (usually in MJ) from the specimen in a certain time exposure to the main burner flames (namely, in the first 600 s), and FIGRA, the fire growth rate (in $W \cdot s^{-1}$), defined as the maximum value of the ratio of HRR from the specimen and the time of its occurrence using a THR threshold (such as 0.2 MJ or 0.4 MJ).

Similarly, SPR is a significant intermediate quantity in the SBI test since it contributes to the determination of two other main output quantities – TSP, the total smoke production (in m^2) from the specimen in a certain time exposure to the main burner flames (namely, in the first 600 s), and SMOGRA, the smoke growth rate (in $m^2 \cdot s^{-2}$), the maximum value of the quotient of smoke production rate from the specimen and the time of its occurrence.

This paper only addresses the measurement uncertainty evaluation of the quantities HRR and SPR, since the subsequent uncertainty propagation from these intermediate key quantities to the output quantities of the SBI test (total heat release, fire growth rate, total smoke production and smoke growth rate) is straightforward and is characterized by simple linear mathematical models.

2.1. *Heat release rate measurement model*

The heat release rate measurement model is derived from the studies performed by [4] in the oxygen consumption calorimetry research field. The measurement principle states that the amount of heat released per unit of consumed oxygen volume, E', during a combustion process (in $MJ \cdot m^{-3}$) is considered constant regardless of the combustion material, which can be expressed by

$$\mathrm{HRR} = E' \cdot x_{\mathrm{O_2}}^0 \cdot \varphi \cdot q_{V_\mathrm{S}}, \tag{1}$$

where $x_{\mathrm{O_2}}^0$ is the oxygen molar fraction in the ambient, φ is the oxygen depletion factor and q_{V_S} is the volumetric flow of air in the ambient (in $m^3 \cdot s^{-1}$).

The amount of heat released per unit of consumed oxygen volume quantity can be determined by the product ofthe oxygen density, $\rho_{\mathrm{O_2}}$, (in $kg \cdot m^{-3}$) and the heat release per unit of consumed oxygen mass, E (in $MJ \cdot kg^{-1}$), i.e.,

$$E' = \rho_{O_2} \cdot E. \tag{2}$$

The oxygen molar fraction in the ambient is

$$x_{O_2}^{0} = x_{O_2}^{\text{initial}} \cdot \left(1 - x_{H_2O}^{0}\right), \tag{3}$$

$x_{O_2}^{\text{initial}}$ being the oxygen molar fraction measured in the initial stage of the SBI test (in the period between 30 s and 90 s), with a gas analyser*, and $x_{H_2O}^{0}$ the water molar fraction in the ambient, determined by the model

$$x_{H_2O}^{0} = \frac{\text{rh}}{100 \cdot p_{\text{atm}}} \exp\left(23.2 - \frac{3816}{T_{\text{initial}} - 46}\right) \tag{4}$$

(derived from the Clausius-Clapeyron equation for water vapour saturation pressure and based on conventional values for the gas constant and the heat vaporization of water), where rh is the relative humidity in moist air (as a percentage), T_{initial} is the initial air temperature inside the exhaust duct (in K) and p_{atm} is the atmospheric pressure (in Pa).

The oxygen depletion factor is calculated by

$$\varphi = \frac{x_{O_2}^{\text{initial}} \cdot \left(1 - x_{CO_2}\right) - x_{O_2} \cdot \left(1 - x_{CO_2}^{\text{initial}}\right)}{x_{O_2}^{\text{initial}} \cdot \left(1 - x_{O_2} - x_{CO_2}\right)}, \tag{5}$$

where $x_{O_2}^{\text{initial}}$ and $x_{CO_2}^{\text{initial}}$ arethe molar fractions of oxygen and carbon dioxide measured in the initial stage of the SBI test with the gas analyser, while x_{O_2} and$x_{CO_2}$are, respectively, the molar fractions of oxygen and carbon dioxide measured with the same equipment in a given time instant after the initial stage.

The volumetric flow of air in the ambient is indirectly measured based on the expression

$$q_{V_s} = \frac{q_{V_c}}{1 + (\alpha - 1) \cdot \varphi}, \tag{6}$$

in which q_{V_c} is the volumetric flow of the gases in the exhaust duct (in $m^3 \cdot s^{-1}$) and α represents the expansion factor. This last quantity reflects the fact that, in a combustion chemical reaction, the amount of substance related to combustion products is not identical to the amount of substance related to the oxygen consumed in the reaction process, i.e.,

$$\alpha = 1 + x_{O_2}^{0} \cdot (\beta - 1). \tag{7}$$

* This equipment receives a gas sample from a normalized exhaust duct in which all water vapour and water-soluble gases are eliminated before measurement.

$x_{O_2}^0$ is obtained from expression (3) and β is the ratio of the amount of substance of combustion products and consumed oxygen. A combustion reaction involving reactants such as hydrocarbons ($C_aH_bO_cN_dX_e$) and oxygen (O_2) originates products such as carbon dioxide (CO_2), water (H_2O), hydrates (HX), carbon monoxide (CO) and nitrogen (N_2). The overall chemical reaction formula is

$$C_aH_bO_cN_dX_e + \left(a - \frac{g}{2} + \frac{b-e}{4} - \frac{c}{2}\right) O_2 \rightarrow (a-c)CO_2 + \frac{b-e}{2} H_2O + eHX + gCO + \frac{d}{2} N_2, \tag{8}$$

where particular constants a to g apply in any specific instance. Therefore, based on expression (8) and by definition, the β ratio is given by

$$\beta = \frac{4a+2b+2e+2d}{4a+b-e-2c-2g}, \tag{9}$$

Depending on the type of hydrocarbon subjected to combustion, several estimates are known for the β ratio, usually values between one and two[†].

The volumetric flow of the gases in the exhaust duct is obtained from

$$q_{V_c} = \frac{k_t}{k_p} \cdot \sqrt{\frac{2\Delta p}{\rho_{T_0}} \cdot \frac{T_0}{T}} \cdot A, \tag{10}$$

where Δp is the differential pressure measured in a bidirectional pressure sensor located inside the exhaust duct (in Pa); ρ_{T_0} is the moist air density[‡] for a reference temperature T_0, equal to 298.15 K; T is the gas temperature in the exhaust duct (in K); A is the area (in m^2) of the exhaust duct circular cross-section; k_p is the differential pressure correction factor; and k_t is the global correction factor.

Since the exhaust duct has a circular cross-section, its area corresponds to

$$A = \frac{\pi}{4} \cdot d^2, \tag{11}$$

where d is the exhaust duct diameter (in m).

For the quantification of the moist air density (considering the reference temperature T_0 in K), the following expression [5] is used:

† Examples of β ratio estimates for the combustion of: carbon (C, $\beta = 1$); ethylene (C_2H_4, $\beta = 1.3$); propene (C_3H_6, $\beta = 1.3$); butane (C_4H_{10}, $\beta = 1.4$); heptane (C_7H_{16}, $\beta = 1.4$); propane (C_3H_8, $\beta = 1.4$); ethane (C_2H_6, $\beta = 1.4$); methane (CH_4, $\beta = 1.5$); hydrogen (H_2, $\beta = 2$).

‡ Since the density of the gas mixture inside the exhaust duct is unknown, this quantity is assumed to be close to the moist air density (expressed in kg·m^{-3}).

$$\rho_{T_0} = \frac{0.34848 \cdot p_{\text{atm}} - 0.009024 \cdot \text{rh} \cdot \exp[0.0612 \cdot (T_0 - 273.15\text{K})]}{T_0}. \tag{12}$$

The differential pressure correction factor is considered in expression (10) due to the use of a bidirectional sensor [6] instead of a conventional Pitot tube (vulnerable to solid particles in the flow). This quantity is defined by

$$k_p = \frac{\sqrt{\frac{\Delta p}{\rho_{T_{\text{amb}}}}}}{v_{\text{c}}}, \tag{13}$$

v_{c} is the flow velocity in the centre of the exhaust duct cross-section (in m·s^{-1}) and $\rho_{T_{\text{amb}}}$ is the moist air density [5] for ambient temperature, T_{amb}, (in K):

$$\rho_{T_{\text{amb}}} = \frac{0.34848 \cdot p_{\text{atm}} - 0.009024 \cdot \text{rh} \cdot \exp[0.0612 \cdot (T_{\text{amb}} - 273.15)]}{T_{\text{amb}}}. \tag{14}$$

The global correction factork_{t} corresponds to the average of three individual corrections, $k_{\text{t, v}}$, $k_{\text{t, propane}}$, $k_{\text{t, heptane}}$, related to the periodic testing of the SBI experimental apparatus aiming, respectively, at the determination of the non-uniformity of the flow velocity in the exhaust duct and the comparison between experimental and theoretical heat release rate values, concerning the combustion of known pure substances such as propane and heptane.

In the case of the $k_{\text{t, v}}$ correction, its quantification is supported by

$$k_{\text{t, v}} = \frac{\sum_{i=1}^{5} v_i}{5 \cdot v_{\text{c}}}, \tag{15}$$

considering the average[§] flow velocities measured in the i radius of the exhaust duct, v_i, and in its centre, v_{c}, all these quantities being expressed in m·s^{-1}.

The $k_{\text{t, propane}}$ correction is expressed by the ratio of the theoretical and the experimental heat release rate values of the propane combustion (in kW) respectively, $\text{HRR}_i^{\text{theoretical}}$ and $\text{HRR}_i^{\text{experimental}}$, i.e.,

$$k_{\text{t, propane}} = k_{\text{t}}' \cdot \frac{\sum \text{HRR}_i^{\text{theoretical}}}{\sum \text{HRR}_i^{\text{experimental}}}, \tag{16}$$

considering the several testing steps indexed by i of this normalized test [1], where k_{t}' is the global correction used in the experimental determination of the heat release rate[**]. The theoretical heat release rate at the ith step is given by

[§] The measurement sample is composed of four velocity measurements in each of five normalized distances from the exhaust duct centre, in addition to four velocity measurements at the centre.

[**] This quantity is also included in $\text{HRR}_i^{\text{experimental}}$; therefore, it can be removed from expression (16).

$$\mathrm{HRR}_i^{\mathrm{theoretical}} = q_{m_i} \cdot \left|\Delta h_{\mathrm{c}}^{\mathrm{l}}\right|_{\mathrm{propane}}. \tag{17}$$

q_{m_i} is the propane mass flow(inkg·s^{-1}) in the ith testing step, and $\left|\Delta h_{\mathrm{c}}^{\mathrm{l}}\right|_{\mathrm{propane}}$ is the low enthalpy of propane combustion per unit of mass (in kJ·kg^{-1}). Note that, in the calculation of $\mathrm{HRR}_i^{\mathrm{experimental}}$ by expression (1), the heat released per unit of consumed oxygen volume adopts a specific estimate and measurement uncertainty known for propane instead of the value mentioned in [4] and used for construction materials in the SBI test.

The$k_{\mathrm{t,heptane}}$ correction is obtained from

$$k_{\mathrm{t,\,heptane}} = k_t' \cdot \frac{\left|\Delta h_{\mathrm{c}}^{\mathrm{l}}\right|_{\mathrm{heptane}} \cdot m_{\mathrm{heptane}}}{\mathrm{THR}}, \tag{18}$$

where k_{t}' is the global correction used in the experimental determination of the total heat release[††], THR (in MJ) during the heptane combustion test, $\left|\Delta h_{\mathrm{c}}^{\mathrm{l}}\right|_{\mathrm{heptane}}$ is the low enthalpy of heptane combustion per unit of mass (in kJ·kg^{-1}) and m_{heptane} is heptane mass used as burning combustible (in kg). As in the case of propane combustion, the estimate and measurement uncertainty of the heat released per unit of consumed oxygen volume adopts known values for heptane, when using expression (1).

2.2. *Smoke production rate measurement model*

The smoke production rate, SPR, is defined in a similar way to the heat release rate quantity. However, its measurement is based on the light attenuation phenomenon resulting from the presence of smoke in an optical path. In this case, the measurement model corresponds to

$$\mathrm{SPR} = \frac{q_{V_{\mathrm{c}}}}{L} \cdot \frac{T}{T_0} \cdot \ln\left(\frac{1}{\tau}\right), \tag{19}$$

where $q_{V_{\mathrm{c}}}$ is the volumetric flow of gases in the exhaust duct (in m^3·s^{-1}), obtained from expression (10); L is the optical path length (in m); the factor T/T_0 is a correction for the temperature difference between the gases in the exhaust duct, T (in K), and the reference temperature, T_0, is equal to 298.15 K; and τ is transmittance, defined as the ratio ofthe luminous intensity measured in a given time instant and in the initial testing stage, I and I_0, respectively. In the SBI test, the luminous intensity that reaches the photodetector installed in the exhaust

[††]This quantity is included in the quantification of THR and can be removed from expression (18).

duct, is considered proportional to the electrical tension between its terminals so that the transmittance quantity is determined by electrical tension measurements.

3. Uncertainty propagation

The propagation of measurement uncertainty comprises two main stages. In the first stage, the input quantities in the mathematical models involved in the measurements are identified. Each input quantity is characterized by a probability density function (PDF) encapsulating knowledge of the quantity, whether obtained by observation or other means. In the second stage, PDFs for the quantities of interest (HRR and SPR) are obtained, based on the propagation of the PDFs for the input quantities through the mathematical models.

In the SBI test, the heat release rate quantity is measured in different test stages, firstly in the preliminary periodic combustion of propane and heptane and, afterwards, during the combustion of the tested specimen. The only significant difference is related to the heat release rate per unit of consumed oxygen mass quantity, which assumes different estimates and measurement uncertainties in each test case (propane, heptane or specimen combustion).

Table 1 shows the adopted probabilistic formulation of the input quantities required for the determination of the total heat release rate related to the combustion of a certain specimen, which already includes (in the global correction factor) the measurement uncertainty of the heat release rate measured in the propane and heptane combustions.

Table 1. Probabilistic formulation of the input quantities related to the heat release rate

Quantity	PDF	Estimate	Standard uncertainty
Relative humidity	Gaussian	60.1 %	1.1 %
Atmospheric pressure	Gaussian	101.4 kPa	0.2 kPa
Initial air temperature inside the exhaust duct	Gaussian	288.3 K	0.1 K
Oxygen density	Gaussian	1.308 $kg{\cdot}m^{-3}$	0.003 $kg{\cdot}m^{-3}$
Heat release per unit of consumed oxygen mass	Gaussian	13.1 $MJ{\cdot}kg^{-1}$	0.3 $MJ{\cdot}kg^{-1}$
Initial molar fraction of oxygen	Gaussian	0.209 5	0.000 04
Molar fraction of oxygen	Gaussian	0.206 7	0.000 2
Initial molar fraction of carbon dioxide	Gaussian	0.000 3	0.000 005
Molar fraction of carbon dioxide	Gaussian	0.001 8	0.000 005
Exhaust duct diameter	Gaussian	0.315 m	0.001 m
Exhaust gas temperature	Gaussian	313.8 K	0.4 K
Ambient temperature	Gaussian	288.6 K	0.7 K
Differential pressure	Gaussian	68.6 Pa	2.1 Pa
Flow velocity in the centre of the duct	Gaussian	9.6 $m{\cdot}s^{-1}$	0.7 $m{\cdot}s^{-1}$
Ratio between the amount of substance of combustion products and of consumed oxygen	Uniform	1.5	0.3
Global correction factor	Gaussian	0.77	0.02

Regarding SPR, Table 2 presents the adopted probabilistic formulation of the input quantities which supported the MCM calculations.

Table 2. Probabilistic formulation of the input quantities related to the smoke production rate

Quantity	PDF	Estimate	Standard uncertainty
Relative humidity	Gaussian	60.1 %	1.1 %
Atmospheric pressure	Gaussian	101.4 kPa	0.2 kPa
Ambient temperature	Gaussian	288.6 K	0.7 K
Exhaust duct diameter	Gaussian	0.315 m	0.001 m
Optical path length	Gaussian	0.315 m	0.001 m
Transmittance	Gaussian	0.974	0.005
Exhaust gas temperature	Gaussian	313.8 K	0.4 K
Differential pressure	Gaussian	68.6 Pa	2.1 Pa
Flow velocity in the centre of the duct	Gaussian	9.6 $m \cdot s^{-1}$	0.7 $m \cdot s^{-1}$
Global correction factor	Gaussian	0.77	0.02

The Monte Carlo method (MCM) is used in the calculation stage [7] [8] and is justified by the multivariate, non-linear and complicated nature of the functional relations between the many input, intermediate and output quantities. For this purpose, a high-quality pseudo-random number generator [9] is used to obtain numerical sequences based on 10^6 trials, in order to obtain robust convergent solutions. In addition, validated computational tools for converting and sorting the generated numerical sequences were used.

Estimates of HRR and SPR and their associated standard uncertainties are given respectively by the expectations and standard deviations of their PDFs [2].

4. Results

The measurement uncertainties of the input quantities shown in Table 1 were propagated by MCM to the intermediate quantities (results shown in Table 3) and then to HRR. With respect to the intermediate quantities related to the output smoke production quantity, the obtained results are shown in Table 4.

Table 5 presents specimen results for a thermal power level of 30 kW for HRR and SPR provided by MCM. Figure 1 shows the PDFs obtained for these quantities, which are visually close to Gaussian PDFs. The PDFs in Figure 1 are in fact approximations given by the MCM in the form of scaled histograms, the quality of which would improve with the number of Monte Carlo trials [7]. Repeating the calculation with a larger number of Monte Carlo trials did not change the values in Table 5. In comparing the results with those from applying the GUM, there was full agreement to the numerical precision given in Table 5.

Additional calculations for higher thermal power levels (up to 250 kW) showed a relative standard uncertainty for HRR between 8.0 % and 9.3 %, while for higher smoke levels (SPR up to 6.8 m^2s^{-1}), the obtained relative standard uncertainty varied between 9.1 % and 12 %, noticing that the 28 % relative standard uncertainty shown in Table 5 corresponds to a much lower smoke level (close to 0,05 $m^2 \cdot s^{-1}$).

Table 3. MCM results for intermediate quantities in the calculation of the heat release rate

Quantity	PDF	Estimate	Standard uncertainty
Water vapour molar faction	Gaussian	0.014 6	0.000 4
Moist air density for ambient temperature	Gaussian	1.180 $kg \cdot m^{-3}$	0.005 $kg \cdot m^{-3}$
Differential pressure correction factor	Gaussian	1.15	0.09
Expansion factor	Gaussian	1.10	0.05
Volumetric flow of gases in the exhaust duct	Gaussian	0.55 $m^3 \cdot s^{-1}$	0.04 $m^3 \cdot s^{-1}$
Heat release per unit of consumed oxygen volume	Gaussian	17.1 $MJ \cdot m^{-3}$	0.4 $MJ \cdot m^{-3}$
Ambient oxygen molar fraction	Gaussian	0.207 4	0.00005
Oxygen depletion factor	Gaussian	0.015	0.001
Ambient volumetric flow	Gaussian	0.55 $m^3 \cdot s^{-1}$	0.04 $m^3 \cdot s^{-1}$

Table 4. MCM results for intermediate quantities related to the smoke release rate

Quantity	PDF	Estimate	Standard uncertainty
Moist air density for ambient temperature	Gaussian	1.220 $kg \cdot m^{-3}$	0.005 $kg \cdot m^{-3}$
Differential pressure correction factor	Gaussian	1.11	0.09
Volumetric flow of gases in the exhaust duct	Gaussian	0.55 $m^3 \cdot s^{-1}$	0.04 $m^3 \cdot s^{-1}$

Table 5. MCM results for the output quantities

Quantity	Estimate	Standard uncertainty	Relative standard uncertainty
HRR	29.8 kW	3.0 kW	10 %
SPR	0.051 $m^2 \cdot s^{-1}$	0.014 $m^2 \cdot s^{-1}$	28 %

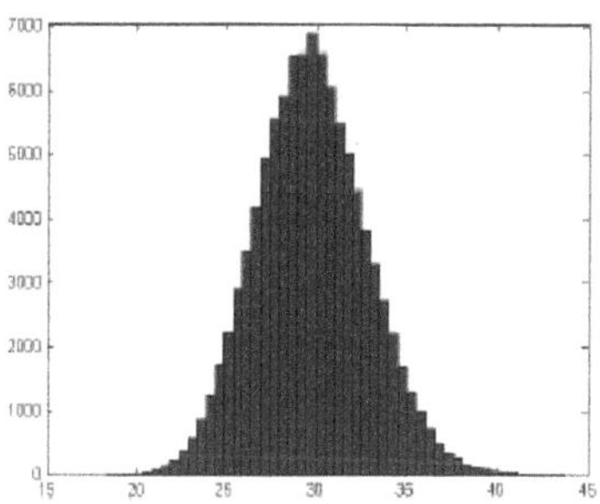

(a)

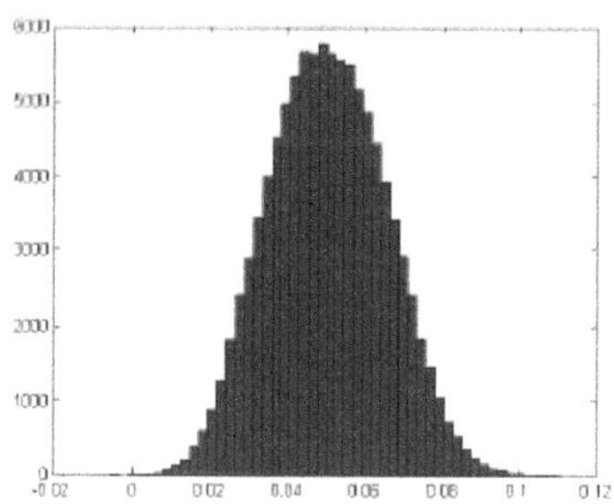

(b)

Figure 1. PDF for the output quantities: (a) heat release rate, HRR; (b) smoke production rate, SPR

5. Conclusions

A Monte Carlo method (MCM),was used to estimate heat release rate (HRR) and smoke production rate (SPR) for the Single Burning Item (SBI) test, which are defined by complicated mathematical models with many input quantities.

MCM also provided uncertainties associated with these estimates. For the typical data used, the GUM method described in [3] is validated to the numerical precision given. A major advantage of using the MCM over the GUM approach is its greater simplicity when dealing with non-linear mathematical models.

Acknowledgment

This work was carried out as part of project 17NRM05 EMUE (Examples of Measurement Uncertainty Evaluation), Advancing measurement uncertainty – comprehensive examples for key international standards, which has received funding from the EMPIR programme co-financed by the Participating States and from the European Union's Horizon 2020 research and innovation programme.

References

[1] EN 13823:2010+A1:2014, *Reaction to fire tests for building products. Building products excluding floorings exposed to the thermal attack by a single item* (CEN, Brussels, Belgium, 2014).

[2] JCGM 100:2008, *Evaluation of measurement data – Guide to the expression of uncertainty in measurement*, (Joint Committee for Guides in Metrology, 2008).

[3] CEN/TR 16988:2016, *Estimation of uncertainty in the single burning item test*, (CEN, Brussels, Belgium, 2016).

[4] W. Parker, *Calculations of the heat release rate by oxygen consumption for various applications*, Internal Report 81-2427-1 (National Bureau of Standards, Washington DC, 1982).

[5] M. Kochsiek and M. Gläser, *Comprehensive Mass Metrology*, Berlin, Wiley-VCH, 2000.

[6] B. McCaffrey, G. Heskestad, *A robust bidirectional low-velocity probe for flame and fire application, Combustion and Flame*, **26**, 1, 125-127, February 1976.

[7] JCGM 101:2008, Supplement 1 to the "*Guide to the expression of uncertainty in measurement" – Propagation of distributions using a Monte Carlo method*, (Joint Committee for Guides in Metrology, 2008).

[8] JCGM 102:2011, Supplement 2 to the "*Guide to the expression of uncertainty in measurement" – Extension to any number of output quantities*, (Joint Committee for Guides in Metrology, 2011).

[9] M. Matsumoto and T. Nishimura, Mersenne Twister: a 623-dimensionally equidistributed uniform pseudorandom number generator, *ACM Transactions on Modeling and Computer Simulation* **8**, 1 (1998).

F Pavese, A B Forbes, N F Zhang, A G Chunovkina (Eds.):
Advanced Mathematical and Computational Tools in Metrology and Testing XII
Series on Advances in Mathematics for Applied Sciences, Vol. 90

https://doi.org/10.1142/9789811242380_0020

Type systems for programs respecting dimensions

Conor McBride and Fredrik Nordvall-Forsberg
Computer and Information Sciences, University of Strathclyde, UK

Type systems can be used for tracking dimensional consistency of numerical computations: we present an extension from dimensions of scalar quantities to dimensions of vectors and matrices, making use of dependent types from programming language theory. We show that our types are unique, and most general. We further show that we can give straightforward dimensioned types to many common matrix operations such as addition, multiplication, determinants, traces, and fundamental row operations.

Keywords: Units of measure; Dimensions; Type systems; Dependent types.

1. Introduction

Measurements of physical quantitites naturally come with *dimensions* such as mass, length and time, and so on. The casual conceptual conflation of physical quantities with mere numbers is a standard computational practice resulting in lamentable errors, notoriously such as colliding with Mars[1]. A common approach to addressing this problem is to bundle numbers with units in compound data structures, then check dynamically whether arithmetic operations make sense before performing them. However, that approach pushes to run time what is much better managed at the time of a program's construction. It is far better to render dimensional nonsense unthinkable than merely to promise to complain if it happens.

In the field of programming languages, *type systems* are used to classify programs by the type of data they manipulate, such as integers, booleans, functions from integers to integers, and so on. Type information is used to guarantee once and for ever that certain classes of errors will not happen when running the program, such as feeding an integer to a function expecting a boolean as an input. Types not only police mistakes, but also mentor development of programs, by structuring the space of meaningful subprograms that can be used to compute a particular intermediate result.

For numerical programs, both dimension checking and type checking are important for the overall trust in computed results. Type systems are natural tools for mechanising the concept of dimensional analysis, yielding a lightweight approach to data integrity for metrology. Indeed, there is already work in this direction, with dimension types implemented in mainstream languages using a plethora of techniques, ranging from static types to dynamic run-time checks (see Bennich-Björkman and McKeever[2] for a comprehensive survey of libraries for dimension and unit checking):

- Microsoft's F#[3] has units of measure built in to static type checking;
- C++'s Boost Units library[4] uses templates to check units statically;
- Java has a proposed API adding classes for dimensioned quantities[5], but run-time casts are inevitable;
- Haskell's type system can now encode basic units of measure as a library[6];
- Python libraries such as Pint[7] cannot do static checking of dimensional correctness, but implement run-time checks instead.

None of the above support static units for compound data structures, such as vectors and matrices, except in overly uniform ways (e.g. matrices of quantities all in the same dimension). Here, we develop a type system for the dimensions of those matrices representing linear transformations between "dimensioned" vector spaces. Hart[8] developed the theory of such matrices, to which we gratefully contribute a rendering as a type system. Since dimensions have an equational theory richer than routinely available in mainstream programming languages (Sec. 2), we will need *dependent types* (Sec. 3) to be accurate about dimensioned matrices and operations thereon (Sec. 4). Griffioen has explored this direction[9], but so far without dependent types, yielding a rigid set of primitive matrix operations.

2. Units of measure, mathematically

When computing with numbers intended as physical quantities, one cheap way to promote virtue over vice is to ascribe *physical dimensions* to those numbers. It is standard (as in the Système Internationale), to fix a set of *fundamental* dimensions such as length, time and mass, then fix standard units (base units) for each: respectively metre, second and kilogram. We may multiply and divide dimensions, so that velocity is length per time and force is mass times distance per time squared. Formally, we define:

Definition 2.1. The **free abelian group** $\mathbb{G}$ on set G has (i) an *identity element*, 1; (ii) *multiplication*, $\_\cdot\_$, satisfying $1 \cdot x = x$, $(x \cdot y) \cdot z = x \cdot (y \cdot z)$ and $x \cdot y = y \cdot x$; (iii) *inversion*, $\_^{-1}$, satisfying $x^{-1} \cdot x = 1$; (iv) *generators*, $G \subset \mathbb{G}$.

We now take the fundamental dimensions as G and obtain the free abelian group $\mathbb{G}$ of all the dimensions. What makes $\mathbb{G}$ *free* is that every homomorphism (structure-preserving interpretation) from $\mathbb{G}$ to any other abelian group is determined exactly by the interpretation of G. We readily obtain a normal form for dimensions by imposing (arbitrarily) a total order on G, e.g. mass M before length L before time T, etc: any dimension may be given as a finite product of distinct fundamental dimensions, in the chosen order, raised to nonzero integral powers. Hence to check equality of dimensions $d \stackrel{?}{=} d'$, we can reduce d and d' to normal forms $d = \mathsf{M}^{n_0} \cdot \mathsf{L}^{n_1} \cdot \mathsf{T}^{n_2}$, $d' = \mathsf{M}^{n'_0} \cdot \mathsf{L}^{n'_1} \cdot \mathsf{T}^{n'_2}$, and then straightforwardly check equality of the exponents $n_i \stackrel{?}{=} n'_i$, rather than applying the group axioms directly.

The crucial step in making dimension checking part of type checking is to allow *abstract* dimensions[10]: addition is not length-specific, but works in one *arbitrary* dimension, which can stand for any dimension in particular. The type system thus does algebra with unknown dimensions in $\mathbb{G}$. In effect, we add variables as generators: ensuring dimensional consistency amounts to computing homomorphisms (if they exist) to instantiate the variables and solve equations. E.g., if we need $x^2 \cdot \mathsf{L} = y \cdot \mathsf{T}^{-1}$, we should take $y = x^2 \cdot \mathsf{L} \cdot \mathsf{T}$, leaving x to vary freely. Recalling that a homomorphism is determined by its action on generators, the above gives a homomorphism from $\mathbb{G}[x, y]$ to $\mathbb{G}[x]$, where brackets show generator extensions. Not all equations have solutions: there is no z such that $z^2 = \mathsf{L}$. If a solution exists, then there is a solution through which all solutions factor — a fact which allows the retention of most general types for a rich class of programs[11].

3. Dependent type systems

Types in programming languages were invented for the benefit of machines, e.g. to support memory layout, but their real payoff is for humans. In typed languages from last century, types are used to detect and avoid basic errors such as dividing an integer by a string. Types are characterised by the *values* they contain, but their job is now to classify *expressions* as meaningful or otherwise. A meaningful expression is expected to compute to a value of its type, when its free variables are substituted by values of their types. For example, matrix multiplication yields a value only when the input sizes (which are numerical values) match appropriately. To do a proper job classifying matrix multiplication as meaningful, our systems must thus let types talk about values. Such type systems are called **dependent type**

systems[12], as types can depend on values. They give us enough language to be honest about meaningfulness as a *relative* notion. If our types of matrices, $\mathsf{Matrix}(n, m)$, specifies their size, then matrix multiplication can be given type

$$\mathsf{Matrix}(n, m) \times \mathsf{Matrix}(m, k) \to \mathsf{Matrix}(n, k)$$

i.e. insisting that the input sizes are compatible, and determining the output size. In Sec. 4, we will generalise the type of matrices from *sizes* to *dimensions*.

This century, types make an active contribution by offering guidance *during* the program construction process, not just criticism afterwards. E.g., only types $\mathsf{Matrix}(n, n)$ admit an identity matrix, and the type already determines which one. Working interactively in a dependently typed language such as Agda[13], we can always see the sizes of the matrix we want, and the components we have available. The space within which we search for programs is correspondingly smaller and better structured.

Notation 3.1. We write $a \in A$ for the assertion that that the expression a is classified as meaningful in type A. We write Set for the type of small types[a]. In general, we write $B \in A \to \mathsf{Set}$ for the assertion that B is a type depending on a value of type A, i.e. given any $a \in A$, $B(a)$ is a type.

With this notation in place, we can now introduce types for dependent functions and dependent pairs — the workhorses of dependently typed programming. Dependent function types give outputs meaning relative to inputs, and dependent pair types reflect the way choices we are offered later can depend on choices we have made earlier.

Definition 3.1. Given a type A and $B \in A \to \mathsf{Set}$, we write $(x \in A) \to B(x)$ for the type of dependent functions, consisting of functions f with domain A, such that $f(a) \in B(a)$ for every $a \in A$. We may just write a standard function type, $A \to B$, for $(x \in A) \to B$ when x does not occur in B. Nested $\to$ groups to the right, i.e. we write $(x \in A) \to (y \in B(x)) \to C(x, y)$ for $(x \in A) \to ((y \in B(x)) \to C(x, y))$.

---

[a]There cannot be a type of all types, for the same reason that there cannot be a set of all sets: this would lead to a paradox. Most ordinary types, such as the type of natural numbers, the type of lists with elements from a small type, the type of functions between small types, etc, are small, but, crucially, Set itself is not a small type. Worried readers are invited think of small types simply as sets of ordinary values.

Definition 3.2. Given a type A and $B \in A \to \mathsf{Set}$, we write $(x \in A) \times B(x)$ for the type of dependent pairs (a, b), where $a \in A$ and $b \in B(a)$. We may just write a standard Cartesian product, $A \times B$, for $(x \in A) \times B$ when x does not occur in B. $\times$ groups more tightly than $\to$, i.e. we write $A \times B \to C$ for $(A \times B) \to C$.

We write $\mathbb{N}$ for the type of natural numbers, and $\mathbb{1}$ for the trivial type with exactly one value. We will make use of two more so-called inductively defined types: the type of lists, and the type of list index positions. We use their description as an opportunity to give informal examples of how types are introduced in dependent type theory.

Example 3.1. Let $X \in \mathsf{Set}$. Then $\mathsf{List}(X)$ is also a type — of lists with elements drawn from X. We always accept that the empty list ε is a value in $\mathsf{List}(X)$. The type $\mathsf{List}(X)$ also admits values of the form "something, somethings", provided that "something" (the head of the list) is one element of X, and the "somethings" (the tail of the list) constitute a further $\mathsf{List}(X)$. That is, $\mathsf{List}(X)$ admits values as follows:

$\varepsilon \in \mathsf{List}(X)$	if	$x \in X$ and $xs \in \mathsf{List}(X)$
	then	$x, xs \in \mathsf{List}(X)$

By explaining how to make values, we have ipso facto explained how to classify expressions. We may read x and xs in the above as *schematic* variables which may stand for arbitrary program expressions, and thus obtain a procedure for classifying expressions which compute to lists by reading the rules from bottom to top. For example, we may check that $7, (2{+}9), 13, \varepsilon \in \mathsf{List}(\mathbb{N})$ by checking that each of 7, 2+9 and 13 are meaningful as natural numbers.

As an example of a dependent type, consider the following type of *evidence* that a given element occurs in a given list.

Example 3.2. Let $X \in \mathsf{Set}$, $x \in X$, and $xs \in \mathsf{List}(X)$. The type $(x \leftarrow xs)$ admits values as follows:

$0 \in (x \leftarrow x, xs)$	if	$i \in (x \leftarrow xs)$
	then	$i + 1 \in (x \leftarrow y, xs)$

That is, the type $(x \leftarrow xs)$ is the type of index positions at which x occurs in xs: 0 is the position of the head of a nonempty list; $i + 1$ is the position in a nonempty list corresponding to position i in its tail. We may check that $1 \in (11 \leftarrow 7, (2{+}9), 13, \varepsilon)$ provided we are willing to do enough work to identify 2+9 with 11. There may be many ways to classify a number (e.g. 1) as meaningful, in accordance with some purpose which we intend. We use $(x \leftarrow xs)$ as a type which gives it additional meaning as an index into a list: we know that this index is bounded by the length of the list, and also know what value sits in the indexed position.

It is not unusual to use numbers as loop indices or as coordinates of matrix entries, and quite often both. By refining our classification of numbers, we can make that connection clear, and begin to characterise the difference between we find at different coordinates.

4. Multidimensional units of measure

In this section, we put our expressive types to work by using them to keep track of units of measure for matrices and matrix operations. We start by summarising the necessary algebraic structure.

4.1. *Monoids and semirings*

Our work will rely heavily on algebraic properties of operations such as addition and multiplication, not only for numbers, but also for dimensions, physical quantities and matrices. Our starting point is the structure of a monoid, being a way to crush (possibly empty) sequences of values into a single value without regard to their grouping.

Definition 4.1. A **monoid** $M = (A, \_\cdot\_, 1)$ consists of

$$A \in \mathsf{Set} \quad \text{with} \quad \_\cdot\_ \in A \times A \to A \quad \text{and} \quad 1 \in A.$$

We call A the carrier of M, and often write $|M| = A$. The multiplication $\_\cdot\_$ is required to be associative, and to absorb the identity element 1 on both sides, i.e.

$$x\cdot(y\cdot z) = (x\cdot y)\cdot z \qquad x\cdot 1 = x \qquad 1\cdot y = y$$

Examples of monoids include $(\mathbb{N}, +, 0)$, $(\mathbb{N}, \times, 1)$, and $(\mathsf{List}(X), +\!\!+, \varepsilon)$, where $+\!\!+$ is list concatenation, and the trivial monoid whose carrier is $\mathbb{1}$. Common refinements of monoids include **commutative monoids**, i.e. where $x\cdot y = y\cdot x$, and **groups**, where there is also a unary inverse $\_^{-1}$ on the carrier, such that $x^{-1}\cdot x = 1$, and **abelian groups**, which combine both.

Definition 4.2. A **monoid homomorphism** between monoids M and N is a function $h \in |M| \to |N|$ which preserves the monoid structure, i.e.

$$h(x\cdot_M y) = h(x)\cdot_N h(y) \qquad h(1_M) = 1_N$$

Examples of monoid homomorphisms include:

- exponentiation k^{-} from $(\mathbb{N}, +, 0)$ to $(\mathbb{N}, \times, 1)$, for a fixed $k \in \mathbb{N}$;
- summation from $(\mathsf{List}(\mathbb{N}), +\!\!+, \varepsilon)$ to $(\mathbb{N}, +, 0)$;
- iterated product from $(\mathsf{List}(\mathbb{N}), +\!\!+, \varepsilon)$ to $(\mathbb{N}, \times, 1)$;
- remappings $[f(x) \mid x \leftarrow \_]$ from $(\mathsf{List}(X), +\!\!+, \varepsilon)$ to $(\mathsf{List}(Y), +\!\!+, \varepsilon)$, for any $f \in X \to Y$, where $[f(x) \mid x \leftarrow xs]$ means the list of the values $f(x)$ for each x drawn in turn from xs.

Once we have additive and multiplicative monoid structure on the same carrier, and they are compatible, we have the machinery for matrix multiplication. This is the structure of a semiring[b].

Definition 4.3. A **semiring** is a type $R \in \mathsf{Set}$, such that: (i) there is a commutative monoid $(R, +, 0)$; (ii) there is a monoid $(R, \_ \cdot \_, 1)$; (iii) $x \cdot \_$ and $\_ \cdot z$ act on $(R, +, 0)$ as monoid homomorphisms, i.e. multiplication distributes:

$$x \cdot 0 = 0 \quad x \cdot (y + z) = x \cdot y + x \cdot z \qquad 0 \cdot z = z \quad (x + y) \cdot z = x \cdot z + y \cdot z$$

As an example, our previously introduced monoids $(\mathbb{N}, +, 0)$ and $(\mathbb{N}, \times, 1)$ neatly combine to form a semiring. Physical quantities in a given dimension, however, do not: if you add two lengths, you get a length, but if you multiply two lengths, you get not a length but an area. We shall need to refine our notion of semiring to account for the dependency of quantities on dimensions. Before we can tackle dimensioned matrices, we need to treat dimensioned scalars.

4.2. *Dimensioned scalars*

Using dependent types, dimensioned scalars can be described elegantly by the structure of a semiring *graded* over the group of dimensions. Grading is a way of inducing relative notions of structure. Let us have a simple example length-respecting concatenation of lists.

Example 4.1. It is standard practice in dependently typed programming to refine lists $\mathsf{List}(X) \in \mathsf{Set}$ with a length index: $\mathsf{ListN}(X, \_) \in \mathbb{N} \to \mathsf{Set}$. $\mathsf{ListN}(X, n)$ is the type of lists of length n with elements from X, but it is not a monoid with respect to concatenation: concatenating two lists of length

[b] To obtain the better known notion of a *ring*, upgrade the additive structure of a semiring to an abelian group.

5 makes a list of length 10. However, dependent pairs in $(n\in\mathbb{N})\times \mathsf{ListN}(X,n)$ recover the monoidal structure of lists; moreover with first projection acting as a monoid homomorphism to $(\mathbb{N},+,0)$. For example,

$$(5,(a,b,c,d,e,\varepsilon)){+\!\!+}(5,(f,g,h,i,j,\varepsilon)) = (10,(a,b,c,d,e,f,g,h,i,j,\varepsilon)).$$

In order to construct this monoid on $(n\in\mathbb{N})\times \mathsf{ListN}(X,n)$, what structure is needed on $\mathsf{ListN}(X,\_)$? We need an "identity element" in $\mathsf{ListN}(X,0)$, and a way to "multiply" a $\mathsf{ListN}(X,n)$ and a $\mathsf{ListN}(X,m)$ to form a $\mathsf{ListN}(X,n+m)$. Intuitively, ε and ${+\!\!+}$ fit the bill: i.e., there is a family of concatenators ${+\!\!+}_{nm} \in \mathsf{ListN}(X,n)\times\mathsf{ListN}(X,m)\to\mathsf{ListN}(X,n+m)$ which respect the ordinary monoid structure of $+$.

Definition 4.4. Let $M = (|M|, \_\cdot\_, 1)$ be a monoid. A **graded monoid over** M is a triple $(L, \_\cdot_{\_\_}\_, 1)$, where

$$L \in |M| \to \mathsf{Set} \quad \text{with} \quad \_\cdot_{ij}\_ \in L(i)\times L(j)\to L(i{\cdot}j) \quad \text{and} \quad 1\in L(1).$$

In truth, the multiplication really has a dependent function type

$$(i\in|M|)\to(j\in|M|)\to L(i)\times L(j)\to L(i{\cdot}j)$$

but in mathematical vernacular, we suppress i and j, and in computational practice, a machine figures them out. The identity and associativity laws are commensurate by the monoid laws of M, and they must hold:

$$x\cdot_{i(jk)}(y\cdot_{jk}z) = (x\cdot_{ij}y)\cdot_{(ij)k}z \qquad x\cdot_{i1}1 = x \qquad 1\cdot_{1j}y = y$$

This is how we make the multiplication structure of physical quantities respect the underlying monoid structure of dimensions. Let us now deploy the same grading technique to the multiplicative structure of semirings: for the additive structure, it is enough to have an ordinary commutative monoid at every individual index.

Definition 4.5. Let $M = (|M|, \_\cdot\_, 1)$ be a monoid. A **semiring graded over** M is a dependent type $R \in |M| \to \mathsf{Set}$, such that: (i) for every $i\in|M|$, there is a commutative monoid, $(R(i), +_i, 0_i)$; (ii) there is a graded monoid $(R, \_\cdot_{\_\_}\_, 1)$; (iii) $x\cdot_{ij}\_ \in R(j)\to R(i{\cdot}j)$ and $\_\cdot_{ij}x \in R(i)\to R(i{\cdot}j)$ act as monoid homomorphisms.

Note that if M is the trivial monoid $M = \mathbb{1}$, this reduces to the ordinary notion of a non-graded semiring. We think of $R(d)$ as the commutative monoid of physical quantities with dimension d: we can add physical quantities of the same dimension — if q, $q'\in R(d)$, then $q +_d q'$ is well typed, and again $q +_d q' \in R(d)$ — and multiplying quantities also multiplies their dimensions: if $q \in R(d)$ and $q' \in R(d')$ then $q \cdot_{dd'} q' \in R(d \cdot d')$. Critically, we do not think of $R(d)$ as a set of *numbers*; both "six foot six" and "two metres" inhabit $R(\mathsf{L})$. Numerical rendering of physical quantities necessitates agreement of *units*!

Definition 4.6. A **unit system** for the graded semiring $R \in |M| \to \mathsf{Set}$ over M is a dependent function $\mathsf{u} \in (d \in |M|) \to R(d)$, satisfying (i) $\mathsf{u}(1) = 1$; (ii) $\mathsf{u}(i \cdot j) = \mathsf{u}(i) \cdot_{ij} \mathsf{u}(j)$; and (iii) $\mathsf{u}(i) \cdot_{i1} x = x \cdot_{1i} \mathsf{u}(i)$ for every $x \in R(1)$.

If M is a free abelian group $M = \mathbb{G}$, most of such a function u is determined by its action on generators — e.g., the SI gives units for all our dimensions, taking $\mathsf{u}(\mathsf{M}) = \mathsf{kg}$, $\mathsf{u}(\mathsf{L}) = \mathsf{m}$ and $\mathsf{u}(\mathsf{T}) = \mathsf{s}$ — but we must still ensure that these are sent to invertible elements which commute with dimensionless quantities. We observe, in this situation, that even though R does not in general support *division*, we have $\mathsf{u}(d) \cdot_{dd^{-1}} \mathsf{u}(d^{-1}) = \mathsf{u}(d \cdot d^{-1}) = \mathsf{u}(1) = 1$. Moreover, the quantities given as $x \cdot_{1d} \mathsf{u}(d)$ constitute a graded semiring in their own right: requirements (ii) and (iii), above, give $(x \cdot_{1i} \mathsf{u}(i)) \cdot_{ij} (y \cdot_{1j} \mathsf{u}(j)) = (x \cdot_{11} y) \cdot_{1(i \cdot j)} \mathsf{u}(i \cdot j)$. Such a u thus allows dimensionless numbers to *encode* physical quantities.

For the rest of the paper, we now assume that we are given a graded semiring $\mathsf{R} \in \mathbb{G} \to \mathsf{Set}$ over our group of physical dimensions.

4.3. *Dimensioned matrices*

What is an appropriate notion of dimensioned matrix? It is not obvious.

On one end of the scale, we could demand that every entry in the matrix is of the same dimension, such as in Fig. 1(a), where we have elided specific quantities for clarity. This is too restrictive in practice: for instance, many elementary applications in physics collect together distance, speed, acceleration, ..., in one matrix — of systematically different dimensions.

On the other end of the scale, we could imagine to let every entry have its own unrelated dimension, as in Fig. 1(d). However most such matrices

$$\begin{pmatrix} \mathsf{L} & \mathsf{L} \\ \mathsf{L} & \mathsf{L} \\ \mathsf{L} & \mathsf{L} \end{pmatrix} \quad \begin{pmatrix} \mathsf{L}\cdot\mathsf{M}^{-1} & \mathsf{T}\cdot\mathsf{M}^{-1} \\ \mathsf{L}\cdot\mathsf{T}^{-1} & \mathsf{T}\cdot\mathsf{T}^{-1} \\ \mathsf{L}\cdot\mathsf{L} & \mathsf{T}\cdot\mathsf{L} \end{pmatrix} \quad \begin{pmatrix} \mathsf{T} & \mathsf{T}\cdot\mathsf{M} \\ \mathsf{L}^{-1}\cdot\mathsf{T} & \mathsf{L}^{-1}\cdot\mathsf{T}\cdot\mathsf{M} \\ \mathsf{M}^{-1}\cdot\mathsf{T} & \mathsf{M}^{-1}\cdot\mathsf{T}\cdot\mathsf{M} \end{pmatrix} \quad \begin{pmatrix} \mathsf{L} & \mathsf{T} \\ \mathsf{T}^{-1} & \mathsf{L}^{3} \\ \mathsf{M}\cdot\mathsf{T} & \mathsf{L}^{-1} \end{pmatrix}$$

(a) Uniform (b) Columns and rows (c) Γ-dimensioned (d) Unrelated

Fig. 1. Possible dimensioned matrices

are without metrological and algebraic meaning — they are mere arrays of dimensioned quantities, rather than representations of linear transformations between dimensioned spaces, as we would expect dimensioned matrices to be. One way to see this is to notice that such dimensioned matrices with unrelated dimensions are not closed under multiplication — see Thm. 4.2 below. Matrix multiplication should correspond to composition of linear transformations, and yield the same closure property.

So instead, we consider more structured representations of such, as in Fig. 1(b), where matrix dimensions are an outer product of dimension vectors, i.e. each entry $M_{i,j}$ has dimension $a_i \cdot b_j$ obtained by multiplying the corresponding entries from a column vector $\vec{a} = a_1, \ldots, a_n$ and a row vector $\vec{b} = b_1, \ldots, b_m$. Be warned that this representation of dimensions is not unique: if M has dimensions given by $\vec{a}$ and $\vec{b}$, it also has dimensions given by $c \cdot \vec{a}$ and $c^{-1} \cdot \vec{b}$ for any group element c, since these perturbations cancel. Such redundancy is bad news if we hope to automate the checking of dimensions, e.g. by reducing it to type checking, since it means that any algorithm must do non-trivial work in order to decide equality.

We remove this redundancy by factoring all dimensions through the dimension b of the top left corner, recording the dimensions of the rest of the first column as and row cs respectively. The dimensions of all other entries (i, j) with $i, j > 1$ are given by multiplying the corresponding entries $a_i \cdot b \cdot c_j$, as in Fig. 1(c). It will yield some handy cancellation if all dimensions in as are inverted — without loss of generality, as $\mathbb{G}$ is a group. This leads to our definition of "Γ-dimensioned matrices".

Definition 4.7. Let $\mathsf{R} \in \mathbb{G} \to \mathsf{Set}$ be a graded semiring, $as, cs \in \mathsf{List}(\mathbb{G})$ lists of elements of $\mathbb{G}$, and $b \in \mathbb{G}$ a group element. We define the type of **Γ-dimensioned matrices** (of the given dimensions) as follows:

$$\Gamma\text{-}\mathsf{Matrix}(as, b, cs) := (a \leftarrow 1, as) \times (c \leftarrow 1, cs) \to \mathsf{R}\ (a^{-1} \cdot b \cdot c)$$

This is making use of the type of "positions" $(x \leftarrow xs)$ from Ex. 3.2. Hence a Γ-dimensioned matrix is a function M such that

$$M(x, y) \in R\ (a^{-1} \cdot b \cdot c) \quad \text{whenever} \quad x \in a \leftarrow 1, as \text{ and } y \in c \leftarrow 1, cs.$$

Notation 4.1. Mnemonically, we *draw* the type of matrices showing each parameter in the place it characterises, making a literal Γ, hence the name:

$$\left[\!\!\left[\begin{array}{cc} b & cs \\ as^{-1} & \end{array}\right]\!\!\right] := \Gamma\text{-}\mathsf{Matrix}(as, b, cs)$$

The $^{-1}$ above as is part of the picture, as a reminder that the as get inverted.

Example 4.2. Assume a unit system based on SI units. The following matrix M of physical quantities can be represented as an array of numbers (by writing the value $M(x, y)$ at numerical coordinates (x, y), as usual), and given the following Γ-dimensioned type:

$$\begin{pmatrix} 5\ \mathsf{m}^2 & 2\ \mathsf{kg}\cdot\mathsf{m}^2 \\ 1\ \mathsf{m}^2/\mathsf{s} & 3\ \mathsf{kg}\cdot\mathsf{m}^2/\mathsf{s} \end{pmatrix} \in \left[\!\!\left[\begin{array}{|cc|} \hline \mathsf{L}^2 & \mathsf{M} \\ \mathsf{T}^{-1} & \\ \hline \end{array}\right]\!\!\right]$$

If $\mathbb{G}$ is the trivial group $\mathbb{G} = \mathbb{1}$, then Def. 4.7 reduces to the familiar notion of $(n+1) \times (m+1)$-sized matrices for natural numbers n and m, as $\mathsf{List}(\mathbb{1}) \cong \mathbb{N}$.

Let us seek solace for the complexity of Γ-dimensioned matrices introduced by eliminating the redundancy from columns-and-rows. We note that for a list $as \in \mathsf{List}(A)$, a position in $(a \leftarrow as)$ is also a position in the remapping $(f(a) \leftarrow [f(x) \mid x \leftarrow as])$ for any function $f \in A \to B$. We can use remapping to normalise dimensions with respect to the corner:

Lemma 4.1. *A columns-and-rows dimensioned matrix with dimensions represented by non-empty lists a, as and b, bs is the same thing as a Γ-dimensioned matrix*

$$\left[\!\!\left[\begin{array}{|cc|} \hline a \cdot b & [b^{-1} \cdot x \mid x \leftarrow bs] \\ [x^{-1} \cdot a \mid x \leftarrow as]^{-1} & \\ \hline \end{array}\right]\!\!\right]$$

A Γ-dimensioned matrix

$$\left[\!\!\left[\begin{array}{|cc|} \hline b & cs \\ as^{-1} & \\ \hline \end{array}\right]\!\!\right]$$

is the same as a columns-and-rows dimensioned matrix with dimensions represented by the lists $[x^{-1} \mid x \leftarrow 1, as]$ and $[b \cdot x \mid x \leftarrow 1, bs]$. □

Hence the concepts of columns-and-rows dimensioned matrices and Γ-dimensioned matrices are interdefinable. Note that if we translate from columns-and-rows to Γ-dimensioned and back again, we may not get back the same lists of dimensions. In contrast, if we translate a Γ-dimensioned matrix forth and back, we preserve dimensions, due to the fusion law $[f(y) \mid y \leftarrow [g(x) \mid x \leftarrow as]] = [f(g(x)) \mid x \leftarrow as]$. This is not a coincidence, or a feat of the translation — Γ-dimensions are unique:

Theorem 4.1. *Assume that quantities have a unique dimension, i.e. if $r \in R(d)$ and $r \in R(d')$ then $d = d'$. Then Γ-dimensioned matrices also have unique Γ-dimensions, i.e. if*

$$M \in \left[\!\!\left[\begin{array}{|cc|} \hline b & cs \\ as^{-1} & \\ \hline \end{array}\right]\!\!\right] \quad \textit{and} \quad M \in \left[\!\!\left[\begin{array}{|cc|} \hline b' & cs' \\ as'^{-1} & \\ \hline \end{array}\right]\!\!\right]$$

then $as = as'$, $b = b'$ *and* $cs = cs'$. □

One might worry that the class of Γ-dimensioned matrices is too small to be useful, but as the following theorem shows, it is actually as good a class of dimensioned matrices as we can hope for, if we believe that matrix multiplication is a fundamental operation (as we do).

Theorem 4.2. *The collection of* Γ*-dimensioned matrices is the largest class of dimensioned matrices closed under multiplication.* □

The proof exploits the fact that multiplication computes a matrix of inner products whose summands must have uniform dimension.

4.4. *Dimension-aware matrix algebra*

Many well known operations on matrices can be given Γ-dimensioned types. Addition lifts straightforwardly, since it acts componentwise:

Theorem 4.3. *Matrix addition and the zero matrix have types*

$$+ \in \left[\begin{array}{c|c} b & cs \\ \hline as^{-1} & \end{array}\right] \times \left[\begin{array}{c|c} b & cs \\ \hline as^{-1} & \end{array}\right] \to \left[\begin{array}{c|c} b & cs \\ \hline as^{-1} & \end{array}\right] \qquad \mathbf{0} \in \left[\begin{array}{c|c} b & cs \\ \hline as^{-1} & \end{array}\right]$$

for any $as, cs \in \mathsf{List}(\mathbb{G})$ *and* $b \in \mathbb{G}$. □

In contrast, multiplication, $\bullet$, is defined only when the matrices involved are compatible, in size, as we are used to, but here also in dimensions:

Theorem 4.4. *Matrix multiplication and the identity matrix have types*

$$\bullet \in \left[\begin{array}{c|c} b & cs \\ \hline as^{-1} & \end{array}\right] \times \left[\begin{array}{c|c} d & es \\ \hline cs^{-1} & \end{array}\right] \to \left[\begin{array}{c|c} b\cdot d & es \\ \hline as^{-1} & \end{array}\right] \qquad \mathbf{1} \in \left[\begin{array}{c|c} 1 & as \\ \hline as^{-1} & \end{array}\right]$$

for any $as, cs, es \in \mathsf{List}(\mathbb{G})$ *and* $b, d \in \mathbb{G}$. □

Each element of the output is given for fixed positions in $(a \leftarrow 1, as)$ and $(e \leftarrow 1, es)$, but computed by an inner product ranging over all positions in $(c \leftarrow 1, cs)$. As we intend the notation to suggest, the row cs cancels the column cs^{-1}, yielding uniform dimension for all summands: $(a^{-1} \cdot b \cdot c) \cdot (c^{-1} \cdot d \cdot e) = a^{-1} \cdot (b \cdot (c \cdot c^{-1}) \cdot d) \cdot e = a^{-1} \cdot (b \cdot d) \cdot e$ as required.

Note that for $A \bullet B$ to be well typed, we require that the row of 'column dimensions' for A match the column of 'row dimensions' for B — this generalises the usual requirement that the *number of* columns of A should match the number of rows of B, and indeed, if $\mathbb{G}$ is the trivial group $\mathbb{G} = \mathbb{1}$, then the former reduces exactly to the latter. Similarly the identity matrix is always a 'square' matrix in the sense that its 'row dimensions' match its 'column dimensions'. This is often the appropriate notion of squareness in a dimensioned setting — many operations on square matrices do not make dimensioned sense on arbitrary matrices of square *size*, but do for square matrices in this stronger sense. A good example is the trace of a square matrix, i.e. the sum of the diagonal elements:

Theorem 4.5. *The trace operator has type*

$$\mathbf{tr} \in \left[\begin{array}{|cc|} b & as \\ as^{-1} & \end{array}\right] \to \mathsf{R}(b)$$

for any $as \in \mathsf{List}(\mathbb{G})$ *and* $b \in \mathbb{G}$. □

However, interestingly, the determinant of a Γ-dimensioned matrix makes sense also for matrices of square *size*. This can be seen by considering the formula for the determinant as a sum over all permutations

$$\mathbf{det}(M) = \sum_{\pi \in S_n} \mathbf{sgn}(\pi) \cdot \prod_{x<n} M(x, \pi(x))$$

and noticing that each summand will always have the same dimension — that of the product of the main diagonal.

Theorem 4.6. *The determinant operator has type*

$$\mathbf{det} \in \left[\begin{array}{|cc|} b & cs \\ as^{-1} & \end{array}\right] \to \mathsf{R}\Big(\big(\prod_{a \leftarrow as} a\big)^{-1} \cdot b^{n+1} \cdot \big(\prod_{c \leftarrow cs} c\big)\Big)$$

for any $b \in \mathbb{G}$, *and* $as, cs \in \mathsf{List}(\mathbb{G})$ *of the same length* $|as| = |cs| = n$. □

4.5. *Elementary row operations*

The important method of Gaussian elimination makes sense for Γ-dimensioned matrices, and so can be used to solve systems of dimensioned equations, compute dimensioned inverses, etc. This can be seen by finding dimensioned types for the elementary row operations used by the method. The proof uses Lem. 4.1 to translate back and forth to the more symmetric columns-and-rows dimensioned matrices. We will explain how the operations use their arguments below:

Theorem 4.7. *The elementary row operations have types thus:*

$$\mathsf{swapFirst} \in (p \in d \leftarrow as) \to \left[\begin{array}{|cc|} b & cs \\ as^{-1} & \end{array}\right] \to \left[\begin{array}{|cc|} d^{-1} \cdot b & cs \\ [d^{-1} \cdot x \mid x \leftarrow as[p \mapsto 1]]^{-1} & \end{array}\right]$$

$$\mathit{multFirst} \in R(d) \to \left[\begin{array}{|cc|}\hline b & cs \\ as^{-1} & \\ \hline\end{array}\right] \to \left[\begin{array}{|cc|}\hline d \cdot b & cs \\ [d \cdot a \mid x \leftarrow as]^{-1} & \\ \hline\end{array}\right]$$

$$\mathit{addMult} \in (d \leftarrow 1, as) \to (e \leftarrow 1, as) \to R(d^{-1} \cdot e) \to \left[\begin{array}{|cc|}\hline b & cs \\ as^{-1} & \\ \hline\end{array}\right] \to \left[\begin{array}{|cc|}\hline b & cs \\ as^{-1} & \\ \hline\end{array}\right]$$

□

The operation swapFirst takes a row to swap with as a position p for dimension d in as, and a matrix, returning a matrix with the dimensions also suitably swapped; the output row dimensions are given by swapping a 1 in place of the d, indicated by $as[p \mapsto 1]$, then renormalising with respect to the $d^{-1} \cdot b$ which is now in the top left corner. The operation multFirst is given a scalar and a matrix, and returns the matrix where the first row has been multiplied by the scalar. Of course, by combining swapFirst and multFirst, we can multiply any row by a scalar. Finally the operation addMult is given a source and a target position as indices of two group elements d and e into the list as, as well as a scalar to multiply by, of dimension $d^{-1} \cdot e$. This is a good example of a refined type being able to help with interactive program construction: the dimension of the multiplier already suggests a specific relationship between the source and target rows.

5. Conclusions and Future Work

Following Hart[8], we have shown how to use dependent types to extend type systems for units of measure from scalars to matrices. The key definition is of Γ-dimensioned matrices, recording the dimensions of matrix entries in a maximally liberal and unique way. We have also shown that many well known operations on matrices can be given Γ-dimension-respecting types. Our present technology, a general-purpose dependently typed language, obliges us to write more explicit proofs (elided here) of the list and group identities we rely upon to make types match. To pay for what we propose to get, we are developing a bespoke type checker with additional domain knowledge[14] about the equational theory of lists and abelian groups. Even to reach 'Base Camp Hart', we type theorists will have to raise our game.

Acknowledgments

We would like to thank Alistair Forbes, Keith Lines, Artem Shinkarov, and Ian Smith for interesting discussions related to this work, and the anonymous referee for comments and suggested improvements to the text. This work is funded by the UK National Physical Laboratory Measurement Fellowship project *Dependent types for trustworthy tools.*

References

1. A. G. Stephenson, L. S. LaPiana, D. R. Mulville, P. J. Rutledge, F. H. Bauer, D. Folta, G. A. Dukeman, R. Sackheim and P. Norvig, *The Mars Climate Orbiter Mishap Investigation Board Phase I Report*, tech. rep., NASA (1999).
2. O. Bennich-Björkman and S. McKeever, The next 700 unit of measurement checkers, in *SLE '18*, (ACM, 2018).
3. A. Kennedy, Programming languages and dimensions, PhD thesis, University of Cambridge, (United Kingdom, 1995).
4. M. C. Schabel and S. Watanabe, Boost C++ libraries, chapter 43 (Boost.Units 1.1.0) https://www.boost.org/doc/libs/1_74_0/doc/html/boost_units.html, (2010).
5. J.-M. Dautelle, W. Keil and O. Santana, JSR 385: Units of measurement https://unitsofmeasurement.github.io/, (2019).
6. T. Muranushi and R. Eisenberg, Experience report: Type-checking polymorphic units for astrophysics research in Haskell, in *Haskell '14*, (ACM, 2014).
7. Pint: makes units easy https://pint.readthedocs.io/, (2020).
8. G. W. Hart, *Multidimensional Analysis* (Springer, 1995).
9. P. Griffioen, A unit aware matrix language and its application in control and auditing, PhD thesis, University of Amsterdam, (the Netherlands, 2019).
10. M. Wand and P. O'Keefe, Automatic dimensional inference, in *Computational Logic: Essays in Honor of Alan Robinson*, eds. J.-L. Lassez and G. Plotkin (MIT Press, 1991) pp. 479–486.
11. A. Gundry, A typechecker plugin for units of measure: Domain-specific constraint solving in GHC Haskell, in *Haskell '15*, (ACM, 2015).
12. A. Bove and P. Dybjer, *Dependent Types at Work*, in *LerNet ALFA Summer School 2008 Revised Tutorial Lectures*, eds. A. Bove, L. S. Barbosa, A. Pardo and J. S. Pinto (Springer, 2009), pp. 57–99.
13. U. Norell, Towards a practical programming language based on dependent type theory, PhD thesis, Chalmers University, (Sweden, 2007).
14. G. Allais, C. McBride and P. Boutillier, New equations for neutral terms: a sound and complete decision procedure, formalized, in *DTP@ICFP 2013*, ed. S. Weirich (ACM, 2013).

F Pavese, A B Forbes, N F Zhang, A G Chunovkina (Eds):
Advanced Mathematical and Computational Tools in Metrology and Testing XII
Series on Advances in Mathematics for Applied Sciences, Vol. 90

https://doi.org/10.1142/9789811242380_0021

Measurement in science: Between evaluation and prediction

F. Pavese
Formerly at CNR-IMGC (then INRiM),
Torino, 10135, Italy
E-mail: frpavese@gmail.com

Measurement, considered a pillar in modern science for acquiring new knowledge, a tool valid from nearly full ignorance up to any stage of progressed knowledge. Stage boundaries cannot be set easily. This reflects onto a wide range of features in expressing cognitive situations. For each of them, in addition to express the level of knowledge on a subject matter, one has to delimitate the areas of ignorance by identifying the limits of the knowledge level on qualifying aspects of the matter under investigation. It is a fact that knowledge is gained in steps. These steps can be classified in different ways. The paper will discuss a step that will be considered preliminary to measurement: evaluation, which is basically a qualitative one. Then, two posterior steps will be illustrated: validation, whose aim is the obtain an inter-subjective agreement about the results of measurement; prediction, that is the true practical aim of science, intending to extrapolate to future observations or behaviours the past and current effects of contingent findings or "laws". This structure of the steps may require a better delimitation of the aims and features of measurement.

Keywords: evaluation; measurement; validation; prediction; simulation.

1. Introduction

The concept of measurement is considered a pillar in modern science for acquiring new knowledge. The following definition is found in the VIM:2012 [1]:

"2.1 *measurement*

Process of experimentally obtaining one or more quantity values that can reasonably be attributed to a quantity

NOTE 1 Measurement does not apply to nominal properties.

NOTE 2 Measurement implies comparison of quantities or counting of entities.

NOTE 3 Measurement presupposes a description of the quantity commensurate with the intended use of a measurement result, a measurement procedure, and a calibrated measuring system operating according to the specified measurement procedure, including the measurement conditions".

Its full validity is presently put in question by the restriction placed by Note 1, and is not intended to apply to quantum measurements. These facts will be shortly commented only in the Final Remarks at the end of this paper.

A measurement *process*—differently from a "measurement" consisting of the trivial reading an instrument indication, case not considered in the following—can be performed only on the basis of models, [1] [2] whose features and implementing *procedures* can be constructed at any level of current knowledge, from nearly full ignorance to any stage of progressed knowledge—being a fact that knowledge is gained and progresses in steps. Boundaries are not easy to be set, and the steps can be classified in different ways.

This reflects onto a wide range of features in modelling cognitive situations. For each of them, in addition to expressing the *level* (or *degree*) of knowledge on a subject matter, one has to delimitate the areas of ignorance by identifying the limits of those levels of knowledge on qualifying aspects of the matter under investigation.

The paper discussion starts from a step that will be considered as preliminary to measurement, *evaluation*, which will basically results to be a *qualitative* one.

Then, after a discussion on the features of the concept of *measurement*, the paper will illustrate two posterior steps: *validation*, whose aim is the obtain an inter-subjective agreement about the results of measurement; *prediction*, that is the true practical aim of science, consisting of an intended temporal extrapolation to future observations or behaviours of the past and present effects of contingent findings or "laws".

2. Evaluation (ex-ante)

The term evaluation will not be considered in this paper in the sense of "evaluation of the uncertainty of a measurement" (clauses 2.28 & 2.29 [1]), except when specified. Nor, it will be considered as "a particular evaluation procedure" in the sense found in [3].

Instead, evaluation will be considered as a cognitive step *preliminary* to measurement, whose foundation embeds at least partially subjective and qualitative reasoning.

---

[1] See term "model" in the Glossary.

That is because it is based on *prior* data and it includes, as its main tool, in addition to mental processes—e.g., analysis of the Community *inter-subjective* state-of-the-art on the subject matter— new *observations of reality*, whatever may be the definition that one prefers to assign to the term "reality".

In fact, any measurement procedure first requires a sufficient understanding of the specific level of knowledge/ignorance about the subject matter. The aim of (scientific) evaluation is to collect just the information necessary to understand the features of the intended measurand and of the available measurement facility, in order to be able to build up a model, necessary to construct the actual measurement model.

These features ranges, in a not necessarily specific order, as follows:

- state-of-the-art of the aimed measurand
- reasons for uncertainty arising from ignorance, and its level
- new *independent* opportunities of observations for advancing knowledge
- aim(s) for the new measurements
- reasons for using a specific measurement method
- construction of a *reference* model [4]
- quantities needed to be measured in it
- aimed uncertainty, and uncertainty allowed by the available instrumentation
- needed calibrations
- partial Type B *ex-ante* uncertainty evaluation. [5]

One of the aims of the above evaluation is to place the correct basis for the measurements to be performed, in such a way that their results can be accepted, in the inter-subjective frame of the relevant Community [6], as 'correct'—i.e., "good" [7]—measurements according to the codes ruling scientific good-practice.

Simulation is also a kind of evaluation: basically it can be considered based on the *dematerialisation* of a cognitive process. Such an evaluation, which is quantitatively based on virtual data, can be used to test a procedure, or an experimental process, generally of a nature that is frequently dynamic. See more in Section 6.

3. Measurement

In modern science, the concept of measurement is conventionally applied to experimental actions (though no more necessarily limited to physical/chemical/ biological quantities) that respect some basic criteria, considered useful and sufficient for allowing *ex-post* evaluation of the measurement results:

- ***independent*** repetitions of the experiment with the same or different methods,
- ***repeatability*** of the results of a single method: ensured for each independent experiment, within a certain uncertainty, as no method can provide exact results. It does not concern (nor affect) *data independence* in distinct experiments,
- ***comparability*** of the data of different *independent* experiments: the *measurand* is required to be the *same* in all compared experiments,
- ***modelling*** of the experiment, also allowing comparability of results between different experiments on the *same measurand*: evaluation *ex-ante* is the needed tools to achieve this goal,
- measurement *original* results must be sufficient for the in-house data elaboration—and more and more required to remain available in time to third-parties (big data issue).

A measurement procedure can be considered *fully-objective* only in the sense that, once it is constructed, is applied—manually or automatically—following fixed steps, irrespective to the imperfect behaviour of the measured system. Fixed steps can even provide a roadmap with several branches, one for each different experimental situation. In general, these features cannot be considered sufficiently reliable yet, so that the full procedure (and its corresponding software) is generally integrated also by human intermediate decisions, instrumental or concerning data pre-processing.

In fact, for instrumental data acquisition, a deterministic procedure tends to freeze out "systematic effects" (also called "systematic error" in VIM:2012, clause 2.17), making them un-detectable; for data processing, the ex-post analysis cannot detect incorrect features of the pre-processed data. In both cases, an evaluation—of the need for the repetition of the measurements by varying strategically features of its implementation—is the only way to provide a higher level of confidence in the correctness of the results: the choice of the procedure variations embeds subjectivity in the measurement process. [8, 9]

The *cause-effect principle* is implied by the process of experimentally acquiring new knowledge. It is a fact that modelling a measurement process should first allow to understand the flow in the chain of the cause-effect features required by the specific measurand, e.g., as processed with the Ishikawa diagram. [4]
A cause-effect chain implies almost invariably a hierarchical order in the relationships between influence quantities and between them and the measurand, in the sense that there is a (often single) correct *ordering* in the logical scheme that must be respected.

This ordering becomes explicit when the model is written as (a set of) mathematical equations describing the functional relationships between quantities, corresponding to the cause-effect relationships. Classically the equations are expressed as continuous functions—which imply causality for their consecutive values (called *autocorrelation*). In this respect, the mathematical modeling is not part of the measurement process but of the evaluation, ex-ante and ex-post, and often represents the basis for the *prediction* step. On the contrary, models that are physical models depict physical relationships. "Data driven" only models often don't define a cause-effect relationship but capture correlation between variables.

When performed completely, the measurement process provides the data that have to be subjected to an *ex-post* evaluation. In fact, how can one distinguish locally between "right" and "wrong" measurements? [4] First, an ex-post evaluation is needed (Section 4).

Finally, measurement can, or must, be performed also in case the *measurand has a stipulated value*, as is the case, for example, for the numerical values used in the definition of the international measurement units (like the SI). In the case of national units realised by the National Metrology Institutes (NMI) at the top of their national traceability chain, it is needed in order to validate the local measurement units, since *no calibration* of the standards is possible [10]. Even when calibration is possible, in these cases the numerical value cannot be changed but any divergence can only reflect on an increase of the associated uncertainty.

4. Evaluation (*ex-post*) and validation

The evaluation *ex-post* is intended here as a *local* evaluation, not the one later arising from the Community. It is almost totally subjective, though basically quantitative, but it has also to develop information suitable for later *inter-*

subjective analysis. It embeds the common problem of the level of ambiguity in public diffusion of new knowledge.

Its aim is to determine the *output values* (numerical or not) of the measurement process: the latter should always end with a set of (more than one) data, and their *associated uncertainty*, on the basis of ex-ante evaluation and of the *ex-post* analysis of the measurement *outputs*.

It consists of several steps:

- noting possible anomalies observed during the measurements
- noting any possible anomaly in the collected data with respect to the ones expected from the *ex-ante* evaluation
- performing all the needed post-elaborations, as/if needed
- full uncertainty evaluation (Type A and Type B) [5]
- new results confirming or contrasting previous data
- analysing the resulting information and comparing it with previous one, when existing
- elaborating on possible observed discrepancies with former data or with third-parties data
- preparing the outcomes as summary numerical tables and/or in graphical form, for external communication to the Community
- indicating the main conclusions drawn from the above, for external communication
- deciding possible need of subsequent changes in the procedure, and/or for in-house repetition of the measurements.

All the above steps are fully or partially subjective, so, consequently, is subjective the communication to the Community. However, the communication must respect a number of rules driving the context of science, which allows inter-communication to be understandable with a minimum of ambiguity.

One basic issue to ensure the above requirements is *comparability*: the adopted measurement model must be such that the resulting information conforms with the *shared format* of the measurand definition, to ensure that the new information can be *compared* with that previously available or with further information arising from third parties.

As detailed in [4], it means that one must take as *reference model the one accepted by the reference Community*—in general, a technical/physical/ chemical/biological… "law"—and construct the corresponding *measurement model* based on the used instrumental apparatus. The latter is always different from the *reference model*—except in very simple cases—as it includes in the corresponding procedure all the influence quantities and parameters necessary to obtain, as the measurement result, an information *compatible with the reference model*.

Compatibility does not mean here existing between quantity values, but the possibility of common understanding. That feature can be incorporated also in an automatic procedure or can be done during the ex-post evaluation. That also means reducing all the measuring-model quantities back to only those included in the reference model—this operation is often called "applying all the corrections".

Finally, one must be aware of the fact that the residual ignorance on the subject matter—which can still be quite extended— can induce *errors*—as distinct from uncertainty— in the inference made on the available results, locally or by the Community.

The acceptance by the relevant Community of the *ex-post* evaluation is called *validation*.

5. Prediction

Prediction is commonly recognized to be the *true practical aim of science*, especially by non-scientists and by the Society: capability to extrapolate to future observations or behaviours past and current effects of contingent findings considered "facts" or "laws". Extrapolation for time-dependent features means that their future trend in time is assumed to follow exactly the rules or laws considered valid at present: their future "changing-point" is difficult of impossible to foresee. If the method is "scientifically-based", it is considered the one providing maximum-reliability to humans. The *lack* of such a capability is often bringing people and media to the conclusion that science is of little use.

However, prediction is a most delicate issue, which can only be verified *a posteriori*. Therefore, it is a *guess* whose foundation requires more care than is usually used—today often the Society pretend to make immediate actions based on long-term predictions.

In principle, a "law" is a prediction, since it pretends to understand the cause-effect relationships between the involved quantities. However, more trivially, it is considered to apply until "it works", i.e. it does not produce conflicting results or applications—at least until the next Kuhn's revolution.

Not all observations outcome a law, as often the cause-effect relationship only rely on probabilistic bases. In this case, the trend in time of the phenomenon would keep on only on such basis.

A third type of prediction is another frequent case: it occurs when the collected dataset is very huge and complex, concerning a physical or other occurrences: typical is the case of local meteorological forecast. A model is built based on the atmosphere overall current laws, and an extrapolation (prediction) is made over a short period of time—from 1 to 15 days—based on collected data.

More ambitious is the further case of prediction on much wider situations, e.g., the whole Earth environment, where the forecast pretends to extrapolate to 30, 50 or even 300 years ahead (for see level). Also in these cases, models are built up, based on data collected globally over past periods—even relatively short and recent, often even shorter than the aimed extrapolation in time. The risk of inaccuracy increases in these case for obvious reasons.

In particular, the mathematical functions used for these calculations suffer of a general problem, which is often overlooked. Any such function (or set of functions) [2] has a *limited adaptability to a change of trend*. In fact, the existing data act as constraints on the function, forcing it to follow, within a given accuracy, their trend. However, when such function is extrapolated forward in time, where no more constraints are available, each type of function will tend to continue a trend strongly determined by its *type* (e.g., in the trivial case of polynomials, there is a distinct difference in case of even or odd polynomials): it may be considered a consequence of the existing *autocorrelation*.

In addition, the more the extrapolation goes far from the initial constraints the more the internal lack of flexibility of the function prevails. For this reason, in the case of critical extrapolations several functions are normally used—so obtaining different behaviours, *very often conflicting* with each other, so not allowing sensible conclusions. At the end of the day, extrapolations are often known to suffer from unreliability. That is often considered intolerable for science: the point is that more and more these extrapolations might lack scientific bases, though rarely admitted.

When prediction means forecast, one must be prepared to admit that the results do not bring to any sensible conclusion: the acquired knowledge can never attain

[2] Except those whose aim to strictly follow the trend implicated by the data (e.g., for *copying* a profile—that cannot be extrapolated).

certainty, and even less perfection, even at a inter-subjective level, which is, at best, a shared convention of a majority. The extent of the majority can itself be affected by non-scientific considerations and can be largely over- or under-estimated.

The decisions to be taken on such a basis, necessary or not, are totally dominated by not-fully scientific factors, and the more, when more important they are. Scientific evidence has often a lifetime validity shorter than those extrapolations, especially with the accelerating increase of knowledge-gaining that is a character of modern science. As a consequence, when decisions must be taken, in addition to the *uncertainty* of the original knowledge, extrapolation may induce *errors* in them—or in the theories.

6. Computer simulation

Complementing what is said in Section 2, simulation can also be considered a particular type of *prediction* (Section 5), generally performed by using mathematical models processed via tools of informatics.

Computer simulation is sometimes considered as a *measurement* and even a form of *observation*. [11]

For sure, one can make a *simulation* only when is able to build up a real *measurement* model. Virtual instruments have been widely developed for the purpose.

However, the quantitative data obtained that way should be considered as pertaining to *qualitative evaluation*, or, if quantitative, be used for *extrapolation*—i.e., *prediction*.

7. Final remarks

As shown in this paper, the cognitive process, of which measurement is a basic step in modern science, has been split into its constitutive steps, in such a way that the measurement step itself is delimited by not less important previous and subsequent steps.

This process structure involves now some delimitation of the measurement operational contents.

It starts from choosing a *reference* model for the measurand intended to be measured, as the basis for the construction of the *measurement* model. It ends by supplying the experimenter the *raw outcomes of the measuring system*—numerical when instrumental, or of different nature otherwise—to allow the experimenter elaborating on, and extracting from, them the final embedded information, to be handled to the Community for review.

Coming now to different aspects of measurement, the concept of measurement in the *quantum frame* has a different meaning and realization features.

It is founded on Born's postulate, whose exact meaning is still the object of debate among physicists, but is used by most of them because "it works"—it is said. A good treatment of that subject matter in relations to the 'classical' measurements can be found, e.g., in [12]. (In all instances, why can wonder why we conceive *extensive infinity* in nature only as a $+\infty$ and not as a $-\infty$?)

Concerning the *objects of measurements*, the one indicated in the VIM:2012 [1] should be considered obsolete, since also *non-quantitative* (often called 'qualitative') quantities have to be included in it, the concept of *magnitude* not necessarily implying only numbers.

One problem for non-quantitative quantities, e.g. for a judgment, is the need, which can be difficult to satisfy, to define the *instrument(s)* used for the measurement.

In addition, a *definition of counting*—a peculiar measurement method (clause 2.1 [1])—must become available, for non-continuous ("granular") quantities.

The two above subject matters will be the object of later author's reflection.

References

1. JCGM 200, BIPM International Vocabulary of Metrology—Basic and General Concepts and Associated Terms (VIM), 3rd edn., 2012, http://www.bipm.org/en/publications/guides/
2. E. Tal, *Naturalness and convention in the International System of Units*, 2018 Measurement 116 631, https://doi.org/10.1016/j.measurement.2017.09.008
3. L. Mari, *Epistemology of measurement*, 2003 Measurement **34**, 17
4. F. Pavese, *On the classification in random and systematic effects*, AMCTM XI, 2018, in A.B. Forbes, N.F. Zhang, A.G. Chunovkina, S. Eichstädt, F. Pavese, (Eds.): Advanced Mathematical and Computational Tools in Metrology and Testing XI, vol.11, Series on Advances in Mathematics for Applied Sciences **89** (Singapore: World Scientific) p 58
5. Guide to the Expression of Uncertainty in Measurement (GUM), 1st edn., International Organization for Standardization, Genève, Switzerland, 1995
6. P. De Courtenay and F. Grégis, *The evaluation of measurement uncertainty and its epistemological ramifications*, 2017 Studies in History and

Philosophy of Science **65-66**, 21, https://doi.org/10.1016/j.shpsa.2017.05.003

7. L. Mari, A. Ravarini, G. Buonanno, R. Ruffini, *Valutare o misurare? Un modello e la sua applicazione alla valutazione delle risorse umane* (in Italian). In English: *To evaluate or to measure: A model and its application for the evaluation of human resources*, 2018 Risorse Umane **6**, 17
8. F. Pavese, *On the degree of objectivity of uncertainty evaluation in metrology and testing*, 2009 Measurement **42**, 1297
9. F. Pavese, *Subjectively vs. objectively-based uncertainty evaluation in metrology and testing*, 2008, IMEKO TC1-TC7 Workshop, on CD-ROM 2008, IMEKO-TC1-TC7-2008-030, https://www.imeko.org/publications/tc7-2008/IMEKO-TC1-TC7-2008-030
10. F. Pavese, *Are the present measurement standards still valid after SI redefinition?*, 2017 Accred. Qual. Assur. **22**, 291, doi 10.1007/s00769-017-1282-8
11. W.S. Parker, *Computer Simulation, Measurement, and Data Assimilation*, 2017 The British Journal for the Philosophy of Science **68**, 273, https://doi.org/10.1093/bjps/axv037
12. A. Neumaier, Born's rule and measurement, 2019 arXiv:1912.09906v1 [quant-ph]
13. T.R. Gruber, G.R. Olsen, *An ontology for engineering mathematics*, 1994 in Principles of Knowledge Representation and Reasoning: Proceedings of the Fourth International Conference KR '94, J. Doyle, E. Sandewall and P. Torasso Eds (San Francisco, California: Morgan Kaufmann Publishers, Inc.) p. 258
14. F. Pavese, *Mathematical and statistical tools in metrological measurement*, 2013, Chapter in: Physical Methods, Instruments and Measurements, [Ed. UNESCO-EOLSS Joint Committee], Encyclopedia of Life Support Systems(EOLSS), Eolss Publishers, Oxford, UK, http://www.eolss.net
15. E.U. Weber, O. Coskunoglu, *Descriptive and Prescriptive Models of decision making: implications for the development of decision aids*, 1990 IEEE Trans. On Systems, Man and Cybernetics, **20**, 310
16. F. Pavese, G.F. Molinar Min Beciet, Modern Gas-Based Temperature and Pressure Measurements, Springer, New York, 2nd Edn., 2012

Glossary (for more, see in [14])

Accuracy: see VIM:2012, clause 2.13. Contrarily to **reproducibility**, it account also for detected "systematic effects".

Bias, method: "difference between the expectation of a test result or measurement result using that method and a true value" (ISO 15725-1)

Comparability, value: Possibility to decide if the value of a measurand is equal to, or lower or larger than a pre-existing value: requires the measurand be the same.

Constant, in physics: Quantity that is defined to have a fixed value. See also *constant, physical value.*

Correction: "compensation for an estimated systematic effect" (VIM:2012, clause 2.53). The compensation is a (best) estimate too. See also *effect, systematic.*

Effect, systematic (measurement): "component of measurement error that in replicate measurement remains constants or varies in a predictable manner" (VIM:2012, clause2.17). In GUM:1995, clause 2.2.3 Note 3 it is said "It is understood that the result of the measurement is the best estimate of the value of the measurand, and that all components of uncertainty, including those arising from *systematic effects*, such as components associated with corrections and reference standards, contribute to the dispersion". Use not advised in [1].

Evaluation: Section 2 (ex-ante) "… cognitive step *preliminary* to measurement, whose foundation embeds at least partially subjective and qualitative reasoning … based on *prior* data"; and 4 (ex post) "… a *local* evaluation, not the one later arising from the Community. It is almost totally subjective, though basically quantitative". Here not VIM:2012, clause 2.28 & 2.29.

Fact: As opposed to "hypothesis" or "supposition" or "inference". Often assumed to be totally "objective". Cambridge dictionary: "Something that is known to have happened or to exist".

Knowledge: as opposed to "ignorance", but always partial, while ignorance can be total. Cambridge dictionary: "understanding of an information about a subject that you get by experience or study, either known by one person or by people generally".

Law: in science, Wikipedia: "Scientific laws or laws of science are statements, based on repeated experiments or observations, that describe or predict a range of natural phenomena" Cambridge dictionary: "a rule, usually made by a government, that is used to order the way in which a society behaves".

Measurement: see Sections 1 and 3.

Model, Descriptive: equivalent to *measurement model.*

Model, Measurement: model including all *influence quantities* estimated to have a significant effect on the *measurand* value.

Model, Prescriptive: Model shared by the relevant Community [6]. Here not a model based on normative (e.g. [15])

Model, Reference: often indicated as "ideal", see *Prescriptive model*

Number: Numerical value in the arithmetical sense, but see also *Value, numerical* and *Constant*. Cambridge dictionary: "(a sign or symbol representing) a unit that forms part of the system of counting and calculating"

Prediction: see Section 5. Prediction is commonly recognized to be the *true practical aim of science*, also by non-scientists and by the Society: capability to extrapolate to future observations or behaviours past and current effects of contingent findings considered "facts" or "laws". Cambridge dictionary: "A statement about what [one] think will happen in the future"

Property, in science: Wikipedia: "A [physical] property is any property that is measurable, whose value describes a state of a physical system". Not the VIM:2012 clause

Property, extensive: additive for non-interactive subsystems, because they increase and decrease as they grow larger and smaller, respectively (IUPAC Green Book 3rd, 1.4)

Property, intensive: independent of the extent of the system (IUPAC, Green Book 3rd, 1.4)

Property, nominal: property of a phenomenon, body or substance that has no magnitude (VIM:2012, clause 1.30)

Quantity: property of a phenomenon, body or substance where the property has a magnitude that can be expressed as a number and a reference (VIM:2012 1.1))

Quantity, Influence: is one estimated to have a significant effect on the *measurand* value

Quantity, Input: here *not* the quantities exempt from *Bias*, like instead in GUM:1995, clause 4.1.2

Reproducibility (condition of measurement): See VIM:2012, clause 2.24. Included "systematic effects" (VIM:2012, clause 2.26, Note 1)

Simulation: see Sections 2 and 6.

Systematic error: See VIM:2012, clause 2.17

Stipulation: (inter-subjective) convention of a Community fixing as exact a numerical value resulting from measurements.

Validation: see Section 4. Acceptance by the relevant Community of the *ex-post evaluation*.

Value, physical constant: constant for a physical constant: a few digits of the numerical value (see below) can often be computed from theoretical considerations.

Value, numerical: number expressing the value of a quantity, depending on the chosen system of units and generally of experimental origin. May not be equivalent to *number* (see above).

Value, conventional: fixed numerical value in a certain system of units assigned by convention (inter-subjective). The fact that it is fixed does not mean that it is a "true" value.

Appendix: An example for measurement of thermodynamic temperature *T*

The following sections are numbered according to the corresponding ones in the main body of the paper.

2. *Evaluation ex-ante*

1.*State-of-the-art of the aimed measurand*: literature on measurements of *T*;

2.*Reasons for uncertainty arising from ignorance, and its level*: pros and cons for the different measuring techniques, in respect to evaluation #10. below;

3.New *independent* opportunities of observations for advancing knowledge: e.g. use of new techniques, like with the use of a cryogenic pressure transducer, or, more recently, of acoustic thermometry; [16]

4.*Aim(s) for the new measurements*: diversification of techniques to help detecting "systematic effects";

5.*Reasons for using a specific measurement method*: e.g., the use of a cryogenic pressure transducer [16] allows to avoid including in the of the measuring equation the most critical part, the modeling (and measurement) of the so-called "dead-volume" (which is customary to call "correction for …");

6.*Construction of a reference model*: as illustrated in [4], the *prescriptive* model is the one adopted by the relevant Community for the measurement of *T*, e.g., concerning the measurement on an Interpolating Constant Volume Gas Thermometer (ICVGT) to realize the ITS-90 between 4.2 K and 24.56 K using ^4He;

7.*Quantities needed to be measured in it*: see below in §**A**);

8.*Aimed uncertainty, and uncertainty allowed by the available instrumentation*: at IMGC the aim was $U < 1$ mK and a choice of instrumentation compatible with that aim, both hardware (calorimeter, resistance bridge for the thermometers, electronic equipment) and software (automatic system for running the experiment and for getting the values of *T*, known to have an intrinsic accuracy equivalent to better than 0.1 mK);

9.*Needed calibrations*: as a metrological good practice, the suitable calibration level must be ensured;

10.*Partial Type B ex-ante uncertainty evaluation* [5]: an uncertainty budget compatible with the aimed uncertainty of the results, must ne compiled as a good practice.

A) *An example of prescriptive measurand model: Thermodynamic temperature measured with a CVGT* (adapted from [4])

A current socially-agreed model of thermodynamic temperature, *T*, for the *ideal gas* is that in Eq. (1):

$$T = pV\ \mathrm{R}^{-1} n^{-1}, \tag{1}$$

where p = pressure, V = volume, n = amount of gas, R = $\mathrm{k_B N_A}$ gas constant, with $\mathrm{k_B}$ = Boltzmann constant, $\mathrm{N_A}$ = Avogadro constant—here upright character is used because k_B and N_A are stipulated exact since 2018.

Therefore, the quantities *prescribed* to be measured are *p, V, n*. Figure 1 shows a usual graphic representation of the model.

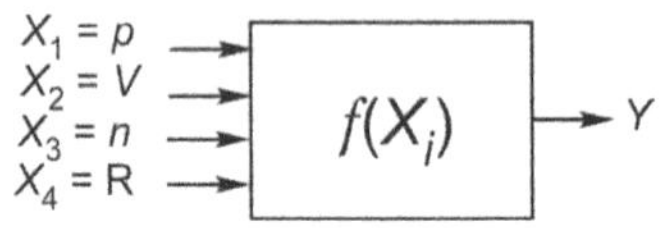

Figure 1. Input-output representation of the *prescriptive* model in Eq. (1). The black box indicates the analytical relationship $f(X_i)$ between the input variables X_i and the output variable $Y = T$.

In a similar generic representation, as that used in [4], the prescriptive model can be shown, using the Ishikawa diagram, as in Fig. 2 (the symbol $^\ominus$ added to a quantity symbol means "reference state" (it is taken from chemical notation for "ideal state"), thus one "socially-agreed.

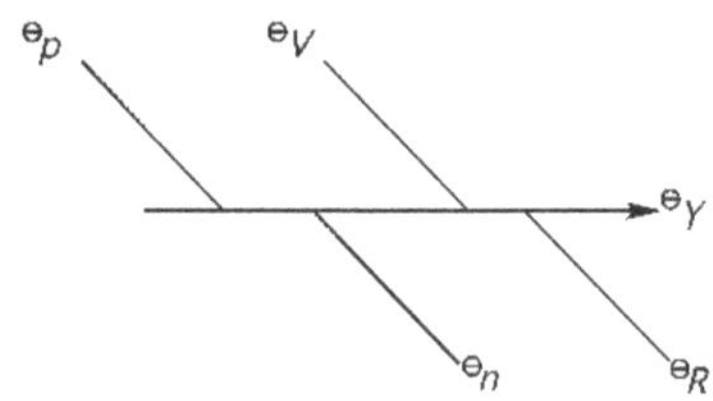

Figure 2. Ishikawa diagram for the *prescriptive* model in Eq. (1). (adapted from [4])

3. *Measurement*

- ***independent*** *repetitions of the experiment with the same (or different) methods*: in each Lab usually only one method is available, but it is good practice to try to make variations in the procedure of the method used, in order to bring to evidence of possible "systematic effects" (see *accuracy* in the Glossary);
- ***repeatability*** *of the results of a single method*: this must be checked for each independent experiment, as no method can provide exact results. It establishes the actual level of the results (*not* the accuracy);
- ***comparability*** of the data of different *independent* experiments: since the *measurand* is required to be the *same* in all compared experiments, each Lab must ensure that since the modeling of the experiment (see *modelling*)
- ***modelling*** of the experiment, also allowing comparability of results between different experiments on the *same measurand*: this is ensure, by definition, starting modelling from the *prescriptive* model shown above in Fig. 2

(see §**A** above). However, each experimental setup needs to start from what is normally called a "measurement model", or *descriptive* model and §**B** below on the CVGT).

***B)** An example of descriptive model for thermodynamic temperature T: constant volume gas thermometry* (CVGT) *(adapted from [4])*

A *descriptive* model for *T*, whose *prescriptive* model is indicated in Eq. (1) and Fig. 2 above for the *ideal* gas, must take into account all the influence quantities of the chosen actual implementation.

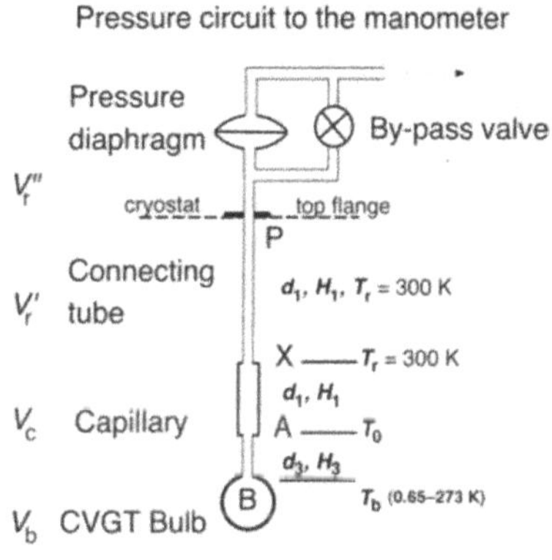

Figure 3. Typical setup of the mechanical part of a CVGT. The gas, whose pressure ${}^{\ominus}p$ is that at ${}^{\ominus}T = T_{\text{bulb}}$ in the CVGT bulb of volume ${}^{\ominus}V$, also fills the above capillary/connecting-tube up to the differential pressure diaphragm and (closed) by-pass valve, generally at room temperature $T_{\text{room}} \neq T_{\text{bulb}}$, where the pressure-measuring manometer is connected to the other side of the differential gauge. The rest of wording in the figure is irrelevant here.

In Fig. 3 [16] a typical material setup is shown.

The influence quantities can be grouped according to the above three categories (detailed description of the symbols is irrelevant here). [16]

(1) *Influence of the physical parameters*:

Substance (*ideal* gas) kinds of non-ideality:

(i) For a low-density gas: $pV = nRT\,[1 + B(T)(n/V) + C(T)(n/V)^2 + \ldots]$, where $B(T)$, $C(T)$ are the *virial coefficients*.

(ii) Non-perfect purity: *different non-ideality of impurities*

Amount of "active" substance , n_{bulb} :

"Active" means that *only* the gas in the isothermal bulb where *T* is measured.

(i) Dead-volume correction, see later: $n_{\text{bulb}} = n - n_{\text{deadvol}}$

(ii) Adsorption of gas on the bulb: $\ln(V/V_0) = \ln(\tau\zeta) + \zeta\Delta_{\text{vap}}H_m(RT\tau + \alpha/RT\tau^2)$

(iii) Impurities: not included in *n*

(iv) Condensation of impurities: "*increases*" *V*.

(1) *Influence of the technical parameters*:

Thermal apparatus (extended, optimised)

Extensions outside isothermal bulb V_{bulb} where $T = T_{\text{bulb}}$ is measured: pressure measuring line (p, n involved)

(i) $(p^*_{\text{int}} + \Delta p_{\text{TM}} + \Delta p_{\text{aer}})V_{\text{bulb}} = RT_{\text{b}}(n_{\text{bulb}} + n_{\text{cap}} + n_{\text{r}})(1+B(n_{\text{bulb}} + n_{\text{cap}} + n_{\text{r}})/V_{\text{bulb}})$,

$n_{\text{cap}} = (\pi/4R)d^2p^*_{\text{int}}(\mathrm{d}h/(T/h))$; $n_{\text{room}} = V_{\text{r}}(p^*/RT_{\text{room}})/(V(p^*/RT)) \propto \approx T^2$

$n_{\text{deadvol}} = n_{\text{cap}} + n_{\text{room}}$

(ii) Pressure p^* is measured at $T_{\text{room}} \neq T_{\text{bulb}}$,

thermomolecular $\Delta p_{\text{TM}} = (C/p_{\text{av}}d^2)[(T_2/T_0)2n{+}2 - (T_1/T_0)^{2n}{+}2]$

aerostatic head $\Delta p_{\text{aer}} \approx ((g/m)/R)p^*_{\text{int}}(\mathrm{d}h/T(h))$

Non-perfect mechanical rigidity of bulb (V_{bulb}):

(i) Pressure difference across bulb walls $\Delta p = (p_{\text{int}} - p_{\text{ext}})$

cylinder $\Delta V/V_0 = (p/E)(d/\tau_{\text{c}} + d/\tau_{\text{c}}/(4 + \tau_{\text{c}}/d))$;

plate $\Delta V/V_0 = (p/E)(55\cdot 10^{-3}\, d^4/L\tau_{\text{p}}^{\;3})$

(ii) Cubic expansion coefficient of the bulb material: $\Delta V_{\text{Cu}}/V_0 = a + bT + cT^2$

(iii)Temperature gradients in the measuring bulb (ΔT)

(2) *Influence of the technical parameters*. *Instruments*: Pressure, Mechanical sensors, Mechanical indication (piston-gage), Electronic indication, Calibrated instruments (*critical*), Volume. *Constant volume gas thermometer*: ideally constant, but deviations (as indicated before). *Acoustic gas thermometer*: microwave (*critical*). *Dielectric, refractive index gas thermometer*: volumetric (*not critical*), Amount of substance. *Indirectly measured quantities*:, Temperature (thermometers usually already calibrated), Electrical resistance measurements, Temperature coefficients of equipment, Thermal controls, Electronic controls.

(3) *Influence of the conditions under which measurements are carried out*

Target accuracy not reached. The conceptual model is directly applicable only for very low requirements (e.g., gas thermometer for the thermostat of a home fridge). Inadequacy of the experimental model. Insufficient capacity of analysis for increasing accuracy. Quality of the experimental set-up. Instruments specifications do not meet the needs. Skill of the operators. Insufficient training in detecting suspicious conditions. Scarce experience in data handling, Insufficient capacity of handling the statistical tools of analysis.

Typically, the number of quantities actually involved in the measurement is *much larger* than those in the initial prescriptive model, and they are *physically localized.* Therefore, e.g., in the example, the generic indications p, V and n are insufficient because actually affected by non-uniqueness. The quantities qualifying the (same) measurand should better be specified for the specific measuring system. In the example, being T the temperature in the bulb:

$p \Rightarrow p_{\text{bulb}}$ for the pressure of the gas in the bulb at temperature T;

$V \Rightarrow V_{\text{bulb},0}$ for the volume of the bulb as it was rigid and at a reference temperature T_0;

$n \Rightarrow n_{\text{bulb}}$ for the amount of "active" gas within the volume $V_{\text{bulb},0}$ of the bulb.

All involved quantities must be taken into account in the measuring system model.

In Fig. 4 a generic descriptive model is shown, again using the Ishikawa diagram. [4] In it, the quantities which have in already in their "reference state" at the moment of the measurement are indicated with the superscript $^{\ominus}$. The others are in a condition that is customary to call "biased", indicated with B in the figure, simply meaning that they are out of their reference state.

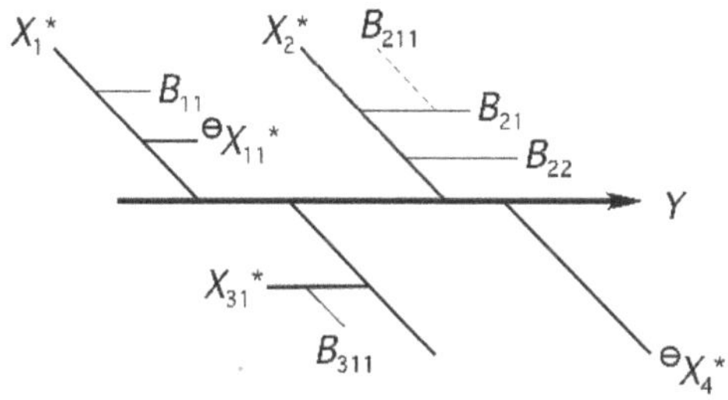

Figure 4. Ishikawa diagram for a generic *descriptive* full model. In it, the superscript * means that those quantities are "localised" in the sense illustrated in [4] and in the example (e.g., for temperature, localisation: that of the "bulb").

However, the model stage in Fig. 4 is *not yet* compatible with the *prescriptive* model.

4. Evaluation (ex-post) and validation

In fact, "bias" requires what is customary to call a "correction", bringing the relevant quantity to its *reference* state. That in [4] is called instead "normalisation". and, in fact, Y in Fig. 4 is not yet the socially-agreed, $^{\ominus}Y$. One get the latter only after all normalisations, $N = -B$, shown in Fig. 5, where N is nothing else than X.

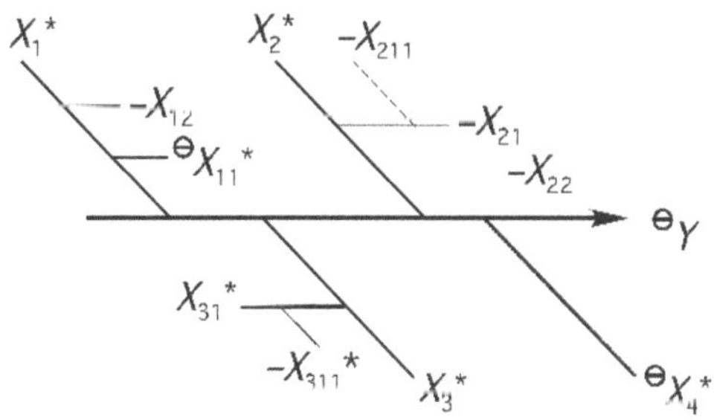

Figure 5. Ishikawa diagram for a generic *final descriptive* full model *compatible* with that in Fig. 2.

This last stage is essential in order to allow validation, since it ensures the comparability of the measurands in different experiments.

(ERRATA-CORRIGE: In [4] Fig. 8 should be corrected, and become as the above Fig. 5)

All aspects of ex-post evaluation and validation are specific of any kind of measurement.

F Pavese, A B Forbes, N F Zhang, A G Chunovkina (Eds):
Advanced Mathematical and Computational Tools in Metrology and Testing XII
Series on Advances in Mathematics for Applied Sciences, Vol. 90

https://doi.org/10.1142/9789811242380_0022

A method based on combinations of forecaster and weighing matrix to detect fault of components in diecasting process

Luca Provezza, Giovanna Sansoni and Matteo Lancini
Department of Industrial and Mechanical Engineering, University of Brescia, Brescia, 25123, Italy
E-mail: luca.provezza@unibs.it
www.unibs.it

Alberto Marini
Engineer Project Manager, Italpresse Gauss, Via Trento n° 178
Capriano Del Colle, 25020, Brescia, Italy
E-mail: alberto.marini@noricangroup.com

This work presents a flexible method to detect the fault of components in a diecasting machine. The core of this method is the combination of sensor-based statistical predictions with the expert knowledge using a series of weights determined in formal interviews. Each feature is extracted from the machine's sensor time history using a least square regression and paired with an uncertainty estimator. Then, each uncertainty estimator is combined with the uncertainty of the relative transducer in order to obtain a combined uncertainty of the two contributions. The final result is a score index representing the distribution of different types of faults in the diecasting machine. A dataset of 451 injections was analyzed to test the method. The historical records of maintenance service recorded 19 events corresponding to a fault of a valve. All the events were correctly detected by the algorithm as well. The uncertainty estimators of the parameters have allowed performing an analysis of the effect of transducers' uncertainty on the final prediction. A higher uncertainty is negligible in the final prediction of fault. This means that the method can work also with transducers with lower accuracy.

Keywords: machine learning; diecasting; uncertainty; technical diagnostics; predictive maintenance.

1. Introduction

Predictions of components fault could lead to great advantages for the manufacturer both, in terms of maintenance and costs. The current trend in fault detection is to employ machine learning algorithms, which represent a powerful tool to detect anomalies, but this approach is generally limited by the large quantity of historical data on failures needed.

Current literature reports many works where different algorithms are used to estimate the fault of components and the Remaining Useful Life (RUL). The approaches are divided into Physics-based approaches and Data-Driven approaches. Physics-based approaches are described by mathematical models to determine the fault of the machinery components. These models usually used structural analysis, contact analysis, cumulative damaged model, cycle fatigue, and so on to determine the fault condition of a component [1,2]. Data-Driven approaches use monitoring data to estimate the degradation of components in time. In this case, no mechanical model is necessary to estimate the fault of a component machine. Data-driven approaches are divided into machine-learning approaches and statistical approaches. Ben Ali et al. [3] proposed a method with a combination of Simplified Fuzzy Adaptive Resonance Map (SFAM) neural network and Weibull Distribution (WD) to determine RUL on rolling element bearings based on vibration signals. As output, they defined seven classes for optimal RUL prediction. Rocha et al. [4] proposed an Internet of Thing (IoT) scheme to remotely manage the remaining useful life of Insulated Gate Bipolar Transistor (IGBT) power modules based on their failure precursors. They adopted Artificial Neural Network (ANN), Support Vector Machine (SVM), Fuzzy, and Wavelet to determine faults of the power converter. Ellefsen et al. [5] provide high RUL prediction accuracy through the combination of unsupervised and supervised learning even with reduced labeling of the data. Recently, hybrid prognostic approaches are used to make prognostic more accurate. There are two different approaches to hybrid prognostic. The first is a physics-based approach combined with data-driven approach [6,7]. The second is based on data-driven approaches combined with data-driven approaches [8-16]. Mazidi et al. [17] proposed a hybrid health condition model where Artificial Neural Networks (ANNs) were created to simulate the normal behavior and then hazards model was applied to the signal to create a health function for wind turbines.

In diecasting, prediction of fault of components is presented in a few works, and represent the only works in literature where the predictive maintenance is applied to the diecasting process. Chen et al. [18] and Liu et al. [19] realize an

IoT to stream online energy data for energy analysis of a diecasting machine. Meanwhile, Yarlagadda [20] proposed a neural network system to generate process parameters for the pressure diecasting process.

Moreover, in diecasting, the fault of components is a rare event and the analysis cannot be performed by investigating large datasets of fault data based on maintenance records. This makes very difficult to detect the fault of components with traditional machine learning algorithms. The analysis must be applied by considering mostly the normal behavior of the machine. A further problem, however, linked with diecasting is in the frequent change in production, which leads to changes in the process parameters. Moreover, sometimes small companies do not update correctly the production identifiers. To solve these problems, a method was created to detect the fault of components in a diecasting machine. The proposed method automatically detects a production change and resets each time the dataset used for training.

2. Material and methods

The method is based on the peculiarity of the diecasting process, that presents different phases equal to each machine and production considered. These phases are the slow motion of the piston to avoid air bubbles inside the die, the stroke with the filling of the die, and the multiplication phase to compensate the shrinkage of the material do to the cooling by giving more pressure in the process. Each phase is interpolated to extract sensitive parameters to perform the prediction of fault. For each parameter, an uncertainty estimator is recorded and combined with the uncertainty of the instrumentation to obtain an uncertainty that takes into account the two contributions.

The core of this method is in the combination of the classical prediction analysis with a weighing matrix given by the experts. The weights are determined in a series of formal interviews for each phase and quantity recorded. The final result is a score index representing the probability of different types of faults in the diecasting machine.

An advantage of the combined uncertainty is to perform a Monte-Carlo analysis to investigate the effect of the uncertainty in the final result through sensitivity analysis by amplifying artificially the matrix of the uncertainty estimators of the parameters.

The details of the proposed method are shown in the flowchart presented in Figure 1.

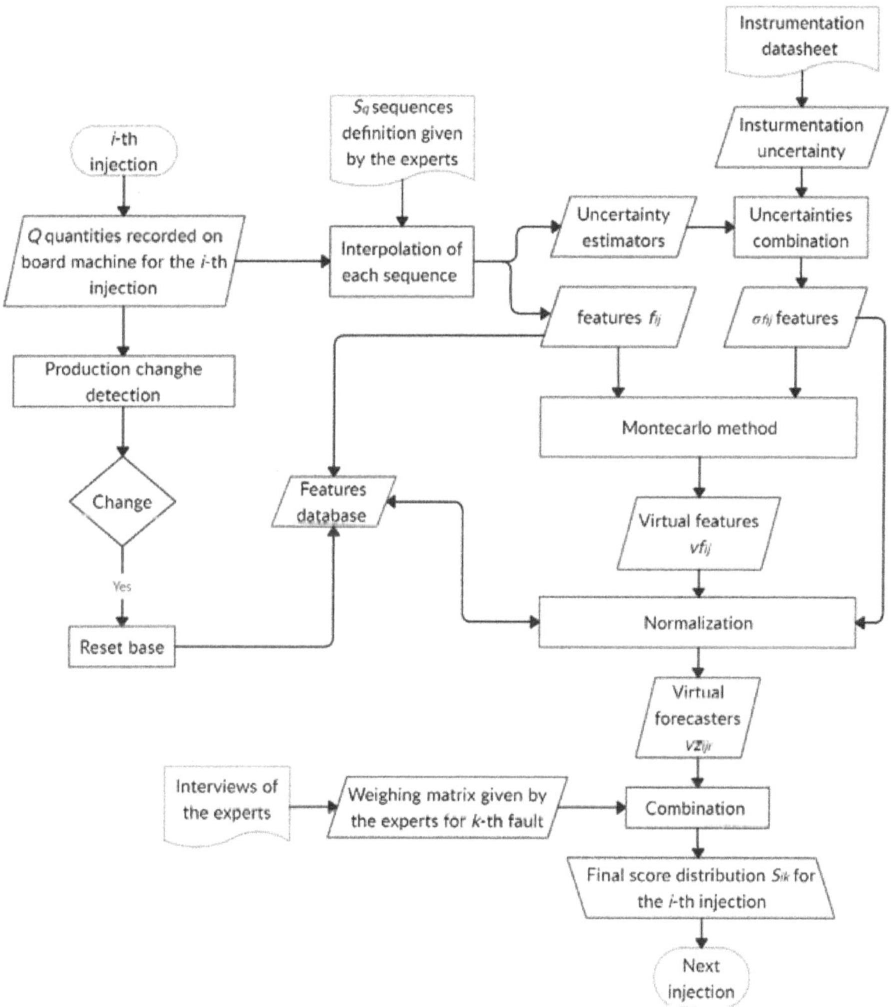

Fig 1. Scheme representation of the method to predict fault of components in the diecasting process.

2.1. *Experts interviews*

The interviews were conducted with the help of six experts chosen from the maintenance and service staff of the machine manufacture for their knowledge of the faults related to the injection group based on 20 years of experience in the field.

The experts were asked to rank from 1 (least significant) to 10 (most significant) each quantity in each phase, indicating the relevance of the quantity for each type of k-th fault.

Then, these ranks are transformed in weights by considering the mean of all ranks given by the experts in each phase in each quantity.

The final function of the weighing matrix is to give more importance to the phases and quantities that are most correlated to a k-th fault according to the opinion the experts.

This matrix is finally combined with the classical machine learning analysis to predict the fault related to a particular component of the machine.

2.2. *Productions detection*

To solve the problem related to the frequent production change, a series of parameters strongly linked to the production recipe are used to detect any production change. These parameters are:

- speed set-points,
- pressure set-points.

These setpoints define the speed and pressure curves that the controller will use as targets during production. A production is identified by these parameters.

The searching technique used to detect a production is presented in Figure 2.

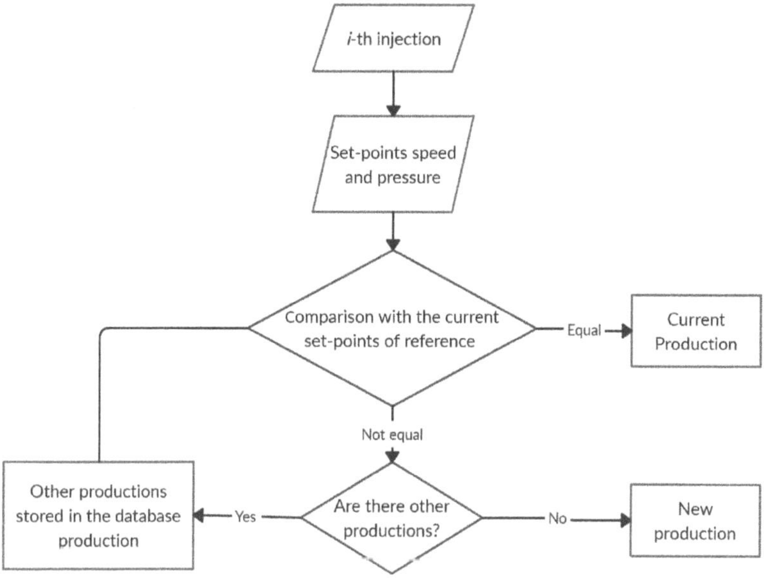

Fig 2. Scheme representation of the productions detections for an *i*-th injection.

The set-points of an injection are compared with the setpoints of the current production stored in the production database.

If the setpoints of the current injection are equal to the setpoints in the database, the injection will belong to that production.

Otherwise, the setpoints of the current injection is compared with the previous setpoints stored in the database until a match is found. If there's no match, a new production is created and the setpoints of the current production are stored in the database production.

2.3. *Predictions analysis*

Q physical quantities are recorded from a number of sensors onboard the diecasting machine such as pressure transducers, electrical command, and feedback of control valves and encoders. These quantities are recorded for each *i*-th injection.

The previous productions detection analysis is considered in the prediction analysis. When a change in production is detected, the algorithm resets the features database used to perform the normalization.

Each q quantity is divided into S_q different sequences, which can be a combination of more production phases (Figures 3a and 3b).

a)

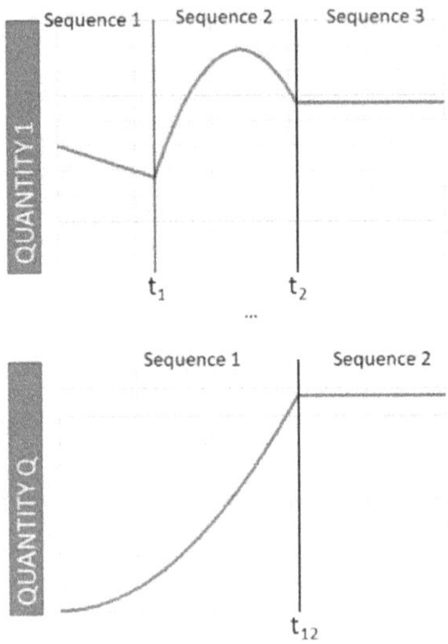

b)

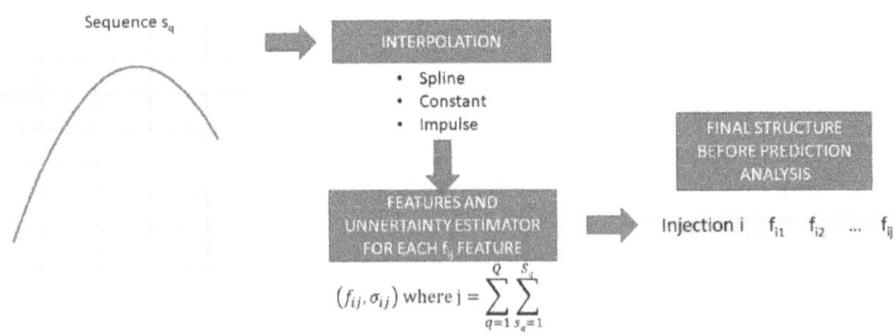

Fig. 3. In Figure 3a, the sequences are determined following the division of times defined by the process. The duration of a sequence is based on what typology of regression is defined for that piece of signal. In Figure 3b is presented the scheme used for the feature extraction for each sequence.

As a matter of fact, a sequence must not necessarily follow the definition given by the process phases. The meaning of a sequence is to follow the trend of the signal recorded that can be a sum of more process phases. The end of a sequence is determined when a signal changes its trend at a particular time division of the process phase.

Once the sequences are defined, the features extraction is performed by considering different typologies of interpolation for each piece of signal in a sequence. The total number of J features is given by the sum of the features of all S_q sequences in all Q quantities. For each sequence, the typologies of interpolation can be:

Spline: the features are defined as the parameters of a b-spline interpolation:

$$\mathcal{B}(t) = \sum_{i=0}^{n} b_i N_{i,p}(t) \tag{1}$$

where b_i is the coefficient of the b-th piecewise b-spline and $N_{i,p}(t)$ represents the basis function of the b-th coefficient. The number of n breakpoints was chosen a priori to have the same number of parameters for all sequences.

The uncertainty estimator of each coefficient of the b-th piecewise spline is the square root of each element of the diagonal of the covariance matrix. The uncertainty combination $\widetilde{\sigma_{ij}^2}$ of the uncertainty estimator with the uncertainty instrumentation is given by the following formulation:

$$\widetilde{\sigma_{i,j}^2} = \frac{\sigma_{ij}^2}{\sigma_0^2}(u^2 + \sigma_0^2) \tag{2}$$

Where σ_{ij} is the uncertainty estimator of the parameter derived from the covariance matrix, u is the uncertainty of the transducer considered for that sequence, and σ_0 is the uncertainty of the least square regression.

Impulse: the maximum value reaches during that sequence represents the parameters of this phase. The uncertainty estimator is calculated with the following formulation:

$$\sigma_{i,j}^2 = \begin{cases} \frac{\sum_{i=1}^{n} q_i^2}{n}, & q_i \neq q_{max} \\ 0, & q_i = q_{max} \end{cases} \tag{3}$$

Where σ_{ij}^2 from equation (3) is a modified variance of the sequence. The quantity q_i is the i-th data recorded in the sequence, and n is the number of the recorded data. The estimated value is zero in the formulation because the value taken as a feature of that sequence is the maximum value of the sequence.

Indeed, the maximum value has contribution zero in the calculation of the variance because is the expected value of the impulse.

Constant: A sequence with a constant trend is defined by the mean of the signal. The uncertainty estimator is represented by the standard deviation of the quantities recorded in the sequence.

Step: when a recorded signal presents different constants trend in steps during the sequence, the parameters used as features are represented by the mean of a number of intervals chosen a priori to have the same number of features for all the injection analyzed. The uncertainty estimator is the standard deviation of each interval of time considered.

For the impulse, constant, and step trend, the combination of the uncertainties is given by:

$$\widetilde{\sigma_{i,j}^2} = (u^2 + \sigma_{ij}^2) \tag{4}$$

where $\widetilde{\sigma_{i,j}^2}$ is the combination of the uncertainties, σ_{ij} is the uncertainty of the fit of that sequence, and u is the uncertainty of the transducer used for that sequence.

The features F_{sq} of the sequence and their uncertainty estimators are then recorded. Since every sequence uses a different curve, the total number of J features is given by the sum of the F_{sq} features in all the S_q sequences and all Q quantities.

As a subsequent step, R =30 000 *virtual features* are generated randomly for each forecaster following a normal distribution centered on its average and with a standard deviation equal to its combined uncertainty. The *virtual features* are normalized by using the mean, the standard deviation of the corresponding feature as recorded in the features database, and by using the combined uncertainty of that features in that injection. This results in a series of $vz_{i,j}$ *virtual forecasters.*

Then, the virtual forecasters $vz_{i,j,r}$ of a sequence are combined in a single parameter VZ_{sq} by considering the r.m.s. value of all parameters of that sequence. The all VZ_{sq} parameters are combined with the weighing matrix determined from the interviews by using the following formulation:

$$VZ_{i,k,r} = \frac{\sum_{sq=1}^{Sq} VZ_{sq} w_{q,s,k}}{\sum_{sq=1}^{Sq} w_{q,s,k}} \tag{5}$$

Where the final $VZ_{i,k,r}$ parameter is computed as the two-tailed central part of a gaussian curve to transform the z-score value in a score index between 0 and 1:

$$S_{i,k,r} = \frac{2}{\sqrt{\pi}} \int_0^{VZ_{i,j,r}} e^{-t^2} dt \quad (6)$$

The final goal of this transformation is to consider a synthetic index that represents the probability to have a deviation from the normal behavior of the component.

The result of all R iterations is this score index that represents a distribution $S_{i,k}$ for the i-th injection. A threshold was set to a value of S equal to 0.75 establishing the boundary between the fault state and normal state.

The Monte Carlo analysis can be used to investigate the effect of the uncertainty parameter in the final prediction making a sensitivity analysis of the uncertainty. The analysis consists of amplifying the combined uncertainties matrix of the parameters, making the Monte Carlo analysis, and see the effect in the final score distribution.

3. Results

An initial application of the method was applied to a dataset of 451 injections to test it. This dataset represents one day of a production where the historical records of maintenance service recorded 19 events corresponding to a problem of the process due to a not correctly work of the multiplication valve. This represents the ground-truth to consider in the final prediction.

This dataset was chosen with the goal to investigate the effect of the amplification of the uncertainty in the final prediction. Different conditions of amplification were considered: without considering the uncertainty of the instrumentation, with the combined uncertainty calculated, 2 times the combined uncertainty, 5 times the combined uncertainty, and 10 times the combined uncertainty.

a)

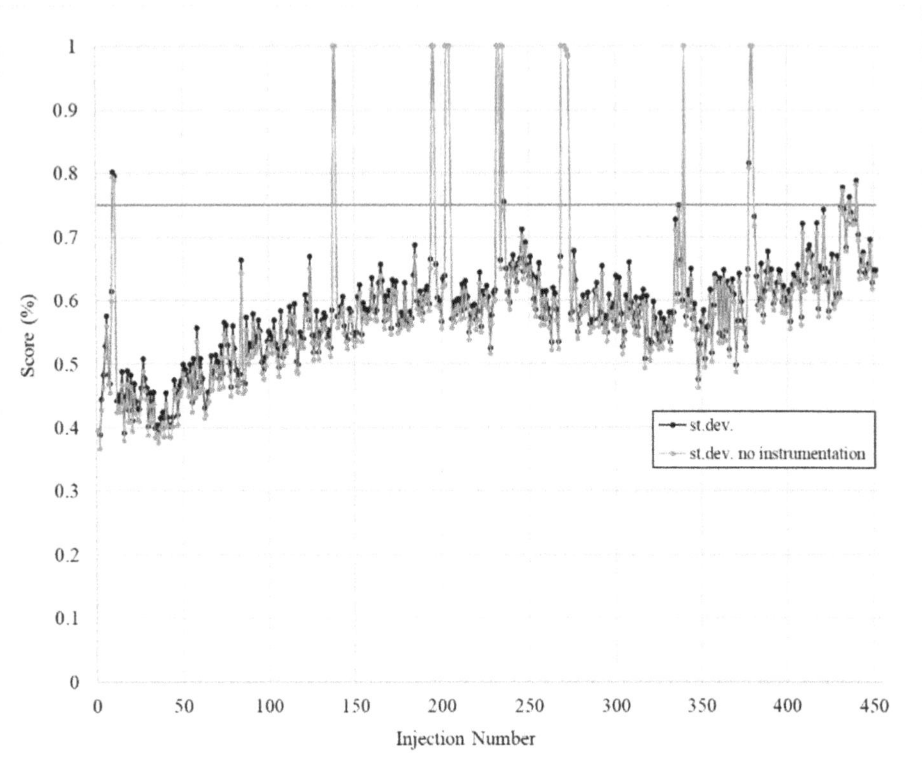
1
0.9
0.8
0.7
0.6
0.5
0.4
0.3
0.2
0.1
0
Score (%)
0
50
100
150
200
250
300
350
400
450
Injection Number
st.dev.
st.dev. no instrumentation

b)

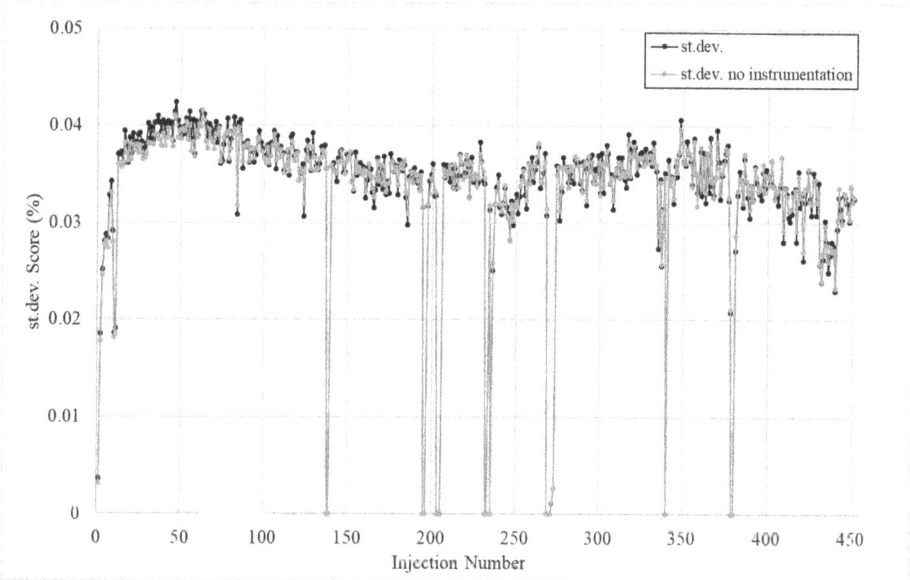

Fig. 4. Analysis performed by considering the comparison between the standard uncertainty and without considering the uncertainty of the instrumentation. Figure 5a presents the mean of the score S distribution and Figure 5b the standard deviation

Figure 4 shows the results of the analysis performed by considering the comparison between the combined uncertainty and without considering the uncertainty instrumentation, in terms of the mean of the score *S* distribution (figure 4a), and in terms of the standard deviation of the score distribution (Figure 4b).

Figure 5 displays the comparison of the analysis carried out for the combined uncertainty, and the amplification of 2 times, 5 times, and 10 times, both in terms of the mean of the score *S* distribution (figure 5a) and in terms of the standard deviation of the score *S* distribution (figure 5b).

a)

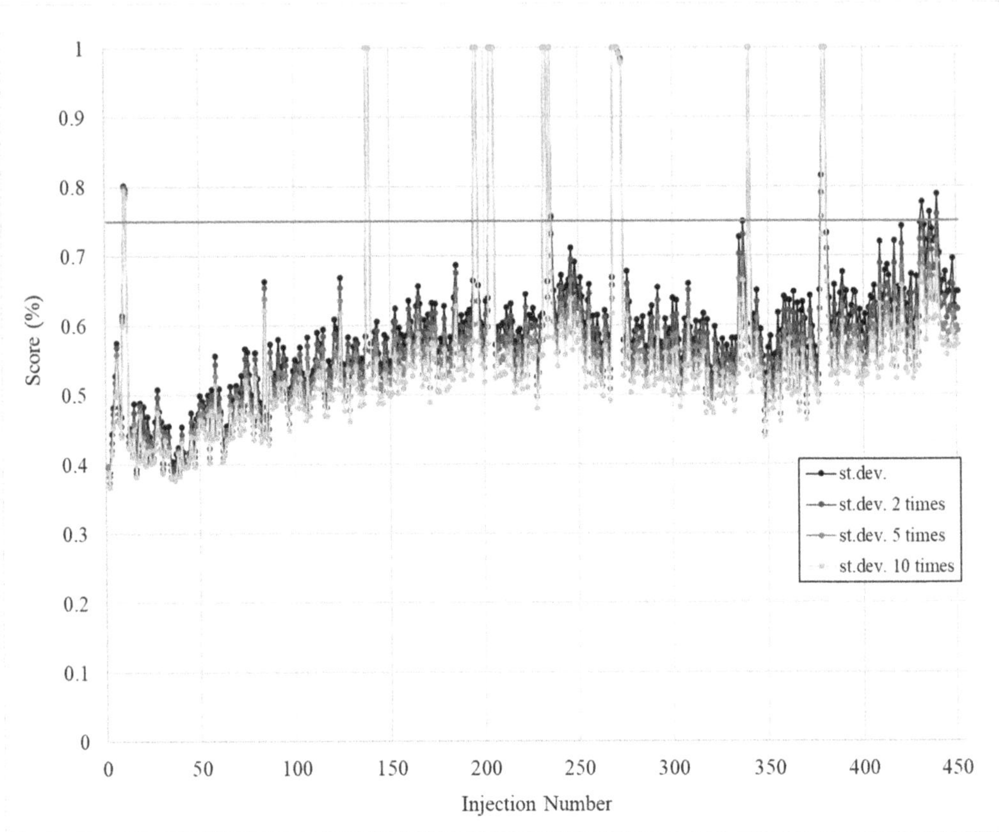
Score (%)
Injection Number
st.dev.
st.dev. 2 times
st.dev. 5 times
st.dev. 10 times

b)

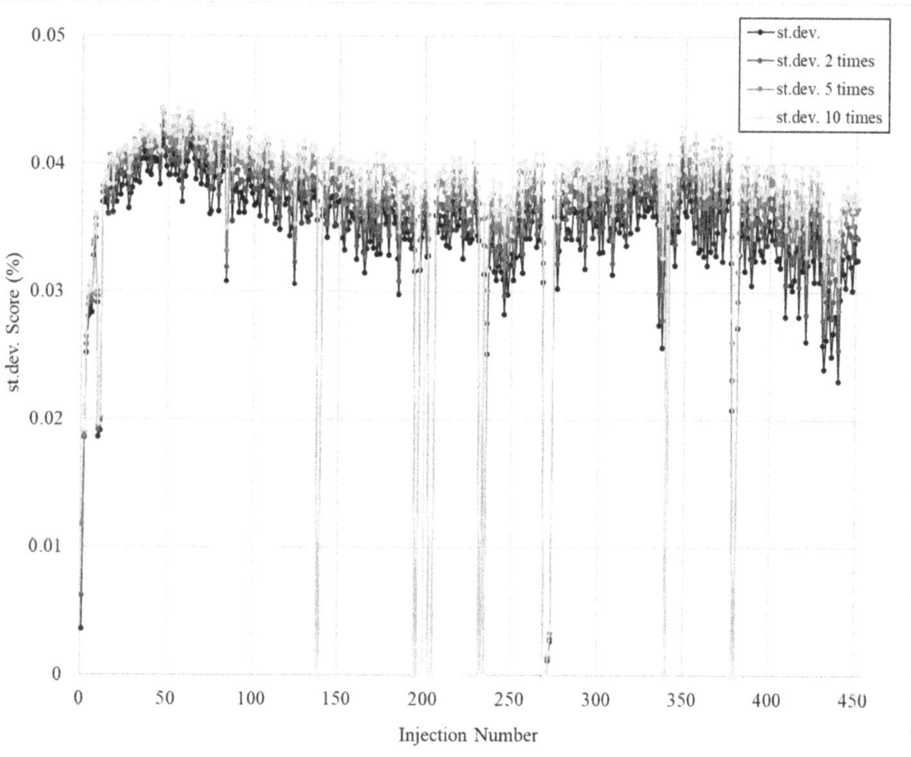

Fig. 5. Analysis performed by considering the comparison between the standard uncertainty and the amplification of the standard uncertainty of 2 times, 5 times, and 10 times. Figure 5a presents the mean of the score S distribution and Figure 5b the standard deviation.

By considering the mean of the score distribution of all analyses performed, all events of fault recorded by the maintenance service were correctly detected by the algorithm, and the 19 events are represented of all injection with a mean of score distribution equal to 1.

The investigation of the uncertainty without considering the uncertainties of the instrumentation (Figures 4a and 4b) returns the same results of the combined uncertainty both in terms of the mean of score distribution and standard deviation of the score distribution.

This result is expected because the values of the uncertainties of the transducers are negligible with respect to the contribution of the uncertainty estimator of the parameters and it has no reflection in the final prediction of the result.

The analysis carried out with the amplification of the combined uncertainty 2 times, 5 times and 10 times (Figures 5a and 5b) presents a result similar to the previous one. The trend of the prediction is similar in all cases of amplification both in terms of the mean and standard deviation of the score distribution. With the increase of the amplification, a decrease in the trend of the score can be seen, but it is negligible in terms of the effect of the final prediction that is the same for all cases of amplification.

This result means that the uncertainty has no reflection in the combination of the forecasters and in the final prediction. The method can be effective even if the transducers have lower accuracy in acquire signals.

4. Conclusion

The proposed method to predict the fault of a component in the diecasting process is based on the combination of z-score and the weighing matrix and then transformed into a score index S between 0 and 1.

The preliminary analysis was performed on a detested of 451 injections. All the events recognized as fault state were correctly detected by the algorithm in all conditions analyzed.

The Montecarlo method has allowed investigating the behavior of the uncertainty in the final result. By considering the uncertainty with no contribution of the instrumentation, the result is the same. The uncertainty of the instrumentation has a negligible contribution to the final result.

The analysis performed with the extended combined uncertainty returns a result similar to all conditions of amplification. A decrease in the trend of the score can be seen with the increase of the amplification of the combined uncertainty, but it is negligible in terms of the final prediction of the fault.

The final results of the investigation of the amplification of the uncertainty are that the uncertainty does not affect the final result of the prediction. This means that the method works even if the transducers are not so accurate in acquire signals.

Acknowledgments

The authors wish to thank Italpresse S.p.a. for the contribution to this work. Another great thanks go to Eng. Roberto Zani, Eng. Andrea Scaratti and Eng. Simone Mutti for their essential contribution to this work.

References

1. Ahmadzadeh F, Lundberg J. Remaining useful life estimation: review. Int J Syst Assur Eng Manage 2014;5(4):461–74.
2. El-Thalji I, Jantunen E. A summary of fault modelling and predictive health monitoring of rolling element bearings. Mech Syst Sig Process 2015;60–61:252–72.
3. Ben, J., Chebel-morello, B., Saidi, L., & Malinowski, S. (2015). Accurate bearing remaining useful life prediction based on Weibull distribution and artificial neural network. Mechanical Systems and Signal Processing, 56–57, 150–172.
4. Rocha, A. V., Melo, D. F., Maia, T. A. C., Ferreira, V. N., & Filho, B. J. C. (2019). IoT-Based Degradation Management for Self-Healing Power Converters. 2019 IEEE Applied Power Electronics Conference and Exposition (APEC), 2802–2809.
5. Ellefsen, A. L., Bjørlykhaug, E., Æsøy, V., Ushakov, S., & Zhang, H. (2019). Remaining useful life predictions for turbofan engine degradation using semi-supervised deep architecture. Reliability Engineering and System Safety, 183(June 2018), 240–251.
6. Qian Y, Yan R, Gao RX. A multi-time scale approach to remaining useful life prediction in rolling bearing. Mech Syst Sig Process 2017;83:549–67.
7. Zhang H, Kang R, Pecht M. A hybrid prognostics and health management approach for condition-based maintenance. IEEE international conference on industrial engineering and engineering management 2009.
8. Du SC, Lv J, Xi LF. Degradation process prediction for rotational machinery based on hybrid intelligent model. Rob Comput Integr Manuf 2012;28(2):190–207.
9. Fang XL, Zhou R, Gebraeel N. An adaptive functional regression-based prognostic model for applications with missing data. Reliab Eng Syst Saf 2015;133:266–74.
10. Xiao Q, Fang Y, Liu Q, Zhou S. Online machine health prognostics based on modified duration-dependent hidden semi-Markov model and high-order particle filtering. Int J Adv Manuf Technol 2018;94(1-4):1283–97.
11. Zheng X, Fang H. An integrated unscented kalman filter and relevance vector regression approach for lithium-ion battery remaining useful life and short-term capacity prediction. ReliabEng Syst Saf 2015;144:74–82.
12. Gebraeel NZ, Lawley M, Liu R, Parmeshwaran V. Residual life predictions from vibration-based degradation signals: a neural network approach. IEEE Trans IndustElectron 2004;51(3):694–700.

13. Mazidi P, Bertling-Tjernberg L, Sanz-Bobi MA. Wind turbine prognostics and maintenance management based on a hybrid approach of neural networks and a proportional hazards model. Proc Inst Mech Eng Part O 2017;231(2):121–9.
14. Tran VT, Pham HT, Yang BS, Nguyen TT. Machine performance degradation as- sessment and remaining useful life prediction using proportional hazard model and support vector machine. Mech Syst Sig Process 2012;32:320–30.
15. Wang Y, Peng Y, Zi Y, Jin X, Tsui KL. A two-stage data-driven-based prognostic approach for bearing degradation problem. IEEE Trans Ind Inf 2016;12(3):924–32.
16. Baptista M, Henriques EMP, de Medeiros IP, Malere JP, Nascimento CL, Prendinger H. Remaining useful life estimation in aeronautics: combining data-driven and Kalman filtering. Reliability Eng Syst Saf 2018.
17. Mazidi P, Du M, Bertling-Tjernberg L, Sanz-Bobi MA. A health condition model for wind turbine monitoring through neural networks and proportional hazard models. Proc Inst Mech Eng Part O 2017;231(5):481–94.
18. Chen, E., Cao, H., He, Q., Yan, J., & Jafar, S. (2019). An IoT based framework for energy monitoring and analysis of die casting workshop. Procedia CIRP, 80, 693–698.
19. Liu, W., Tang, R., & Peng, T. (2018). An IoT-enabled Approach for Energy Monitoring and Analysis of Die Casting Machines. Procedia CIRP, 69(May), 656–661.
20. Yarlagadda, P. K. D. V. (2000). Prediction of die casting process parameters by using an artificial neural network model for zinc alloys. International Journal of Production Research, 38(1), 119–139.

F Pavese, A B Forbes, N F Zhang, A G Chunovkina (Eds):
Advanced Mathematical and Computational Tools in Metrology and Testing XII
Series on Advances in Mathematics for Applied Sciences, Vol. 90

https://doi.org/10.1142/9789811242380_0023

Practical experimental design of task-specific uncertainty evaluation for coordinate metrology*

O. Sato† and T. Takatsuji

National Institute of Advanced Industrial Science and Technology,
AIST Central 3, Umezono 1-1-1,
Tsukuba, Ibaraki 305-8563, Japan
†E-mail: osm-satou@aist.go.jp
www.aisto.go.jp

A. Balsamo

Istituto Nazionale di Ricerca Metrologica, Strada. delle Cacce, 91,
Torino, 10135, Italy
E-mail: a.balsamo@inrim.it

This research describes a strategy for evaluating the measurement uncertainty of the feature dimensions and forms of products using a coordinate measuring machine, which is widely used in manufacturing industry. The proposed strategy outputs a task-specific measurement uncertainty, where the task includes the specification of the distribution of measurement points. Being executable with a small number of measurement trials, the strategy is practical for industrial use.

Keywords: Dimensional metrology, Measurement uncertainty, Design of experiment, Analysis of variance, Coordinate measuring machine.

1. Introduction

When sharing product information among the concerned parties in connected industries, the products must be unambiguously defined. Product shapes are specified by their dimensions and geometrical tolerances [1].

Coordinate measuring machines (CMMs) and other three-dimensional measuring systems validate the actual forms of products against their respective designed forms [2]. When verifying the dimensions of products, the measurement uncertainties must be considered [3]. A task-specific measurement uncertainty is

* This work is funded by the European Metrology Programme for Innovation and Research (EMPIR). The EMPIR initiative is co-funded by the European Union's Horizon 2020 research and innovation programme and the EMPIR Participating States.

evaluated by combining the standard uncertainties of the contributors [4]. However, it is hard to estimate the uncertainty contributions of individual dimensional measurements using a CMM, because measurement results are affected by many factors, such as deformations of the machine frame and environmental temperature changes [5][6]. In some studies, the uncertainty is instead assessed by design of experiment (DOE) [7-9], but this method is also difficult to perform on actual industrial floors.

The authors of the present study are developing an experimental procedure for task-specific measurement uncertainty evaluation using a CMM. The proposed uncertainty estimation applies a DOE and the analysis of variance (ANOVA) approach. The development is part of the EURAMET project titled Standards for the Evaluation of the Uncertainty of Coordinate Measurements in Industry (EUCoM).

This article describes the evaluation of the measurement uncertainties of the feature dimensions and forms of products, which are derived from the designated distributions of measurement points on the product surfaces. From a statistical viewpoint, the distribution of the measurement points used for feature extraction [10] should be non-patterned to reduce the systematic-error contribution to the measurement. Unfortunately, applying a random-points sampling strategy on typical industrial CMMs is impractical, because these machines are numerically controlled by a computer. The proposal strategy outputs a task-specific measurement uncertainty, where the task includes the specification of the distribution of measurement points.

The experiment designed in this article is similar to Taguchi's L4 orthogonal array method [11].

2. Uncertainty contributors of a CMM measurement task

Essentially, a CMM measures single-point coordinates. The dimensions and forms of the features are then computed from the set of measurement points sampled by the CMM. Therefore, the uncertainty in the dimensions or forms is derived from the uncertainty in the single-point coordinates: u_P. The uncertainty u_P is a combination of various uncertainty contributors, which are difficult to evaluate individually because of the complex factors in a CMM's measurement task.

In practice, combined uncertainties of measurements are dominated by random errors and systematic errors. The contributions of these respective factors on the measurement uncertainty propagate and are modularized to their contributors such as the items of an uncertainty budget. By proper design of the measurement experiment, the unknown systematic errors can be treated as random errors.

2.1. *Randomizing the unknown systematic-error sources*

When a product is measured on a CMM in one orientation, the measured coordinates of a single point in the product's coordinate system (x, y, z) include random and systematic errors. The errors derived from the contributors are propagated to the machine coordinates of the CMM. The systematic-error contribution depends on the location of the single point in the machine coordinate system and on the orientation of the product coordinate system relative to the machine coordinate system.

If the measurement is made in a specified position and orientation, the measurement result is biased by systematic error. Even if the measurement is repeated and the measurement results are averaged, the effect of the bias is neither eliminated nor reduced. To evaluate the magnitude of the bias, extra measurements are necessary.

In the present study, a product is measured at different positions and orientations on a CMM. This approach randomizes the systematic errors in the respective measurements by varying the magnitude and coefficient of the error propagation to the measured results. The effect of the systematic error is then reduced by averaging the repeated measurement results. The respective contributions of the systematic errors are evaluated from the multi-orientation measurements using the ANOVA technique.

3. Design of experiment

To accurately estimate the systematic-error effect, the measurements should be performed in the maximum possible number of positions and orientations. Moreover, the positions and orientations should be randomly distributed. However, increasing the number of measurement repetitions and trials raises the measurement cost. Therefore, the DOE should provide the uncertainty contribution of the unknown systematic error in the CMM within a small number of measurement trials. In this research, the effect of the systematic error was evaluated from the measurement results of only four significantly different orientations on a CMM.

The systematic error in the CMM is observed as the deformation of the cubic region of interest (ROI) of the CMM, which holds the workpiece to be measured [12][13]. The influence of the ROI deformation is approximated by a linear transformation function. To evaluate the effect of the systematic error in the ROI, it is sufficient to perform the measurements for uncertainty estimation in a few but essential number of orientations.

Figure 1 illustrates the four essential orientations for randomizing the systematic error on a CMM [14]. The primal position (home position) is the ordinal orientation of the workpiece selected by the operator. The other three positions are obtained after rotating the coordinate system of workpiece by 90° about the 1st, 2nd, and 3rd axes of the CMM machine's coordinate system.

The systematic errors caused by the X-, Y-, and Z-axis of the CMM in the ROI are unchanged throughout the measurements, but their propagations onto the coordinates of the workpiece coordinate system differ among the four positions. After averaging the measurement results at the four positions, the effect of the unknown systematic errors is reduced. The contribution of the systematic error in the averaged output is evaluated along with the variance of the measurement results at the four positions.

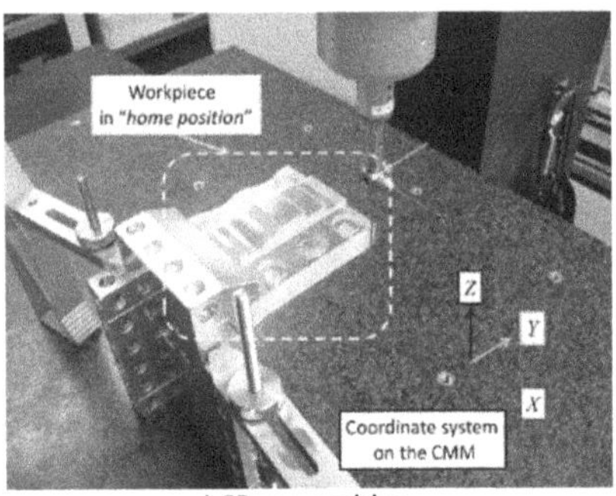

a) Home position

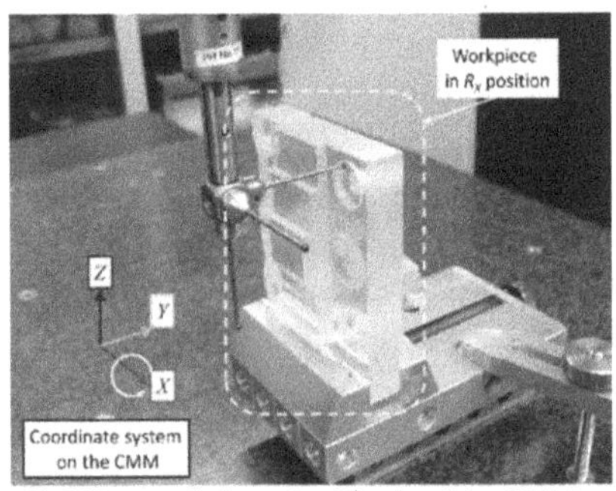

b) 90° rotated about the 1st axis of the CMM

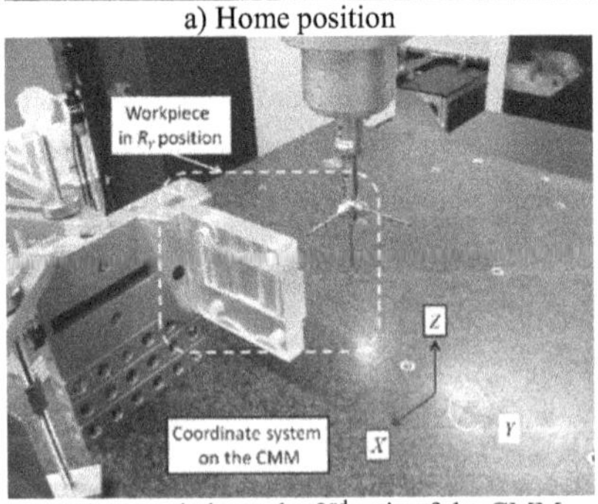

c) 90° rotated about the 2nd axis of the CMM

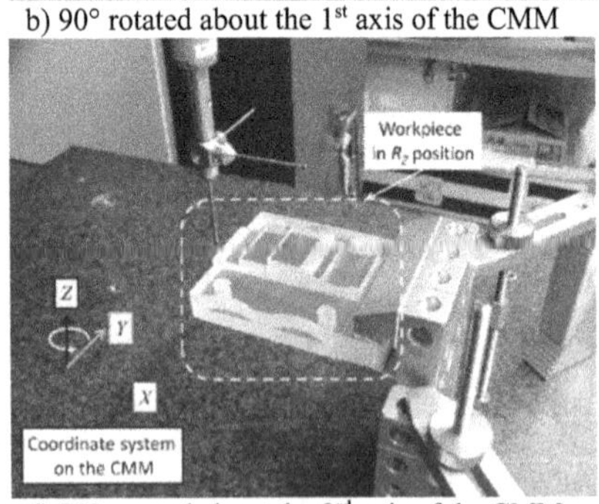

d) 90° rotated about the 3rd axis of the CMM

Fig. 1. Setup for randomizing and measuring the unknown systematic error.

4. Simulation

To demonstrate that the uncertainty evaluated by the measurement results at the four orientations closely approximates the true uncertainty, a Monte Carlo Simulation (MCS) was performed. As mentioned in Sec. 2, the essential measurements by a CMM are single-point coordinates. To simplify the measurement task, the measurement workpiece in the following simulation was a tetrahedron.

The coordinates of the four vertices of the tetrahedron were measured on the CMM in several orientations. The workpiece coordinate system was constructed from the coordinates of three vertices (the 1st, 2nd, and 3rd vertices) of the tetrahedron. The coordinates of the remaining vertex (the 4th vertex) were reserved for checking the simulation.

Table 1 shows the MCS conditions. The validation set consisted of 500 randomly generated tetrahedron sets and deformed ROIs. Each side of each tetrahedron was about100 mm long. In each ROI, the tetrahedron was measured in 200 random orientations and four specified orientations (see Fig. 1).

Table 1. Conditions of the Monte Carlo simulations

Measurement object	Vertices of tetrahedrons (side length about 100 mm)
Orientations in one ROI	200 at random
Trials of ROI deformation	500 at random
Deformation of the ROI	
Scale factor	5 μm/m
Perpendicularity between axes	2" (10 μrad)

Figure 2 presents a simulation result in one ROI. The red dots show the deviations between the averaged coordinates of the 4th vertex and the measured coordinates of the 4th vertices of the 200 randomly orientated tetrahedrons in the workpiece coordinate system. The blue crosses show the deviations between the averaged coordinates of the 4th vertex and the measured coordinates of the 4th vertices of the four specified orientated tetrahedrons. As shown in Fig. 2, the measured positions of the 4th vertex in the ROI deviated more widely among the orientations than the distribution of vertex positions given by the random orientations in the MCS.

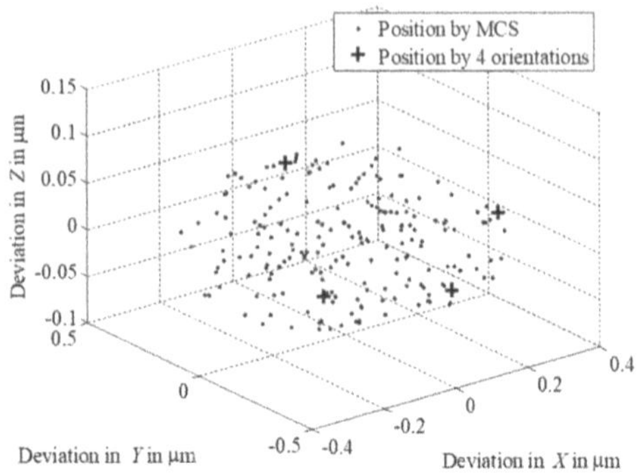

Fig. 2. Distributions of 4th vertex position of the tetrahedron in a single deformed ROI.

Figure 3 relates the expanded uncertainties in the coordinate measurements of the 4th vertex evaluated from the 200 random orientation measurements ($U_{geo,\ MCS}$) to those obtained from the four specified orientations ($U_{geo,\ 15530\text{-}2}$)[15] in the 500 corresponding ROIs. The standard uncertainties in the simulated 4th vertex coordinates, $u_{geo,\ MCS}$ and $u_{geo,\ 15530\text{-}2}$, respectively, were computed as the standard deviations of the Euclidean distances between the mean point of the repeated measurements and the corresponding measured point. Here $u_{geo,\ MCS}$ was the standard uncertainty in the 200 measurement results in a given ROI and $u_{geo,\ 15530\text{-}2}$ was that of the four measurement results in the same ROI. The coverage factors ($k = 2$ for $U_{geo,\ MCS}$ and $k = 3$ for $U_{geo,\ 15530\text{-}2}$) were determined by considering their effective degrees of freedom. In most cases, the proposed strategy overestimated the precisely estimated uncertainty, meaning that $U_{geo,\ 15530\text{-}2} \geq U_{geo,\ MCS}$. In the remaining 5% of cases in the MCS trial, the proposed strategy under-estimated the precise uncertainty.

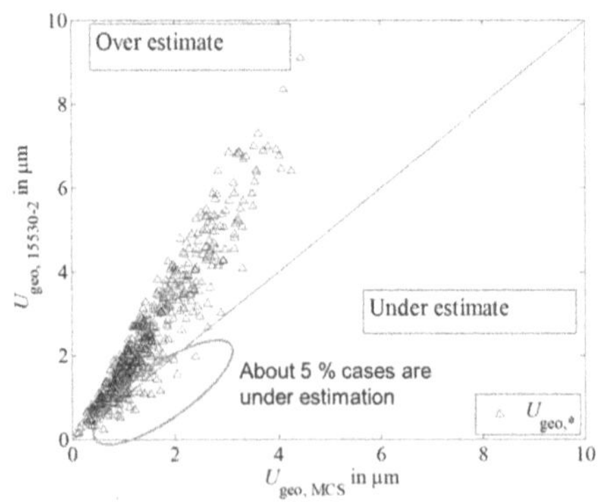

Fig. 3. Comparison of expanded uncertainties derived from the two evaluations.

5. Discussion

In industry, the measurement uncertainty is used to make decisions whether a product meets its specification requirements [3]. To avoid the risk of false acceptance or false rejection, the uncertainty approximated by the easy-to-use approach should over-estimate or under-estimate the actual uncertainty by as little as possible.

Figure 4 shows the degree of under estimation by the proposed method. The degree of under estimation under the given conditions was unacceptable in only one case. Therefore, the risk of false acceptances and rejections using the proposed method is deemed sufficiently small.

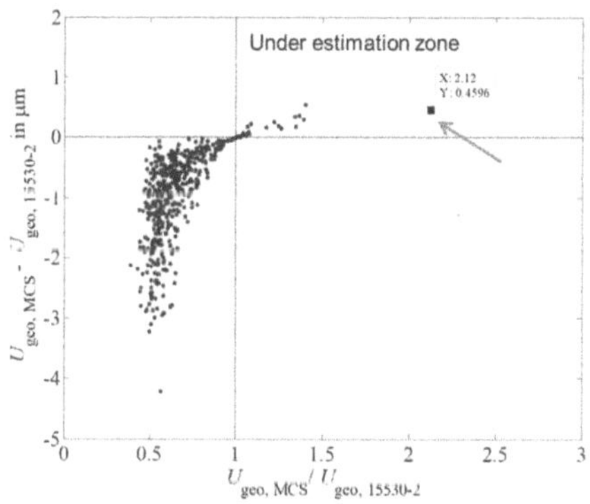

Fig. 4. Degree of under estimation in the proposed strategy.

6. Summary

Single-point coordinate measurements are essential in dimensional metrology; therefore, uncertainty evaluation of coordinate measurements is the basic process to estimate the complex task-specific uncertainties using CMMs. In this article, the authors proposed and demonstrated a simple method for estimating measurement uncertainties. After combining the measurement results in four specified orientations on a CMM, the unknown systematic errors in the CMM system are sufficiently randomized, and their contributions to the measurement results can be approximated with small risk of under estimation.

Acknowledgments

This work is funded by the European Metrology Programme for Innovation and Research (EMPIR). The EMPIR initiative is co-funded by the European Union's Horizon 2020 research and innovation programme and the EMPIR Participating States.

References

1. ISO 5459:2011, Geometrical product specifications (GPS) -- Geometrical tolerancing -- Datums and datum systems.
2. S. Carmignato, et al., Dimensional artefacts to achieve metrological traceability in advanced manufacturing, *CIRP Annals Manufacturing Technology*, **69** 693(2020).
3. ISO 14253-1:2017, Geometrical product specifications (GPS) -- Inspection by measurement of workpieces and measuring equipment -- Part1: Decision rules for verifying conformity or nonconformity with specifications.
4. JCGM 100:2008, Evaluation of measurement data -- Guide to the expression of uncertainty in measurement.
5. R. Wilhelm, et al., Task specific uncertainty in coordinate measurement, *CIRP Annals Manufacturing Technology*, **50** 553 (2001).
6. K. Takamasu, NEDO International joint research project international standard development of virtual CMM (2002)
7. M. Kritikos, et al., A random factorial design of experiments study on the influence of key factors and their interactions on the measurement uncertainty: a case study using the ZEISS CenterMax, *Applied Science*, **10** 37 (2020).
8. C. J. Feng, et al., Design and analysis of experiments in CMM measurement uncertainty study, *Precision Engineering*, **31** 94 (2007).
9. A. Piratelli, B. Di Giacomo, CMM uncertainty analysis with factorial design, *Precision Engineering*, **27** 283 (2003).
10. ISO 14406:2010, Geometrical product specifications (GPS) -- Extraction.
11. G. Taguchi, *Taguchi on Robust Technology Development: Bringing Quality Engineering Upstream*, (ASME Press, 1992).
12. V. Nardelli, et al., A simple solution to interim check of coordinate measuring machines, *Proc. XVIII IMEKO WORLD CONGRESS*, (Rio de Janeiro, Brazil, 2006).
13. R. Furutani, et al., Comparison of daily check artefacts of CMM, *Proc. 11th Int. Symposium on Measurement and Quality Control 2013*, (Cracow-Kielce, Poland, 2013).

14. Q. C. Dang, et al., Complete 3-D self-calibration of coordinate measuring machines, *CIRP Annals Manufacturing Technology*, **55** 527 (2006).
15. ISO/DTS 15530-2:2007, Geometrical product specifications (GPS) -- Coordinate measuring machines (CMM): Technique for evaluation of the uncertainty of measurement -- Part 2: Use of multiple measurement strategies in measurements of artefacts.

F Pavese, A B Forbes, N F Zhang, A G Chunovkina (Eds.):
Advanced Mathematical and Computational Tools in Metrology and Testing XII
Series on Advances in Mathematics for Applied Sciences, Vol. 90

https://doi.org/10.1142/9789811242380_0024

How to separate absolute and relative error components: Interval case

C. Servin

Computer Science and Information Technology Systems Department,
El Paso Community College (EPCC),
919 Hunter Dr. Room A-1720,
El Paso, TX 79915, USA
E-mail: cservin1@epcc.edu

O. Kosheleva

Department of Teacher Education,
University of Texas at El Paso,
500 W. University,
El Paso, TX 79968, USA
E-mail: olgak@utep.edu

V. Kreinovich

Department of Computer Science,
University of Texas at El Paso,
500 W. University,
El Paso, TX 79968, USA
E-mail: vladik@utep.edu

Usually, measurement errors contain both absolute and relative components. To correctly gauge the amount of measurement error for all possible values of the measured quantity, it is important to separate these two error components. For probabilistic uncertainty, this separation can be obtained by using traditional probabilistic techniques. The problem is that in many practical situations, we do not know the probability distribution, we only know the upper bound on the measurement error. In such situations of interval uncertainty, separation of absolute and relative error components is not easy. In this paper, we propose a technique for such a separation based on the maximum entropy approach, and we provide feasible algorithms – both sequential and parallel – for the resulting separation.

Keywords: absolute error, relative error, interval uncertainty, Maximum entropy approach

1. Need to Separate Absolute and Relative Error Components: General Case

Measurement uncertainty: a brief reminder. Most information about the world comes from measurements. Computers that process this information treat all the numerical inputs as exact numbers, but, of course, measurements are never 100% accurate. Each measurement result $\widetilde{x}$ is, in general, different from the actual (unknown) value x of the measured quantity: in other words, the *measurement error* $\Delta x \stackrel{\text{def}}{=} \widetilde{x} - x$ is, in general, different from 0; see, e.g., Rabinovich[9].

Because of this measurement uncertainty, the results of data processing are also, in general, different from what we would get if we knew the exact values of the corresponding quantities. To make adequate decisions based on the results of data processing, we need to gauge this uncertainty. For this, we need to know how accurate were the corresponding measurements.

How measurement uncertainty is usually described. Situations change. As a result, in slightly different situations, when we use the same measuring instrument several times to measure the same quantity, in general, we get slightly different results. We cannot predict which result we will get for each measurement. At best, what we can do is find the frequency with which different measurement errors appear in such a measurement. In other words, at best, we can determine, for each measured value x, the probability distribution – as characterized, e.g., by the probability density function (pdf) ρ_x that describes the probability of different values of the measurement error Δx: for each interval $\left[\underline{\delta}, \overline{\delta}\right]$, the probability that the measurement error Δx belongs to the interval is equal to

$$\int_{\underline{\delta}}^{\overline{\delta}} \rho_x(\Delta x)\, d\Delta x.$$

In many cases, the measurement error is the result of a joint action of the large number of small independent factors. In such cases, the measurement error is the sum of several independent variables of about the same size. It is known – this is called the Central Limit Theorem – that the distribution of such a sum is close to Gaussian; see, e.g., Sheskin[10]. Gaussian distributions are indeed frequently observed in metrology; see, e.g., Rabinovich[9].

In general, a Gaussian distribution is described by two parameters:

- its mean and
- its standard deviation σ.

If the mean is different from 0 – i.e., if we have a *bias* – then we can simply re-calibrate this measurement instrument by subtracting this bias from all the measurement results. Thus, we can safely assume that the bias is 0, and that the only characteristic describing the measurement error of measuring the quantity x is the standard deviation $\sigma(x)$.

Case of interval uncertainty. To find (and compensate for) the bias and to find the standard deviation $\sigma(x)$, we need several times to compare:

- the results of a measurement performed by the analyzed measuring instrument and
- the results of measuring the same quantity with a much more accurate ("standard") measuring instrument.

Such standard instruments are expensive. Thus, determining the mean and standard deviation is usually a very expensive and time-consuming procedure. Because of this, in many practical situations, this procedure is not performed. In such situations, the only information that we have about the measurement error Δx is the upper bound $\Delta(x)$ on its absolute value:

$$|\Delta x| = |\widetilde{x} - x| \le \Delta(x). \tag{1}$$

This upper bound has to be provided by the manufacturer of the measuring instrument. Indeed, if no such upper bound is provided, then, based on the measurement results, we cannot say anything at all about the actual value, it can be arbitrarily large or it can be arbitrarily small. In other words, this would be a wild guess, not a measurement.

The upper bound is obtained by the manufacturer of the measuring instrument by analyzing the whole bunch of similar mass-produced measuring instruments, with different biases and different values of other characteristics. This is a much cheaper and much less time-consuming process than individual calibration of each individual measuring instrument – which is needed for the probability uncertainty description. The use of this process allows the manufacturer to produce relatively cheap and affordable sensors and other measuring instruments.

In this case, once we know the measurement result $\widetilde{x}$, what can we conclude about the measurement error? The dependence of $\Delta(x)$ on x is usually smooth: relatively small changes of x lead to relatively small changes in $\Delta(x)$. Measurements are usually reasonably accurate, so the measurement result $\widetilde{x}$ is very close to the actual value x. Thus, the value $\Delta(x)$ is very close to the value $\Delta(\widetilde{x})$. The relative difference $\dfrac{\Delta(\widetilde{x}) - \Delta(x)}{\Delta(x)}$

between these two bounds is proportional to the relative difference $\frac{\widetilde{x}-x}{x} = \frac{\Delta x}{x}$. Thus, the absolute difference between these two bounds is of the order $(\Delta(x))^2$ – quadratic in terms of $\Delta(x)$:

$$\Delta(\widetilde{x}) - \Delta(x) = \frac{\Delta(\widetilde{x}) - \Delta(x)}{\Delta(x)} \cdot \Delta(x) \sim \frac{\Delta x}{x} \cdot \Delta(x) \sim$$

$$\frac{\Delta(x)}{x} \cdot \Delta(x) = \frac{1}{x} \cdot (\Delta(x))^2.$$

Thus, this difference is much smaller than $\Delta(x)$. So, from the practical viewpoint, this difference can be safely ignored. For example, if the upper bound on the absolute value of the measurement error is 5%, then its square is 0.25%, and is indeed negligible in comparison.

Since, from the practical viewpoint, we have $\Delta(\widetilde{x}) \approx \Delta(x)$, the inequality (1) takes the form $|\widetilde{x} - x| \leq \Delta(\widetilde{x})$. Thus, once we know the measurement result $\widetilde{x}$, the only information that we gain about the actual (unknown) value x of the measured quantity is that this value must satisfy the inequality

$$\widetilde{x} - \Delta(\widetilde{x}) \leq x \leq \widetilde{x} + \Delta(\widetilde{x}), \tag{2}$$

i.e., that this value x is in the interval $[\widetilde{x} - \Delta(\widetilde{x}), \widetilde{x} + \Delta(\widetilde{x})]$. Because of this, situations when we only know the upper bounds on the measurement errors are known as situations of *interval uncertainty*; see, e.g., Jaulin et al.[2], Moore et al.[6], and Mayer[5].

Need for extrapolation and interpolation. In practice, during any time interval, we can only perform a finite number of tests. In these tests, we estimate the values $\sigma(x_i)$ or $\Delta(x_i)$ for finitely many different values

$$x_1 < x_2 < \ldots < x_n.$$

In practice, we will encounter other values of the quantity x. We therefore need to extrapolate and/or interpolate the estimates $\sigma(x_i)$ or $\Delta(x_i)$ to other values of the quantity x.

How to extrapolate and interpolate. The dependence of $\sigma(x)$ and $\Delta(x)$ on the corresponding value x is usually smooth. Thus, a natural idea is to expand this dependence in Taylor series and keep only a few first terms in this expansion. As a result, the dependence $\sigma(x)$ or $\Delta(x)$ becomes a polynomial. So:

- we can estimate the coefficients of this polynomial based on the known values $\sigma(x_i)$ or $\Delta(x_i)$, and then
- we can use the resulting polynomial formula to estimate the corresponding metrological characteristic $\sigma(x)$ or $\Delta(x)$ for all x.

Absolute and relative error components. The simplest polynomial is a linear function. In the vicinity of each value, each smooth function can be reasonably well approximated by a linear function; see, e.g., Feynman et al.[1]. So, for some reasonable-size range, can safely use linear expressions $\sigma(x) = a + b \cdot x$ and $\Delta(x) = a + b \cdot x$.

What is meaning of the two terms in this expansion?

- When we have only the constant term, i.e., when $b = 0$, the corresponding characteristic of the measurement error is *absolute* in the sense that it does not change – it is the same for all the values of the measured quantity x. This error component is therefore known as the *absolute* error component.
- When we only have the linear term, i.e., when $a = 0$, then the absolute value of the measurement-error characteristic changes, but the *relative* error $\dfrac{\sigma(x)}{x}$ or $\dfrac{\Delta(x)}{x}$ remains constant. This component is thus called a *relative* error component.

Comment. The possibility of a linear approximation assumes a relatively limited range of values x. What can we do in situations when this linearity assumption does not hold for the whole range of possible values of x¿

- A first natural idea is to divide the range into sub-ranges on each of which we can approximate $\sigma(x)$ and/or $\Delta(x)$ by a linear function. For each of these sub-ranges, we can then apply the linear-approximation algorithm described in the following text.
- Alternatively, we can consider quadratic (or higher order) dependence of $\sigma(x)$ or $\Delta(x)$ on x. In this case, we need to modify our algorithm; the main idea behind this modification is described as a comment in the section where this algorithm is described.

Need to separate absolute and relative error components. It is therefore desirable to estimate the coefficients of the above linear dependence based on the known values $\sigma(x_i)$ or $\Delta(x_i)$. In other words, we need to *separate* the overall measurement errors into absolute and relative error components.

Comment. In general, the larger the value of the measured quantity, the larger the measurement error. However, we do not want to exclude the opposite situations, when larger values of the quantity can be measured with higher accuracy – and thus, with smaller measurement errors. In such situations, the corresponding coefficient b is negative.

2. How to Separate Absolute and Relative Error Components: Probabilistic Case

Towards a formal description of the problem. Suppose that we have several situations i with, in general, different values of the quantity of interest x – e.g., if we measure mass, we have several objects with different masses. Let x_i denote the true (unknown) value of the quantity x in situation i.

For each situation i, we perform several (s_i) measurements and get several values $\widetilde{x}_{ik}$, $k = 1, 2, \ldots, s_i$. In line with the above description, we assume that all the measurement errors are normally distributed with mean 0 and standard deviation $a + b \cdot x_i$, for some unknown values x_i, a, and b, and that measurement errors corresponding to different measurements are independent.

In this case, each value $\Delta x_{ik} = \widetilde{x}_{ik} - x_i$ of the measurement error is normally distributed with mean 0 and standard deviation $a + b \cdot x_i$. Thus, the corresponding probability density has the form

$$\frac{1}{\sqrt{2\pi} \cdot (a + b \cdot x_i)} \cdot \exp\left(-\frac{(\widetilde{x}_{ik} - x_i)^2}{2(a + b \cdot x_i)^2}\right).$$

Since all measurement errors are independent, the overall probability density is equal to the product of these expressions:

$$\prod_i \prod_{k=1}^{s_i} \frac{1}{\sqrt{2\pi} \cdot (a + b \cdot x_i)} \cdot \exp\left(-\frac{(\widetilde{x}_{ik} - x_i)^2}{2(a + b \cdot x_i)^2}\right). \tag{3}$$

A natural idea to select the values of the corresponding parameters x_i, a, and b which is the most probable, i.e., for which the corresponding probability (3) is the largest possible. This is known as the *Maximum Likelihood Method*; see, e.g., Sheskin[10]. Thus, we arrive at the following solution.

Solution. Based on the measurement results $\widetilde{x}_{ik}$, we find the values x_i, a, and b that maximize the expression (3).

3. How to Separate Absolute and Relative Error Components in the Interval Case: Analysis of the Problem

Towards a precise formulation of the problem: what we want and what we have. We want to find the expression $\Delta(x) = a + b \cdot x$ that describes the desired upper bounds on the measurement error for all the values x from some range $[\underline{x}, \overline{x}]$.

To find this expression, we can use the values $\Delta(x_i)$ provided for several values $\underline{x} < x_1 < x_2 < \ldots < x_n < \overline{x}$ of the measured quantity x.

Important conditions. We want to make sure that we do not underestimate the measurement errors. So, for all i, we must have

$$\Delta(x_i) \le a + b \cdot x_i. \tag{4}$$

Another constraint is that the value $\Delta(x) = a + b \cdot x$ must be non-negative for all values x from the interval $[\underline{x}, \overline{x}]$. It is easy to see that this condition is equivalent to requiring that this value is non-negative at the two endpoints of this interval, i.e., that we have:

$$0 \le a + b \cdot \underline{x} \tag{5}$$

and

$$0 \le a + b \cdot \overline{x}. \tag{6}$$

These conditions are not sufficient. These conditions are necessary, but not sufficient to determine the desired values a and b. Indeed, if we have some values a and b that satisfy inequalities (4)–(6), and we increase the value a, then all these inequalities will still be preserved.

So, we need to have a criterion that would allow us to select one pair of the values a and b from all the pairs that satisfy the above inequalities.

Idea: use Maximum Entropy Approach. In general, when we make a selection, we decrease the uncertainty of the situation: instead of all possible dependencies $\Delta(x)$ which are consistent with the known values $\Delta(x_i)$, we select one specific dependence. A natural idea is to not invent certainty where there is none, i.e., to select an alternative that maximally preserves the original uncertainty.

In the statistical approach, uncertainty of the probability distribution with the probability density $\rho(x)$ is naturally measured by entropy

$$S = -\int \rho(x) \cdot \ln(\rho(x))\, dx,$$

which is proportional to the smallest average number of binary (yes-no) questions that we need to ask to determine the value of the random variable with given accuracy; see, e.g., Jaynes & Bretthorst [3] and Nguyen at al. [7]. So, the above natural idea leads to maximum entropy approach – when from all possible distributions, we select the one with the largest entropy.

Interval uncertainty corresponds to a *class* of distributions – namely, the class of all the distributions located on the given interval. The average number of binary questions depends on the distribution. In this case, a natural measure of uncertainty is the largest possible average number of questions – i.e., largest possible entropy of all the distributions from this class. We will call this value the *entropy* of the given class of probability distributions.

Thus, a reasonable idea is to select the values a and b for which thus defined entropy is the closest to the original entropy value.

From the idea to the exact optimization criterion. What is the entropy in the interval case?

We consider n measurements corresponding to n different values $x_1, \ldots, x_n$. To provide a full description of each combination of measurement outcomes, we need to describe n corresponding measurement errors Δx_i. Thus, to get a complete description of measurement accuracy, we need to provide a probability distribution of the set of the corresponding tuples $(\Delta x_1, \ldots, \Delta x_n)$.

Originally, for each i, the only information that we have about each measurement error Δx_i is that it belongs to the interval $[-\Delta(x_i), \Delta(x_i)]$. Thus, the set of all possible values of the tuple is the box

$$B = [-\Delta(x_1), \Delta(x_1)] \times \ldots \times [-\Delta(x_i), \Delta(x_i)] \times \ldots \times [-\Delta(x_n), \Delta(x_n)].$$

So, we need to consider the class of all possible probability distributions on this box.

The entropy of this class, as we have mentioned earlier, can be described as the largest possible entropy of all the distributions located on this box. It is known (see, e.g., Jaynes & Bretthorst [3]) that this largest value is attained when we consider the uniform distribution on this box, with the constant probability density $\rho(x) = \text{const}$. This constant can be determined from the condition that the overall probability be equal to 1: $\int \rho(x)\, dx = 1$. For the constant, this implies that $\rho \cdot V = 1$, where V denotes the volume of the box. Thus, the probability density is equal to $\rho = \dfrac{1}{V}$. So, the entropy takes the form

$$S = -\int \rho(x) \cdot \ln(\rho(x))\, dx = -\int \frac{1}{V} \cdot \ln\left(\frac{1}{V}\right) dx =$$

$$-V \cdot \frac{1}{V} \cdot \ln\left(\frac{1}{V}\right) = -\ln\left(\frac{1}{V}\right).$$

Since the logarithm of the ratio is equal to the difference between logarithms, we conclude that

$$\ln\left(\frac{1}{V}\right) = \ln(1) - \ln(V) = -\ln(V),$$

thus $S = \ln(V)$.

The volume of the box is equal to the product of the lengths $2\Delta(x_i)$ of its sides:

$$V = \prod_{i=1}^{n} (2\Delta(x_i)) = 2^n \cdot \prod_{i=1}^{n} \Delta(x_i).$$

The logarithm of the product is equal to the sum of the logarithms, so we conclude that the original entropy is equal to

$$S = \ln(V) = n \cdot \ln(2) + \sum_{i=1}^{n} \ln(\Delta(x_i)).$$

When we replace each value $\Delta(x_i)$ with a linear expression $a + b \cdot x_i$, we similarly get the values

$$V' = \prod_{i=1}^{n} (2(a+b) \cdot x_i) = 2^n \cdot \prod_{i=1}^{n} (a + b \cdot x_i), \tag{7}$$

and

$$S' = \ln(V') = n \cdot \ln(2) + \sum_{i=1}^{n} \ln(a + b \cdot x_i). \tag{8}$$

We want to find the values a and b for which the entropy S' is the closest to the original entropy S.

For each i, due to inequality (4), we have $0 \le \Delta(x_i) \le a + b \cdot x_i$, so the product V of the left-hand sides is always smaller than or equal to the product V' of the right-hand sides. Thus, we always have

$$S = \ln(V) \le S' = \ln(V').$$

So, selecting the value S' which is the closest to S means selecting the smallest possible value S'.

Adding a constant $n \cdot \ln(2)$ to all the values of the entropy does not change which value is larger. Thus, finding the values a and b that minimize the entropy S' is equivalent to finding the values Δ_i for which the sum

$$s \stackrel{\text{def}}{=} \sum_{i=1}^{n} \ln(a + b \cdot x_i) \tag{9}$$

attains the smallest possible value.

So, we arrive at the following exact formulation of the problem.

Formulation of the problem in precise terms.

- *Given:*
  - values $\underline{x} < x_1 < x_2 < \ldots < x_n < \overline{x}$ and
  - values $\Delta(x_1), \ldots, \Delta(x_n)$.
- *What we want to compute:* Among all the pairs of values a and b that satisfy the inequalities (4)–(6), we need to find the values a and b for which the expression (9) attains the smallest possible value.

Towards the algorithm. First, to somewhat simplify the formulation of the problem, let us notice that the conditions (5) and (6) can be formulated in the form (4).

- For the condition (5), it is sufficient to take $x_0 = \underline{x}$ and $\Delta(x_0) = 0$.
- For the condition (6), it is sufficient to take $x_{n+1} = \overline{x}$ and $\Delta(x_{n+1}) = 0$.

So, instead of the conditions (4)–(6), we can consider the condition (4) for $i = 0, 1, \ldots, n, n+1$.

Second, let us take into account that since we have constraints of the non-strict inequality type, then for the minimizing pair of a and b, some inequalities will be equalities and some strict inequalities.

When at least two inequalities become strict equalities, e.g., equalities j and k, i.e., when $a + b \cdot x_j = \Delta(x_j)$ and $a + b \cdot x_k = \Delta(x_k)$, then we have two linear equations for the two unknowns a and b. From these two equations, we can find the values of both unknowns:

$$b = \frac{\Delta(x_k) - \Delta(x_j)}{x_k - x_j} \tag{10}$$

and

$$a = \frac{x_k \cdot \Delta(x_j) - x_j \cdot \Delta(x_k)}{x_k - x_j}. \tag{11}$$

The objective function (9) is smooth. Thus, if none of the inequalities (4) are equalities (or if only one inequality is an equality), then the value of the objective function is a (local or global) minimum of the objective function (or a local or global minimum under a single constraint). In both cases, this minimum is minus infinity – corresponding to the case when one of the values $a + b \cdot x_i$ is 0 and thus, its logarithm is equal $s = -\infty$. This cannot be the solution to our problem since, due to inequalities (4), we have $\ln(\Delta(x_i)) \le \ln(a + b \cdot x_i)$ hence $\sum_{i=1}^{n} \ln(\Delta(x_i)) \le s$. Here, $\Delta(x_i) > 0$ for all i, so $-\infty < \ln(\Delta(x_i))$ and $-\infty < s$.

Thus, at least two inequalities (4) must be equalities. Hence, the minimizing values a and b are determined by the formulas (10) and (11) corresponding to some indices j and k. So, we arrive at the following algorithm that always solves the above problem.

4. How to Separate Absolute and Relative Error Components in the Interval Case: Algorithm

Formulation of the problem: reminder. *We are given:*

- the values $\underline{x} < x_1 < x_2 < \ldots < x_n < \overline{x}$ and
- the values $\Delta(x_1), \ldots, \Delta(x_n)$.

What we want: among all the pairs (a, b) that satisfy the following inequalities:

$$\Delta(x_i) \le a + b \cdot x_i \tag{4}$$

for all $i = 1, \ldots, n$,

$$0 \le a + b \cdot \underline{x}, \tag{5}$$

and

$$0 \le a + b \cdot \overline{x}, \tag{6}$$

we want to find the pair (a, b) for which the expression

$$s = \sum_{i=1}^{n} \ln(a + b \cdot x_i) \tag{9}$$

attains the smallest possible value.

Algorithm. To find the desired values a and b, we do the following:

- First, we compute $x_0 = \underline{x}$, $x_{n+1} = \overline{x}$, and $\Delta(x_0) = \Delta(x_{n+1}) = 0$.

- Then, for each pair of indices $0 \le j < k \le n+1$:
  - First, we use the formulas
$$b = \frac{\Delta(x_k) - \Delta(x_j)}{x_k - x_j} \quad (10)$$
and
$$a = \frac{x_k \cdot \Delta(x_j) - x_j \cdot \Delta(x_k)}{x_k - x_j}. \quad (11)$$
to compute the corresponding values a and b.
  - Then, we check whether for the resulting values a and b, the inequality (4) is satisfied for all $i = 0, 1, \ldots, n+1$.
  - If the inequality (4) is satisfied for all i, then we compute the value (9).
- We then compare all the resulting values of the expression (9) and return the values a and b for which the expression (9) attains its smallest possible value.

Comment. In geometric terms, for each pair $j < k$, we draw a straight line $a + b \cdot x$ through the points $(x_j, \Delta(x_j))$ and $(x_k, \Delta(x_k))$ and select only those pairs for which all other points $(x_i, \Delta(x_i))$ are located beneath or on this line.

This algorithm is feasible. For each of $O(n^2)$ pairs (j, k):

- we check $O(n)$ inequalities;
- if inequalities are satisfied, we compute the sum (9) of n numbers;
- then, we compare the sum (9) with largest-sum-so-far, and if the new sum is larger, replace the largest-so-far with this sum.

So, for each of $O(n^2)$ pairs, we need $O(n)$ computational steps, to the total of $O(n^2) \times O(n) = O(n^3)$. Cubic-time algorithms are feasible (see, e.g., Papadimitriou[8] and Kreinovich at al.[4]), so we indeed have a feasible algorithm.

This algorithm is parallelizable. If we have an unlimited number of processors, then operations corresponding to each pair (j, k) can be performed on a separate group of processors. For each group, checking all inequalities can be done in parallel by $n + 2$ processors in one step. So, overall, we need $O(n^2) \cdot O(n) = O(n^3)$ processors.

Computing the sum can be done, as usual: in one step we add pairs of values, then add sums of pairs into sums of 4 values, etc. At the end, in

$O(\log(n))$ steps, we compute the sum. Thus, the overall time for processing each pair in parallel is $O(\log(n))$.

Now, we have $O(n^2)$ values, we need to find the smallest of these values. This can also done in $O(\log(n^2)) = O(\log(n))$ steps: first, we find the minimum of pairs, then minima of groups of 4, etc. So overall, we need $O(\log_2(n))$ time on $O(n^3)$ processors. The fact that this problem can be solved in polylog (even log) time on polynomial number of processors means that our problem belongs to the class NC of parallelizable problems; see, e.g., Papadimitriou[8].

Comment. In this paper, we concentrated on the case when we approximate the dependence $\Delta(x)$ by a linear functions – i.e., by the first two terms in the Taylor expression. Similar ideas and algorithms can be used if we approximate it by a quadratic, cubic, and other fragments of the Taylor series. In this case, instead of pairs (j, k), we will have to consider triples, quadruples, etc. The algorithm will be more time-consuming but still feasible.

5. Numerical Example

Description of the situation. Let us assume that the quantity x takes values from the interval $[\underline{x}, \overline{x}] = [0, 4]$, and that the actual (unknown) dependence of the upper bound $\Delta(x)$ on the measurement error on x has the form $\Delta^{\text{act}}(x) = 1 + x$, i.e. the form $\Delta^{\text{act}}(x) = a^{\text{act}} + b^{\text{act}} \cdot x$ with $a^{\text{act}} = b^{\text{act}} = 1$. This means, in particular, that for three values $x_1 = 1$, $x_2 = 2$, and $x_3 = 3$, we have $\Delta^{\text{act}}(1) = 1$, $\Delta^{\text{act}}(2) = 3$, and $\Delta^{\text{act}}(3) = 4$.

Suppose that the manufacturer of the measuring instrument provides slightly larger bounds on the measurement errors: $\Delta(x_1) = 2.1$, $\Delta(x_2) = 3.1$, and $\Delta(x_3) = 4.2$.

Resulting input. In this case, we have $n = 3$,

$$\underline{x} = 0 < x_1 = 1 < x_2 = 2 < x_3 = 2 < \overline{x} = 4,$$

and

$$\Delta(x_1) = 2.1, \quad \Delta(x_2) = 3.1, \quad \Delta(x_3) = 4.2.$$

Let us apply the algorithm. Following the algorithm, we first compute the values

$$x_0 = \underline{x} = 0, \quad x_4 = \overline{x} = 4, \quad \Delta(x_0) = \Delta(x_4) = 0.$$

Then, as we have mentioned in the previous section, in geometric terms, for each pair $j < k$, we draw a straight line $a + b \cdot x$ through the points $(x_j, \Delta(x_j))$ and $(x_k, \Delta(x_k))$ and select only those pairs for which all other points $(x_i, \Delta(x_i))$ are located beneath or on this line.

Let us consider all pairs $j < k$; we will consider them in lexicographic order.

- For $j = 0$ and $k = 1$, we get the line $a + b \cdot x = 2.1 \cdot x$, for which $a + b \cdot x_1 = 2.1$, $a + b \cdot x_2 = 4.2$, $a + b \cdot x_3 = 6.3$. The inequality (4) is satisfied for all i. Thus, we compute the expression (9). In our case, this expression has the form

$$s = \ln(2.1) + \ln(4.2) + \ln(6.3) = \ln(2.1 \cdot 4.2 \cdot 6.3) = \ln(55.566).$$

- For $j = 0$ and $k = 2$, we get the line $a + b \cdot x = 1.55 \cdot x$. For $x_1 = 1$, we have

$$\Delta(x_1) = 2.1 > a + b \cdot x_1 = 1.55,$$

so the inequality (4) is violated for $i = 1$. The pair $(j.k) = (0, 2)$ is therefore dismissed.
- Similarly, for $j = 0$ and $k = 3$, we get the line $a + b \cdot x = 1.4 \cdot x$, so the inequality (4) is also violated for $x_1 = 1$.
- For $j = 0$ and $k = 4$, we get the line $a + b \cdot x = 0$, so the inequality (4) is violated for x_1.
- For $j = 1$ and $k = 2$, we get $a + b \cdot x = 1.1 + x$. Here,

$$\Delta(x_3) = 4.2 > a + b \cdot x_3 = 4.1,$$

so this pair is also dismissed.
- For $j = 1$ and $k = 3$, we get $a + b \cdot x = 1.05 + 1.05 \cdot x$. The inequality (4) holds for all i, and

$$s = \ln(2.05 \cdot 3.15 \cdot 4.2) = \ln(27.1215).$$

- For $j = 1$ and $k = 4$, the line $a + b \cdot x$ decreases and thus, does not cover the points $(x_2, \Delta(x_2))$ and $(x_3, \Delta(x_3))$. This pair is dismissed.
- For $j = 2$ and $k = 3$, we get $a + b \cdot x = 0.9 + 1.1 \cdot x$. Here,

$$\Delta(x_1) = 2.1 > a + b \cdot x_1 = 2.0,$$

so this pair is also dismissed.
- For $j = 2$ and $k = 4$, the line $a + b \cdot x$ decreases and thus, does not cover the point $(x_3, \Delta(x_3))$. This pair is dismissed

- Finally, for $j = 3$ and $k = 4$, we get $a + b \cdot x = 16.8 - 4.2 \cdot x$. Here, the inequality (4) is satisfied for all i, and

$$s = \ln(12.6 \cdot 8.4 \cdot 4.2) = \ln(476.28).$$

We are left with three pairs $(j, k) = (0, 1)$, $(j, k) = (1, 3)$, and $(j, k) = (3, 4)$.

The value s is the smallest for the pair $(j, k) = (1, 3)$, so we conclude that

$$a = b = 1.05.$$

This result is reasonable. The values a and b that we obtained are indeed close to the actual values $a^{\text{act}} = b^{\text{act}} = 1$.

Acknowledgments

This work was supported in part by the National Science Foundation grants 1623190 (A Model of Change for Preparing a New Generation for Professional Practice in Computer Science), and HRD-1834620 and HRD-2034030 (CAHSI Includes).

The authors are thankful to the anonymous referees for valuable suggestions, and to Franco Pavese for his encouragement.

References

1. R. Feynman, R. Leighton, and M. Sands, *The Feynman Lectures on Physics* (Addison Wesley, Boston, Massachusetts, 2005).
2. L. Jaulin, M. Kiefer, O. Didrit, and E. Walter, *Applied Interval Analysis, with Examples in Parameter and State Estimation, Robust Control, and Robotics* (Springer, London, 2001).
3. E. T. Jaynes and G. L. Bretthorst, *Probability Theory: The Logic of Science* (Cambridge University Press, Cambridge, UK, 2003).
4. V. Kreinovich, A. Lakeyev, J. Rohn, and P. Kahl, *Computational Complexity and Feasibility of Data Processing and Interval Computations* (Kluwer, Dordrecht, 1998).
5. G. Mayer, *Interval Analysis and Automatic Result Verification* (de Gruyter, Berlin, 2017).
6. R. E. Moore, R. B. Kearfott, and M. J. Cloud, *Introduction to Interval Analysis* (SIAM, Philadelphia, 2009).

7. H. T. Nguyen, V. Kreinovich, B. Wu, and G. Xiang, *Computing Statistics under Interval and Fuzzy Uncertainty* (Springer Verlag, Berlin, Heidelberg, 2012).
8. C. Papadimitriou, *Computational Complexity* (Addison-Wesley, Reading, Massachusetts, 1994).
9. S. G. Rabinovich, *Measurement Errors and Uncertainties: Theory and Practice* (Springer, New York, 2005).
10. D. J. Sheskin, *Handbook of Parametric and Non-Parametric Statistical Procedures* (Chapman & Hall/CRC, London, UK, 2011).

F Pavese, A B Forbes, N F Zhang, A G Chunovkina (Eds):
Advanced Mathematical and Computational Tools in Metrology and Testing XII
Series on Advances in Mathematics for Applied Sciences, Vol. 90

https://doi.org/10.1142/9789811242380_0025

Proficiency testing with ordinal categorical data

Tomomichi Suzuki† and Momoko Isono
Department of Industrial Engineering, Tokyo University of Science,
2641 Yamazaki, Noda, Chiba, 2788510, Japan
†E-mail: szk@rs.tus.ac.jp

Jun-ichi Takeshita
Research Institute of Science for Safety and Sustainability,
National Institute of Advanced Industrial Science and Technology,
16-1 Onogawa, Tsukuba, Ibaraki, 3058569, Japan
E-mail: jun-takeshita@aist.go.jp

Proficiency testing is carried out to evaluate the performance of participating laboratories by summarizing and comparing the measurement results. This paper discusses methodologies related to measurements with ordinal categorical data. It investigates in detail three methods of statistical significance test with ordinal categorical data, from the viewpoint of whether they can be applied to proficiency testing procedures. These methods are: the cumulative test proposed by Taguchi, the cumulative chi-square test proposed by Takeuchi and Hirotsu, and ORDANOVA proposed by Bashkansky, Gadrich, and Kuselman. Our results suggest that the ORDANOVA method is applicable to proficiency testing and proposes a procedure using these statistics to analyze proficiency testing with ordinal categorical data.

Keywords: proficiency testing, ordinal categorical data, cumulative test, cumulative chi-square test, ORDANOVA

1. Introduction

Proficiency testing is carried out to evaluate the performance of participating laboratories by summarizing and comparing the measurement results. Statistical methods to support the analysis of proficiency testing are provided by ISO 13528 Statistical methods for use in proficiency testing by interlaboratory comparison [1]. ISO 13528 focuses on quantitative data that assume continuous distributions, namely normal distributions.

However, in many cases, non-quantitative data are obtained through measurements. In this paper, we consider cases of measurements with ordinal categorical cases, such as grades. Literature on proficiency testing with ordinal categorical data is limited. This paper investigates the existing statistical significance test methods on ordinal categorical data from the viewpoint of their applicability to proficiency testing procedures. We then propose a method to analyze proficiency testing with ordinal categorical data.

2. Statistical Methods for Analyzing the Variation of Ordinal Categorical Data

2.1. *Data format*

The data format discussed in this paper is shown in Table 1.

Table 1. Data format of cumulative test

A \ B	B_1	...	B_k	...	B_K	Total
A_1	n_{11}	...	n_{1k}	...	n_{1K}	$n_{1.}$
:	:		:		:	:
A_m	n_{m1}	...	n_{mk}	...	n_{mK}	$n_{m.}$
:	:		:		:	:
A_M	n_{M1}	...	n_{Mk}	...	n_{MK}	$n_{M.}$
Total	$n_{.1}$	...	$n_{.k}$	...	$n_{.K}$	N

This is an $M \times K$ contingency table with two variables A and B. Variable A has M categories (A_1, A_2, ..., A_M) and variable B has K *ordinal* categories (B_1, B_2, ..., B_K). Each item belongs to one category of A and one category of B. In proficiency testing, variable A represents laboratory and variable B represents measurement result in ordinal category.

Number of items classified as A_m and B_k is denoted as n_{mk}. The total number of items classified as A_m regardless of variable B (that is, marginal frequency) is denoted as $n_{m.}$. In the same manner, the total number of items classified as B_{mk} regardless of variable A is denoted as $n_{.k}$. Finally, the overall number of items in the table is denoted as N.

The cumulative total number of items classified as B_1 up to B_k for a particular A_m is denoted as N_{mk}. Therefore, p_{mk}, the probability of an item classified as B_k for a particular A_m can be estimated by $n_{mk}/n_{m.}$, and P_{mk}, the cumulative probability of an item classified as B_1 up to B_k for a particular A_m can be estimated by $N_{mk}/n_{m.}$. In addition, the marginal probabilities $p_{.k}$ and $P_{.k}$ are estimated by $n_{.k}/N$ and $N_{.k}/N$, respectively.

2.2. *ORDANOVA*

ORDANOVA is proposed by Bashkansky, Gadrich, and Kuselman [4]. The data format and the null hypothesis of the ORDANOVA are the same as that of the cumulative test and the cumulative chi-square test.

In ORDANOVA, variable A is set to the laboratory because its main focus is on measurement evaluation. The number of observations within a laboratory A_m is considered to be the same across all laboratories and is denoted as n, which means

$$n_{1.} = n_{2.} = \cdots = n_{M.} = n. \tag{1}$$

In ORDANOVA, an estimate of P_{mk} is denoted by $\hat{F}_{km}$ as in the original paper. (Note the exchange of subscripts). In addition, $\hat{F}_{k\cdot}$ is introduced, which is expressed as

$$\hat{F}_{k\cdot} = \frac{1}{M}\sum_{m=1}^{M} \hat{F}_{km} \ \left(= \hat{P}_{.k}\right). \tag{2}$$

ORDANOVA utilizes the h^2 statistic suitable for evaluating the variation of ordinal categorical data, as discussed by Blair and Lacy [5], which is expressed as

$$h^2 = \frac{1}{(K-1)/4}\sum_{k=1}^{K-1} F_k(1 - F_k). \tag{3}$$

The total variation of the data, $\hat{h}^2_{(T)}$, the variation within laboratory m, $\hat{h}^2_{m(W)}$, and the variation between laboratories for category k, $\hat{S}^2_{k(B)}$, are expressed as

$$\hat{h}^2_{(T)} = \frac{1}{(K-1)/4}\sum_{k=1}^{K-1} \hat{F}_{k\cdot}\left(1 - \hat{F}_{k\cdot}\right), \tag{4}$$

$$\hat{h}^2_{m(W)} = \frac{1}{(K-1)/4}\sum_{k=1}^{K-1} \hat{F}_{km}\left(1 - \hat{F}_{km}\right), \tag{5}$$

$$\hat{S}^2_{k(B)} = \frac{1}{M}\sum_{m=1}^{M}\left(\hat{F}_{km} - \hat{F}_{k\cdot}\right)^2. \tag{6}$$

Using these equations, the total variation can be decomposed into within-laboratory variation, $\hat{h}^2_{(W)}$, and between-laboratory variation, $\hat{S}^2_{(B)}$, as follows:

$$\hat{h}^2_{(T)} = \hat{h}^2_{(W)} + \hat{S}^2_{(B)}$$
$$= \frac{1}{M}\sum_m^M \hat{h}^2_{m(W)} + \frac{1}{(K-1)/4}\sum_{k=1}^{K-1} \hat{S}^2_{k(B)}. \quad (7)$$

The test statistic I is calculated by

$$I = \frac{\hat{S}^2_{(B)}/df_B}{\hat{h}^2_{(T)}/df_T}, \quad (8)$$

where $df_B = M - 1$ and $df_T = N - 1$. The distribution of I is also discussed in Bashkansky, Gadrich, and Kuselman [4], and it follows a chi-square distribution when the number of categories $K = 2$.

3. Searching for Measures of Proficiency Testing

3.1. *Measures of proficiency testing*

The statistical significance test procedures introduced in Section 2 are investigated from the view of using their statistics in proficiency testing. As a result, three measures are proposed all of which are from ORDANOVA. These measures are explained below.

3.2. *a) Sum of cumulative proportions*

The first of the three measures is the sum of cumulative proportions F_{km}, which is $\sum_{k=1}^K F_{km}$. The idea of this measure is illustrated in Fig. 1. The shaded area expresses the value of $\sum_{k=1}^K F_{km}$ which shows the performance of a laboratory.

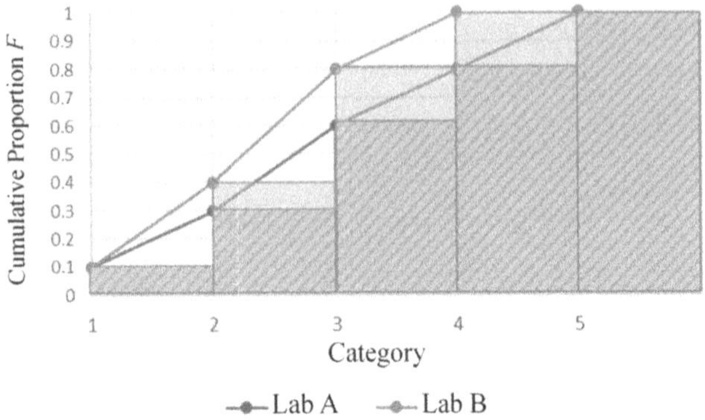

Fig. 1. Sum of cumulative proportions.

3.3. *b) Adjusted value of category where cumulative proportion becomes 0.5*

The second of the proposed measures is the adjusted value of the category where the cumulative proportion F_{km} becomes 0.5. The idea of this measure is illustrated in Fig. 2. Local linearity is assumed between the cumulative proportion and the category number. The value of the category that gives a cumulative proportion value of 0.5 is considered as a measure. In case F_{1m} becomes larger than 0.5, F_{0m}=0 is used to calculate the appropriate measure.

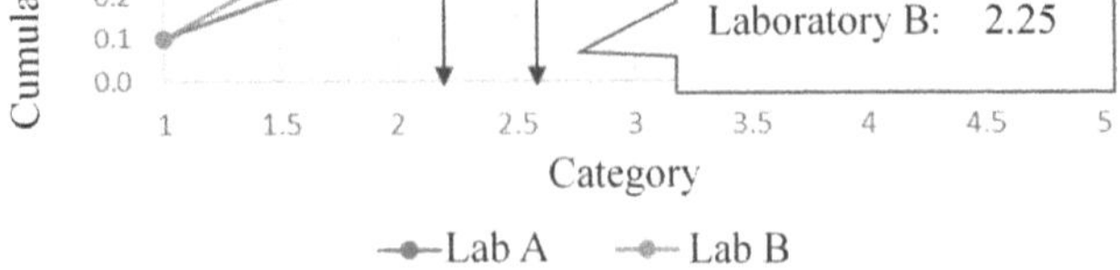

Fig. 2. Category value with cumulative proportion of 0.5.

3.4. *c) Laboratory component of the total variation given by ORDANOVA*

The last of the proposed measures is the laboratory component of the total variation given by ORDANOVA. The idea of this measure is based on Eq. (9), which is the arrangement of Eq. (7).

$$\hat{h}^2_{(T)} = \frac{1}{M}\frac{1}{(K-1)/4}\sum_{m=1}^{M}\sum_{k=1}^{K-1}\{\hat{F}_{km}(1-\hat{F}_{km}) + (\hat{F}_{km} - \hat{F}_{k.})^2\}. \qquad (9)$$

Here, $\sum_{k=1}^{K-1}\{\hat{F}_{km}(1-\hat{F}_{km}) + (\hat{F}_{km} - \hat{F}_{k.})^2\}$ is used as a measure for evaluating laboratory performance.

3.5. *Notes on the proposed measures*

The proposed measures are intended to represent the specific characteristics of the ordinal categorical data. However, we need to note that the measures a) and b) implicitly assume equidistance between categories and also implicitly assume linear latent variable in measure b).

Another point to make is these measures do not make a clear distinction between location and dispersion measures of ordinal distribution. It is, however, very difficult to make a clear distinction between them in ordinal distribution, even in the very simple case which is the binary data.

These aspects give rise to search other measures that might be more preferable in evaluating the precision of the ordinal measurements in the future research.

4. Simulation

4.1. *Procedures*

If we can assume a latent continuous distribution behind ordinal categorical data, it would be easy to set the scene for the simulation. However, in this paper, no latent distribution is assumed, so a small framework is proposed to simulate realistic scenarios. The number of categories K is set to 5, the number of laboratories is set to 21, the number of repetitions n is set to 50, and the number of iterations of the simulations is set to 1000. The procedure is as follows:

Procedure 1. Setting up base probabilities (scenarios)
The base probabilities of the multinomial distribution are set. The values of the probabilities are listed in Table 2. Scenario 1 reflects a normal distribution type, Scenario 3 reflects uniform distribution type and Scenario 2, very unusual distributions (U-shaped and biased) in order to verify the results in various cases.

Table 2. Base probabilities

Category	Scenario 1	Scenario 2	Scenario 3
1	0.1	0.4	0.2
2	0.2	0.2	0.2
3	0.4	0.1	0.2
4	0.2	0.1	0.2
5	0.1	0.2	0.2

Procedure 2. Setting up the probabilities of each laboratory
The probabilities of the multinomial distribution are set such that the resulting 10–100 % of the categories are moved one category higher/lower than the original category. There will be 10 laboratories that give higher categories, and another 10 laboratories that give lower categories. A 100% higher category means every classification will be one category higher than the base scenarios. These settings are intended to realize laboratory bias. The probabilities of a multinomial distribution of laboratories in Scenario 1 are shown in Table 3 as an example.

Table 3. Probabilities for laboratories (Senario 1)

Category	Lab 1 (-100%)	Lab 2 (-90%)	....	Lab 11 (base)	....	Lab 20 (+90%)	Lab 21 (+100%)
1	0.30	0.28		0.10		0.01	0.00
2	0.40	0.38		0.20		0.11	0.10
3	0.20	0.22		0.40		0.22	0.20
4	0.10	0.11		0.20		0.38	0.40
5	0.00	0.01		0.10		0.28	0.30

Procedure 3. Simulating the data
Data sets are obtained using the multinomial distribution given in Procedure 2. For each scenario, 1000 virtual proficiency testing are conducted. In other words, 1000 realizations of multinomial distribution is obtained for each laboratory for each scenario.

Procedure 4. Calculation of the proposed measures
The measures of evaluating laboratory performance proposed in Section 3 are calculated using the data obtained in Procedure 3.

4.2. *Results*

The results of the simulation are discussed by examining and comparing the distribution of the measures proposed in Section 3. The results discuss within-laboratory and between-laboratory effects. Below are the simulation results that show the distributions of the proposed measures for the 21 laboratories that participated.

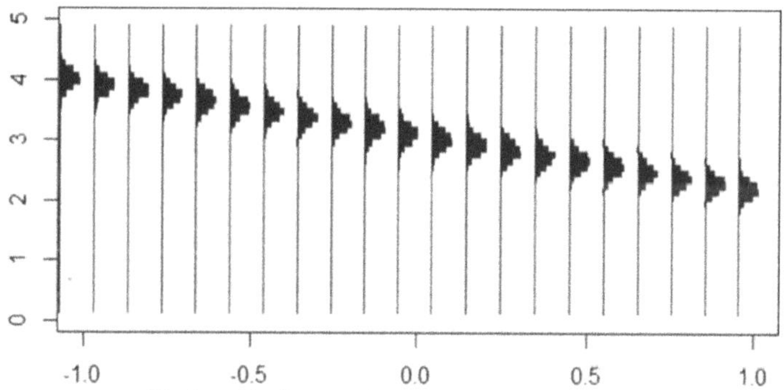

Fig. 3a Distribution of measure a) for Scenario 1

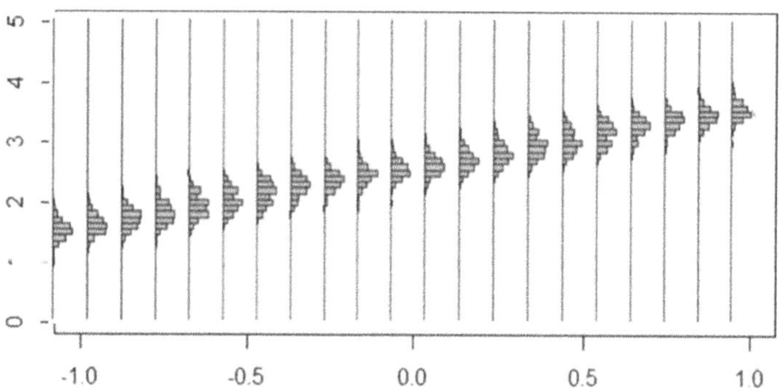

Fig. 3b Distribution of measure b) for Scenario 1

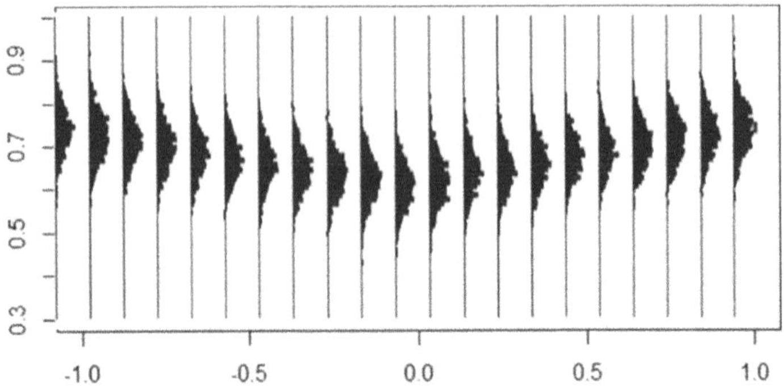

Fig. 3c Distribution of measure c) for Scenario 1

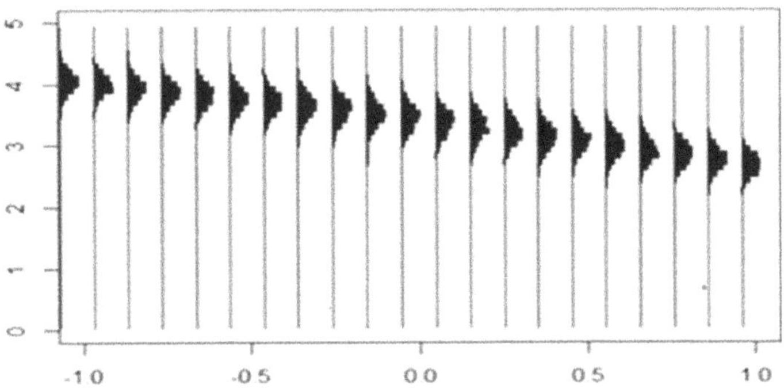

Fig. 4a Distribution of measure a) for Scenario 2

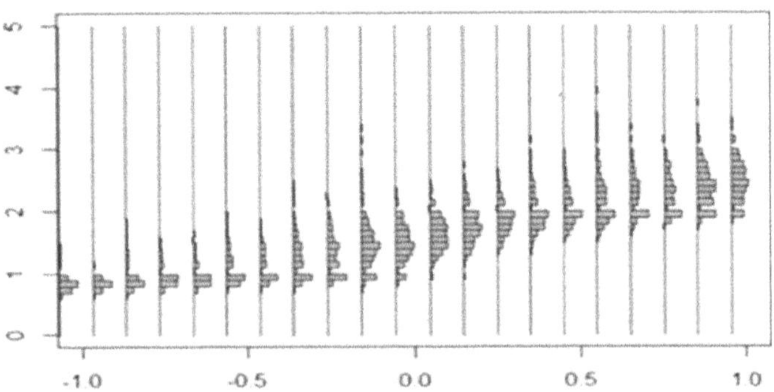

Fig. 4b Distribution of measure b) for Scenario 2

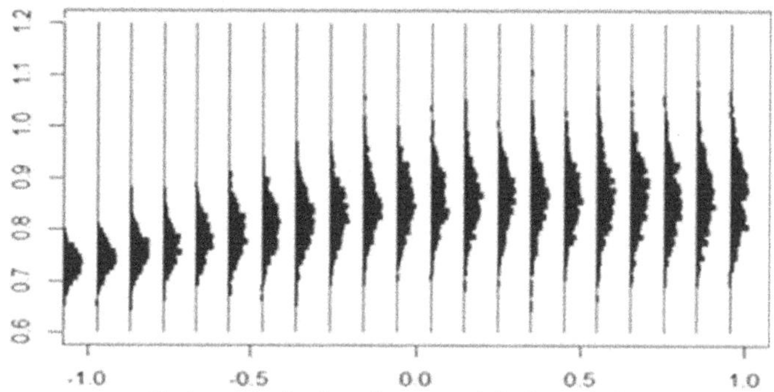

Fig 4c. Distribution of measure c) for Scenario 2

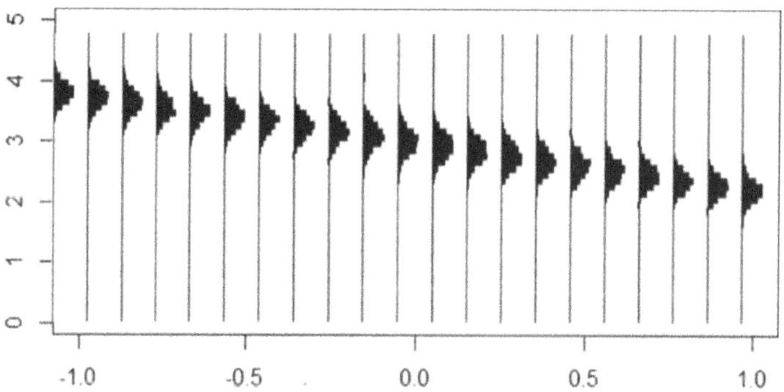

Fig 5a. Distribution of measure a) for Scenario 3

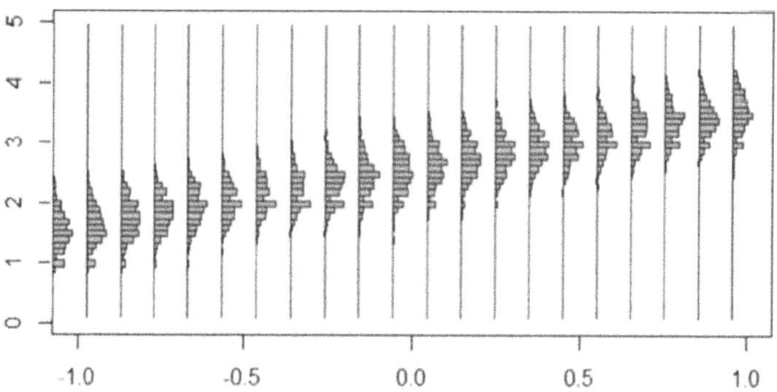

Fig 5b. Distribution of measure b) for Scenario 3

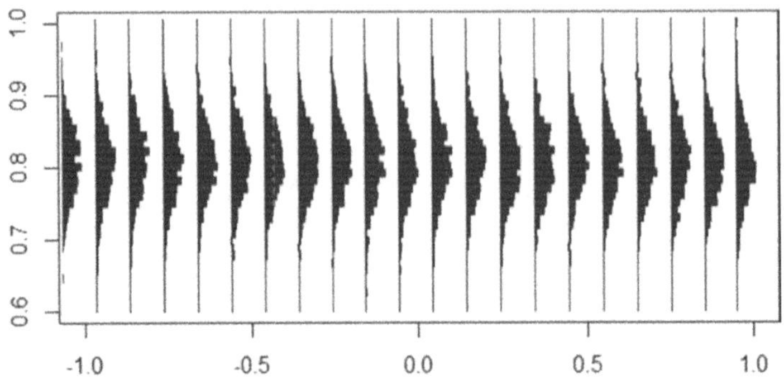

Fig 5c. Distribution of measure c) for Scenario 3

We discuss the results from two perspectives. The first is the relation between the intended laboratory bias and the resulting proposed measures. In this regard, the proposed measures a) and b) seem applicable, and measure a) seems robust compared to measure b). Measure c) does not seem to reflect the intended laboratory bias.

The second perspective is the distribution of measures in each case. Most of the distributions have a single peak, although the distribution is distorted in some cases with measure b). The dispersion of the measures compared to the amount of laboratory bias is small for measure a) and is considerably larger for measure c). From the results, it can be concluded that measure a) is the most appropriate, and measure b) may be applicable for the purpose. Measure c) is not ideal for evaluating laboratory performance.

5. Conclusions

A methodology for analyzing proficiency testing using ordinal categorical data is proposed. Three measures for evaluating laboratory performance were proposed and examined. The procedure as a whole can be considered appropriate. The proposed measures can be applied, although a significant improvement is possible. We believe this can be a starting point for further research on the topic.

Acknowledgments

The authors would like to thank the anonymous referees for the suggestions on improving the quality of the paper. This study was supported JSPS KAKENHI Grant Number JP20K12203.

References

1. International Organization for Standardization, *ISO 13528 Statistical methods for use in proficiency testing by interlaboratory comparison*, (ISO, 2015).
2. International Organization for Standardization, *ISO 5725-1 Accuracy (trueness and precision) of measurement methods and results –Part 1: General principles and definitions,* (ISO, 1994).
3. International Organization for Standardization, *ISO 5725-2, Accuracy (trueness and precision) of measurement methods and results – Part 2: Basic method for the determination of repeatability and reproducibility of a standard measurement method*, (ISO, 1994).
4. E. Bashkansky, T. Gadrich, I. Kuselman, Interlaboratory comparison of test results of an ordinal or nominal binary property: analysis of variation, *Accreditation and Quality Assurance* **17**, 239–243 (2012).
5. J. Blair, M. G. Lacy, Statistics of ordinal variation, *Sociological Methods & Research* **28**, 251–280 (2000).

F Pavese, A B Forbes, N F Zhang, A G Chunovkina (Eds.):
Advanced Mathematical and Computational Tools in Metrology and Testing XII
Series on Advances in Mathematics for Applied Sciences, Vol. 90

https://doi.org/10.1142/9789811242380_0026

How to describe measurement errors: A natural generalization of the Central Limit Theorem beyond normal (and other infinitely divisible) distributions

J. C. Urenda

Departments of Computer Science and Mathematical Sciences
University of Texas at El Paso
500 W. University
El Paso, TX 79968, USA
E-mail: jcurenda@utep.edu

O. Kosheleva

Department of Teacher Education
University of Texas at El Paso
500 W. University
El Paso, TX 79968, USA
E-mail: olgak@utep.edu

V. Kreinovich

Department of Computer Science
University of Texas at El Paso
500 W. University
El Paso, TX 79968, USA
E-mail: vladik@utep.edu

When precise measurement instruments are designed, designers try their best to decrease the effect of the main factors leading to measurement errors. As a result of this decrease, the remaining measurement error is the joint result of a large number of relatively small independent error components. According to the Central Limit Theorem, under reasonable conditions, when the number of components increases, the resulting distribution tends to Gaussian (normal). Thus, in practice, when the number of components is large, the distribution is close to normal – and normal distributions are indeed ubiquitous in measurements. However, in some practical situations, the distribution is different from Gaussian. How can we describe such distributions? In general, the more parameters we use, the more accurately we can describe a distribution. The class of Gaussian distributions is 2-dimensional, in the sense that each distribution from this family can be uniquely determined by 2 parameters: e.g., mean and standard deviations. Thus, when the approximation of the measurement error by a normal distribution is insufficiently accurate, a natural idea is to consider families with more parameters. What are 3-, 4-, 5-, n-dimensional limit families of this type? Researchers have considered 3-dimensional classes of distributions, which can – under weaker assumptions – be used to describe similar limit cases; distributions from these families are known as infinitely divisible ones. A natural next question is to describe all possible n-dimensional families for all n. Such a description is provided in this paper.

Keywords: measurement error, Central Limit Theorem, infinitely divisible distributions

1. Central Limit Theorem and Distributions of Measurement Error: A Brief Reminder and Formulation of the Problem

Specifics of precise measuring instruments: main idea. Measurements are never absolutely accurate: the measurement result $\widetilde{x}$ is, in general, different from the actual (unknown) value x of the measured quantity; see, e.g., Novitskii and Zograph[1], Orlov[2], and Rabinovich[3].

For most measurement instruments, we can usually identify several main factors that contribute to this measurement error. For many instruments, these factors include thermal noise, interference of nearby electric lines, etc. To increase the measurement accuracy, we need to decrease the effect of these factors. For example:

- to decrease the effect of thermal noise, we need to cool down the measuring instrument;
- to decrease the effect of electric lines, we can place the instrument in a conducting box, etc.

As a result of this thorough decrease, all *major* factor affecting measurement uncertainty have been decreased. Thus, the remaining measurement error is the joint effect of many *small* independent error components.

This idea can help describe the probability distribution of measurement errors. Interestingly, the above seemingly *qualitative* idea can help us describe, in *quantitative* terms, the probability distribution of the corresponding measurement errors.

Specifically, in many cases, there are theorems – they are called *limit theorems* – that state that when the number of components increases, the distribution of the sum of that many independent components tends to distributions from a certain family. Thus, in practice, when the number of components is large, the actual distribution of the measurement error is close to the corresponding limit distribution.

Central Limit Theorem. Historically the first – and most well-known – limit theorem is the Central Limit Theorem that states that under some reasonable conditions, the sum of many similar-size independent random variables tends to Gaussian (normal) distribution; see, e.g., Sheskin[4]. So, under these conditions, the distribution of the resulting measurement error is close to Gaussian; see, e.g., Rabinovich[3].

For many measuring instruments, the distribution of the measurement error is indeed close to Gaussian.

Need for limit theorems beyond Central Limit Theorem. For some measuring instruments, however, the distribution of the measurement error is different from Gaussian – and even when it is close to Gaussian, it is not exactly equal to Gaussian. To describe such distributions, we need to go beyond normal distributions.

How can we do that? The more parameters we use – i.e., in other words, the higher the dimension of the corresponding family – the more accurately we can describe the corresponding distributions. A general Gaussian distribution can be described by two parameters: mean μ and standard deviation. A natural idea is thus to consider more general – e.g., 3-parametric – families of distributions.

First natural property of the class of limit distributions: closeness under addition. Suppose that we have two different families of independent small random variables:

- the first one tends to a random variable X, and
- the second one tends to the random variable Y.

Then, when we combine variables from both families, the resulting limit random variable is simply equal to the sum $X + Y$ of the two limit random variables. So, the probability distribution corresponding to this sum can also appear in the limit.

Thus, the desired family of limit probability distributions must satisfy the following property:

- if X and Y are independent random variables from this family,
- then their sum $X + Y$ also belongs to this same family.

This property can be easily described in terms of the distribution's *characteristic function* $\chi_X(\omega) \stackrel{\text{def}}{=} E[\exp(\mathrm{i}\cdot\omega\cdot X)]$, where X is the corresponding random variable, $\mathrm{i} \stackrel{\text{def}}{=} \sqrt{-1}$, and $E[Z]$ denotes the expected value of the random variable Z. Indeed, here

$$\exp(\mathrm{i} \cdot \omega \cdot (X+Y)) = \exp(\mathrm{i} \cdot \omega \cdot X) \cdot \exp(\mathrm{i} \cdot \omega \cdot Y), \tag{1}$$

thus the expected values of both sides are also equal:

$$E[\exp(\mathrm{i} \cdot \omega \cdot (X+Y))] = E[\exp(\mathrm{i} \cdot \omega \cdot X) \cdot \exp(\mathrm{i} \cdot \omega \cdot Y)]. \tag{2}$$

Since the variables X and Y are independent, the variables $\exp(\mathrm{i} \cdot \omega \cdot X)$ and $\exp(\mathrm{i} \cdot \omega \cdot Y)$ are also independent. Hence, the expected value of their product is equal to the product of their expected values, so

$$E[\exp(\mathrm{i} \cdot \omega \cdot (X+Y))] = E[\exp(\mathrm{i} \cdot \omega \cdot X)] \cdot E[\exp(\mathrm{i} \cdot \omega \cdot Y)], \tag{3}$$

i.e., indeed

$$\chi_{X+Y}(\omega) = \chi_X(\omega) \cdot \chi_Y(\omega). \tag{4}$$

Thus, in terms of the characteristic function, the above property takes a very simple form: the family of the characteristic functions must be closed under multiplication.

Examples. This property is definitely true for the characteristic functions of the normal distribution $\chi(\omega) = \exp(\mathrm{i} \cdot \mu \cdot \omega - \sigma^2 \cdot \omega^2/2)$. Indeed:

- if we have two independent normally distributed random variables X_1 and X_2 with means μ_i and variances $V_i = \sigma_i^2$,
- then, as one can easily check, the product of their characteristic functions

  $$\chi_1(\omega) = \exp(\mathrm{i} \cdot \mu_1 \cdot \omega - V_2 \cdot \omega^2/2) \tag{5}$$

  and

  $$\chi_2(\omega) = \exp(\mathrm{i} \cdot \mu_2 \cdot \omega - V_2 \cdot \omega^2/2) \tag{6}$$

  also had the same form

  $$\chi(\omega) = \exp(\mathrm{i} \cdot \mu \cdot \omega - V \cdot \omega^2/2), \tag{7}$$

  with $\mu = \mu_1 + \mu_2$ and $V = V_1 + V_2$.

In general, this property is similarly satisfied for families of the type

$$\chi(\omega) = \exp(C_1 \cdot f_1(\omega) + \ldots + C_n \cdot f_n(\omega)), \tag{8}$$

where the functions $f_i(\omega)$ are fixed, but the parameters C_i can take any value. Indeed:

- if we have two distributions X_1 and X_2 from such a family, with characteristic functions

$$\chi_1(\omega) = \exp(C_{11} \cdot f_1(\omega) + \ldots + C_{1n} \cdot f_n(\omega)) \tag{9}$$

  and

$$\chi_2(\omega) = \exp(C_{21} \cdot f_1(\omega) + \ldots + C_{2n} \cdot f_n(\omega)), \tag{10}$$

- then the characteristic sum of the sum $X_1 + X_2$ has a similar characteristic function

$$\chi(\omega) = \exp(C_1 \cdot f_1(\omega) + \ldots + C_n \cdot f_n(\omega)), \tag{11}$$

  with $C_i = C_{1i} + C_{2i}$.

Comment. It is sufficient to describe the values of the characteristic function only for $\omega > 0$. Indeed, for $\omega < 0$, we have

$$\exp(\mathrm{i} \cdot \omega \cdot X) = \exp(-\mathrm{i} \cdot |\omega| \cdot X) = [\exp(\mathrm{i} \cdot |\omega| \cdot X)]^*, \tag{12}$$

where for each complex number $z = a + \mathrm{i} \cdot b$, the notation $z^* \stackrel{\text{def}}{=} a - \mathrm{i} \cdot b$ denotes its complex conjugate. By taking the expected value of both sides, we conclude that for $\omega < 0$, we have

$$\chi(\omega) = [\chi(|\omega|)]^*. \tag{13}$$

Thus, it is indeed sufficient to consider the values of the characteristic function only for $\omega > 0$.

Towards other natural properties. Other properties of the limit family are related to the fact that numerical values of a physical quantity depend:

- on the choice of a measuring unit and,
- (for many quantities like time or temperature) on the selection of the starting point.

Scale-invariance.

- If we replace the original measuring unit by a new unit which is λ times smaller,
- then all the numerical values of this quantity are multiplied by λ.

Comment. This idea is well-known in metrology: it is, e.g., an important feature to the so-called *interval scales*; see, e.g., Rabinovich[3].

Scale-invariance (cont-d). In particular, if we replace the original measuring unit by a new unit which is λ times smaller, then, instead of the original random variable X, we get the random variable $X' = \lambda \cdot X$. In the new units, the characteristic function has the form

$$\chi_\lambda(\omega) = E[\exp(\mathrm{i} \cdot \omega \cdot X')] = E[\exp(\mathrm{i} \cdot \omega \cdot (\lambda \cdot X))] =$$

$$E[\exp(\mathrm{i} \cdot (\omega \cdot \lambda) \cdot X)] = \chi(\lambda \cdot \omega). \tag{14}$$

We are interested in the universally applicable limit family of distributions, a family that should not depend on the choice of the measuring unit. Thus, we should require that:

- if a function $\chi(\omega)$ belongs to the limit family,
- then, for every $\lambda > 0$, the function $\chi_\lambda(\omega) = \chi(\lambda \cdot \omega)$ should also belong to this family.

This property is called *scale-invariance.*

Comment. One can easily see that the family of characteristic functions corresponding to normal distributions has this property. Indeed:

- if we have a characteristic function
$$\chi(\omega) = \exp(\mathrm{i} \cdot \mu \cdot \omega - V \cdot \omega^2/2) \tag{15}$$
corresponding to normal distribution,
- then, for each $\lambda > 0$, we have
$$\chi_\lambda(\omega) = \chi(\lambda \cdot \omega) = \exp(\mathrm{i} \cdot \mu \cdot \lambda \cdot \omega - V \cdot \lambda^2 \cdot \omega^2/2) =$$
$$\exp(\mathrm{i} \cdot \mu_\lambda \cdot \omega - V_\lambda \cdot \omega^2/2), \tag{16}$$
where we denoted $\mu_\lambda \stackrel{\text{def}}{=} \lambda \cdot \mu$ and $V_\lambda \stackrel{\text{def}}{=} \lambda^2 \cdot V$.

Shift-invariance.

- If we replace the original starting point with a new one which is x_0 units smaller,
- then to all numerical values, we add x_0.

In particular, instead of the original random variable X, we get the random variable $X' = X + x_0$. In the new units, the characteristic function has the form

$$\chi_{x_0}(\omega) = E[\exp(\mathrm{i} \cdot \omega \cdot X')] = E[\exp(\mathrm{i} \cdot \omega \cdot (X + x_0))] =$$

$$\exp(\mathrm{i} \cdot \omega \cdot x_0) \cdot E[\exp(\mathrm{i} \cdot \omega \cdot X)] = \exp(\mathrm{i} \cdot \omega \cdot x_0) \cdot \chi(\omega). \tag{17}$$

We are interested in the universally applicable limit family of distributions, a family that should not depend on the choice of the starting point. Thus, we should require that:

- if a function $\chi(\omega)$ belongs to the limit family,
- then, for every x_0, the function $\chi_{x_0}(\omega) = \exp(\mathrm{i} \cdot \omega \cdot x_0) \cdot \chi(\omega)$ should also belong to this family.

This property is called *shift-invariance.*

Comment. One can easily see that the family of characteristic functions corresponding to normal distributions has this property. Indeed:

- if we have a characteristic function
$$\chi(\omega) = \exp(\mathrm{i} \cdot \mu \cdot \omega - V \cdot \omega^2/2) \tag{18}$$
corresponding to normal distribution,
- then, for each x_0, we have
$$\chi_{x_0}(\omega) = \exp(\mathrm{i} \cdot \omega \cdot x_0) \cdot \exp(\mathrm{i} \cdot \mu \cdot \omega - V \cdot \omega^2/2) =$$
$$\exp(\mathrm{i} \cdot (\mu + x_0) \cdot \omega - V \cdot \omega^2/2), \tag{19}$$
i.e., the same form with $\mu + x_0$ instead of μ.

What is known: infinitely divisible distributions. For each positive real number a, there is a family of distributions that satisfy all three above-described properties. For $\omega > 0$, the corresponding characteristic functions have the form $\chi(\omega) = \exp(\mathrm{i} \cdot \omega \cdot x_0 + c \cdot \omega^a)$ for a complex value $c = c_r + \mathrm{i} \cdot c_i$.

Distributions described by these formulas are known as *infinitely divisible.*

Comment. It should be mentioned that normal distributions are a particular case of this family corresponding to $a = 2$ and $c_i = 0$.

Remaining problem. Distributions from the 3-parametric family of infinitely divisible distributions – defined by the above types of characteristics functions – do not always provide a precise description of how the measurement errors are distributed; see, e.g., a detailed empitical analysis in Novitskii and Zograph[1] and in Orlov[2]. A natural idea is thus to try 4-parametric, 5-parametric, etc. families.

A natural question is: which n-parametric families satisfy the three above-described properties?

What we do in this paper. In this paper, we provide a full description of all the families of this type.

2. Definition and the Main Result

Analysis of the problem. We are looking for families of characteristic functions of the type

$$\chi(\omega) = \exp(C_1 \cdot f_1(\omega) + \ldots + C_n \cdot f_n(\omega)), \tag{20}$$

where functions $f_i(\omega)$ are fixed, and the parameters C_i can take any value. It is reasonable to assume that the functions $f_i(\omega)$ are smooth for $\omega > 0$.

We want to find families which are scale- and shift-invariant. So, we arrive at the following definition:

Definition 1. *By a* limit family *we mean the family F of the functions (20) – corresponding to some functions differentiable $f_i(\omega)$, which satisfies the following two properties:*

- *if a function $\chi(\omega)$ belongs to the family F, then, for every $\lambda > 0$, the function $\chi(\lambda \cdot \omega)$ also belongs to the family F; and*
- *if a function $\chi(\omega)$ belongs to the family F, then, for every x_0, the function $\exp(\mathrm{i} \cdot \omega \cdot x_0) \cdot \chi(\omega)$ also belongs to the family F.*

Proposition 1. *For each limit family, each function $\chi(\omega)$ from this family has the form $\exp(\ell(\omega))$, where $\ell(\omega)$ is a linear combination of the functions $(\ln(\omega))^k \cdot \omega^a$ for some non-negative integer k and some complex value a.*

Comment. For $k = 0$ and real a, we get the characteristic functions of the normal distribution and of the infinitely divisible distributions.

Proof. Instead of the characteristic functions of the type (20), it is convenient to consider their logarithms

$$\ell(\omega) = \ln(\chi(\omega)) = C_1 \cdot f_1(\omega) + \ldots + C_n \cdot f_n(\omega). \tag{21}$$

Let us denote the class of all the logarithms corresponding to all the characteristic functions from the class F by L. By S, let us denote the set of all linear combinations of functions from the set L.

In terms of these logarithms, the scale-invariance property has the similar form:

- if a function $\ell(\omega)$ belongs to the family L
- then the re-scaled function $\ell(\lambda \cdot \omega)$ should also belong to the family L.

From this, we can conclude that:

- if a function $\ell(\omega)$ belongs to the family S
- then the re-scaled function $\ell(\lambda \cdot \omega)$ should also belong to the family S.

In particular, since each function $f_i(\omega)$ belongs to the family S, the re-scaled function $f_i(\lambda \cdot \omega)$ also belongs to the family S, i.e., has the form

$$f_i(\lambda \cdot \omega) = C_{i1}(\lambda) \cdot f_1(\omega) + \ldots + C_{in}(\lambda) \cdot f_n(\omega), \tag{22}$$

for some coefficients C_{ij} which are, in general, depending on λ.

Let us fix i and fix λ and consider n different values of ω:

$$\omega_1, \ldots, \omega_j, \ldots, \omega_n. \tag{23}$$

Then, we have n linear equations for n unknowns $C_{i1}(\lambda), \ldots, C_{in}(\lambda)$:

$$f_i(\lambda \cdot \omega_1) = C_{i1}(\lambda) \cdot f_1(\omega_1) + \ldots + C_{in}(\lambda) \cdot f_n(\omega_1),$$

$$\ldots$$

$$f_i(\lambda \cdot \omega_j) = C_{i1}(\lambda) \cdot f_1(\omega_j) + \ldots + C_{in}(\lambda) \cdot f_n(\omega_j), \tag{24}$$

$$\ldots$$

$$f_i(\lambda \cdot \omega_n) = C_{i1}(\lambda) \cdot f_1(\omega_n) + \ldots + C_{in}(\lambda) \cdot f_n(\omega_n).$$

Due to Cramer's rule, the solution to the system of linear equations can be represented:

- as the ratio of two polynomials depending on the coefficients and on the right-hand sides, i.e.,
- as a differentiable function of the coefficients and of the right-hand sides.

Here:

- The coefficients do not depend on *lambda* at all and are, thus, differentiable (namely, constant-valued) functions of λ.
- The right-and sides $f_i(\lambda \cdot \omega_i)$ are also differentiable functions of λ – since the functions f_i are differentiable.

Thus, the values $C_{ij}(\lambda)$ are also differentiable function of λ.

So, in the equation (22), all the functions are differentiable. Thus, we can differentiate both sides with respect to λ and get the following equation:

$$\omega \cdot f_i'(\lambda \cdot \omega) = C_{i1}'(\lambda) \cdot f_1(\omega) + \ldots + C_{in}'(\lambda) \cdot f_n(\omega), \tag{25}$$

where f_i' and C_{ij}', as usual, denotes the derivative of the corresponding function f_i or C_{ij}.

In particular, for $\lambda = 1$, we get

$$\omega \cdot f_i'(\omega) = c_{i1} \cdot f_1(\omega) + \ldots + c_{in} \cdot f_n(\omega), \tag{26}$$

where we denoted $c_{ij} \stackrel{\text{def}}{=} C_{ij}'(1)$.

The equation (26) can be further simplified if instead of the original variable ω, we introduce:

- a new variable $w = \ln(\omega)$ for which $\omega = \exp(w)$, and
- the corresponding new functions $F_i(w) \stackrel{\text{def}}{=} f_i(\exp(w))$ for which

$$f_i(\omega) = F_i(\ln(\omega)). \tag{27}$$

In these terms,

$$F_i'(w) = \frac{dF_i}{dw} = \frac{df_i(\exp(w))}{dw} = f_i'(\exp(w)) \cdot \exp(w) = f_i'(\omega) \cdot \omega, \tag{28}$$

which is exactly the left-hand side of the equation (26). Thus, in terms of the new variables and new functions, the equations (26) corresponding to $i = 1, \ldots, n$ take the form

$$F_1'(w) = c_{11} \cdot F_1(w) + \ldots + c_{1j} \cdot f_j(w) + \ldots + c_{1n} \cdot F_n(w),$$

$$\ldots$$

$$F_i'(w) = c_{i1} \cdot F_1(w) + \ldots + c_{ij} \cdot f_j(w) + \ldots + c_{in} \cdot F_n(w), \tag{29}$$

$$\ldots$$

$$F_b'(w) = c_{n1} \cdot F_1(w) + \ldots + c_{nj} \cdot f_j(w) + \ldots + c_{nn} \cdot F_n(w).$$

So, the functions $F_1(w), \ldots, F_n(w)$ satisfy a system of linear differential equations with constant coefficients. It is know that a general solution to such a system is a linear combination of the functions $w^k \cdot \exp(a \cdot w)$, where:

- a is an eigenvalue of the matrix $\|c_{ij}\|$, and

- k is a non-negative integer which is smaller than the multiplicity of the eigenvalue.

Thus, each function $f_i(\omega) = F_i(\ln(\omega))$ is a linear combination of the functions $(\ln(\omega))^k \cdot \exp(a \cdot \ln(\omega))$. Here,

$$\exp(a \cdot \ln(\omega)) = (\exp(\ln(\omega)))^a = \omega^a, \tag{30}$$

so we conclude that each function $f_i(\omega)$ is a linear combination of the expressions $(\ln(\omega))^k \cdot \omega^a$.

Since each function $\ell(\omega) = \ln(\chi(\omega))$ is a linear combination of the functions $f_i(\omega)$, it is also equal to the linear combination of the expressions $(\ln(\omega))^k \cdot \omega^a$. The proposition is thus proven.

Acknowledgments

This work was supported in part by the National Science Foundation grants 1623190 (A Model of Change for Preparing a New Generation for Professional Practice in Computer Science), and HRD-1834620 and HRD-2034030 (CAHSI Includes).

The authors are greatly thankful to the anonymous referees for their valuable suggestions and to Dr. Franco Pavese for his encouragement and help.

References

1. P. V. Novitskii and I. A. Zograph, *Estimating the Measurement Errors*, (Energoatomizdat, Leningrad, 1991), in Russian.
2. A. I. Orlov, How often are the observations normal?, *Industrial Laboratory* **57**(7), 770–772 (1991).
3. S. G. Rabinovich, *Measurement Errors and Uncertainties: Theory and Practice* (Springer, New York, 2005).
4. D. J. Sheskin, *Handbook of Parametric and Non-Parametric Statistical Procedures* (Chapman & Hall/CRC, London, UK, 2011).

F Pavese, A B Forbes, N F Zhang, A G Chunovkina (Eds.):
Advanced Mathematical and Computational Tools in Metrology and Testing XII
Series on Advances in Mathematics for Applied Sciences, Vol. 90

https://doi.org/10.1142/9789811242380_0027

Modelling of the dynamic gravimetric preparation of calibration gas mixtures using permeation for trace gas analysis

Adriaan M. H. van der Veen* and Heleen Meuzelaar
VSL, Department of Chemistry, Mass, Pressure and Viscosity, Delft, The Netherlands
**E-mail: avdveen@vsl.nl, www.vsl.nl*

Merima Čaušević
Institute of Metrology of Bosnia and Herzegovina, Sarajevo, Bosnia and Herzegovina
E-mail: merima.causevic@met.gov.ba

Maurice G. Cox
National Physical Laboratory, Teddington, United Kingdom
E-mail: maurice.cox@npl.co.uk

Permeation is a technique for realising a primary measurement standard for gas composition. A permeation tube is suspended in a permeation chamber and emits a constant mass flow of the nominally pure substance contained in it. This flow is combined with a flow of a carrier gas to obtain a calibration gas mixture with known composition. We used an automated weighing system to monitor the mass loss of the tube, and prepared mixtures of ammonia (NH_3) in nitrogen. The advantage of such dynamic gas standards is that unlike static standards they do not have stability issues and the composition of the calibration gas mixture can be chosen more rapidly than with static mixtures.

We revisited and extended existing measurement models describing the standard. We show how dependencies between the input quantities in the extended model can be taken into account. We describe the effect of temperature fluctuations on the permeation rate, temperature and pressure effects on the dispersion of the weighing data. For evaluating the linearity of the balance, a simple Bayesian model was established that takes into account the repeatability and resolution of the balance. We also show that the use of ordinary least squares regression to obtain the permeation rate is justified.

Keywords: Permeation; gas analysis; Bayesian inference; law of propagation of uncertainty, implicit measurement model.

1. Introduction

Permeation[1] is a technique for preparing dynamically calibration gas mixtures[2]. The method is a dynamic-gravimetric method; mass flow rates are used, together with information concerning the purity of the materials[3] and the molar masses of the components to calculate the composition. One of the mass flow rates originates from the permeation tube ("permeation rate"), the other from a thermal mass flow controller. In this example, we describe the calculation of the composition of the calibration gas mixture expressed in amount fractions, as used in many high-end applications.

2. Principle

The method, in ISO 6145-10[1], concerns the preparation of a gas mixture of ammonia (NH_3) in high-purity nitrogen (N_2; grade 6.0). The measurand is the amount fraction ammonia, the most abundant component present. A permeation tube, containing NH_3 of known purity was hooked inside the temperature- and pressure-controlled permeation chamber to a magnetic suspension balance, continuously measuring the mass of the permeation tube. The permeation chamber is purged at a known flow rate by the carrier gas, regulated by a thermal mass flow controller (MFC), operated as in[4] (figure 1). The permeation rate of the component of interest through the membrane depends on the properties of the component and the membrane permeability: the chemical nature and structure of the membrane, the area and thickness of the membrane, and the temperature and pressure gradient of the calibration component across the membrane. Different amount frac-

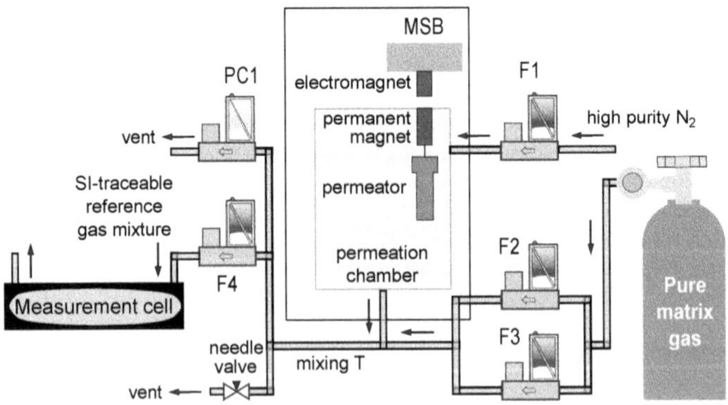

Fig. 1. Schematic overview of the permeation system and the auxiliary equipment to prepare one-stage and two-stage calibration gas mixtures

tions are realised by changing the carrier gas or permeation flow, in turn changed by adjusting the temperature in the permeation chamber.

The measurement model for the amount fraction of component k is[1]

$$y_k = \frac{\frac{\dot{m}_1}{\bar{M}_1} x_{1k}}{\frac{\dot{m}_1}{\bar{M}_1} + \frac{\dot{m}_2}{\bar{M}_2}}. \qquad (1)$$

$\dot{m}_1 = \mathrm{d}m_1/\mathrm{d}t$ is the permeation rate (mass flow rate from the permeation tube)[1], m_1 is the mass of the permeation tube and t time, $\dot{m}_2$ is the mass flow rate measured by the MFC. $\bar{M}_1$ and $\bar{M}_2$ are molar masses of the parent gases, and x_{1k} the amount fraction of the component of interest in parent gas 1. As nominally pure substances are used, $x_{1k} \approx 1$, but a purity analysis according to ISO 19229[3] is used to estimate x_{1k} and evaluate the associated standard uncertainty. Equation (1) presumes that the component of interest is not present in the carrier gas, a reasonable assumption for ammonia in nitrogen. The most general form of model is that of ISO 6142-1[5,6], applicable when the assumptions underlying equation (1) are not met.

Effects incorporated in the measurement model include temperature fluctuations in the permeation chamber, which affect the permeation rate of the tube and the performance of the weighing, and pressure fluctuations in the permeation chamber, which affect the weighing.

The law of propagation of uncertainty (LPU) is used to evaluate the standard uncertainty associated with y_k. We obtain the expressions for the sensitivity coefficients (first partial derivatives of y_k with respect to the input quantities) by using differentials[7]. Equation (1) can be written as

$$y_k = \frac{\dot{n}_1 x_{1k}}{\dot{n}_1 + \dot{n}_2}, \qquad \dot{n}_j = \dot{m}_j/\bar{M}_j.$$

Expressions for the partial derivatives with respect to the inputs are

$$\mathrm{d}y_k = \frac{\dot{n}_1}{\dot{n}_1 + \dot{n}_2}\mathrm{d}x_{1k} + \frac{\dot{n}_2 x_{1k}}{(\dot{n}_1 + \dot{n}_2)^2}\mathrm{d}\dot{n}_1 - \frac{\dot{n}_1 x_{1k}}{(\dot{n}_1 + \dot{n}_2)^2}\mathrm{d}\dot{n}_2,$$

where

$$\mathrm{d}\dot{n}_j = \frac{1}{\bar{M}_j}\mathrm{d}\dot{m}_j - \frac{\dot{m}_j}{\bar{M}_j^2}\mathrm{d}\bar{M}_j.$$

The standard uncertainty $u(y_k)$ associated with y_k, taking account of the covariances between $\bar{M}_1$ and $\bar{M}_2$ as well as between $\bar{M}_1$ and x_{1k} is given by

$$u^2(y_k) = \left(\frac{\partial y_k}{\partial x_{1k}}\right)^2 u^2(x_{1k}) + \left(\frac{\partial y_k}{\partial \dot{m}_1}\right)^2 u^2(\dot{m}_1) + \left(\frac{\partial y_k}{\partial \bar{M}_1}\right)^2 u^2(\bar{M}_1)$$
$$+ \left(\frac{\partial y_k}{\partial \dot{m}_2}\right)^2 u^2(\dot{m}_2) + \left(\frac{\partial y_k}{\partial \bar{M}_2}\right)^2 u^2(\bar{M}_2)$$
$$+ 2\frac{\partial y_k}{\partial \bar{M}_1}\frac{\partial y_k}{\partial \bar{M}_2} u(\bar{M}_1, \bar{M}_2) + 2\frac{\partial y_k}{\partial x_{1k}}\frac{\partial y_k}{\partial \bar{M}_1} u(x_{1k}, \bar{M}_1),$$

where

$$\frac{\partial y_k}{\partial x_{1k}} = \frac{\dot{n}_1}{\dot{n}_1 + \dot{n}_2}, \qquad \frac{\partial y_k}{\partial \dot{m}_1} = \frac{\dot{n}_2 x_{1k}}{(\dot{n}_1 + \dot{n}_2)^2}\frac{1}{\bar{M}_1}, \qquad \frac{\partial y_k}{\partial \bar{M}_1} = -\frac{\dot{n}_2 x_{1k}}{(\dot{n}_1 + \dot{n}_2)^2}\frac{\dot{m}_1}{\bar{M}_1^2}$$
$$\frac{\partial y_k}{\partial \dot{m}_2} = -\frac{\dot{n}_1 x_{1k}}{(\dot{n}_1 + \dot{n}_2)^2}\frac{1}{\bar{M}_2}, \qquad \frac{\partial y_k}{\partial \bar{M}_2} = \frac{\dot{n}_1 x_{1k}}{(\dot{n}_1 + \dot{n}_2)^2}\frac{\dot{m}_2}{\bar{M}_2^2}.$$

Assuming the amount fraction x_{1k} is calculated by difference (ISO 19229[3]),

$$u(x_{1k}, \bar{M}_1) = -\sum_{i \neq k} M_i u^2(x_{1i})$$

3. Dynamic gas mixture preparation

3.1. *Weighing*

Resolution, repeatability and linearity of the balance affect the uncertainty of weighing. Eccentric loading of the balance does not occur: the weighted tube is fixed throughout the measurement. Since repeatability of the weighing affects the readings used to calculate the permeation rate, this effect is taken into account when calculating the uncertainty of the permeation rate. The resolution of the balance is 1 µg, affecting the linearity assessment.

Fluctuations of temperature, pressure and convection are reflected in the dispersion of the weighing data and so not evaluated separately. Temperature and pressure fluctuations give rise to fluctuations in the density of the carrier gas (N_2 in this example), which can be calculated from an equation of state[8] or a simplified formula such as in ISO 6976[9].

The evaluation of balance repeatability was combined with a linearity assessment. A 20 g weight was suspended in the permeation chamber for a prolonged period of time. The balance reading was recorded in the same way as for a permeation tube. After allowing for a stabilisation period and taking a sufficient number of readings, a weight of 100 mg was added. Again after a stabilisation period, a further number of readings was taken. The results of the weighings are shown in figure 2, from which it can be seen that there are at least two effects at play: (1) the repeatability of weighing and (2) the resolution of the balance. To evaluate balance linearity, the recorded

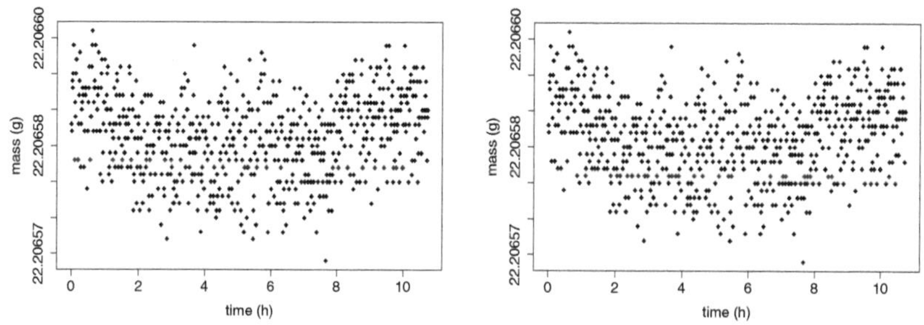

Fig. 2. Results of the weighings performed for a linearity assessment of the balance over a range of 100 mg

mean difference of the two weighings was compared with the conventional mass of the 100 mg weight. The recorded mass difference was

$$d = \mu_2 - \mu_1, \tag{2}$$

the μ_i denoting the mean balance readings. From both weighings, 700 observations were used. A naive evaluation would use the Type A evaluation method of the GUM[10], based on the normal distribution, or GUM Supplement 1 (GUM-S1)[11], based on the t distribution. Neither of these two methods addresses the two effects (finite resolution and repeatability) duly however, as in the recorded readings these effects are inseparable. A Bayesian model was constructed that addresses both effects. It is based on the R[12] package `RStan`[13]. A Bayesian model is given for evaluating the standard deviation of the mean of a series of observations. The likelihood is

$$z_i \sim N(\mu, \sigma^2),$$

with unknown mean μ and unknown variance σ^2. The z_i denote raw (unrounded) observations. The joint prior takes the form of an (uninformative) Jeffreys' prior $p(\mu, \sigma^2) \propto \sigma^{-2}$[14]. The raw observations are distributed as

$$z_i \sim \mathrm{R}(y_i, r),$$

where R denotes the rectangular distribution with mean y_i and semi-width $r = 0.5\,\mu\mathrm{g}$ (the resolution of the balance). Save for using z_i instead of

y_i, this model is well covered in the literature[14] and also revisited in the framework of evaluating measurement uncertainty[15].

Fitting the data shown in figure 2 yields the following results (all in µg):

```
Inference for Stan model: Model3.
4 chains, each with iter=10000; warmup=5000; thin=1;
post-warmup draws per chain=5000, total post-warmup draws=20000.

mean se_mean     sd         2.5%        97.5% n_eff Rhat
mu1      100151.305   0.001 0.217 100150.880 100151.737 41351     1
sigma1        5.674   0.001 0.153      5.385      5.981 38617     1
mu2           2.268   0.001 0.201      1.871      2.662 39273     1
sigma2        5.242   0.001 0.140      4.978      5.522 39697     1
diff     100149.036   0.001 0.298 100148.454 100149.628 40528     1

Samples were drawn using NUTS(diag_e) at Wed Jan 29 12:51:54 2020.
For each parameter, n_eff is a crude measure of effective sample size,
and Rhat is the potential scale reduction factor on split chains (at
convergence, Rhat=1).
```

Four chains were used with 10 000 iterations, of which the first 5000 were used to warm-up the sampler. The definitions and interpretation of the parameters $\widehat{R}$ and the effective chain length n_{eff} in the Markov Chain Monte Carlo method have been discussed elsewhere[15] and are not repeated here. From the values of these parameters, it can be concluded that the fit of the model is satisfactory. Convergence is confirmed by inspecting the trace plot of the four parameters (not shown). The calculated mass difference is 100 149.036 µg with an associated standard uncertainty 0.298 µg. The mass of the 100 mg weight is 100.181 mg with associated standard uncertainty 0.005 mg (true mass). The relative deviation is 0.03 %, negligible in view of the uncertainty on the permeation rate. Hence, neither a correction was made nor was the uncertainty incorporated in the calculations.

3.2. *Uncertainty of the mass flow rate of the dilution gas*

The volume flow of the carrier gas is constant and equals 250 mL min^{-1} at reference conditions (0 °C and 101 325 Pa), which corresponds to a mass flow rate of 0.3126 g min^{-1}. The required density of the gas can be computed using the method in ISO 6976[9]. The standard uncertainty of the volume flow rate is 0.2 %. For the mass flow rate, the same relative standard uncertainty applies, or an absolute standard uncertainty 0.0006 g min^{-1}. The uncertainty of the gas density at reference conditions is ignored, as it is $\approx$ 0.02 %[16], which is negligible in view of the standard uncertainty associated with the mass flow rate.

3.3. *Permeation rate*

Permeation data were recorded at two temperatures: see figure 3. Following ISO 6145-10 [1], a number of observations were taken so that $u(\dot{m})/|\dot{m}| \approx 1\,\%$. This requirement corresponds to observations taken at an interval of $\approx$ 20 min. The calculated permeation rate is the average taken over the time span.

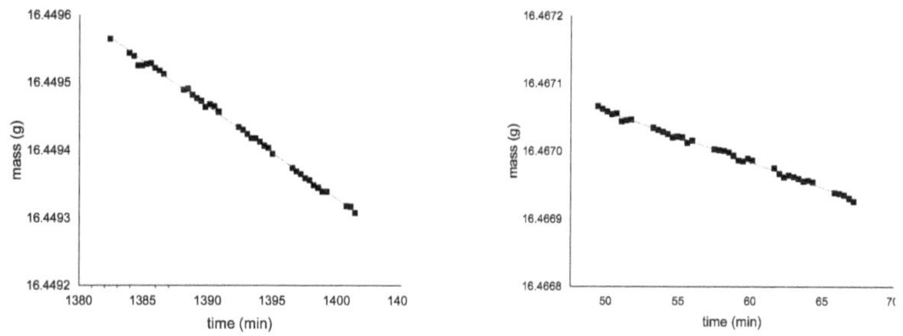

Fig. 3. Mass of the permeation tube of NH_3 at nominally 62 °C (left) and 52 °C (right) as a function of time

The resolution of the time measurement is 1 s. Hence, modelling this effect with the rectangular distribution gives a standard uncertainty of $1\,\mathrm{s}/\sqrt{12} = 0.29\,\mathrm{s}$. The resolution of the balance is 1 µg. The standard deviation of weighing a 10 g weight for approximately 40 hours is 2.9 µg. These standard uncertainties were used in errors-in-variables (EIV) regression [17], which is appropriate since it takes into account uncertainties associated with both independent and dependent variables. Results of fitting the recorded mass of the tube by a straight line as a function of time include the permeation rate (the slope of the straight line). Practically all residuals meet the consistency criteria of Deming regression [18], namely that their absolute value does not exceed the expanded uncertainty [17]. From the EIV regression, $\dot{m} = -13.599\,\mu\mathrm{g\,min}^{-1}$ with standard uncertainty $u(\dot{m}) = 0.088\,\mu\mathrm{g\,min}^{-1}$. Using ordinary least squares (OLS) [19], $\dot{m} = -13.599\,\mu\mathrm{g\,min}^{-1}$ with standard uncertainty $u(\dot{m}) = 0.108\,\mu\mathrm{g\,min}^{-1}$. To the last stated digit, the calculated permeation rates concur, which implies that it is justified to use OLS (although the standard uncertainty is some 20 % larger than that from EIV). OLS can be readily performed using the built-in function `LINEST` in MS Excel, which provides the above values for the slope and associated standard uncertainty. To obtain both the estimate and standard uncertainty, `LINEST` should be called with the last two arguments equal to `TRUE`; the first to allow for a non-zero intercept, the second for calculating additional statistics.

3.4. *Temperature dependence of the permeation rate*

The temperature dependence of the permeation rate follows Arrhenius' law[20]:

$$\dot{m} = k_0 \exp\left(-\frac{E_\mathrm{A}}{RT}\right). \tag{3}$$

E_A denotes the activation energy and k_0 is a constant, dependent on, e.g., the component and the polymer used for the permeation tube. From expression (3),

$$\frac{\partial \dot{m}}{\partial T} = \frac{E_\mathrm{A}}{RT^2}\dot{m}. \tag{4}$$

Values for k_0 and E_A were obtained using the measured permeation rate at two temperatures. At $T = 52.45\,°\mathrm{C}$, OLS gave $\dot{m} = -7.4921\,\mathrm{\mu g\,min^{-1}}$ and $u(\dot{m}) = 0.0795\,\mathrm{\mu g\,min^{-1}}$. The activation energy is $59.164\,\mathrm{kJ\,mol^{-1}}$ and $k_0 = 43.069\,\mathrm{\mu g\,min^{-1}}$. The sensitivity coefficient at $T = 62.00\,°\mathrm{C}$ is $4.81 \times 10^{-7}\,\mathrm{\mu g\,min^{-1}\,K^{-1}}$. The standard uncertainty contribution due to temperature fluctuations on the permeation rate is $0.0022\,\mathrm{\mu g\,min^{-1}}$.

4. Composition of the calibration gas mixture

4.1. *Purity of the materials*

Material purity affects the molar masses $\bar{M}_1$ and $\bar{M}_2$ of the parent gases in equation (2). The only restriction in the modelling is that the component of interest does not occur in the carrier gas. As NH_3 is not a common impurity in high-purity N_2, this assumption is reasonable. Both the amount fractions NH_3 in parent 1 (the permeation tube) and N_2 in parent 2 (dilution gas) are calculated by difference[3]. The composition of the materials used is summarised in table 1. The amount fractions methane, carbon monoxide, carbon dioxide and oxygen in nitrogen stem from the limit of quantification of the respective methods. The derivation of the values and standard uncertainties has been performed using the rectangular distribution in accordance with ISO 19229.

Table 1. Purity information of ammonia (Parent 1) and nitrogen (Parent 2), expressed in amount fractions (mol mol^{-1})

Component	Parent 1 x	Parent 1 $u(x)$	Parent 2 x	Parent 2 $u(x)$
Argon			0.000 005	0.000 003
Methane			8.00×10^{-9}	5.00×10^{-9}
Carbon monoxide			1.50×10^{-8}	9.00×10^{-9}
Carbon dioxide			1.00×10^{-8}	6.00×10^{-9}
Hydrogen			2.50×10^{-8}	1.50×10^{-8}
Water	0.010	0.002	1.00×10^{-8}	6.00×10^{-9}
Nitrogen			0.999995	0.000 003
Oxygen			5.00×10^{-9}	3.00×10^{-9}
Ammonia	0.990	0.002		

4.2. *Molar masses*

The molar masses of the components are calculated from the molecular formulæ and the standard atomic weights[21] as described in ISO 6142-1[5]. Correlations, arising due to the fact that some elements occur in multiple components are taken into account[22]. The molar masses of the parent gases are computed using

$$\bar{M}_1 = \sum_{i=1}^{q} x_{i1} M_i, \qquad \bar{M}_2 = \sum_{i=1}^{q} x_{i2} M_i. \tag{5}$$

Noting that the molar masses are the shared variables in expressions (5),

$$u(\bar{M}_1, \bar{M}_2) = \sum_i x_{i1} x_{i2} u^2(M_i) + \sum_i \sum_k x_{i1} x_{k2} u(M_i, M_k).$$

gives an expression for the covariance between $\bar{M}_1$ and $\bar{M}_2$, obtained by applying equation F.1 in the GUM[10] or using LPU for the explicit, multivariate measurement model of GUM-S2[23]. Expressions (5) form the multivariate measurement model. The covariance matrix contains $u^2(\bar{M}_1)$, $u^2(\bar{M}_2)$ and $u(\bar{M}_1, \bar{M}_2)$. The molar masses of parent gases were calculated as $\bar{M}_1 = 17.040\,63\,\mathrm{g\,mol^{-1}}$ for ammonia and $\bar{M}_1 = 28.013\,77\,\mathrm{g\,mol^{-1}}$ for nitrogen. The associated squared standard uncertainties and covariance were computed as $u^2(\bar{M}_1) = 3.99 \times 10^{-6}\,\mathrm{g^2\,mol^{-2}}$, $u^2(\bar{M}_2) = 2.42 \times 10^{-7}\,\mathrm{g^2\,mol^{-2}}$ and $u(\bar{M}_1, \bar{M}_2) = 1.19 \times 10^{-7}\,\mathrm{g^2\,mol^{-2}}$. Hence, the correlation coefficient is $r(\bar{M}_1, \bar{M}_2) = 0.121$, implying weak correlation between the molar masses of the two parent gases.

4.3. *Composition*

The amount fraction NH_3 is 70.8 µmol mol^{-1} with standard uncertainty 0.6 µmol mol^{-1}. For the components added intentionally (in this example NH_3 and N_2), it is appropriate to assume the normal distribution or t distribution to obtain a coverage factor to calculate the expanded uncertainty. Hence, the result can be stated as (70.8 ± 1.2) µmol mol^{-1} $(k = 2)$ with 95 % probability. The full composition of the calibration gas mixture is needed for a multistage preparation (e.g., the further dilution of the calibration gas mixture by means of MFCs)[22].

5. Conclusions

From our studies, as far as the amount fractions of abundant components are concerned, the GUM[10] can be used to obtain an estimate and standard uncertainty of component amount fractions. Also, the determination of coverage intervals for these amount fractions using the normal or t distribution is appropriate. For impurities, it can be necessary to use another probability density function for establishing coverage intervals, such as the beta distribution[3,24].

We have shown how the use of OLS to obtain the permeation rate can be justified. The principal question was whether the uncertainty associated with the recorded time measurements could be ignored in the calculation of the permeation rate. The resolution of the time measurement is 1 s. The comparison between the permeation rates obtained by EIV and OLS shows that OLS can be used.

The method of total differentials used in section 2 is a convenient way to obtain expressions for the partial derivatives. It is particularly useful in multistage measurement models, as it eases the use of the chain rule of differentiation, which can be readily implemented by substituting differentials[7] giving compact expressions for the sensitivity coefficients. When applying this method to a multivariate measurement model (such as the model in ISO 6142-1[5] to compute the amount fractions of all components in the calibration gas mixture), the chain rule of differentiation can be implemented in the form of a matrix multiplication[22].

References

1. *ISO 6145-10 Gas analysis – Preparation of calibration gas mixtures using dynamic volumetric methods – Part 1: Permeation methods,*

ISO 6145–10, International Organization for Standardization (Geneva, Switzerland, 2002).

2. *ISO 6145 Gas analysis – Preparation of calibration gas mixtures using dynamic methods*, ISO, International Organization for Standardization (Geneva, Switzerland, 2002-2018), 11 parts.
3. *ISO 19229 Gas analysis – Purity analysis and the treatment of purity data*, ISO, International Organization for Standardization (Geneva, Switzerland, 2018), Second edition.
4. *ISO 6145–7 Gas analysis – Preparation of calibration gas mixtures using dynamic volumetric methods – Part 7: Thermal mass-flow controllers*, ISO 6145–7, International Organization for Standardization (Geneva, Switzerland, 2018).
5. *ISO 6142–1 Gas analysis – Preparation of calibration gas mixtures – Gravimetric method for Class I mixtures*, ISO, International Organization for Standardization (Geneva, Switzerland, 2015), First edition.
6. *ISO 14167 Gas analysis – General quality aspects and metrological traceability of calibration gas mixtures*, ISO, International Organization for Standardization (Geneva, Switzerland, 2018), First edition.
7. M. L. Boas, *Mathematical Methods in the Physical Sciences* (John Wiley & Sons Inc, 2005).
8. A. M. H. van der Veen, Evaluating measurement uncertainty in fluid phase equilibrium calculations, *Metrologia* **55**, S60 (feb 2018).
9. *ISO 6976 Natural gas – Calculation of calorific values, density, relative density and Wobbe indices from composition*, ISO, International Organization for Standardization (Geneva, Switzerland, 2016).
10. BIPM, IEC, IFCC, ILAC, ISO, IUPAC, IUPAP and OIML, *Guide to the Expression of Uncertainty in Measurement, JCGM 100:2008, GUM 1995 with minor corrections* (BIPM, 2008).
11. BIPM, IEC, IFCC, ILAC, ISO, IUPAC, IUPAP and OIML, *Supplement 1 to the 'Guide to the Expression of Uncertainty in Measurement' – Propagation of distributions using a Monte Carlo method, JCGM 101:2008* (BIPM, 2008).
12. R Core Team, *R: A Language and Environment for Statistical Computing*. R Foundation for Statistical Computing, Vienna, Austria, (2019).
13. Stan Development Team, RStan: the R interface to Stan (2018), R package version 2.18.2.
14. A. Gelman, J. Carlin, H. Stern, D. Dunson, A. Vehtari and D. Rubin, *Bayesian Data Analysis*, 3rd edn. (Chapman and Hall/CRC, Boca Raton, Florida, USA, 2013).

15. A. M. H. van der Veen, Bayesian methods for type A evaluation of standard uncertainty, *Metrologia* **55**, 670 (2018).
16. *ISO/DTR 29922 Natural gas – Supporting information on the calculation of physical properties according to ISO 6976*, ISO 29922, International Organization for Standardization (Geneva, Switzerland, 2016), First edition.
17. *ISO 6143 Gas analysis – Comparison methods for determining and checking the composition of calibration gas mixtures*, ISO, International Organization for Standardization (Geneva, Switzerland, 2001), Second edition.
18. W. Deming, *Statistical Adjustment of Data* (J. Wiley & Sons, Incorporated, 1943).
19. N. R. Draper and H. Smith, *Applied Regression Analysis* (Wiley-Blackwell, 1998).
20. A. D. McNaught and A. Wilkinson, *Compendium of Chemical Terminology – IUPAC Recommendations (IUPAC Chemical Data) – Gold Book* (Wiley, 2014).
21. J. Meija, T. B. Coplen, M. Berglund, W. A. Brand, P. D. Bièvre, M. Gröning, N. E. Holden, J. Irrgeher, R. D. Loss, T. Walczyk and T. Prohaska, Atomic weights of the elements 2013 (IUPAC technical report), *Pure and Applied Chemistry* **88**, 265 (2016).
22. A. M. H. van der Veen and K. Hafner, Atomic weights in gas analysis, *Metrologia* **51**, p. 80 (2014).
23. BIPM, IEC, IFCC, ILAC, ISO, IUPAC, IUPAP and OIML, *Supplement 2 to the 'Guide to the Expression of Uncertainty in Measurement' – Extension to any number of output quantities, JCGM 102:2011* (BIPM, 2011).
24. A. M. H. van der Veen and G. Nieuwenkamp, Revision of ISO 19229 to support the certification of calibration gases for purity, *Accreditation and Quality Assurance* **24**, 375 (2019).

F Pavese, A B Forbes, N F Zhang, A G Chunovkina (Eds):
Advanced Mathematical and Computational Tools in Metrology and Testing XII
Series on Advances in Mathematics for Applied Sciences, Vol. 90

https://doi.org/10.1142/9789811242380_0028

Adaptive measuring system with dynamic error estimation of the first-order sensor

A. S. Volosnikov

School of Electrical Engineering and Computer Science,
South Ural State University (National Research University),
Chelyabinsk, Russian Federation
E-mail: volosnikovas@susu.ru
www.susu.ru/en

A problem of the dynamic measurement error estimation and correction is considered. The adaptive measuring system with the dynamic measurement error estimation is developed. Correction of the dynamic error is achieved by simultaneous restoring and filtering of the sensor input signal being measured. Signal processing consists in determining of the FIR-filter optimal order by the standard deviation minimization of the dynamic measurement error estimation. Computer simulation of the proposed measuring system for the first-order sensor is carried out. The purpose of the research is to demonstrate the approach efficiency. A good proximity between input and restored sensor signals as well as between true and estimated dynamic measurement errors is shown. Future works consist in the approach expansion on high-order measuring systems.

Keywords: Dynamic Measurement; Adaptive Systems; Adaptive Signal Processing; FIR Filters; Dynamic Error; Error Estimation; Dynamic Uncertainty.

1. Introduction

The bibliographic review on the theory of dynamic measurements demonstrates the interest of researchers in the problem of estimating and correcting the dynamic error (uncertainty) of measurements.

It is noted in [1] and [2] that compensation of the dynamics of linear time-invariant systems and an estimate of the corresponding uncertainty arouse a growing interest in metrology. In [3] it is shown that modeling is useful in the design of measuring systems and in the quantification of uncertainties. The role of digital processing, mathematical and simulation modeling of the process of dynamic measurement is illustrated.

Approaches for restoring dynamically distorted signals of measuring systems based on classical and modern methods of automatic control theory are considered in [4]-[7].

2. Background

The present paper continues the work presented in [8], where a measuring system (MS) with dynamic measurement error (DME) estimation were proposed. The MS design is based on convolution of an inverse sensor model and low-pass FIR-filters. This provides simultaneous restoring and filtering of the sensor input signal, and therefore DME correcting.

Optimal adjustment of FIR-filters orders with specified window function is performed in view of additive noises existed at the sensor output. The optimization criterion is the minimum of the standard deviation (SD) of the DME estimation. This estimation is obtained from an additional channel of the MS during the signal processing of the measured sensor output. The approach implies formation a single DME estimation for a whole realization of the measured sensor signal. The estimate remains constant regardless of the instantaneous signal change. Thus, it is aimed at post-processing of the sensor output data.

3. Measuring System in Adaptive Mode

In the present work the adaptive MS is investigated. The adaptation procedure consists in determining of MS optimal parameters at each discrete observation in measured sensor output. Along with it, the proposed approach provides the DME estimation at each discrete point of restored input signal.

Consider a sensor model as an LTI-system (*i.e.* linear time-invariant system) described by the first-order transfer function (TF) as follows:

$$W_S(p) = \frac{Y(p)}{U(p)} = \frac{1}{pT_1 + 1}, \tag{1}$$

where $U(p)$ and $Y(p)$ are Laplace transforms of input and output signals of the sensor; T_1 is the sensor time constant, p is the complex frequency.

The discrete equivalent of TF(1) can be represented on the basis of the *z*-transform as follows:

$$W_S(z) = \frac{1}{\alpha_1 z + \alpha_0}, \tag{2}$$

where for the sampling period T

$$\begin{cases} \alpha_1 = \left(1 - \exp\left(-T/T_1\right)\right)^{-1} \\ \alpha_0 = -\exp\left(-T/T_1\right)\left(1 - \exp\left(-T/T_1\right)\right)^{-1} \end{cases}. \tag{3}$$

Block diagram of the adaptive MS with DME estimation for the first-order sensor TF(2) is shown in Fig. 1.

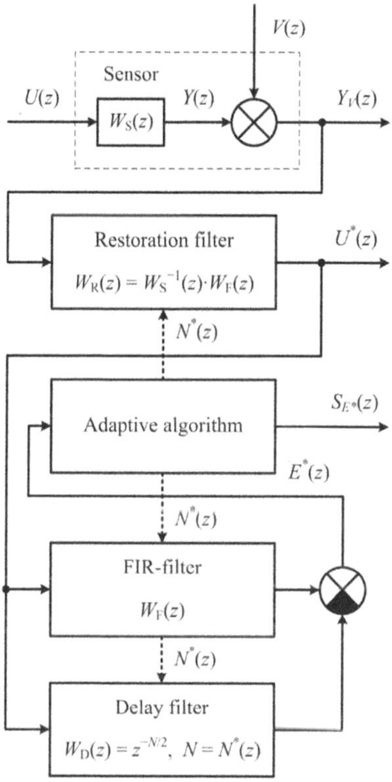

Fig. 1. Block diagram of the adaptive measuring system with DME estimation.

In the block diagram above $Y_V(z) = Y(z) + V(z)$ is the sensor measured output, where $V(z)$ is the additive noise. The restored sensor input $U^*(z)$ is the output of the restoration filter described by the following TF:

$$W_R(z) = W_S^{-1}(z)\cdot W_F(z), \tag{4}$$

where

$$W_F(z) = \sum_{i=0}^{N^*} \beta_i z^{-i} \tag{5}$$

is the TF of a low-pass FIR-filter with symmetric impulse response, i.e. $\beta_i = \beta_{N^*-i}$ for $i = 0, 1, 2, \ldots, N^*$.

A low-pass FIR-filter can be represented as a window of width N^*+1 (that defines the filter cut-off frequency) with the following vector of coefficients:

$$\mathrm{B_F} = \begin{bmatrix} \beta_{N^*} & \beta_{N^*-1} & \beta_{N^*-2} & \dots & \beta_2 & \beta_1 & \beta_0 \end{bmatrix}. \tag{6}$$

Then, the vector of coefficients for the restoration filter is evaluated by the following convolution:

$$\mathrm{B_R} = \mathrm{A_S} \circ \mathrm{B_F} = \begin{bmatrix} \alpha_1 & \alpha_0 \end{bmatrix} \circ \begin{bmatrix} \beta_{N^*} & \beta_{N^*-1} & \beta_{N^*-2} & \dots & \beta_2 & \beta_1 & \beta_0 \end{bmatrix}, \tag{7}$$

where vector $\mathrm{A_S} = \begin{bmatrix} \alpha_1 & \alpha_0 \end{bmatrix}$ represents the sensor inverse TF $W_\mathrm{S}^{-1}(z)$;

In the block diagram above the DME signal is estimated as follows [8]:

$$E^*(z) = U^*(z)W_\mathrm{F}(z) - U^*(z)W_\mathrm{D}(z) = U^*(z)\left[W_\mathrm{F}(z) - W_\mathrm{D}(z)\right], \tag{8}$$

while the true DME is defined given a known sensor input $U(z)$ as follows:

$$E(z) = U^*(z)W_\mathrm{F}(z) - U(z)W_\mathrm{D}(z), \tag{9}$$

where $W_\mathrm{D}(z) = z^{-N^*/2}$ is the TF of the group delay compensation filter.

The mean and SD of the DME estimation (8) in adaptive mode within the window of width N^*+1 are respectively as follows:

$$\begin{cases} \overline{E^*} = \left(\sum_{i=0}^{N^*} E^*(iT) \right) \Big/ \left(N^*+1\right) \\ \sigma_{E^*} = \sqrt{\left(\sum_{i=0}^{N^*} \left(E^*(iT) - \overline{E^*} \right)^2 \right) \Big/ N^*} \end{cases}. \tag{10}$$

The minimum value of (9) corresponds to the optimal order N^*_{opt} of the FIR-filter (5):

$$S_{E^*}(k) = \min_{N^*(k) \to N^*_{opt}(k)} \left[\sigma_{E^*}(N^*) \right], \tag{11}$$

where k are samples of discrete time $t_k = k \cdot T$.

The estimation (11) is a function of discrete time, as well as the MS adjustable parameter $N^*(k)$. The optimization of this parameter is achieved with a minimum of the DME estimation (10), which is the criterion (11) for the adaptive algorithm. Thus, the DME is estimated and corrected with each new observation of measured signal on the basis of knowledge about preceding states of the MS. This mode besides post-processing is also aimed at real-time processing of sensor output data.

4. Simulation Study

Computer simulation of the described MS in adaptive mode is performed. The results are shown in Figs. 2–5. Harmonic and pulse signals of unity amplitude are used as typical input impacts of the sensor. There are two types of additive noise at the sensor output. The first one is harmonic of 5 % amplitude, and the second one is random with Gaussian distribution of 3 % SD.

In the upper diagram graphs of FIR-filter optimal orders changing in time are shown (the light gray curve is for the true order defined given a known sensor input and the black curve for the estimated one defined in compliance with (11)). In the middle diagram SD graphs of the DME changing in time are shown (the light gray curve is for the true DME defined given a known sensor input and the black solid curve for the estimated one defined in compliance with (11)). Graphs of the MS signals are shown in Figs. 2(a)–5(a). These diagrams demonstrate a good proximity between true and estimated values of FIR-filter optimal orders as well as between true and estimated values of the DME. True values are determined on the difference between true and restored input signals of the sensor (9) in the same way as estimated ones. In the lower diagram graphs of the sensor signals are shown (the black dashed curve is for the measured output, the solid one is for the restored input and the light gray curve behind the black one is for the delayed input to be measured). These diagrams demonstrate almost complete match between true and restored input signals of the sensor. The difference between all estimated and true values is mainly determined by noises at the sensor output and residual errors in its dynamics compensation.

Graphs of the DME estimation signals within the window function of the FIR-filter at certain times are shown in Figs. 2(b)–5(b). These signals are associated with respective types of input signals and additive noises specified in Figs. 2(a)–5(a). Each diagram in Figs. 2(b)–5(b) presents a series of curves defined in compliance with (8) within the window function at certain times. Each of curves in a series is associated with a certain FIR-filter order. Black solid curves at each diagram show graphs of the DME estimation signals with minimal SD. Such a value of SD in compliance with (11) is defined for each discrete observation of the sensor measured output on the basis of the several preceding observations within the window function of the FIR-filter.

It means that for each successive sample in the discrete time domain, a function is constructed, which is the DME estimation as the minimized SD of the absolute dynamic error. This estimation yields the optimal order of the restoration FIR-filter within its window function. And as it shown below all these estimates are adaptive to additive noises at the sensor output.

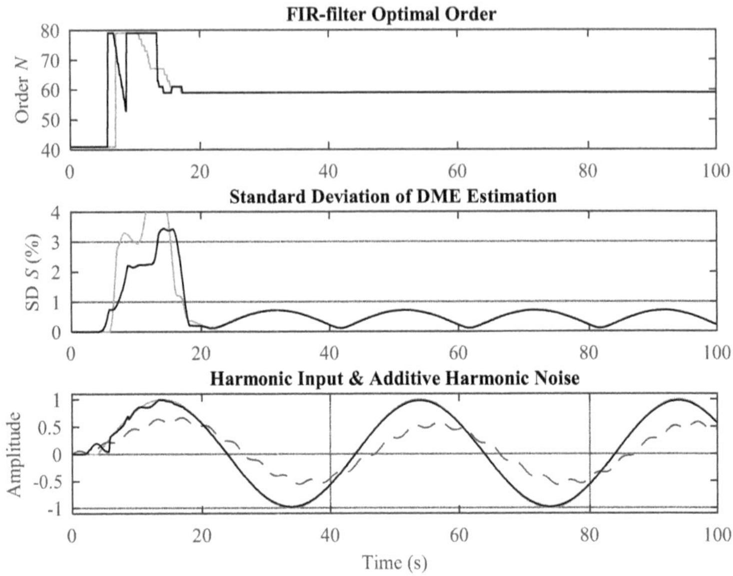

Fig. 2(a). Computer simulation of the MS in adaptive mode for harmonic sensor input and additive harmonic noise at the sensor output.

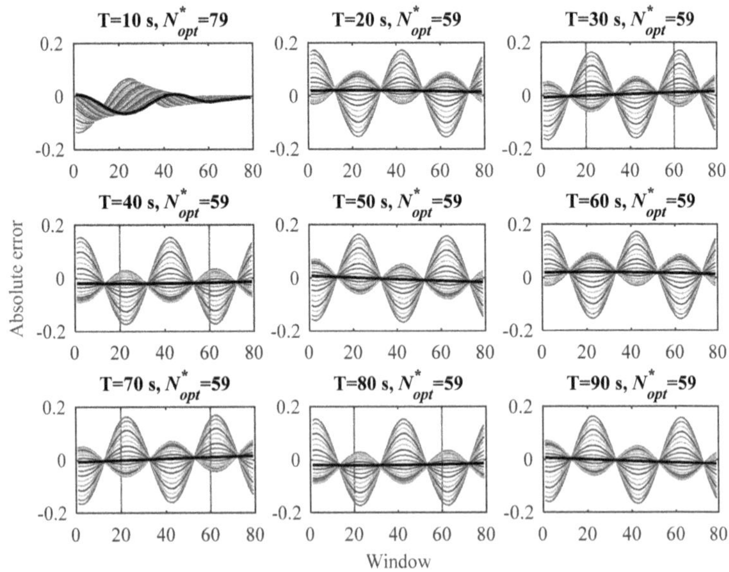

Fig. 2(b). DME estimation signals within the window function of the FIR-filter at certain times for the MS signals shown in Fig. 2(a).

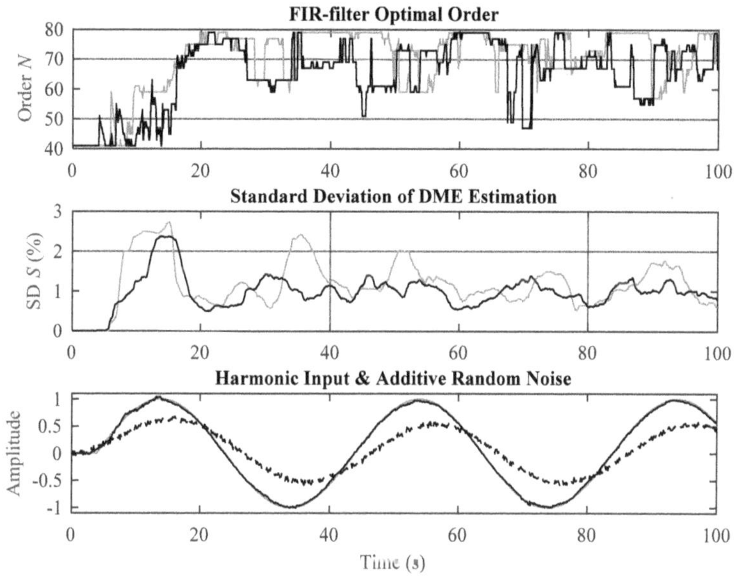

Fig. 3(a). Computer simulation of the MS in adaptive mode for harmonic sensor input and additive random noise at the sensor output.

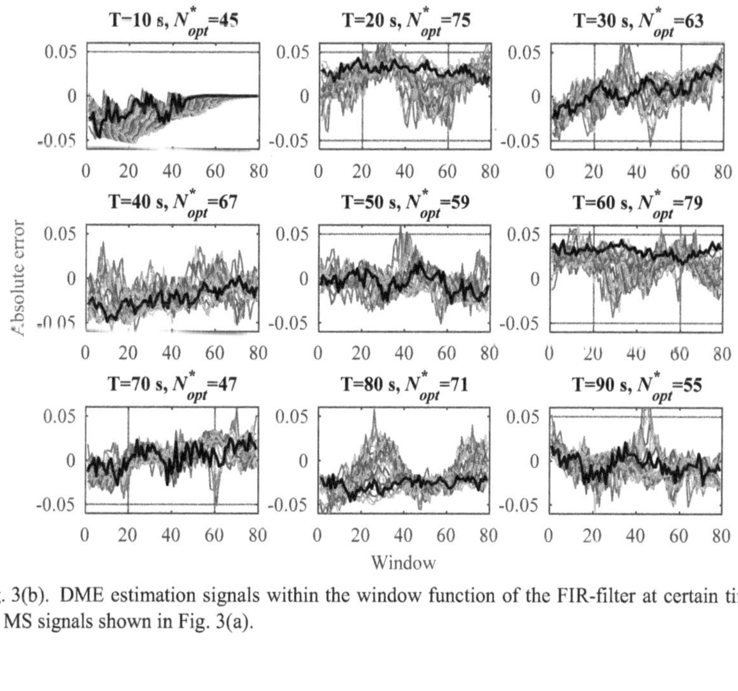

Fig. 3(b). DME estimation signals within the window function of the FIR-filter at certain times for the MS signals shown in Fig. 3(a).

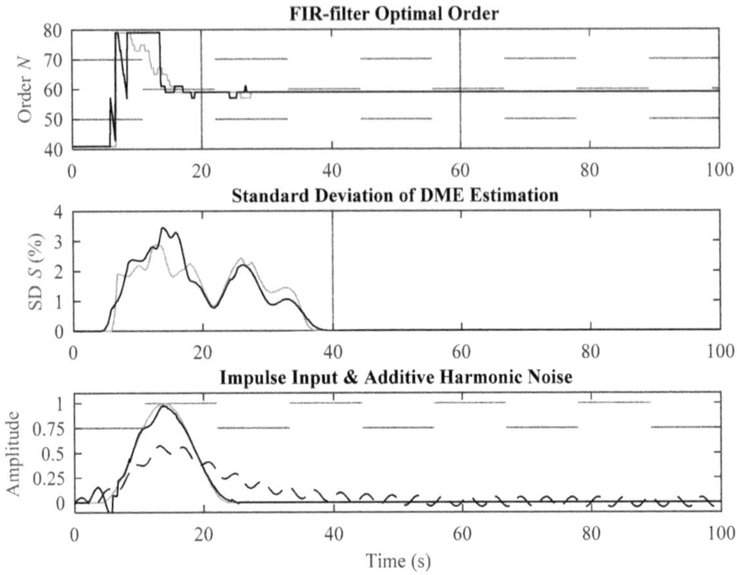

Fig. 4(a). Computer simulation of the MS in adaptive mode for impulse sensor input and additive harmonic noise at the sensor output.

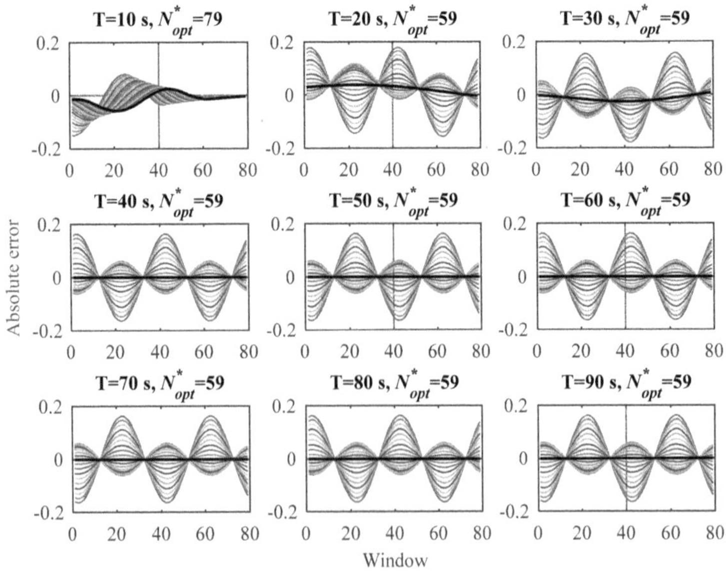

Fig. 4(b). DME estimation signals within the window function of the FIR-filter at certain times for the MS signals shown in Fig. 4(a).

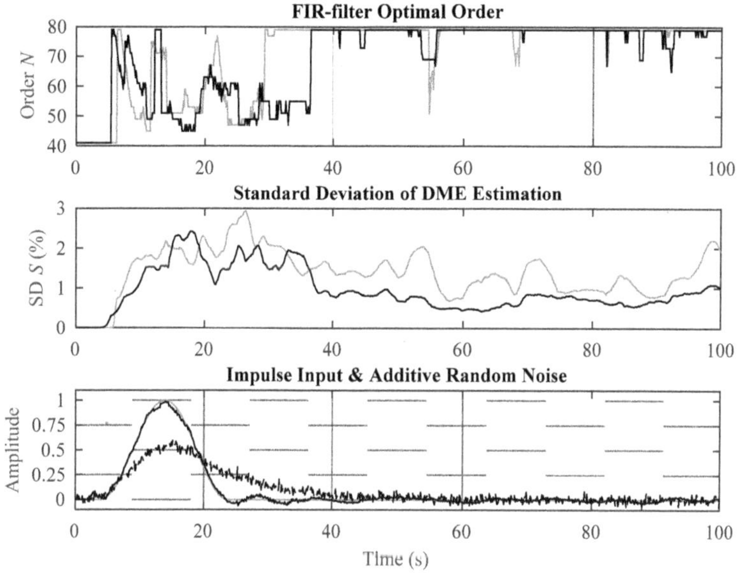

Fig. 5(a). Computer simulation of the MS in adaptive mode for impulse sensor input and additive random noise at the sensor output.

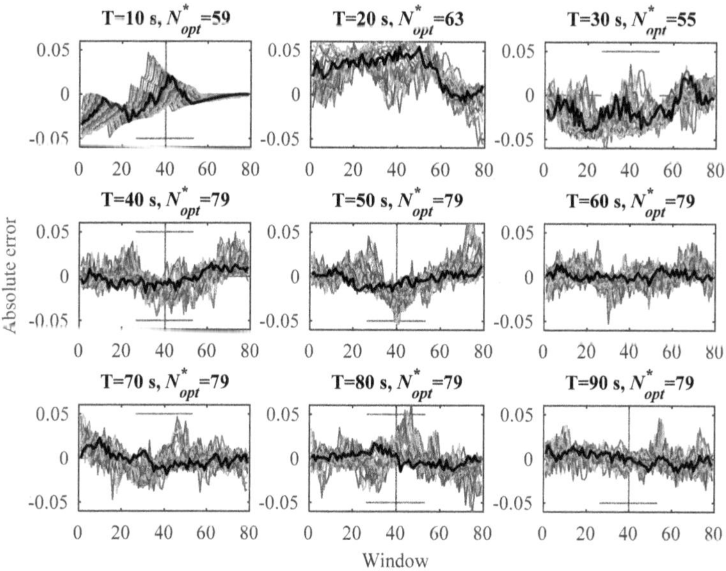

Fig. 5(b). DME estimation signals within the window function of the FIR-filter at certain times for the MS signals shown in Fig. 5(a).

5. Summary

The research presents an approach to adaptive estimation and correction of the DME. The approach consists in simultaneous filtering and restoring of a sensor input signal by means of the restoration filter which is the convolution of the inverse sensor model and a low-pass FIR-filter. The criterion for the adaptive adjustment of the FIR-filter order is the minimum of the DME estimation.

The performed computer simulation demonstrates an acceptable match between true and restored input signals. Optimum estimates of the FIR-filter order as a function of time for input signals of various types have been obtained in the presence of additive noises at the output of the first-order sensor. In addition, the minimized SD of the DME estimation changing in time have been calculated. All these estimates are adaptive to additive noises of a sensor.

The proposed approach to the DME estimating and correcting is aimed at post-processing as well as real-time processing of the sensor measured signal. This can serve as a basis for the uncertainty evaluation of dynamic measurements adaptive to the sensor output instantaneous change. Future works consist in the approach expansion on high-order MSs in accordance with [8].

References

1. S. Eichstädt, C. Elster, T.J. Esward and J.P. Hessling, Deconvolution filters for the analysis of dynamic measurement processes: a tutorial, *Metrologia* **47** (5), 522 (2010).
2. S. Eichstädt, A. Link and C. Elster, Dynamic Uncertainty for Compensated Second-Order Systems, *Sensors* **10** (8), 7621 (2010).
3. T.J. Esward, Investigating dynamic measurement applications through modelling and simulation, *Tech. Mess.* **83** (10), 557 (2016).
4. A.L. Shestakov, Dynamic measurements based on automatic control theory approach, in *Advanced Mathematical and Computational Tools in Metrology and Testing X*, eds. F. Pavese, W. Bremser, A. Chunovkina, N. Fischer and A.B. Forbes, (World Scientific, Singapore, 2015), pp. 66-77.
5. E.V. Yurasova, M.N. Biziaev and A.S. Volosnikov, Approaches in a sensor model of error correction in dynamic measurements, *Procedia Engineering* **129**, 764 (2015).
6. A.S. Volosnikov and A.L. Shestakov, Neural network approach to reduce dynamic measurement errors, *ACTA IMEKO* **5** (3), 24 (2016).
7. A.L. Shestakov, Dynamic measuring methods: A review, *ACTA IMEKO* **8** (1), 64 (2019).
8. A.S. Volosnikov and E.V. Yurasova, Dynamic measurement error evaluation and minimization based on FIR-filter, in *2018 Global Smart Industry Conference (GloSIC'2018)*, (Chelyabinsk, Russia, 2018).

F Pavese, A B Forbes, N F Zhang, A G Chunovkina (Eds):
Advanced Mathematical and Computational Tools in Metrology and Testing XII
Series on Advances in Mathematics for Applied Sciences, Vol. 90

https://doi.org/10.1142/9789811242380_0029

Method of estimation uncertainties of indirect multivariable measurements including the accuracy of processing function as extension of GUM-S2

Zygmunt Lech Warsza
Research net Lukasiewicz - Industrial Research Institute of Automation and Measurements (PIAP-L), Warszawa, Poland
E-mail: zlw1936@gmail.com

Jacek Puchalski
Central Office of Measures (GUM), Warszawa, Poland
E-mail: jacek.puchalski@gum.gov.pl

The new extended mathematical model for evaluation uncertainties of the multivariable indirect measurements, which upgrades the method of GUM-Supplement 2 is presented. In this model the uncertainties and correlations of the processing function parameters are also taken into account. It can be used for multivariable measurements and also to describe the accuracy of instruments and systems that perform such measurements. An example is included on estimated uncertainties of indirectly-measured output voltage and current of a twoport circuit, by considering uncertainties and correlations of its elements.

Keywords: Indirect multivariable measurements; Measurand; Uncertainty; Processing function; Correlation coefficient; Covariance matrix.

Designations used in paper

$\boldsymbol{X}, \boldsymbol{Y}$ - input and output vector measurands; $\boldsymbol{X_0}, \boldsymbol{Y_0}$ - their initial values; $\overline{\boldsymbol{X}}, \overline{\boldsymbol{Y}}$ – vectors of estimators of n values $\overline{x}_i$ and of m values $\overline{y}_j$; u_{xi}, u_{yi}, u_x and $\delta_{(xi-xi0)}$, $\delta_{(yi-yi0)}$ – absolute and relative standard uncertainties; $\boldsymbol{F}(\boldsymbol{X})$, $\boldsymbol{F}(\boldsymbol{X},\boldsymbol{P})$ – ideal and model of the real multi-variable functions of processing $\boldsymbol{X}$ to $\boldsymbol{Y}$; $\boldsymbol{U_F}$, $\boldsymbol{U_{\delta F}}$ - its covariance matrices and $\boldsymbol{S} = \frac{\partial Y}{\partial X}$, $\boldsymbol{S_\delta} = \frac{\partial Y}{Y} / \frac{\partial X}{X}$ – sensitivity matrices of absolute and relative uncertainties; $\boldsymbol{U_X}$, $\boldsymbol{U_Y}$, $\boldsymbol{U_{\delta(X-X0)}}$; $U_{\delta(Y-Y0)}$, $\boldsymbol{U_{YP}}$ – covariance matrices of: X, Y ; $\boldsymbol{U_P}$, $\boldsymbol{U}(\boldsymbol{P},\boldsymbol{X})$ – covariance and correlation matrices of $\boldsymbol{P}$ parameters of $\boldsymbol{F}(\boldsymbol{X},\boldsymbol{P})$.

1. Introduction – actual state

In the fundamental and technical research, in control, monitoring, and technical diagnostics, many physical quantities and parameters have to be measured for characterizing objects under the test. In many cases there is any possibility to carrying out direct measurements of their values and the indirect methods must be applied. On the measurement system's input is *n* joined quantities $\boldsymbol{X}$, and from their results or signals is processed the output *m*-variable measurand $\boldsymbol{Y}$, i.e.:

$$Y = F(X) \tag{1}$$

The method of evaluating $\boldsymbol{Y}$ values' estimators, their covariance matrix with uncertainties and correlation coefficients is given in the GUM Supplement 2 [1]. These recommendations cover only exceptional cases with an ideal situation when uncertainties of processing $\boldsymbol{X}$ by the multivariable function $\boldsymbol{F}(\cdot)$ are negligible. As an example is the calculation of the active component $P=U\,I\cos\varphi$ of the AC power by the exact multiplication of its separately measured factors.

Even if the basic relation (1) is nonlinear, in the most cases, errors and uncertainties of $\boldsymbol{X}$ elements are small, and linear relations of covariance matrices describe the propagation of absolute and relative variances [1-S2], [7], i.e.:

$$\boldsymbol{U_Y} = \boldsymbol{S}\,\boldsymbol{U_X}\boldsymbol{S}^{\mathrm{T}} \tag{2a}$$

$$\boldsymbol{U}_{\delta(\boldsymbol{Y}-\boldsymbol{Y_0})} = \boldsymbol{S_\delta}\,\boldsymbol{U}_{\delta(\boldsymbol{X}-\boldsymbol{X_0})}\,\boldsymbol{S}_{\boldsymbol{\delta}}^{\mathrm{T}}, \tag{2b}$$

Where: $\boldsymbol{U_X} = \boldsymbol{U_{X_0}} + \boldsymbol{U_{X-X_0}}$, $\boldsymbol{U_Y} = \boldsymbol{U_{Y_0}} + \boldsymbol{U_{Y-Y_0}}$, $\boldsymbol{S} = \frac{\partial \boldsymbol{Y}}{\partial \boldsymbol{X}}$, $\boldsymbol{S_\delta} = \frac{\partial \boldsymbol{Y}}{\partial \boldsymbol{X}}\frac{\boldsymbol{X}}{\boldsymbol{Y}} = \boldsymbol{S}\cdot\left(\frac{\boldsymbol{X}}{\boldsymbol{Y}}\right)$.

For example matrices: $\boldsymbol{U_X}$ of the size 3D (3x3) and $\boldsymbol{U}_{\delta(\boldsymbol{Y}-\boldsymbol{Y_0})}$ for $\boldsymbol{X_0} = \boldsymbol{0}$ are

$$\boldsymbol{U_X} = \begin{bmatrix} u_{x_1}^2 & \rho_{12}u_{x_1}u_{x_2} & \rho_{13}u_{x_1}u_{x_3} \\ \rho_{12}u_{x_1}u_{x_2} & u_{x_2}^2 & \rho_{23}u_{x_2}u_{x_3} \\ \rho_{13}u_{x_1}u_{x_3} & \rho_{23}u_{x_2}u_{x_3} & u_{x_3}^2 \end{bmatrix} = \boldsymbol{u}(\boldsymbol{X})\cdot\boldsymbol{R_X}\cdot\boldsymbol{u}^T(\boldsymbol{X}) = \begin{bmatrix} u_{x_1} \\ u_{x_2} \\ u_{x_3} \end{bmatrix}\begin{bmatrix} 1 & \rho_{12} & \rho_{13} \\ \rho_{12} & 1 & \rho_{23} \\ \rho_{13} & \rho_{23} & 1 \end{bmatrix}\begin{bmatrix} u_{x_1} \\ u_{x_2} \\ u_{x_3} \end{bmatrix}^{\mathrm{T}}$$

$$\boldsymbol{U_{\delta Y}} = \begin{bmatrix} \delta_{y_1}^2 & \cdots & \rho_{y_{1m}}\delta_{y_1}\delta_{y_m} \\ \cdots & \cdots & \cdots \\ \rho_{y_{1m}}\delta_{y_1}\delta_{y_m} & \cdots & \delta_{y_m}^2 \end{bmatrix} = \boldsymbol{S_\delta}\begin{bmatrix} \delta_{x_1}^2 & \cdots & \rho_{x_{1n}}\delta_{x_n}\delta_{x_1} \\ \cdots & \cdots & \cdots \\ \rho_{x_{1n}}\delta_{x_n}\delta_{x_1} & \cdots & \delta_{x_n}^2 \end{bmatrix}\boldsymbol{S}_{\boldsymbol{\delta}}^{\mathrm{T}};\quad \boldsymbol{S_\delta} = \begin{bmatrix} \frac{x_1}{y_1}\frac{\partial y_1}{\partial x_1} & \cdots & \frac{x_n}{y_1}\frac{\partial y_1}{\partial x_n} \\ \cdots & \cdots & \cdots \\ \frac{x_1}{y_m}\frac{\partial y_m}{\partial x_1} & \cdots & \frac{x_n}{y_m}\frac{\partial y_m}{\partial x_n} \end{bmatrix}$$

Their scatter regions are defined by a multidimensional normal probability distributions. The *n*- and *m*-dimensional covariance matrices $\boldsymbol{U_X}$ and $\boldsymbol{U_Y}$ are positive definite, i.e. det($\boldsymbol{U_X}$) >0, det($\boldsymbol{U_Y}$) >0, and det($\boldsymbol{R_X}$) >0, det($\boldsymbol{R_Y}$) >0 as equal for the matrix of correlation coefficients, called the correlator $\boldsymbol{R}$. Then cover regions, e.g. of probability p_0 =(.95, .99) used in measurements for extended uncertainties [1], are described by the *n*- and *m*-size hyper-ellipsoids with centers at ends of vectors $\overline{\boldsymbol{X}}, \overline{\boldsymbol{Y}}$ and components as averages of random variables $\boldsymbol{X}$, $\boldsymbol{Y}$.

All or some components of the vector $\overline{\boldsymbol{Y}}$ of output results can be next used separately or jointly. In the latter case, it is necessary to find and consider not only uncertainties u_{yi}, but also the correlations between pairs of the variables y_i. In many cases the indirect measurements of *m*–components of the measurand $\boldsymbol{Y}$ are made now by the automatic measurement system. If not, then for obtain correlation coefficients, a collection of individual quantities x_i of $\boldsymbol{X}$ should be synchronically measured, and results $\overline{\boldsymbol{X}}$, $\overline{\boldsymbol{Y}}$ and covariance matrices $\boldsymbol{U_X}$ or $\boldsymbol{U}_{\delta(\boldsymbol{X}-\boldsymbol{X_0})}$ and $\boldsymbol{U_Y}$, $\boldsymbol{U}_{\delta(\boldsymbol{Y}-\boldsymbol{Y_0})}$ externally calculated by formulas (1), (2a, b).

2. Basic formulas of the method extended the GUM-S2

After the correction of known systematic errors in the measurement systems, other errors are described by uncertainties. They are inside the measured object, in measurements of input values ***X***, and in the implementation of processing function ***Y***=***F***(***X***) computationally or instrumentally way. There are uncertainties type A depended on the random errors during the experiment and also uncertainties type B of unknown randomized systematic errors of the measured object's model and of the instrumental part of the measurement system. The actual GUM-S2 method [1] and earlier literature, e.g., [2-5] consider only cases of the ideal function ***F***(***X***). Situations of not accurate and correlated parameters ***P*** of the real multivariable function ***F***(***X, P***) are not yet covered. There are additional uncertainties due to approximations in the model of measured object and the instrumental system's limitations, and also, when measurements of variables ***Y*** are possible only indirectly via other internal parameters of the tested object's. In the high precise measurements, even the rounding of calculations is also essential, including ones from the precision of the digital part of the circuit. Then in above cases for the evaluation of uncertainties of indirect measurements of the multivariable output measurand ***Y*** the processing function have to be described as

$$\boldsymbol{Y} = \boldsymbol{F}(\boldsymbol{X}, \boldsymbol{P}) \tag{3}$$

The basic equations for descriptions of the accuracy by absolute errors and uncertainties with the processing function ***F***(***X, P***) are given in Table 1.

In the measurement practice, including electrical measurement systems, there is usually a more straightforward case when the directly measured variables ***X*** of the tested object and internal parameters ***P*** of the measurement system are not correlated. For example it is when ***X*** is differently located then ***P*** and do not affect themselves, and their external influences are also not related. Then the matrix ***U*** in (5), (5b) does not occur and $\boldsymbol{V} = \boldsymbol{V}^{\mathrm{T}} = 0$. The propagation equation of variances has then only a two-component form (6).

The first component of (6), which depends on the uncertainties and correlations of the input measurand ***X***, is similar as in the GUM-S2 [1]. The second component of (6), depending on the processing function's uncertainties, appeared only in given here the extended method. It expresses the influence of uncertainties and correlation coefficients of ***P*** parameters of function ***F***(***X, P***).

From (2b) and (6) follows the covariance matrix $\boldsymbol{U}_{\delta(\boldsymbol{Y}-\boldsymbol{Y}_0)}$ – formula (7).for the relative uncertainties $\delta_{(yi\text{-}yi0)}$ of $\boldsymbol{Y} - \boldsymbol{Y}_0$ increments, or of ***Y***, if uncertainties of the initial values $\boldsymbol{X}_0$ of $\boldsymbol{X}$ are negligible, $(||\boldsymbol{u}_X(\boldsymbol{X})|| \gg ||\boldsymbol{u}_X(\boldsymbol{X}_0)|| \approx \boldsymbol{0}$ as measurement of values in the beginning of range are usually avoided) and $||\boldsymbol{U}_{X-X_0}|| \approx ||\boldsymbol{U}_X(\boldsymbol{X})||$. New relations (6), (7) should be included in the GUM-S2.

Table 1. Formulas of the extended method of estimate uncertainties of indirect multivariable measurements

General formula		$Y = F(X,P) \quad (u_F \neq 0)$ (3) where: $Y = [y_1, \ldots y_m]^T$, $X = [x_1, \ldots x_n]^T$, $P = [p_1, p_2, \ldots p_k]^T$ (3a,b,c)
Deviations (absolute errors)		$\Delta Y = \Delta F(X,P) = S_{X,P}[\Delta X, \Delta P]^{\mathrm{T}} = S_X \Delta X + S_P \Delta P$ (4a, b) where: $S_{X,P}$ - sensitivity matrix of function $F(X,P)$ of dimensions $[(n+k)\times m]$, $S_X \equiv S$, S_P and S_δ - sensitivity matrices of influence of deviations: ΔX, ΔP and $\delta_X = \frac{\Delta X}{X}$, $\delta_P = \frac{\Delta P}{P}$
Sensitivity matrices of $\Delta X, \Delta P, \delta_X$		$S_X = \frac{\partial Y}{\partial X} = \begin{bmatrix} \frac{\partial y_1}{\partial x_1} & \cdots & \frac{\partial y_1}{\partial x_n} \\ \cdots & \cdots & \cdots \\ \frac{\partial y_m}{\partial x_1} & \cdots & \frac{\partial y_m}{\partial x_n} \end{bmatrix}$, $S_p = \frac{\partial Y}{\partial P} = \begin{bmatrix} \frac{\partial y_1}{\partial p_1} & \cdots & \frac{\partial y_1}{\partial p_k} \\ \cdots & \cdots & \cdots \\ \frac{\partial y_m}{\partial p_1} & \cdots & \frac{\partial y_m}{\partial p_k} \end{bmatrix}$, $S_\delta = \begin{bmatrix} \frac{x_1}{y_1}\frac{\partial y_1}{\partial x_1} & \cdots & \frac{x_n}{y_1}\frac{\partial y_1}{\partial x_n} \\ \cdots & \cdots & \cdots \\ \frac{x_1}{y_m}\frac{\partial y_m}{\partial x_1} & \cdots & \frac{x_n}{y_m}\frac{\partial y_m}{\partial x_n} \end{bmatrix}$ (4c,d,e)
Propagation of variances in general case $V \equiv SUS_P^T \neq 0$		$U_Y = U_Y(X,P) = S_{X,P}\, U_{X,P} = [S, S_p] \begin{bmatrix} U_X & U \\ U^T & U_p \end{bmatrix} \begin{bmatrix} S^T \\ S_P^T \end{bmatrix}$ (5) or $U_Y = U_{YX} + U_{YF} = S_X U_X S_X^T + U_{YF}$ (5a) where $U_{YF} = S_p U_p S_P^T + S_X U S_P^T + (S_X U S_P^T)^T \equiv U_{YP} + V + V^T$ (5b)
Covariance matrices: U_Y, U_X, U_P and matrix U of correlation X and P		$U_Y = \begin{bmatrix} u_{y1}^2 & \cdots & \rho_{y1m} u_{y1} u_{ym} \\ \cdots & \cdots & \cdots \\ \rho_{y1m} u_{ym} u_{y1} & \cdots & u_{ym}^2 \end{bmatrix}$, $U_X = \begin{bmatrix} u_{x1}^2 & \cdots & \rho_{x1n} u_{x1} u_{xn} \\ \cdots & \cdots & \cdots \\ \rho_{x1n} u_{xn} u_{x1} & \cdots & u_{xn}^2 \end{bmatrix}$ (5c, d) $U_p = \begin{bmatrix} u_{p1}^2 & \cdots & \rho_{p1k} u_{p1} u_{pk} \\ \cdots & \cdots & \cdots \\ \rho_{p1k} u_{pk} u_{p1} & \cdots & u_{pk}^2 \end{bmatrix}$, $U = \begin{bmatrix} \rho_{x1p1} u_{x1} u_{p1} & \cdots & \rho_{x1pk} u_{x1} u_{pk} \\ \cdots & \cdots & \cdots \\ \rho_{xnp1} u_{xn} u_{p1} & \cdots & \rho_{xnpk} u_{xn} u_{pk} \end{bmatrix}$ (5e, f)
Propagation of variances for $V = 0$	**absolute uncertainties**	$U_Y = S\, U_X\, S^T + S_P U_P S_P^T$ (6)
	relative uncertainties	$U_{\delta(Y-Y_0)} = S_\delta U_{\delta(X-X_0)} S_\delta^T + S_{\delta P} U_{\delta P} S_{\delta P}^T$ (7)
	Covariance matrices of type A and type B uncertainty components	$U_Y = U_{YA} + U_{YB} = \left(S U_{XA} S^T + S_P U_{PA} S_P^T\right) + \left(S U_{XB} S^T + S_P U_{PB} S_P^T\right)$ (8) $U_{XA} = \begin{bmatrix} u_{x1A}^2 & \cdots & \rho_{Ax1,n} u_{x1A} u_{xnA} \\ \cdots & \cdots & \cdots \\ \rho_{Ax1,n} u_{xnA} u_{x1A} & \cdots & u_{xnA}^2 \end{bmatrix}$, $U_{XB} = \begin{bmatrix} u_{x1B}^2 & \cdots & \rho_{Bx1,n} u_{x1B} u_{xnB} \\ \cdots & \cdots & \cdots \\ \rho_{Bx1,n} u_{xnB} u_{x1B} & \cdots & u_{xnB}^2 \end{bmatrix}$ $U_{PA} = \begin{bmatrix} u_{p1A}^2 & \cdots & \rho_{Ap1k} u_{p1A} u_{pkA} \\ \cdots & \cdots & \cdots \\ \rho_{Ap1k} u_{pkA} u_{p1A} & \cdots & u_{pkA}^2 \end{bmatrix}$, $U_{PB} = \begin{bmatrix} u_{p1B}^2 & \cdots & \rho_{Ap1k} u_{p1B} u_{pkB} \\ \cdots & \cdots & \cdots \\ \rho_{Bp1k} u_{pkB} u_{p1B} & \cdots & u_{pkB}^2 \end{bmatrix}$ Uncertainties and correlation coefficient of two variables: (8a-d) $u_1^2 = u_{1A}^2 + u_{1B}^2$; $u_2^2 = u_{2A}^2 + u_{2B}^2$; $\rho_{1,2} = \frac{\rho_A u_{1A} u_{2A} + \rho_B u_{1B} u_{2B}}{\sqrt{u_{1A}^2 + u_{1B}^2}\sqrt{u_{2A}^2 + u_{2B}^2}}$ (8e, f, g)

In papers [6-10] is stated that only sets of deviations with uncertainties of one type, i.e. only of u_A or only of u_B, can correlate with each other for variables as components of the same or of different multi-measurands. Similarly to the variance $u^2 = u_A^2 + u_B^2$ of the single measured variable, the covariance matrix of multi-measurand also is the sum of two component covariance matrices of type ***A*** and ***B***, i.e. $U_X = U_{XA} + U_{XB}$, $U_Y = U_{YA} + U_{YB}$ - formula (8). Elements of the component covariance matrices type A and B are given in (8a)-(8d) and the method of their calculations in (8e, f, g). Only the correlation coefficients ρ_{xA} in the U_{XA} matrix of the input measurand X, can be experimentally determined by synchronous measurements of all X variables. On the other hand, the coefficients

ρ_{Bij} in the $\boldsymbol{U_{XB}}$ matrix, similarly to type B uncertainties, which have to be estimated heuristically as well as by some non-statistical approach and even based on the other data, which are not current measurement data. If two quantities are measured with the same, or similar instrument, and are in similar conditions, then the correlation coefficient ρ_{xB} equal to 1 can be assumed [6]-[10]. For measurements in different operating conditions or by different instruments, the correlation coefficient is closer to 0. The correlation coefficient –1 is also possible. It occurs, when changes of both variables from common interactions have opposite signs.

Type A and type B uncertainties for individual quantities of the output measurand $\boldsymbol{Y}$ should be carried out separately from the component $\boldsymbol{U_{YA}}$, $\boldsymbol{U_{YB}}$ of the covariance matrix $\boldsymbol{U_Y}$, according to formulas (8), (8a-g). If, in the period of time of measurement experiment, the values of $\boldsymbol{P}$ system parameters are constant, the matrix $\boldsymbol{U_{PA}}$ does not occur, and the $\boldsymbol{U_P}=\boldsymbol{U_{PB}}$ matrix, which is estimated for the possible long period changes of deviations $\Delta\boldsymbol{P}$. However, if $\Delta\boldsymbol{P}$ is changing randomly in the duration of the experiment, then the elements of their component matrix $\boldsymbol{U_{PA}}$ must also be estimated heuristically based on technical data and own knowledge. It is also valuable to perform additional measurements in specially created conditions to estimate $\boldsymbol{P}$ parameters' short random changes.

In several papers, the authors described how to use this extended method. A few examples of implementation this model to indirect measurement of a two-terminal circuit parameters through a four-terminal T type network, considering the uncertainties and correlation of its impedances in general case and for $U = 0$ are presented in detail in [8, 10]. One example is the indirect measurements of voltage and current of the branch, which loaded at the output a four-terminal circuit with twoport type T structure as its equivalent scheme. It is given below.

3. Uncertainties of indirect measurement of twoport variables

Let us consider as the example, the indirect observation of the voltage U_2 and current I_2 of an inaccessible branch based on direct measurements of voltage U_1 and current I_1 on other pair of network's terminals of a tested object. If this network is linear and passive then for the description of such experiment, the twoport of structure type T as the equivalent circuit can be used (Fig.1).

For the passive T-type twoport (Fig 1), from Kirchhoff's laws is: $I = I_1 + I_2$, $U_1 - Z_1 I_1 - Z_3 I = 0$; $U_2 - Z_2 I_2 - Z_3 I = 0$, and following relations are obtained

$$U_2 = \left(1 + \frac{Z_2}{Z_3}\right) U_1 - \left(Z_1 + Z_2 + \frac{Z_1 Z_2}{Z_3}\right) I_1$$

$$I_2 = \frac{1}{Z_3} U_1 - \left(1 + \frac{Z_1}{Z_3}\right) I_1 \qquad \text{(9a,b)}$$

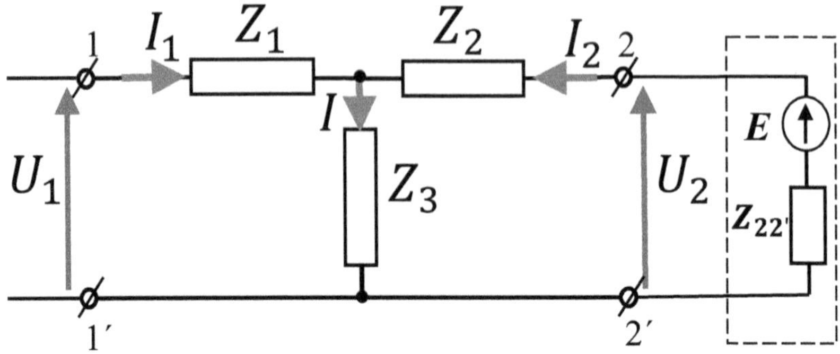

Fig 1. Diagram of a passive twoport T-type circuit

Then the relation between output measurand $\boldsymbol{Y} = [U_2, I_2]^{\mathrm{T}}$ and input measurand as $\boldsymbol{X}=[U_1, -I_1]^{\mathrm{T}}$ of the T-type twoport has the form of matrix function $\boldsymbol{Y}=\boldsymbol{B}\cdot\boldsymbol{X}$:

$$\begin{bmatrix} U_2 \\ I_2 \end{bmatrix} = \boldsymbol{B} \begin{bmatrix} U_1 \\ -I_1 \end{bmatrix} = \begin{bmatrix} B_{11}U_1 - B_{12}I_1 \\ B_{21}U_1 - B_{22}I_1 \end{bmatrix} \tag{10}$$

$$\text{Where } \boldsymbol{B} = \begin{bmatrix} B_{11} & B_{12} \\ B_{21} & B_{22} \end{bmatrix} = \begin{bmatrix} 1 + \frac{Z_2}{Z_3} & Z_1 + Z_2 + \frac{Z_1 Z_2}{Z_3} \\ \frac{1}{Z_3} & 1 + \frac{Z_1}{Z_3} \end{bmatrix} \tag{10a}$$

For the passive and reversible linear twoport, the determinant of the matrix $\boldsymbol{B}$ is

$$\det(\boldsymbol{B}) = B_{11}B_{22} - B_{12}\, B_{21} = 1 \tag{10b}$$

and only three of matrix $\boldsymbol{B}$ elements are independent, the fourth one follows from (10b). In the opposite situation, when estimated are variables on the twoport input from measurements on its output, the matrix $\boldsymbol{A}$ is used. For the twoport T it is like $\boldsymbol{B}$ with replaced impedances Z_2 and Z_1.

3.1 Covariance matrix $\boldsymbol{U}_Y$ when $\boldsymbol{U}_P = \boldsymbol{0}$ (GUM-S2 case)

The accuracy of the variables U_2, I_2 on output terminals 2, 2' is determined by values, uncertainties and correlations of internal impedances Z_1, Z_2, Z_3 and of variables U_1, I_1 measured on input terminals 1, 1'. If uncertainties u_{Z_i} of twoport impedances Z_i (i =1, 2, 3) are negligible, then in formula (6) from Table 1 the matrix $\boldsymbol{U}_P = \boldsymbol{0}$ and the covariance matrix of output measurand $\boldsymbol{Y}$ is as in the GUM-S2 method [1], i.e. $\boldsymbol{U}_Y = \boldsymbol{S}_X \boldsymbol{U}_X \boldsymbol{S}_X^T$.

If the twoport input variables U_1, $-I_1$ are measured by independent instruments and are uncorrelated, i.e. $\rho_1 = 0$, their covariance and the sensitivity matrices are

$$\boldsymbol{U}_{X0} = \begin{bmatrix} u_{U_1}^2 & 0 \\ 0 & u_{I_1}^2 \end{bmatrix}, \qquad \boldsymbol{S}_X = \frac{\partial Y}{\partial X}\Big|_{\substack{U_P = 0 \\ \rho_1 = 0}} = \begin{bmatrix} B_{11} & -B_{12} \\ B_{21} & -B_{22} \end{bmatrix}. \tag{11}$$

From (6) and (11), if $U_{YP}=0$, $\rho_1 = 0$, the covariance matrix $\boldsymbol{U_{Y0}} = \boldsymbol{S_X U_{X0} S_X^T}$ is:

$$\boldsymbol{U_{Y0}} = \begin{bmatrix} u_{U_2}^2 & \rho_2 u_{U_2} u_{I_2} \\ \rho_2 u_{U_2} u_{I_2} & u_{I_2}^2 \end{bmatrix} = \begin{bmatrix} B_{11}^2 u_{U_1}^2 + B_{12}^2 u_{I_1}^2 & B_{11}B_{21}u_{U_1}^2 + B_{12}B_{22}u_{I_1}^2 \\ B_{11}B_{21}u_{U_1}^2 + B_{12}B_{22}u_{I_1}^2 & B_{21}^2 u_{U_1}^2 + B_{22}^2 u_{I_1}^2 \end{bmatrix} \quad (12)$$

If, as in GUM-S2, the assumption is made that the processing function $\boldsymbol{F}=\boldsymbol{B}\cdot\boldsymbol{X}$ is exact, variances of the voltage and current and their correlation coefficient ρ_2 at output of the passive T-type twoport are:

$$u_{U_2}^2 = \frac{1}{Z_3^2}[(Z_2 + Z_3)^2 u_{U_1}^2 + (Z_1Z_2 + Z_3Z_2 + Z_1Z_3)^2 u_{I_1}^2],$$

$$u_{I_2}^2 = \frac{1}{Z_3^2}[u_{U_1}^2 + (Z_1 + Z_3)^2 u_{I_1}^2], \qquad \rho_2 = \frac{B_{11}B_{21}u_{U_1}^2 + B_{12}B_{22}u_{I_1}^2}{u_{I_2}u_{U_2}} \quad (12a, b, c)$$

Correlation coefficient $\rho_2 \neq 0$ and is always positive as $B_{11}B_{21} > 0$, $B_{12}B_{22} > 0$.

For $Z_2 = 0$, the twoport T circuit is changed to a voltage divider Z_1, Z_3, and above patterns simplify to:

$$u_{U_{2D}}^2 = u_{U_1}^2 + Z_1^2 u_{I_1}^2, \qquad u_{I_{2D}}^2 = \frac{1}{Z_3^2}\left[u_{U_1}^2 + (Z_1 + Z_3)^2 u_{I_1}^2\right]$$

$$\rho_{2D} = \frac{1}{Z_3}\frac{u_{U_1}^2 + (Z_1 + Z_3)u_{I_1}^2}{u_{I_2}u_{U_2}} \quad (13a, b, c)$$

For correlated input variables U_1,$-I_1$, the covariance matrix $\boldsymbol{U_X}$ is:

$$\boldsymbol{U_X} = \begin{bmatrix} u_{U_1}^2 & \rho_1 u_{U_1} u_{I_1} \\ \rho_1 u_{U_1} u_{I_1} & u_{I_1}^2 \end{bmatrix} \quad (14)$$

Their uncertainties are now modified by elements of correlation:

$$u_{U_{2\rho}}^2 = u_{U_2}^2 - 2B_{11}B_{12}\rho_1 u_{U_1}u_{I_1}, \quad u_{I_{2\rho}}^2 = u_{I_2}^2 - 2B_{21}B_{22}\rho_1 u_{U_1}u_{I_1} \quad (14a,b)$$

and correlation coefficient:

$$\rho_{2\rho} = \frac{B_{11}B_{21}u_{U_1}^2 + B_{12}B_{22}u_{I_1}^2 - (B_{11}B_{22} + B_{12}B_{21})\rho_1 u_{U_1}u_{I_1}}{u_{I_{2\rho}}u_{U_{2\rho}}} \quad (14c)$$

In cases $\rho_1 = \pm 1$, these absolute and relative uncertainties are:

$$u_{U_{2\rho}} = |B_{11}u_{U_1} \mp B_{12}u_{I_1}|, \qquad u_{I_{2\rho}} = |B_{21}u_{U_1} \mp B_{22}u_{I_1}|, \quad (15a,b)$$

$$\delta_{U_{2\rho}} = \left|\frac{u_{U_{2\rho}}}{U_2}\right| = \frac{|B_{11}u_{U_1} \mp B_{12}u_{I_1}|}{|B_{11}U_1 - B_{12}I_1|}, \qquad \delta_{I_{2\rho}} = \left|\frac{u_{I_{2\rho}}}{I_2}\right| = \frac{|B_{21}u_{U_1} \mp B_{22}u_{I_1}|}{|B_{21}U_1 - B_{22}I_1|}, \quad (16a, b)$$

and correlation coefficient:

$$\rho_{2\rho} = \frac{(B_{11}u_{U_1} \mp B_{12}u_{I_1})\cdot(B_{21}u_{U_1} \mp B_{22}u_{I_1})}{|(B_{11}u_{U_1} \mp B_{12}u_{I_1})\cdot(B_{21}u_{U_1} \mp B_{22}u_{I_1})|} = \begin{cases} 1 \text{ for } \rho_1 = -1 \\ \pm 1 \text{ for } \rho_1 = 1 \end{cases} \quad (16c)$$

If variables U_1, I_1 are uncorrelated, i.e. $\rho_1 = 0$, then $\rho_{2\rho} = \rho_2$ is given by (12c).

3.2 Component $\boldsymbol{U_{YP}}$ of $\boldsymbol{U_Y}$ for uncorrelated impedances of twoport T

For $V=0$, and uncorrelated impedances Z_1, Z_2, Z_3, from equations (5b) and (6) in the Table 1 follows that the covariance matrix $\boldsymbol{U_{P0}}$ and the sensitivity matrix $\boldsymbol{S_P}$ of the twoport T take forms:

$$\boldsymbol{U_{P0}} = \begin{bmatrix} u_{Z_1}^2 & 0 & 0 \\ 0 & u_{Z_2}^2 & 0 \\ 0 & 0 & u_{Z_3}^2 \end{bmatrix}, \quad \boldsymbol{S_P} = \begin{bmatrix} -I_1\left(1+\frac{Z_2}{Z_3}\right), & \frac{U_1}{Z_3} - I_1\left(1+\frac{Z_1}{Z_3}\right), & -\frac{Z_2}{Z_3^2}(U_1 - I_1 Z_1) \\ -\frac{I_1}{Z_3} & 0 & -\frac{1}{Z_3^2}(U_1 - I_1 Z_1) \end{bmatrix} \tag{17a,b}$$

If in (17b) the current I of Z_3 is used, i.e. $I = I_1 + I_2 = (U_1 - I_1 Z_1)/Z_3$, then the sensitivity matrix $\boldsymbol{S_P}$ simplifies to:

$$\boldsymbol{S_P} = \frac{-1}{Z_3}\begin{bmatrix} (Z_2+Z_3)I_1 & Z_3(I-I_1) & Z_2 I \\ I_1 & 0 & I \end{bmatrix} \tag{18}$$

The equation (6) in Table 1 shows that the $\boldsymbol{U_{YP}}$ component of the covariance matrix $\boldsymbol{U_Y} = \boldsymbol{U_{YX}} + \boldsymbol{U_{YP}}$ of the output measurand $\boldsymbol{Y}$, as function of the uncertainties of uncorrelated impedances of the T twoport, in this case is

$$\boldsymbol{U_{YP}} = \boldsymbol{S_P U_{P0} S_P^T}$$
$$= \frac{1}{Z_3^2}\begin{bmatrix} I_1^2(Z_2+Z_3)^2 u_{Z_1}^2 + (I-I_1)^2 Z_3^2 u_{Z_2}^2 + I^2 Z_2^2 u_{Z_3}^2, & -I_1^2(Z_2+Z_3)u_{Z_1}^2 + I^2 Z_2 u_{Z_3}^2 \\ -I_1^2(Z_2+Z_3)u_{Z_1}^2 + I^2 Z_2 u_{Z_3}^2, & I_1^2 u_{Z_1}^2 + I^2 u_{Z_3}^2 \end{bmatrix} \tag{19}$$

Despite the non-correlated impedances Z_i, when their $\boldsymbol{U_P}$ matrix is diagonal, the voltage U_2 and current I_2 as variables of the output measurand $\boldsymbol{Y}$, will be correlated. Then for uncorrelated variables $\boldsymbol{X}$, variances of the output voltage and current are

$$u_{U_2}^2 = \frac{1}{Z_3^2}[(Z_2+Z_3)^2 u_{U_1}^2 + (Z_1Z_2 + Z_3Z_2 + Z_1Z_3)^2 u_{I_1}^2 + I_1^2(Z_2+Z_3)^2 u_{Z_1}^2 + I^2 Z_2^2 u_{Z_3}^2$$
$$+ (I-I_1)^2 Z_3^2 u_{Z_2}^2]$$

$$u_{I_2}^2 = \frac{1}{Z_3^2}[u_1^2 + (Z_1+Z_3)^2 u_{I_1}^2 + I_1^2 u_{Z_1}^2 + I^2 u_{Z_3}^2] \tag{20a, b}$$

When $Z_2 = 0$, the T-type twoport becomes a divider Z_1, and $Z_3 U_2 = U_1 - Z_1 I_1$. Then for the variance $u_{I_2}^2$ is still (20b), but $u_{U_2}^2$ is changed and it is:

$$u_{U2D}^2 = u_{U_1}^2 + Z_1^2 u_{I_1}^2 + I_1^2\, u_{Z_1}^2 \tag{21}$$

3.3 Matrix $\boldsymbol{U_Y}$ for correlated impedances and uncorrelated input variables $\boldsymbol{X}$

In the case $\boldsymbol{U} = \boldsymbol{0}$, $V = \boldsymbol{0}$, from (6) and (18), for the accuracy of function $\boldsymbol{F}$ is now responsible the covariance matrix $\boldsymbol{U_{YP}}$, i.e.:

$$\boldsymbol{U_{YP}} = \boldsymbol{S_P U_P S_P^T}$$
$$= \frac{1}{Z_3^2}\begin{bmatrix} (Z_2+Z_3)I_1 & Z_3(I_1 - I) & Z_2 I \\ I_1 & 0 & I \end{bmatrix} \boldsymbol{U_P} \begin{bmatrix} (Z_2+Z_3)I_1 & Z_3(I_1 - I) & Z_2 I \\ I_1 & 0 & I \end{bmatrix}^T \tag{22}$$

where: the covariance matrix $\boldsymbol{U_P}$ of impedances has now all elements

$$\boldsymbol{U_P} = \begin{bmatrix} u_{Z_1}^2 & \rho_{Z12}u_{Z_1}u_{Z_2} & \rho_{Z13}u_{Z_1}u_{Z_3} \\ \rho_{Z12}u_{Z_1}u_{Z_2} & u_{Z_2}^2 & \rho_{Z23}u_{Z_2}u_{Z_3} \\ \rho_{Z13}u_{Z_1}u_{Z_3} & \rho_{Z23}u_{Z_2}u_{Z_3} & u_{Z_3}^2 \end{bmatrix} \quad (23)$$

If measurements of U_1 and I_1 are independent, their correlation factor is $\rho_1 = 0$, and after calculation $\boldsymbol{U_{YP}}$, from (14a) follows the variance of output voltage

$$u_{U_2}^2 = B_{11}^2 u_{U_1}^2 + B_{12}^2 u_{I_1}^2 + u_{UF}^2 \quad (24)$$

where: $B_{11} = 1 + \frac{Z_2}{Z_3}$; $B_{12} = Z_1 + Z_2 + \frac{Z_1 Z_2}{Z_3}$; $u_{UF}^2 = W_{UZ_1}^2 u_{Z_1}^2 + W_{UZ_2}^2 u_{Z_2}^2 + W_{UZ_3}^2 u_{Z_3}^2$
$+2(\rho_{Z12}W_{UZ_1}W_{UZ_2}u_{Z_1}u_{Z_2} + \rho_{Z23}W_{UZ_2}W_{UZ_3}u_{Z_2}u_{Z_3} + \rho_{Z13}W_{UZ_1}W_{UZ_3}u_{Z_1}u_{Z_3})$;
$W_{UZ_1} = I_1\left(1 + \frac{Z_2}{Z_3}\right)$; $W_{UZ_2} = I_1 - I$; $W_{UZ_3} = I\frac{Z_2}{Z_3}$ (24a-f)

For $u_{U2}^2 \geq 0$ from (2a), (24a-f) results the condition for the permissible values of correlation coefficients, that the determinant of correlator $\boldsymbol{R}$ cannot be negative:

$$det(\boldsymbol{R}) = \begin{vmatrix} 1 & \rho_{Z12} & \rho_{Z13} \\ \rho_{Z12} & 1 & \rho_{Z23} \\ \rho_{Z13} & \rho_{Z23} & 1 \end{vmatrix} = 1 - \rho_{Z12}^2 - \rho_{Z13}^2 - \rho_{Z23}^2 + 2\rho_{Z12}\rho_{Z13}\rho_{Z23} \geq 0 \quad (25)$$

Thus, the minimal correlation coefficients of three impedance pairs $\rho_{Z12} = \rho_{Z13} = \rho_{Z23} = -1$ cannot coincide, as $det(\boldsymbol{R}) = -4 \leq 0$.

If one of above coefficients, e.g. $\rho_{Z12} = \pm 1$, then should be $\rho_{Z23} = \pm \rho_{Z13}$.

It has to be analyzed when the correlations of the T-type twoport impedances and determinants of matrices $\boldsymbol{R}$ and $\boldsymbol{U_P}$ are equal to zero. The boundary values of correlation coefficients for $det(\boldsymbol{R}) = 0$ result from the equation (25), e.g. such values of the correlation coefficient ρ_{Z23} can be obtained from the expression

$$\rho_{Z23} = f(\rho_{Z12}, \rho_{Z13}) = \rho_{Z12}\rho_{Z13} \pm \sqrt{(1-\rho_{Z12}^2)(1-\rho_{Z13}^2)} \quad (26)$$

Condition: $1 - \rho_{Z12}^2 - \rho_{Z13}^2 - \rho_{Z23}^2 + 2\rho_{Z12}\rho_{Z13}\rho_{Z23} \geq 0$ means that the value of ρ_{Z23} belongs to the range defined as follows

$$\rho_{Z12}\rho_{Z13} - \sqrt{(1-\rho_{Z12}^2)(1-\rho_{Z13}^2)} \leq \rho_{Z23} \leq \rho_{Z12}\rho_{Z13} + \sqrt{(1-\rho_{Z12}^2)(1-\rho_{Z13}^2)} \quad (27)$$

It is the interior of a solid resembling tetrahedron with modified walls (fig. 2).

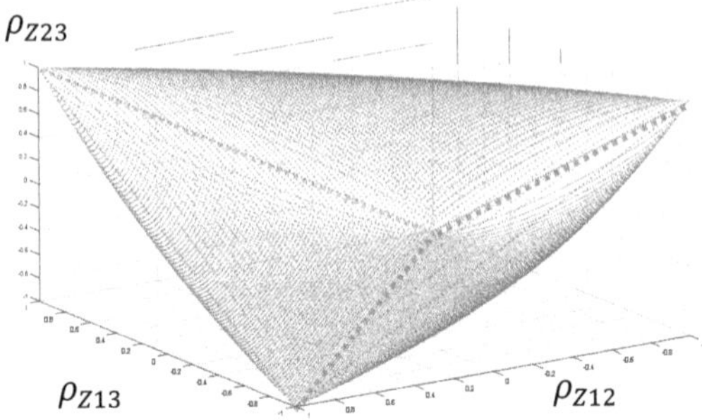

Fig 2. The space of three correlation coefficients for det $(\boldsymbol{R}) \geq 0$

If impedances are correlated only, the uncertainty of current I_2 can be find from its variance $u_{I_2}^2$ in the formula (28) obtained from (22), which is:

$$u_{I_2}^2 = \frac{1}{Z_3^2}\left[u_{U_1}^2 + (Z_1 + Z_3)^2 u_{I_1}^2 + I_1^2 u_{Z_1}^2 + I^2 u_{Z_3}^2 + 2I_1 I \rho_{Z13} u_{Z_1} u_{Z_3}\right] \tag{28}$$

For fully correlated impedances Z_1, Z_3, $\rho_{Z13} = 1$, the formula (28) simplify to

$$u_{I_2}^2 = \frac{1}{Z_3^2}\left[u_{U_1}^2 + (Z_1 + Z_3)^2 u_{I_1}^2 + \left(I_1 u_{Z_1} + I u_{Z_3}\right)^2\right] \tag{29}$$

In the general case, the correlation coefficient between U_2 and I_2 is:

$$+\frac{1}{Z_3^2}\frac{\begin{bmatrix} I_1^2(Z_2 + Z_3)u_{Z_1}^2 + I_1 I(Z_3 + 2Z_2)\rho_{Z13}\, u_{Z_1} u_{Z_3} + \\ +(I_1 - I)I_1 Z_3 \rho_{Z12} u_{Z_1} u_{Z_2} + I(I_1 - I)Z_3 \rho_{Z23} u_{Z_2} u_{Z_3} + Z_2 I^2 u_{Z_3}^2 \end{bmatrix}}{u_{U_2} u_{I_2}} \tag{30}$$

For the voltage divider, i.e. when $Z_2 = 0$, and uncorrelated input variables U_1, I_1

$$\rho_{2D} = \frac{u_{U_1}^2 + Z_1(Z_1 + Z_3)u_{I_1}^2 + I_1^2 u_{Z_1}^2 + I_1 I \rho_{Z13} u_{Z_1} u_{Z_3}}{Z_3 u_{U_2} u_{I_2}} \tag{31}$$

3.4. Numerical example

The output variables $\boldsymbol{Y} = [U_2, I_2]^{\mathrm{T}}$ of the linear and passive twoport have to be find indirectly from measurements of its input variables $\boldsymbol{X}=[U_1, I_1]^{\mathrm{T}}$. The type T scheme of twoport is used (Fig. 1). Given are its impedances $Z_i=(Z_i, Z_2, Z_3)$ $=100\ \Omega$, known with relative uncertainties $\delta_{Z\,i}= .2\%$ ($u_{Z\,i}=2$ mΩ), and should be consider the correlation of their pairs with values coefficients: $\rho_{ZiZj}\equiv \rho=0$; .5; 1.0. Then elements of the matrix $\boldsymbol{B}$ (10a) of processing function $\boldsymbol{Y}=\boldsymbol{B}\cdot\boldsymbol{X}$ of this T type twoport are: $B_{11}=2$, $B_{12}=300\ \Omega$, $B_{21}=\ .01$S, $B_{22}=2$.

The input variables U_1, I_1 have initial uncertainties $u_{U10} = u_{I10} = 0$ and relative ones $\delta_{U1}=\delta_{I1}= .2\%$, both uncorrelated.The external branch, connected to output terminals 2 2' is unknow and its equivalent circuit can be passive or active. e.g.

uncertainties u_{U_2}, u_{I_2} and ρ_2 are found from (24) and (28)-(30). U_1=100 V, ranges of other variables are: U_2= (20 ÷ 80)V; I_1= (.6 ÷ .4)A; I_2 = (-0.2 ÷ 0,2) A.

Results of calculations of formulas are given in tables 2a, b, c.

Table 2a. Data of input and output variables of T-twoport, uncertainties of U_1, I_1, Z_i, and sensitivity coefficients of the output voltage uncertainty.

Parameter / No	U_1 [V]	u_{U_1} [V]	I_1 [A]	u_{I_1} [mA]	U_2 [V]	I_2 [A]	$u_{Z_1} = u_{Z_2} = u_{Z_3}$ [Ω]	$W_{UZ_1} = I_1\left(1+\frac{Z_2}{Z_3}\right)$ [A]	$W_{UZ_2} = I_1 - I$ [A]	$W_{UZ_3} = I\frac{Z_2}{Z_3}$ [A]
1			0,6	1,2	20	-0.2		1,2	0,2	0,4
2			0,55	1,1	35	-0.1		1,1	0,1	0,45
3	100	0,2	0,5	1	50	0	0,2	1	0	0,5
4			0,45	0,9	65	0.1		0,9	-0,1	0,55
5			0,4	0,8	80	0.2		0,8	-0,2	0,6

Table 2b. Coefficients of equation: $u_{U_2}^2 = B_{11}^2 u_{U_1}^2 + B_{12}^2 u_{I_1}^2 + W_{UZ_1}^2 u_{Z_1}^2 + W_{UZ_2}^2 u_{Z_2}^2 + W_{UZ_3}^2 u_{Z_3}^2 + +2(\rho_{Z12} W_{UZ_1} W_{UZ_2} u_{Z_1} u_{Z_2} + \rho_{Z23} W_{UZ_2} W_{UZ_3} u_{Z_2} u_{Z_3} + \rho_{Z13} W_{UZ_1} W_{UZ_3} u_{Z_1} u_{Z_3})$ (24)

Para-meter No	B_1^2 [1]	B_{12}^2 [Ω²]	$W_{UZ_1}^2$ [A²]	$W_{UZ_2}^2$ [A²]	$W_{UZ_3}^2$ [A²]	$2\rho_{Z12}W_{UZ_1}W_{UZ_2}$ $2\rho_{Z23}W_{UZ_2}W_{UZ_3}$ $2\rho_{Z13}W_{UZ_1}W_{UZ_3}$ ρ=0	$2\rho_{Z12}W_{UZ_1}W_{UZ_2}$ [A²]		$2\rho_{Z23}W_{UZ_2}W_{UZ_3}$ [A²]		$2\rho_{Z13}W_{UZ_1}W_{UZ_3}$ [A²]	
							ρ=0,5	ρ=1	ρ=0,5	ρ=1	ρ=0,5	ρ=1
1		90000	1,44	0,04	0,16		0,24	0,48	0,08	0,16	0,48	0,96
2		90000	1,21	0,01	0,2025		0,11	0,22	0,045	0, 09	0,495	0,99
3	4	90000	1	0	0,25	0	0	0	0	0	0,5	1
4		90000	0,81	0,01	0,3025		-0,09	-0,18	-0,055	-0,11	0,495	0,99
5		90000	0,64	0,04	0,36		-0,16	-0,32	-0,12	-0,24	0,48	0,96

Table 2c. Sensitivity coefficients of the current I_2 variance elements:

$$u_{I_2}^2 = B_{21}^2 u_{U_1}^2 + B_{22}^2 u_{I_1}^2 + \frac{1}{Z_3^2}\left[I_1^2 u_{Z_1}^2 + I^2 u_{Z_3}^2 + 2 I_1 I \rho_{Z13} u_{Z_1} u_{Z_3}\right] \quad (28)$$

Parameter No	B_{21}^2 [S^2]	B_{12}^2 [1]	I_1^2/Z_3^2 [A^2S^2]	I^2/Z_3^2 [A^2S^2]	$2I_1I\rho_{Z13}$ ρ=0	$2I_1I\rho_{Z13}/Z_3^2$ [A^2S^2] ρ=0,5	ρ=1
1			$3,6\cdot10^{-5}$	$1,6\cdot10^{-5}$		$2,4\cdot10^{-5}$	$4,8\cdot10^{-5}$
2			$3,03\cdot10^{-5}$	$2,02\cdot10^{-5}$		$2,475\cdot10^{-5}$	$4,95\cdot10^{-5}$
3	10^{-4}	4	$2,5\cdot10^{-5}$	$2,5\cdot10^{-5}$	0	$2,5\cdot10^{-5}$	$5\cdot10^{-5}$
4			$2,03\cdot10^{-5}$	$3,02\cdot10^{-5}$		$2,475\cdot10^{-5}$	$4,95\cdot10^{-5}$
5			$0,2\cdot10^{-5}$	$3,6\cdot10^{-5}$		$2,4\cdot10^{-5}$	$4,8\cdot10^{-5}$

Figures 3a-e show: a, b - uncertainties u_{U_2}, u_{I_2} of output variables $\underline{U}_2$, I_2, c, d - their relative values $\delta_{U_2} \approx \frac{u_{U2}}{U_2}$ (as $u_{U_2}(0) \ll u_{U_2}$), $\delta_{I_2} = |\frac{u_{I2} - u_{I2=0}}{I_2}|$ and 3e - the correlation coefficient ρ_2, all as function of output current I_2 for three correlation coefficients of the twoport impedance pairs ρ = (0, 0.5, +1).

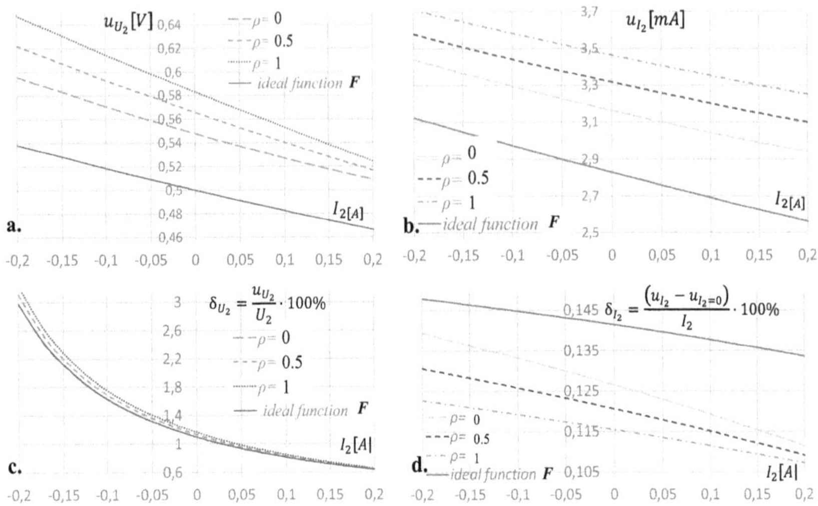

Fig. 3: **a**, **b**. Absolute uncertainties u_{U_2}, u_{I_2}; **c**, **d**. relative uncertainties $\delta_{U_2}, \delta_{I_2}$.

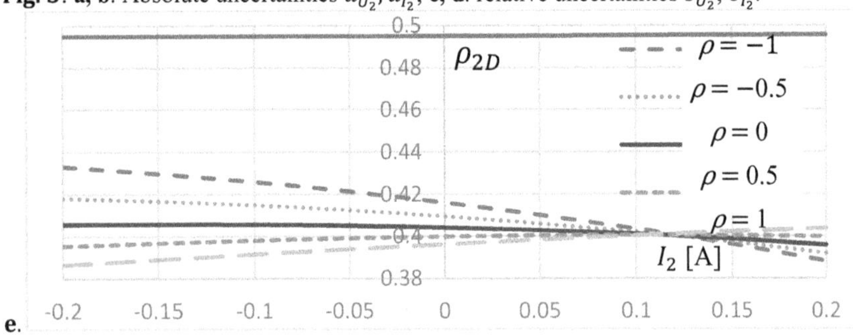

Fig 3e. Correlation coefficient ρ_2, all of output variables U_2, I_2 of T-type twoport as function of current I_2 for three equal correlation coefficients ρ of impedances $Z_i = 100\ \Omega$ of relative uncertainty $\delta_{Z_i} = \delta_{U_1} = \delta_{I_1} = 0.2\%$.

From the above graphs results that the absolute uncertainties of voltages and current at the output of twoport increase with the value of positive correlation coefficient. The equivalent circuit of the load branch of twoport could be the impedance $Z_{22'}$ and voltage source E_2 in series (fig.1). Their values can be find from two measurements of U_1, I_1 for different supply of the input net.

Summary and conclusions

The currently used method for assessing the accuracy of indirect multivariable measurements is discussed in Supplement 2 of the GUM [1]. Its use is for cases of the ideal multivariable processing function ***Y***=***F***(***X***), i.e. without uncertainties.

The paper discusses a model with basic formulas listed in Table 1, which extends the GUM-S2 method. The covariance matrix elements, which describe the accuracy of output measurand ***Y***, were derived considering uncertainties and correlations of the multivariable input measurand ***X*** and of the parameters ***P*** of the real processing functions ***Y***=***F***(***X***, ***P***) performed in the measurement system.

As an example, the application of this extended method in indirect 2D measurements of the voltage and current of the two-terminal circuit branch, is presented. This branch is available for measurements only indirectly changing input variables of other pairs of network terminals through the equivalent circuit of the twoport T structure. Such measurement occurs in the identifying parameters of inaccessible elements and branches of the electrical circuit, in multi-sensor measurements, in the technical diagnostics and in control of industrial processes.

The formulas of increased total uncertainty of the estimated voltage and current due to uncertainties of the impedances of this system are found. It did not exceed the sum of uncertainties of input variables and twoport impedances.

In the presented twoport, the estimated uncertainties of the voltage and current at its output depend on the value and sign of correlation coefficients of input quantities and circuit parameters. The minimizing of uncertainties is possible.

The proposed method can be usefully used to evaluate of uncertainties of indirect multi-parameter measurements made with a set of instruments, as well as for the assessment of the accuracy of measuring instruments and systems with an integrated measurement circuit for direct reading of multivariable measurements.

This method can be the basis for a new version of the Supplement 2 to the guide GUM, or should be included in the new version of this guide. It will allows also the assessment of the accuracy of multivariable instrumental measurement systems through uncertainties. Other partial methods are discussed in [4], [6-10].

It is also possible by presented method to analyze the accuracy of several other multi-parameter measurement systems, e.g., equivalent circuits of AC elements and AC bridges, networks as in [2], [3], [5], [7, 8, 10], components of power in AC networks with different waveforms, statistical properties of multi-variable systems with non-Gaussian probability distributions and various processing functions of multivariable measurands.

References

1. JCGM100:2008, Evaluation of measurement data - Guide to the expression of uncertainty in measurement. +S2: Supplement 2. Extension to any number of output quantities.JCGM102:2011
2. D'Antona, G.: Measurement Data Processing Using Random Matrices: A Generalized Formula for the Propagation of Uncertainty. IEEE Trans. of Instrumentation and Measurement v. 53, n. 2, April 2004, p. 537-545

3. Augustyn, J.: Uncertainty evaluation of the processing algorithm of a time-variable quantities in multi-channel measurement systems, Metrology and Measurement Systems, v. 13, n. 3, p. 303 – 313 (2006)
4. Warsza Z.L.: About evaluation of multivariate measurements results. *JAMRIS* vol. 6, n. 4, p. 27 -32 (2012)
5. Ramos P.M., Janeiro F.M., Girão P.M.: Uncertainty evaluation of multivariate quantities: A case study on electrical impedance, Measurement, vol. 78, n. 1, pp. 397 – 411(2016)
6. Warsza, Z., Puchalski, J.: Method of the estimation of uncertainties in multiparameter measurements of correlated quantities. Proc. of AMSA V (2019) 18-20 Sept. Novosibirsk NSTU pp.47-59 (amsa.conf.nstu.ru)

7, 8. Warsza, Z.L., Puchalski, J.: Extended method for evaluation uncertainties of indirect multi-parameter measurements and circuits for these measurements (in Polish)
Part 1. Influence of correlation and uncertainties of processing function - basic formulas. *PAR* v. 23, n.3 p.55-63 (2019), DOI:10.14313/PAR_233/55
Part 2. Application on the example of a system for indirect two-parameter measurements *PAR* v.23, n.4 p.87-100 (2019), DOI:10.14313/PAR_234/87

9,10. Warsza, Z.L., Puchalski, J. Estimation of uncertainty in indirect multi-variable measurements Part 1. Case of correlated quantities. Part 2. Influence of the processing function accuracy. in: R. Szewczyk et al. (Eds.), Proc. of Automation 2020, Towards Industry of the Future. pp. 309–344. Springer Nature Switzerland AG (2020) doi.org/10.1007/978-3-030-40971-5_29 and_30

F Pavese, A B Forbes, N F Zhang, A G Chunovkina (Eds.):
Advanced Mathematical and Computational Tools in Metrology and Testing XII
Series on Advances in Mathematics for Applied Sciences, Vol. 90

https://doi.org/10.1142/9789811242380_0030

Incompatibility of types of measurement uncertainty: A digital paradox and other examples

R. Willink[1,2]

[1] *Measurement Standards Laboratory of New Zealand, PO Box 31310, Lower Hutt 5040, New Zealand*
[2] *University of Otago (Wellington), PO Box 7343, Wellington 6242, New Zealand*
E-mail: robin.willink@measurement.govt.nz

Aleatory measurement uncertainty (the uncertainty of randomness) can be distinguished from epistemic measurement uncertainty (the uncertainty of doubt). Unacceptable results in several example measurement problems show that these forms of uncertainty cannot always be combined using the idea of probability. Associated principles of Bayesian statistics are discussed. One conclusion is that a prior distribution on a nuisance parameter governing the quality of a measurement does not adequately represent its contribution to the uncertainty statement. The implications for the evaluation of measurement uncertainty and for the utility of Bayesian statistical methods are far-reaching.

Keywords: Aleatory uncertainty; Epistemic uncertainty; Nuisance parameter; Bayesian statistics; Subjective probability.

1. Introduction

Typically, a measurement problem involves the combination of estimates of many quantities that are related either to the measurand or to the experimental procedure. Consequently, the analysis involves the combination of different components of measurement uncertainty. Some components might be associated with the presence of random data, in which case they are treated using the concepts of frequency and probability in a statistical analysis. Other components are assessed by expert opinion, and these might be expressed using probability distributions to describe strength of belief, even without experimental data. One current approach to the analysis of measurement uncertainty advocates that components of these two different types can and should be combined using the laws of probability, in particular using Bayes' Theorem.

Many scientists find such hybrid methodology intuitive, perhaps not considering it to be hybrid at all. However, there are a number of instances in which incorporating the degree-of-belief concept into the analysis leads to unwanted behaviour in the results,[1,2] and this suggests that the methodology is not as reliable as might be hoped. In this paper, we consider several new examples of unacceptable behaviour with this approach to the analysis of measurement uncertainty. We argue that such behaviour is linked to an error in the methodology: uncertainty associated with randomness in the data and uncertainty associated with a lack of knowledge about a constant are different in nature and cannot always be combined. Some of the examples are realistic and some are contrived but informative. Section 2 presents three examples and then Section 3 seeks to provide explanations. Section 4 presents a realistic example of a different kind, which raises further questions. Section 5 gives suggestions and conclusions.

2. Three measurement problems

2.1. *A digital paradox*

Consider the standard problem where there is an error with unknown standard deviation σ and another error caused by subsequent discretization. Results given elsewhere[3] for repeated measurement in this situation imply that, when belief about σ is represented by a probability distribution, the uncertainty associated with the sample mean as the measured value can *decrease* when this distribution is broadened to accommodate *larger* values of σ. This behaviour is observed in more detail here.

Suppose that a quantity θ is measured using a device having Gaussian random error and that the result is displayed after rounding to the nearest integer x. If the Gaussian distribution has mean zero and standard deviation σ then x is the outcome of a random variable $X = \lfloor Y + 0.5 \rfloor$ with $Y \sim \mathrm{N}(\theta, \sigma^2)$, where $\lfloor \cdot \rfloor$ indicates the 'greatest integer not greater than'. Therefore $\Pr(X = x) = \Pr(Y \leq x + 0.5) - \Pr(Y \leq x - 0.5)$. So the probability mass function of X for $x = \pm 0, 1, 2, \ldots$ is

$$\mathrm{P}(x|\theta, \sigma) \equiv \Pr(X = x) = \Phi\left(\frac{(x + 0.5) - \theta}{\sigma}\right) - \Phi\left(\frac{(x - 0.5) - \theta}{\sigma}\right),$$

with $\Phi(\cdot)$ the standard normal distribution function, e.g. $\Phi(1.96) = 0.975$. Suppose that θ is to be estimated from a set of n observations and that σ is unknown. Some metrologists might carry out an analysis that attributes the unknowns θ and σ probability distributions, as in the Bayesian and fiducial paradigms of inference. The analysis here applies to the Bayesian approach, because it is the 'prior distribution' of σ that is relevant. Here σ is a *nuisance parameter* because it is unknown but not related to θ.

For simplicity, suppose that – before observing the data – the Bayesian analyst considers each value in a wide range to be equally likely for θ and each value between 0 and some maximum value σ_{max} to be equally likely for σ. The prior distributions that the analyst attributes to θ and σ must be independent because the true value of θ is unrelated to the behaviour of the device that is used to estimate it. Therefore, the joint prior density function attributed to the parameters is $f_{\text{prior}}(\theta, \sigma) \propto 1$ for $0 < \sigma < \sigma_{\text{max}}$. The uniformity of this distribution means that the joint posterior density function of θ and σ has the form of the likelihood function. So the posterior density function of θ when the vector of data is $(x_1, \ldots, x_n)$ is $f_{\text{post}}(\theta) = \int_0^{\sigma_{\text{max}}} \prod_{j=1}^n \mathrm{P}(x_j|\theta, \sigma)\, d\sigma$. Suppose that the data all take the same value, which we assume to be $x = 0$ without loss of generality. Then $\prod_{j=1}^n \mathrm{P}(x_j|\theta, \sigma) = \mathrm{P}(0|\theta, \sigma)^n$. The corresponding posterior density of θ is shown for $\sigma_{\text{max}} = 10, 0.1, 0.001$ and for $n = 4, 8, 24$ in Fig. 1a, b and c. The mean of the posterior distribution, which is 0 in each case, would be taken as the measured value. Also, the standard deviation of the distribution would be taken as the standard uncertainty of measurement, u.

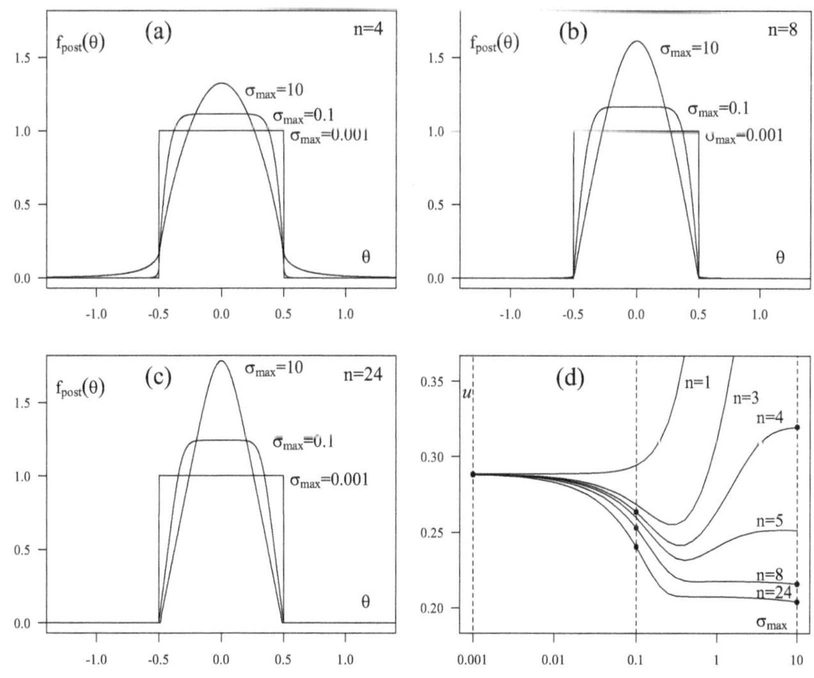

Fig. 1. Posterior distributions and standard uncertainties u for different sample sizes and different maximum values thought possible for σ.

Obviously, a larger value of $\sigma_{\max}$ represents greater uncertainty on the part of the Bayesian analyst about σ. Yet Fig. 1 shows that a larger value of $\sigma_{\max}$ can give a narrower posterior distribution for θ and a smaller standard uncertainty u, as clearly seen in Fig. 1d. This result is accurate mathematically, but how can it be meaningful or correct as a representation of measurement uncertainty? Being less sure about one thing, σ, cannot increase our sureness or strengthen our knowledge about something that is unrelated, θ, either before or after observation of the data. Yet Fig. 1d clearly shows that, for some $\sigma_{\max}$ and n, the posterior standard deviation decreases as $\sigma_{\max}$ increases. This phenomenon is not linked to the forms of the distributions: the same qualitative behaviour would be seen with non-Gaussian sampling distributions and non-uniform prior distributions. Nor is it linked to the use of *improper* prior distributions. So this phenomenon will be of concern to those who favour the subjective Bayesian approach, as well as to those who employ 'objective' Bayesian methods.

In this example, the behaviour of concern is counter-intuitive. We now present two other examples where this behaviour can be found, albeit in a different form. In these auxiliary examples, the behaviour is intuitive, not counter-intuitive. The examples are contrived, but they illustrate that there are theoretical problems with the methodology. As shall be explained in Section 2.4, the three examples have a common feature.

2.2. *Doubt about the standard deviation of the error*

Consider measuring a voltage θ using a meter for which the specification sheet suggests an accuracy of $\pm 1\,\text{mV}$. Suppose that this is interpreted to mean that the error is drawn from a Gaussian distribution with mean zero and standard deviation $\sigma = 1\,\text{mV}$. The standard deviation would be used as the standard uncertainty, so $u = 1\,\text{mV}$. This would be the standard uncertainty obtained in a frequentist analysis, and it would also be the standard uncertainty obtained in a Bayesian analysis with a uniform prior distribution for θ and a degenerate prior distribution at $1\,\text{mV}$ for σ. But suppose that subsequent, closer, inspection of the specification sheet indicates that the printed figure might be $\pm .1\text{mV}$ instead. The Bayesian analysis must change to take this new piece of information into account. The distribution for σ becomes a distribution on two points, $0.1\,\text{mV}$ and $1\,\text{mV}$, and so the posterior distribution of θ becomes a mixture of Gaussian distributions with the same mean but with different standard deviations. The standard uncertainty calculated is smaller – by approximately 30% if the two possibilities are weighted equally. Thus, doubt introduced about σ appears to improve the measurement! This is unacceptable.

2.3. *Doubt about the sample size*

The sample size in repeated observation plays an important role in determining the uncertainty of measurement, so doubt about the sample size will influence the figure of uncertainty stated. Suppose that two analysts work together to make a number of repeated Gaussian observations and that, as usual, they base their calculation of the standard uncertainty on the sample standard deviation s. Let u_n be the figure of uncertainty appropriate with sample size n, so $u_{n+1} < u_n$, all other things being equal. Imagine that the analysts omit to record the number of measurements made, and that one believes there were 6 or 7 while the other is confident there were only 6. If the first analyst attributes n the probabilities p and $1-p$ on its possible values 6 and 7 then the standard uncertainty obtained is $u = pu_6 + (1-p)u_7$. This is necessarily smaller than the standard uncertainty u_6 that is obtained by the analyst who was more sure of an important figure, n. This would not make any practical, metrological, sense.

2.4. *A common feature*

Before proceeding, we note a relevant feature of the three examples. In each case there is an uncertain nuisance parameter that is linked to the scale of the error in the measured value and to which the analyst must attribute a probability distribution if the analysis is to be Bayesian: in the first example, the parameter σ must be given a prior distribution; in the second example, σ must be attributed a probability distribution, albeit only on two points; and in the third example, the unknown sample-size n, which acts as a nuisance parameter, must be considered to have a distribution.

3. Different types of uncertainty

In each of the three examples, the Gaussian probability distribution represents the random generation of data in a repeatable process. And in each of the three examples, a second probability distribution was proposed to represent something different – a lack of knowledge about something that is fixed and not repeated. So, in each example, the Gaussian probability distribution describes *aleatory* uncertainty,[4] which is uncertainty associated with the idea that a process is unpredictable, as in the phrase "the outcome is uncertain". And, in each example, the proposed additional distribution of the Bayesian describes *epistemic* uncertainty,[4] which is uncertainty of not knowing a fact, as with "I am uncertain about that". Because of unfamiliarity, we will refer to aleatory uncertainty and epistemic uncertainty as *stochastic* uncertainty and *personal* uncertainty, respectively. The claim that a Bayesian analysis of the measurement can give correct results takes as a premise the idea that these different types of uncertainty can be combined in the same analysis. The phenomena observed call into question the validity of this premise, so we shall now discuss these ideas more fully.

3.1. *The natures of probability*

Much has been written about different interpretations of probability, two of which are the *frequency* interpretation, which is associated with the concept of stochastic uncertainty, and the *subjective* interpretation, in which the uncertainty represented is personal. But these are interpretations of *statements* of probability, not interpretations of probability itself. Probability itself is not interpreted – probability is merely an abstract mathematical construct that obeys Kolmogorov's axioms. Rather, probability interprets. That is, stochastic uncertainty and personal uncertainty are not interpretations *of* probability: they are interpreted or represented *by* probability. Probability represents frequency behaviour when called upon to do so, and it can represent strength of belief when called upon to do so.

But do the principles of scientific inference always allow probability to represent both of these different things in the same analysis? The existence of the anomalous behaviour seen here suggests that they do not. Consider the first example. One of the probability distributions involved describes frequency behaviour, $\mathrm{N}(\theta, \sigma^2)$, while another describes belief about a constant, $\mathrm{U}(0, \sigma_{\max})$. We suggest that, in this case, the two representations cannot be combined in Bayes' Theorem if the final distribution calculated for the measurand θ, the posterior distribution, is to have its intended meaning. The same may be said of the two representations in each of the other two examples. Clearly, an examination of Bayes' Theorem is appropriate.

3.2. *Bayes' Theorem as frequency and belief*

To see a successful interaction of the two types of uncertainty in Bayes' Theorem, we consider the archetypal example of a medical diagnosis. Imagine that a screening programme is put into place to detect a serious condition known to affect 0.1% of the population. Individuals with no symptoms are tested systematically, with the response '+' suggesting the presence of the condition, C, and the response '−' suggesting its absence, $\overline{\text{C}}$. The test has a *sensitivity* of 91% and a *specificity* of 96%, i.e. $\Pr(+|\text{C}) = 0.91$ and $\Pr(-|\overline{\text{C}}) = 0.96$. Jimmy is tested and receives a positive response. Dismayed, he asks the doctor "what is the probability I have the condition?" There is no other information to suggest that Jimmy has or has not got the condition, so the appropriate *positive predictive value* (PPV) is

$$\begin{aligned}\Pr(\text{C}|+) &= \frac{\Pr(+|\text{C})\Pr(\text{C})}{\Pr(+)} = \frac{\Pr(+|\text{C})\Pr(\text{C})}{\Pr(+|\text{C})\Pr(\text{C}) + \Pr(+|\overline{\text{C}})\Pr(\overline{\text{C}})} \qquad (1)\\ &= \frac{0.91 \times 0.001}{0.91 \times 0.001 + 0.04 \times 0.999} = 0.02.\end{aligned}$$

Thus, the doctor would inform Jimmy that, of all the people who would test positive for the condition, only 2% would actually have the condition. The prevalence of the condition and the specificity of the test are low enough to make the PPV quite small: Jimmy can be reassured despite the positive test result.

The analysis above was carried out using Bayes' Theorem, which for this problem is (1). Here, this equation expresses frequency behaviour: every individual probability seen in (1) represents a relative frequency among some reference set, (i.e. some denominator). Let θ_{J} denotes Jimmy's true condition-status, with $\theta_{\text{J}} = 1$ meaning he has the condition and $\theta_{\text{J}} = 0$ meaning he does not have the condition. The doctor observes that the relative frequency of the condition in the population is 0.1%, and so regards θ_{J} as being randomly drawn from the Bernoulli distribution with parameter $\Pr(\text{C}) = 0.001$, i.e. the doctor regards θ_{J} as the outcome of a random variable, $\Theta_{\text{J}} \sim \text{Bernoulli}(0.001)$. Likewise, the doctor takes positive tests to occur with relative frequencies of 91% and $100 - 96 = 4\%$ among those with and without the condition. The result, 2%, is a relative frequency of those with the condition among those who would return a positive test.

The analysis just given would appeal to a frequentist statistician. But it could be reformulated as a Bayesian analysis if θ_{J} itself, not Θ_{J}, were regarded as a random variable with a Bernoulli distribution, i.e. if, prior to observing the result of Jimmy's test, the doctor considered that $\Pr(\theta_{\text{J}} = 1) = p$ for some specific figure $p \equiv \Pr(\text{C})$. The figure p might not be the overall prevalence, 0.001, but it would be formed as a subjective judgment of the doctor based on expertise, (perhaps because Jimmy is a member of a recognizable sub-population with a different prevalence). For that Bayesian doctor, the figure p is the prior probability that Jimmy has the disease, and *a priori*, $\theta_J \sim$ Bernoulli(p). The probability Pr(C) would be a personal attribute of the doctor, and so a better notation would be $\Pr_{\rm doc}$(C). (Note the difference in the frequentist and Bayesian understandings of θ_J. For the frequentist, the constant θ_J can (if required) be seen as *the outcome of* a random variable Θ_J. For the Bayesian, θ_J *is* a random variable.)

So this example can illustrate both frequentist and Bayesian practice. For the frequentist doctor, Pr(C) represents stochastic uncertainty associated with randomly drawing Jimmy's status from a frequency distribution. For the Bayesian doctor, $\Pr_{\rm doc}$(C) represents personal uncertainty associated with the doctor's knowledge of Jimmy's circumstances. In both cases, Bayes' Theorem (1) is used appropriately to combine the stochastic uncertainty of the testing process with the 'prior' uncertainty described by Pr(C), and in both cases the figure of probability obtained can be given a frequency interpretation. So how is it that the Bayesian concept of 'degree of belief' as probability works here but fails in, say, the first example?

3.3. *Explanations*

One clue is that the personal distribution of the Bayesian doctor can be given a useful frequency interpretation (but the prior distribution on σ in the discretization example cannot). The reference set for the prior probability $\Pr_{\rm doc}$(C) is the set of statuses of people with the same characteristics as Jimmy, as perceived by the doctor, and the distribution Bernoulli(p) can be seen as the corresponding frequency distribution. Subsequently, the doctor's posterior distribution of θ_J can be seen as a statement of the frequency of disease in a group of people who would test positive and have the same characteristics as Jimmy.

Another clue is that in the discretization problem there is more that one unknown parameter, the measurand θ and the nuisance parameter σ. Both parameters must be given prior distributions if the analysis is to be Bayesian. The distribution that is influential in permitting the phenomenon is the prior distribution of σ, and it is simply the fact that the upper limit of this distribution, $\sigma_{\max}$, might vary that gives rise to the unwanted effect. This suggests that the rogue element in the analysis is the requirement that the nuisance parameter be given a prior distribution at all. This suggestion is consistent with the behaviour observed in the other two examples.

Consider now the example with the voltmeter. The stark simplicity of this example helps identify another important idea: an academic Bayesian statistician would find no fault in the behaviour! If the input distributions were accurate representations of the Bayesian's beliefs, the posterior distribution would faithfully represent his or her (theoretical) resulting belief about θ. The difficulty arises here because the analyst is working for someone else, a client. The belief of the Bayesian metrologist, which is intrinsically subjective and inescapably personal, has no scientific relevance to the client. The metrologist cannot pretend that increased doubt about his or her process-parameter σ diminishes the client's doubt about the client's measurand θ. The personal, non-scientific, nature of Bayesian statistics is exposed: a result deemed relevant to one person is not relevant to another.

3.4. *Re-examining Bayes' Theorem*

The results that have been described lead us to re-examine the meaning of Bayes' Theorem. Bayes' Theorem is an identity in conditional probability that is accepted by all statisticians: the truism $\Pr(B \cap A) = \Pr(A \cap B)$, which has complete symmetry, is simply being re-expressed as

$$\Pr(B|A) - \frac{\Pr(A|B)\Pr(B)}{\Pr(A)}. \tag{2}$$

So how can problems arise with it? The answer seems to be that A and B are often entities of different types: A is usually an event governed by stochastic uncertainty and B is usually an hypothesis subject to personal uncertainty. So a symmetric system is being expected to deal with asymmetric inputs while conforming to its symmetric nature. As applied here, A stands for the potential occurrence of an event E that – if it were given the opportunity to occur through a sampling process – would be subject to stochastic uncertainty in the future, while B stands for the potential truth of an hypothesis H about which there is personal uncertainty that actually exists in the present. Thus, Bayes' Theorem (2) expresses the probability Pr(H is true | E occurs if given the chance to occur) in the form

$$\Pr(\text{H is true} \mid \text{E occurs if given the chance to occur}) = \\ \frac{\Pr(\text{E occurs if given the chance to occur} \mid \text{H is true})\Pr(\text{H is true})}{\Pr(\text{E occurs if given the chance to occur})}.$$

The brevity of the symbolism in (2) obscures the fact that the two types of propositions differ in the nature of what they describe. Bayes' Theorem (2) is a reversible transformation of a symmetric relationship. Without further evidence, is there anything about it that says it can be applied when A and B are qualitatively different, i.e. when the notion of probability is being employed to represent two different concepts at the same time?

4. Another phenomenon

The examples given so far involve both personal uncertainty and stochastic uncertainty. However, a similar effect can be observed in a problem that involves only personal uncertainty.[5] Consider measuring the real part of a complex quantity with magnitude Γ and phase α. So the measurand is $\theta \equiv \Gamma \cos \alpha$. Suppose that the magnitude is known to be equal to b but that nothing is known about the phase. Before observation of any data, the distribution attributed to θ will be the arc-sine distribution with half width b, the arc-sine distribution being the distribution of the cosine of a uniform random angle. So the corresponding standard deviation is $b/\sqrt{2}$. But imagine that the magnitude is not known and that the figure b is an upper bound on Γ. It is not immediately clear how to define a corresponding uniform distribution of points on the disc, but if it is uniform with respect to Cartesian coordinates then the distribution being attributed to θ will have mean 0 and standard deviation $b/2$ while if it is uniform with respect to magnitude then the distribution being attributed to θ will have mean 0 and standard deviation $b/\sqrt{6}$. Whichever way is chosen, the standard deviation is markedly less than $b/\sqrt{2}$. The same analysis holds for the imaginary components. So knowing less about the magnitude of the complex quantity is interpreted as reducing uncertainty about both its real and imaginary components! This example involves only personal uncertainty, and the phenomenon it exhibits deserves further study. It suggests that the problem for Bayesian analyses is more extensive.

5. Principles and conclusions

There are many remarks that could be made about the issues raised in this paper, and an examination of every aspect of this problem is not possible here. So this section concludes the paper with a number of conclusions (C), principles (P) and suggestions (S), all given in the form of assertions.

5.1. *General principles*

The Bayesian approach to the analysis of measurement uncertainty uses the term 'uncertainty' in at least one of its senses, this being 'doubt about a potential fact' or 'lack of confidence in an hypothesis', i.e. personal uncertainty (epistemic uncertainty). This accords with the idea that post-data statements of assurance about the measured quantity are more valuable than pre-data statements of assurance about the measurement procedure. With this in mind, we suggest that:

C. Personal uncertainty is different in nature from stochastic uncertainty, and it cannot necessarily be treated the same way.
P. An increase in personal uncertainty about any quantity must not be permitted to decrease the stated measurement uncertainty. (Conversely, epistemic certainty that the stochastic variability is large means that the measurement uncertainty is surely large.)
S. Attributing a probability distribution to a varying quantity is a statement of knowledge (about a frequency distribution). Attributing a probability distribution to a constant is a statement of doubt (about the constant). The two ideas are fundamentally different.

5.2. *Nuisance parameters*

Difficulties can arise when a probability distribution is attributed to a parameter ϕ that is not functionally related to the measurand θ but is instead related to the stochastic uncertainty in the measurement, e.g. $\phi = \sigma$. The following statements might be made about such a nuisance parameter, ϕ.

C. Assigning a probability distribution to ϕ is not always meaningful for the task at hand.
P. A change in personal uncertainty about ϕ does not necessarily lead to a change in the measurement uncertainty of the estimate of θ.
S. The component of measurement uncertainty associated with uncertainty about ϕ should be expressed differently. The appropriate representation would employ worst-case values!

5.3. *Bayes' Theorem*

Bayes' Theorem is not controversial in itself, but there is debate about its use with unobservable probability distributions.

C. Bayes' Theorem is a re-expression of a symmetric relationship. There is no reason why it must be meaningful with asymmetric inputs.
P. With regard to the concept of probability, the calculus of probability, and Bayes' Theorem in particular, the question to ask is not "how do we interpret them?" but "what can they properly represent?"
S. Bayes' Theorem can be applied when probability distributions are constructed to represent uncertainties *of the same type*, e.g. only frequency distributions for data or only personal uncertainties about constants.

5.4. *Bayesian statistics*

Bayesian analyses involve considering each relevant unknown quantity to have a probability distribution, notably every unknown constant. In general, these distributions are unobservable and are intrinsically subjective.

C. One premise of a Bayesian analysis is that frequency behaviour and prior opinion can always be considered simultaneously. The examples given in this paper show that – if a Bayesian analysis conducted by one person is to be meaningful to another – this premise is invalid.

P. If one of the distributions appearing in Bayes' Theorem is a frequency distribution then all the other distributions must admit a frequency interpretation. In that case, each prior distribution must be interpretable as a frequency distribution of parameter values in many experiments.

S. Bayesian analysis is formulaic and deductive in style (e.g. "if the prior is this and the likelihood is this then the posterior is that"). However, its accuracy depends on its premises and the inputs to the formula. Inference that is based on abductive reasoning, e.g. a frequentist confidence interval, is not constrained by a single formula or its requirements.

5.5. *Final comments*

If the principles of probability and statistics are to be used to describe the accuracy of measurements then the concept of 'measurement uncertainty' must be amenable to this. Probability distributions are models of behaviour, either the behaviour of data in experiments, as with stochastic uncertainty, or the behaviour of individuals when choosing between possibilities, as with personal uncertainty. We cannot expect to successfully combine the behaviour of data with the behaviour of people by calling them both 'uncertainty' and then treating them as if they were the same thing.

References

1. B. D. Hall, *Metrologia* **45**, L5 (2008).
2. R. Willink, *Measurement Uncertainty and Probability* (Cambridge University Press, 2013).
3. R. B. Frenkel, *Metrologia* **46**, 661 (2009).
4. G. Ülkümen, C. R. Fox and B. F. Malle, *Journal of Experimental Psychology: General* **145**, 1280 (2016).
5. B. D. Hall, *Metrologia* **44**, L62 (2007).

F Pavese, A B Forbes, N F Zhang, A G Chunovkina (Eds.):
Advanced Mathematical and Computational Tools in Metrology and Testing XII
Series on Advances in Mathematics for Applied Sciences, Vol. 90

https://doi.org/10.1142/9789811242380_0031

Invertible calibration curves: Hyperbolic segments and hyperbolic splines

R. Willink[1,2]

[1] *Measurement Standards Laboratory of New Zealand,*
PO Box 31310, Lower Hutt 5040, New Zealand
[2] *University of Otago (Wellington), PO Box 7343, Wellington 6242, New Zealand*
E-mail: robin.willink@measurement.govt.nz

A function with the form $y = (a+ex)/(1+cx)$ is monotonic for all values of the parameters and can be inverted easily. This function describes a rectangular hyperbola, a segment of which might be used as a calibration curve for reading from x to y and, more importantly, from y to x. Methods are given for finding the hyperbolic segment that minimizes the sum of squared residuals to a set of points $\{(x, y)\}$ and for finding a spline of hyperbolic segments that minimizes an objective function, e.g. the sum of squared residuals. A method is also given for constructing a spline of hyperbolic segments that interpolates a set of points. An analysis of model error is given.

Keywords: Curve fitting; Spline; Least squares; Rectangular hyperbola.

1. Introduction

Calibrating an instrument that produces a reading y for a stimulus x involves recording values of y at known values of x, fitting a curve to the data, and inverting the curve to obtain a function $x(y)$ that can be used to estimate the stimulus from a future reading. The curve is fitted as $y(x)$ but is used as $x(y)$, so it would be advantageous to employ a function that permits inversion. Sometimes a quadratic function $y = a+bx+cx^2$ is used, with c allowing some non-linearity in the form of a parabola. Inversion gives the function $x = -b \pm \sqrt{\{[b^2 - 4(a-y)c]/(2c)\}}$, which is multi-valued and has a different form. However, unless theory suggests that the underlying function is quadratic, there is no reason in principle to use such an equation. Here, we describe instead the use of a segment of a hyperbola, for which the inverse is given explicitly by a function with the same form. We demonstrate the use of the hyperbola as a single function and as the form of the segment in a spline, in which there is greater flexibility and the possibility of points of inflexion.

2. A hyperbolic segment

A calibration curve might be chosen to have the form

$$y = \frac{a + ex}{1 + cx}, \tag{1}$$

in which non-linearity enters through a non-zero value of c. This function has the property that x is expressible in terms of y in the same form: the inverse of (1) is

$$x = \frac{a' + e'y}{1 + c'y} \qquad a' = -a/e \qquad e' = 1/e \qquad c' = -c/e.$$

The curve could therefore be used in both forward and inverse directions without difficulty. Rearranging these equations gives a single equation of the form $Ax + By + Cxy = 1$, which describes a rectangular hyperbola with asymptotes parallel to the coordinate axes. Thus, over the range of the data, the curve we consider is an arc of a hyperbola. We will say that, when only a finite range of values is involved, equation (1) describes a *hyperbolic segment*. Examples of hyperbolic segments are given in Figure 1a.

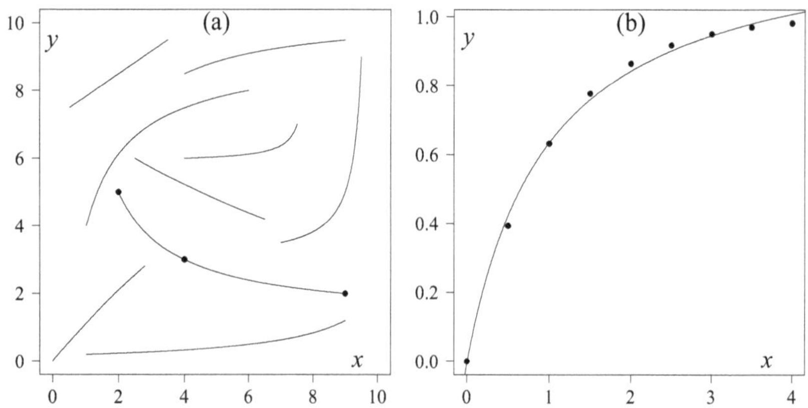

Fig. 1. (a) Example hyperbolic segments and (b) a least-squares fit to theoretical data

In the sense that a rectangular hyperbola and a quadratic function are both defined by three parameters, they are as flexible as each other. But this does not mean that they are equally suited to approximating the kinds of monotonic relationships that underlie calibration problems. The ability of a quadratic function to model non-monotonic behaviour will be wasted when that function is used as a calibration curve, while this is not the case for the hyperbolic function (1). The hyperbolic function has a pole at $x = -1/c$, which could be problematic if the curve is to be extrapolated. However, this point seems likely to be a long way from the x-values of interest, and the behaviour over the relevant range will be entirely satisfactory. Either side of the pole the function is monotonic, and far from the pole y approaches the value e/c. Thus, a hyperbolic segment can model behaviour asymptotic to a constant. Figure 1a illustrates the flexibility of this function.

In what follows we (i) give an explicit expression for the unique hyperbolic segment that passes through any three points, (ii) describe a simple way of finding the least-squares hyperbolic segment that fits $n > 3$ points, (iii) give an explicit expression for a hyperbolic spline that interpolates $n > 3$ points and has first derivatives matched at the points, allowing points of inflexion in the calibration curve, and (iv) describe the fitting of a least-squares spline of fewer segments to a set of points. Subsequently, we consider some relevant aspects of an appropriate error analysis.

An equation describing a hyperbolic segment can be written in various ways. The form given in (1) facilitates inversion and shows how the standard linear expression $y = a + ex$ is modified by non-zero c. However, when fitting to data in Sections 3, 5 and 6 we will use the representation

$$y = y_* + \frac{b(x - x_*)}{1 + c(x - x_*)}$$

for some point (x_*, y_*) through which the curve must pass, while in Section 4 we will use the representation $y = a + bx/(1 + cx)$.

3. The interpolating hyperbolic segment

There is a unique hyperbolic segment interpolating any three points that are strictly monotonic in both x and y. Let the points be (x_1, y_1), (x_2, y_2), (x_3, y_3) with $x_1 < x_2 < x_3$ and $y_1 < y_2 < y_3$. Then the hyperbola is

$$y = y_1 + \frac{b(x - x_1)}{1 + c(x - x_1)}$$

$$\text{with} \quad b = \frac{y_3 - y_1}{x_3 - x_1} \cdot \frac{y_2 - y_1}{x_2 - x_1} \cdot \frac{x_3 - x_2}{y_3 - y_2} \quad \text{and} \quad c = \frac{\frac{y_3 - y_1}{x_3 - x_1} - \frac{y_2 - y_1}{x_2 - x_1}}{y_2 - y_3}.$$

For example, the marked segment on Figure 1a connects the points $(2, 5)$, $(4, 3)$ and $(9, 2)$, and has $b = -15/7$ and $c = 4/7$.

4. The least-squares hyperbolic segment

Let us fit a hyperbolic segment to data $(x_1, y_1), \ldots, (x_n, y_n)$ by weighted least squares with residuals in y. The representation of (1) used is

$$y = a + \frac{bx}{1+cx} \qquad b \equiv e - ac. \tag{2}$$

Unless otherwise stated, all summation is over $i = 1, \ldots, n$.

We seek the values of the parameters a, b and c that minimize

$$O = \sum w_i \left\{ y_i - \left(a + \frac{bx_i}{1+cx_i} \right) \right\}^2$$

where w_i is the weight to be associated with the squared residual at the point (x_i, y_i). No analytical solution seems available, so the method of solution will involve a numerical search. However, we only require a line search because for any value of c the solution to the resulting two-parameter problem can be found analytically if we write

$$z_i = z_i(c) = \frac{x_i}{1+cx_i}$$

to obtain a linear function $y = a + bz$. So $O = \sum w_i \{y_i - (a + bz_i)\}^2$ and, for any set of z_i values, the values of b and a that minimize O are the weighted least-squares solutions for the straight line

$$b = \frac{\sum w_i (y_i - \tilde{y})(z_i - \tilde{z})}{\sum w_i (z_i - \tilde{z})^2} \qquad \tilde{y} \equiv \frac{\sum w_i y_i}{\sum w_i} \qquad \tilde{z} \equiv \frac{\sum w_i z_i}{\sum w_i}$$

$$a = \tilde{y} - b\tilde{z}.$$

Thus, we find the value of c minimizing $O = O(c)$, record the corresponding values of b and a, and set $e = b + ac$. A suitable starting value for searching on c would be the value of c minimizing the weighted sum of squared residuals with the quadratic function $y = a + bx(1 - cx)$, which describes a curve very similar to (2) when $|c|$ is small. This value is

$$c = \frac{S(w,y,x)S(w,x,x^2) - S(w,y,x^2)S(w,x,x)}{S(w,y,x)S(w,x^2,x^2) - S(w,y,x^2)S(w,x,x^2)}$$

where $S(w,u,v) \equiv \sum w_i (u_i - \tilde{u})(v_i - \tilde{v})$ with $\tilde{u} \equiv \sum w_i u_i / \sum w_i$ and $\tilde{v} \equiv \sum w_i v_i / \sum w_i$. Alternatively, a simple starting value of $c = 0$ would suffice. Figure 1b shows the least-squares fit to the points (x_i, y_i) given by $x_i = (i-1)/2$ and $y_i = 1 - e^{-x_i}$ for $i = 1, \ldots, 9$. The parameters of the hyperbolic segment are $a = -0.0082$, $b = 1.310$ and $c = 1.040$.

5. An interpolating hyperbolic spline

Now we consider the construction of a spline of hyperbolic segments that is continuous, has continuous first derivative, and passes through the points $(x_1, y_1), \ldots, (x_n, y_n)$ with $x_1 < \cdots < x_n$ and $y_1 < \cdots < y_n$. This *hyperbolic spline*[1] will be made up of $n-1$ segments between the pairs of adjacent x_k values. Similarly, the inverse function will be a hyperbolic spline made up of $n-1$ hyperbolic segments between the pairs of adjacent y_k values. Both functions will be strictly monotonic. The points at which the segments of a spline meet are called knots. (The subscript k will be used instead of i when a point is a knot.) So we seek a function $y(x)$ of the form

$$y = y_k + \frac{b_k(x - x_k)}{1 + c_k(x - x_k)} \qquad x_k \le x \le x_{k+1} \qquad k = 1, \ldots, n-1 \qquad (3)$$

such that the function is continuous and has continuous first derivative,

$$\frac{dy}{dx} = \frac{b_k}{\{1 + c_k\,(x - x_k)\}^2} \qquad x_k \le x \le x_{k+1} \qquad k = 1, \ldots, n-1.$$

Let us define

$$s_k \equiv \frac{y_{k+1} - y_k}{x_{k+1} - x_k} \qquad k = 1, \ldots, n-1,$$

so s_k is the average slope in the interval from (x_k, y_k) to (x_{k+1}, y_{k+1}). Equating the first derivatives at the knots gives the simple but unusual recursion $b_{k+1} = s_k^2/b_k$. Putting $x = x_{k+1}$ and $y = y_{k+1}$ in (3) gives $c_k = b_k/(y_{k+1} - y_k) - 1/(x_{k+1} - x_k)$, so

$$c_k\,(x_{k+1} - x_k) = \frac{b_k}{s_k} - 1.$$

Thus, for any value of the derivative at the initial point $b_1 = dy/dx|_{x=x_1}$, the unique interpolating hyperbolic spline is given by

$$y = y_k + \frac{b_k(x - x_k)}{1 + \left(\frac{b_k}{s_k} - 1\right)\frac{x - x_k}{x_{k+1} - x_k}} \qquad x_k \le x \le x_{k+1} \qquad b_{k+1} = \frac{s_k^2}{b_k} \qquad (4)$$

for $k = 1, \ldots, n-1$. By symmetry about the line of slope 1, the inverse of this function is the hyperbolic spline

$$x = x_k + \frac{(y - y_k)/b_k}{1 + \left(\frac{s_k}{b_k} - 1\right)\frac{y - y_k}{y_{k+1} - y_k}} \qquad y_k \le y \le y_{k+1} \qquad (5)$$

for $k = 1, \ldots, n-1$. It remains to choose b_1.

A similar method for constructing such a spline has been obtained by Martin *et al*,[1] who give their analysis by referring to a *Möbius transformation* $f(t) = (A + Bt)/(C + Dt)$ and associated 2×2 matrices. They consider $n+1$ points numbered from $k = 0, \ldots, n$ and they write λ_k where we have b_{k+1}/s_{k+1}. So the dimensionless quantity λ describes the ratio of the slope at the beginning of a segment to the average slope over that segment, and setting $\lambda_0 = 1$ means that the first segment is straight, i.e. $c_1 = 0$. Martin *et al* suggest that if the derivative at x_1 is not known then it might be chosen to minimize the curvature of the function, and they measure this by the quantity $\Lambda \equiv \max\{\lambda_0, \ldots, \lambda_{n-1}, \lambda_0^{-1}, \ldots, \lambda_{n-1}^{-1}\}$ because curvature is associated with values of λ that differ from 1. In our notation, their suggestion is to set $b_1 = s_1\sqrt{\max\{v_k\} \times \min\{v_k\}}$ with $v_1 \equiv 1$, $v_k \equiv v_{k-1}(s_{k-1}/s_k)^{(-1)^k}$ for $k = 2, \ldots, n-1$. This measure has the property that the same function would be obtained whether the independent variable was x or y. The function that would result from swapping the x and y values therefore describes the same relationship as the inverse function $x(y)$ given by (5). This approach strengthens the concept of invertibility to one that involves a concept of symmetry and an invariance.

Here, we suggest (i) scaling to obtain a variable $\acute{y}$ that equalizes the extent of the function in the x and y directions and then (ii) minimizing the sum of a suitable statistic of the function $\acute{y}(x)$ and the same statistic of the inverse function $x(\acute{y})$. The result will be a function with the same properties of uniqueness and inversion as the spline of Martin *et al*. We will minimize the integrated squared second derivative, as is common in the regularization of curves. This measure of smoothness responds to all of the function, not just the single values of λ.

The spline with the y-axis scaled so that the average slope is 1 uses the values $\acute{y}_k = y_1 + r(y_k - y_1)$ for $k = 1, \ldots n$ with

$$r \equiv \frac{x_n - x_1}{y_n - y_1}.$$

The corresponding value of b_k is $\acute{b}_k \equiv rb_k$. For the spline with scales equalized in this way, $\acute{y}(x)$, the integrated squared second derivative is

$$\begin{aligned}\phi_{\acute{y}x} &\equiv \int_{x_1}^{x_n} \left(\frac{-2\acute{b}_k c_k}{\{1 + c_k(x - x_k)\}^3}\right)^2 \\ &= \frac{4}{5}\sum_{k=1}^{n-1} \frac{r^2 b_k^2}{x_{k+1} - x_k}\left(\frac{b_k}{s_k} - 1\right)\left\{1 - \left(\frac{s_k}{b_k}\right)^5\right\}.\end{aligned}$$

For the inverse spline, $x(\acute{y})$, the quantity $\acute{b}_k$ is replaced by its reciprocal, and the corresponding integral is found to be

$$\phi_{x\acute{y}} = \frac{4}{5}\sum_{k=1}^{n-1} \frac{1}{r^3(y_{k+1}-y_k)b_k^2}\left(\frac{s_k}{b_k}-1\right)\left\{1-\left(\frac{b_k}{s_k}\right)^5\right\}.$$

Therefore, the quantity minimized to obtain the unique smoothest spline is $\phi \equiv \phi_{\acute{y}x} + \phi_{x\acute{y}}$, and the associated value of b_1 is found numerically.

Figure 2 gives a set of data and the parameters of the smoothest hyperbolic spline interpolating these data. Figure 2 also shows this spline and two other interpolating hyperbolic splines with different initial slopes. (For our spline, $b_1 = 2.84$. The corresponding value for the spline of Martin *et al* is $b_1 = 2.98$. The two splines are not distinguishable on Figure 2.)

k	x_k	y_k	b_k	s_k	c_k
1	11	0.31	2.84	2.07	0.37
2	12	2.38	1.51	1.00	0.51
3	13	3.38	0.66	0.73	-0.09
4	14	4.11	0.80	0.91	-0.12
5	15	5.02	1.03	1.11	-0.07
6	16	6.13	1.20	1.48	-0.19
7	17	7.61			

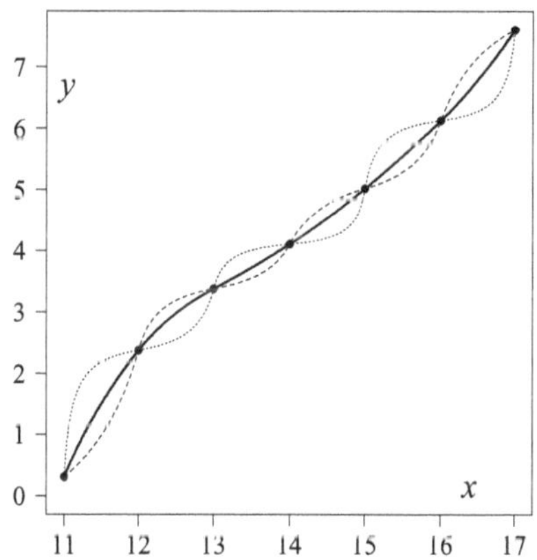

Fig. 2. The dataset, the parameters of the smoothest interpolating hyperbolic spline, the spline (solid) and those with initial gradient $b_1 = 1$ (dashed) and $b_1 = 20$ (dotted)

6. The least-squares hyperbolic spline

Now we consider the construction of a monotonic least-squares hyperbolic spline to n data points $(x_1, y_1), \ldots, (x_n, y_n)$ using m segments with knots at $x_{(1)}, \ldots, x_{(m+1)}$, where $x_{(k)}$ does not need to be an x_i value. As before, $x_1 < \cdots < x_n$, but now we no longer require that $y_1 < \cdots < y_n$: monotonicity is imposed on the fitted function, not on the y-data. The knots are free, but the spline obtained over $[x_1, x_n]$ will not depend on the x-coordinates of the end knots, so without loss of generality we set $x_{(1)} = x_1$ and $x_{(m+1)} = x_n$. The data are indexed by $i = 1, \ldots, n$ and the knots by $k = 1, \ldots, m+1$.

The spline will be given by

$$y = y_{(k)} + \frac{b_k(x - x_{(k)})}{1 + c_k(x - x_{(k)})} \qquad x_{(k)} \le x \le x_{(k+1)} \qquad k = 1, \ldots, m.$$

A hyperbolic spline on m segments with knots having fixed x-coordinates has $m+2$ parameters. So, with the x-coordinates of the interior knots being free, our spline is defined by a vector of $2m+1$ unknowns, $\mathbf{z} \equiv (z_1, \ldots, z_{2m+1})$, which we can regard as a vector of parameters. These parameters can unambiguously define (i) the derivative at $x_{(1)}$, (ii) the values of x at the interior knots $x_{(2)}, \ldots, x_{(m)}$ and (iii) the values of the function at the knots $y(x_{(1)}), \ldots, y(x_{(m+1)})$. The corresponding spline can then be constructed using (4). For any candidate vector of parameters $\mathbf{z}$, the spline can be evaluated at $x_1, \ldots, x_n$ to give the fitted values $\hat{y}_1(\mathbf{z}), \ldots, \hat{y}_n(\mathbf{z})$ from which the objective function

$$O(\mathbf{z}) = \sum \{w_i\,(y_i - \hat{y}_i(\mathbf{z}))\}^2$$

can be evaluated. A numerical search on $\mathbf{z}$ gives the least-squares spline.

For efficiency and numerical stability, the value of each parameter will be of order 1 and will affect the spline monotonically. For simplicity in obtaining a solution, the parameters are unconstrained. However, the x-coordinates of the knots and the y-coordinates at these knots must be increasing, so unbounded parameter values are transformed into positive quantities using exponential functions. Suitable choices are parameters $z_1, \ldots, z_{2m+1}$ that define the spline through the equations

$$b_1 = \exp(z_1)\frac{y_n - y_1}{x_n - x_1}$$

$$x_{(k)} = x_{(k-1)} + t_k\left(x_{(m+1)} - x_{(1)}\right) \qquad k = 2, \ldots, m$$

$$y_{(1)} = y_1 + z_{m+1}(y_n - y_1) \qquad t_j \equiv \frac{\exp(z_j)}{1 + \exp(z_j)}$$

$$y_{(m+1)} = y_{(1)} + \exp(z_{m+2})(y_n - y_1)$$

$$y_{(k)} = y_{(k-1)} + t_{m+1+k}\left(y_{(m+1)} - y_{(1)}\right) \qquad k = 2, \ldots, m.$$

Also, a suitable starting vector for estimating these parameters is

$$(z_1, \ldots, z_{2m+1}) = (0, q_2, q_3, \ldots, q_m, 0, 0, q_2, q_3, \ldots, q_m)$$

with $q_k \equiv -\log(m+1-k)$. These equations show that the derivative at the first knot is kept positive, the x-coordinates of the interior knots are kept in order and their starting positions are evenly spaced between the limits $x_{(1)} = x_1$ and $x_{(m+1)} = x_n$. Also, the values of y at the knots are kept in order and their starting values are evenly spaced over $[y_1, y_n]$.

A set of 7 data points was generated by the mechanism '$y = 0.5 + 0.5 \log x$ + random error' and a set of 9 points was generated by the mechanism '$y = \{1 + \exp(6 - x)\}^{-0.5}$ + random error'. Figure 3 shows the underlying true logarithmic and sigmoidal functions (dashed lines), and gives the data. Figure 3 also shows the least-squares hyperbolic splines with $m = 2$ segments and $m = 3$ segments respectively (solid lines), and gives the values of the parameters of these splines, $\{z_j\}$. (Spline (b) exhibits some overfitting. In practice, there would be more data-points per knot.)

(a)

x_i	1.00	2.00	3.00	4.00	5.00	6.00	7.00
y_i	0.47	0.84	1.00	1.17	1.32	1.44	1.48

z_i	3.558	-1.616	0.000	0.023	-0.636

b_1	$x_{(2)}$	$y_{(1)}$	$y_{(2)}$	$y_{(3)}$
0.599	1.995	0.470	0.828	1.504

(b)

x_i	1.00	2.00	3.00	4.00	5.00	6.00	7.00	8.00	9.00
y_i	0.06	0.08	0.18	0.31	0.51	0.74	0.86	0.95	1.00

z_i	0.401	0.044	-1.190	-0.015	0.002	0.061	-0.270

b_1	$x_{(2)}$	$x_{(3)}$	$y_{(1)}$	$y_{(2)}$	$y_{(3)}$	$y_{(4)}$
0.047	5.088	6.000	0.046	0.538	0.739	1.002

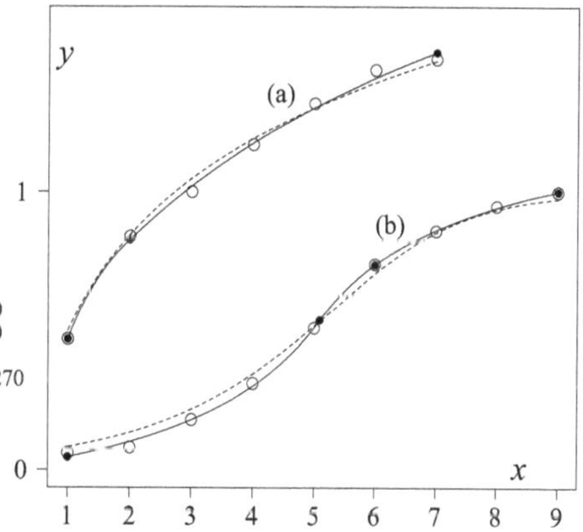

Fig. 3. Exact datasets and calculated figures of least-squares hyperbolic splines (solid) and the true functions (dashed) for (a) $y = 0.5 + 0.5 \log x$ and (b) $y = 1/\sqrt{(1 + e^{6-x})}$.

7. Error analysis – for interpolating functions

There are at least two sources of error to consider: one relating to error in the data and the other to error associated with the form of the interpolating function, which can be called *model-form error*. The sensitivity of the fitted function to errors in the data can be assessed by perturbing the data. So here we only consider the model-form error, and we show how it might be assessed with the interpolating hyperbolic spline. The principles apply for other forms of interpolating function also. Interpolated data are treated as being exact, in which case there is no model-form error at the data-points. The error is then restricted to the intervals between the points, and it can be expected to be greatest midway between the adjacent x_k values. The following method for assessing the size of the error is proposed.

Suppose that $y(x)$ is a spline interpolating the data $(x_1, y_1), \ldots, (x_n, y_n)$, and imagine being given an additional datum $(\breve{x}, \breve{y})$ that is consistent with and has the same status as an original datum. The quantity $y(\breve{x}) - \breve{y}$ must be interpretable as an error caused solely by the choice of the form of the interpolating function. This suggests that, to assess the size of the model-form error in $y(x)$ in the broad vicinity of a point x_k, we might remove the point (x_k, y_k) from the dataset, obtain a new spline $y_{\backslash k}(x)$ and calculate a deviation function $g_{\backslash k}(x) \equiv y_{\backslash k}(x) - y(x)$. This function would slowly vary between zeroes at the remaining x-values and would tend to be too large in the interval $[x_{k-1}, x_{k+1}]$, especially at x_k, where the model-form error is zero. With these things in mind, we suggest that each datum be held out in turn and that the function $f(x) = \max_k\{|g_{\backslash k}(x)| : k = 1, \ldots, n\}$ be calculated. Clearly, this would be a conservative guide to the model-form error near the x_k values, so it would be appropriate to take into account the anchoring property of each datum. Therefore, the function suggested for describing the standard uncertainty in $y(x)$ at x due to model-form error is

$$u\left(y(x)\right) \equiv \left\{4\left(\frac{x - x_k}{x_{k+1} - x_k}\right)\left(1 - \frac{x - x_k}{x_{k+1} - x_k}\right)\right\} f(x) \qquad x_k \le x \le x_{k+1}$$

for $k = 1, \ldots, n-1$. The multiplier of $f(x)$ is a quadratic function that is equal to zero at x_k and x_{k+1} and has its greatest value, 1, at $(x_k + x_{k+1})/2$. So $u\left(y(x)\right)$ takes the value zero at the knots but rises to $f(x)$ midway between each pair of adjacent knots. Figure 4a shows the functions $u\left(y(x)\right)$ and $f(x)$ for the data of Figure 2.

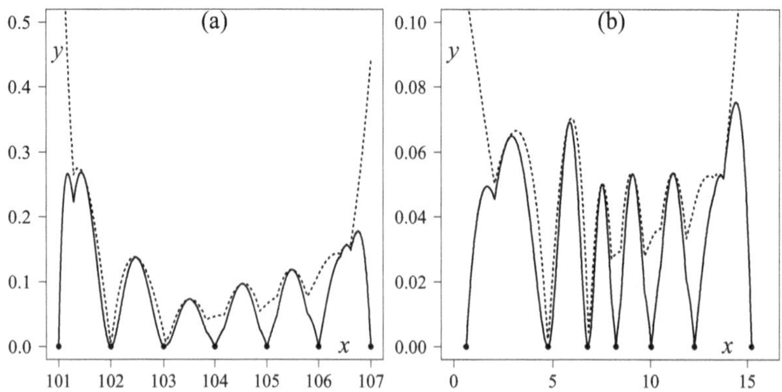

Fig. 4. Functions describing the standard uncertainty due to error from using a hyperbolic spline as a calibration curve with the data of Figure 2: (a) solid line $u\left(y(x)\right)$ and dotted line $f(x)$, (b) solid line $u\left(x(y)\right)$ and dotted line $f(y)$

The function $u(y(x))$ defined in this way is intended to represent the component of standard uncertainty in y at individual values of x point by point. (It is not to be interpreted as the border of an uncertainty region applicable to every point simultaneously.) Because the calibration curve will often be used as $x(y)$, it will often be more appropriate to calculate $u(x(y))$. This function can be formed in exactly the same way but using the inverse spline $x(y)$ and the y_k values. Figure 4b shows the functions $u(x(y))$ and $f(y)$ for the data of Figure 2.

8. A practical study – inversion of a high-order polynomial

A hyperbolic spline can be constructed to mimic a reference function used in the International Temperature Scale of 1990 (ITS-90).[2] The definition of the scale in the region from 273.15 K to 1234.73 K involves the use of a platinum resistance thermometer. A ratio of resistances W is related to temperature T by a reference function that is a ninth-order polynomial $W(T)$. Use of the scale requires application of the inverse function $\acute{T}(W)$, which is a ninth-order polynomial consistent with $W(T)$ to within 0.13 mK, i.e. $|\acute{T}(W(T)) - T| \leq 0.13$ mK, as shown by the dashed line in Figure 5.

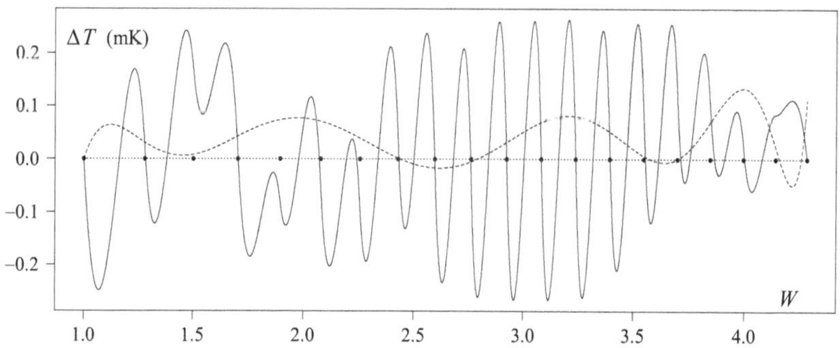

Fig. 5. Deviations in measured temperatures with hyperbolic spline and ITS-90 reference functions: $\Delta T = T_h(W) - \acute{T}(W)$ (solid) and $\Delta T = \acute{T}(W(T)) - T$ (dashed)

An optimization method was configured to approximate these functions by a hyperbolic spline $T_{\mathrm{h}}(W)$, which can be inverted exactly. Over this range, the difference between $T_{\mathrm{h}}(W)$ and $\acute{T}(W)$ is less than 0.3 mK, i.e. $|T_{\mathrm{h}}(W) - \acute{T}(W)| \leq 0.3$ mK, as shown by the solid line. The figure of 0.3 mK is comparable to the summary figure of 0.5 mK that can be associated with non-uniqueness in ITS-90 from 'subrange inconsistency'.[2] Table 1 gives defining parameters of this spline and the temperatures at the knots.

Table 1. Figures for a spline $T_{\rm h}(W)$ approximating the reference function $\acute{T}(W)$.

k	x_k (W) (exact)	z_k (exact)	y_k (T) (rounded)	k	x_k (W) (exact)	z_k (exact)	y_k (T) (rounded)
1	1.000	0.00000000	273.1600	11	2.923	0.54338023	795.7669
2	1.281	0.07405322	344.3822	12	3.082	0.59280680	843.3039
3	1.502	0.13345332	401.5114	13	3.239	0.64247327	891.0716
4	1.704	0.18868003	454.6268	14	3.394	0.69240328	939.0928
5	1.895	0.24175344	505.6712	15	3.547	0.74261846	987.3882
6	2.078	0.29341389	555.3567	16	3.698	0.79313641	1035.9749
7	2.255	0.34416604	604.1686	17	3.847	0.84396923	1084.8644
8	2.428	0.39455230	652.6286	18	3.995	0.89547411	1134.4002
9	2.596	0.44425918	700.4352	19	4.141	0.94731316	1184.2575
10	2.761	0.49386675	748.1463	20	a	1.00000000	1234.9301

Note: $a = 4.28642053$ (exact)

9. Concluding comments

A calibration curve is often formed as $y(x)$ but used as $x(y)$. The monotonic behaviour of a rectangular hyperbola and the ability to invert this function make the hyperbolic segment suitable for use in calibration tasks. This article has described methods for forming a calibration curve using a single hyperbolic segment or a spline of segments, either interpolating the data or fitted to the data by the principle of least squares, with residuals in the y dimension. The analysis could easily be adapted to use residuals in the x dimension instead.

The underlying equation, $Ax + By + Cxy = 1$, can be generalized to accommodate more variables, in particular, with the equation $Ax + By + Cz + Dxy + Exz + Fyz + Gxyz = 1$. This new equation describes a rectangular hyperboloid, with each pair of variables being related in the way described in this paper. For example, if z is fixed, the relationship between x and y is the relationship described here. Therefore, calibration functions in more than two dimensions can be constructed using the same principles. Also, an invertible spline with more derivatives matched might be built around the more difficult equation $y^g = (a + ex^h)/(1 + cx^h)$.

References

1. R. Martin, K. Lever and J. McCarthy, Hyperbolic splines and non-linear distortion, *IEEE Trans. Signal Process.* **48**, 1825 (2000).
2. J. V Nicholas and D. R .White, *Traceable Temperatures*, 2nd edn. (Wiley, Chichester, 2001).

F Pavese, A B Forbes, N F Zhang, A G Chunovkina (Eds.):
Advanced Mathematical and Computational Tools in Metrology and Testing XII
Series on Advances in Mathematics for Applied Sciences, Vol. 90

https://doi.org/10.1142/9789811242380_0032

A logical contradiction in the representation of constants by probability distributions: Premises and implications

R. Willink[1,2]

[1] *Measurement Standards Laboratory of New Zealand,
PO Box 31310, Lower Hutt 5040, New Zealand*
[2] *University of Otago (Wellington), PO Box 7343, Wellington 6242, New Zealand*
E-mail: robin.willink@measurement.govt.nz

A probability distribution is sometimes said to encode a set of information about a constant quantity. If that is the case then there will be some probability distribution that represents the sum of such sets of information. A derivation is given of the appropriate method for obtaining this distribution when its components represent disjoint sets of information. This derivation avoids an incomplete step that was part of an earlier analysis (*Metrologia*, **43**, 12–20, 2006). The result leads to a logical contradiction: the combined distribution is sensitive to the arbitrary ordering of two legitimate steps in its generation. The only cause of this can be fault in the premise of the procedure, which is the premise that a set of information or 'state of knowledge' about a quantity can be represented by a unique probability distribution. This result serves as a conclusive counter-example to the notion of 'logical probability'. It has implications for the development of a scientific method for the evaluation of uncertainty in measurement. The argument is extended verbally to also disqualify the notion that a distribution can represent the state of knowledge approximately.

Keywords: Logical probability; Bayesian statistics; Combination of data.

1. Introduction

One current approach to the analysis of experimental data permits the analyst to attribute a probability distribution to an unknown quantity with a unique constant true value, as in a Bayesian statistical analysis. This practice is appealing and, for many, it accords with everyday thinking. For example, a probability distribution is being attributed to the year of manufacture of an antique clock when someone says "For me, the probability that the clock was made before 1900 is about 0.4 and the probability that it was made before 1800 is about 0.1." Here, two probability statements are being made about the unknown year, Y, which is a quantity with a fixed but unknown value. So the constant Y is being regarded as a random variable: it is being considered to have a known probability distribution, albeit one that is described in minimal detail.

The idea that constant physical quantities can be treated in this way is an essential part of methods of uncertainty analysis advocated in literature surrounding the revision of the *Guide to the Expression of Uncertainty in Measurement.*[1] The 'input quantities' to a calculation are identified with probability distributions, and these distributions are combined to obtain the distribution for the measurand. The implied claim is that each contributing distribution is meaningful and adequate for the purpose at hand and that, in any given situation, there is some final distribution that would be accurate, in some sense. This idea finds voice in statements such as[2]

> A probability distribution (on the set of possible values of the measurand) provides a complete characterization of measurement uncertainty ...

A statement with this implication adheres to the idea of 'logical probability', where a probability is seen as an objective degree to which one proposition implies another, so it might be regarded as a rational degree of belief.[3]

The scientific basis of this practice of attributing a probability distribution to a constant is unclear. Instead, its acceptance appears to be based on the use of probability in other spheres of life. In fact, there seems to be an assumption in metrological literature that the notion of probability not just *can be* but *should be* employed in this way, and that it is normal to do so. If such an assumption be scientifically correct then the subsequent application of the laws of probability *cannot lead to a logical contradiction.* In this paper, we observe such a contradiction: two equivalent methods for combining probability distributions for a constant give results that differ, both quantitatively and qualitatively. It seems difficult to escape the conclusion that the concept of 'logical probability' is invalid.

Before proceeding, we distinguish two assumptions that relate to this practice of attributing a distribution to an unknown constant quantity X:

(i) "The information available to someone about X can be adequately represented as a probability distribution for X."
(ii) "People with the same information about X should assign the same probability distribution to X."

These assumptions constitute respective claims that (i) there is a probability distribution that adequately represents a set of information and that (ii) this representation is complete and unique. There might be some debate in metrology as to whether the methodology in question, the 'uncertainty approach',[4] makes the second claim. However, we proceed on the basis that it does. Nevertheless, we shall relax this assumption in Section 4, where it will be seen that the essential conclusion is unaffected. So the relevant premise is taken to be that *observers with the same information about an unknown quantity can and should assign the same pdf to that quantity.* This is the essence of 'logical probability' and the 'uncertainty approach'.

2. Combining unrelated probability distributions for a constant

The combination of measurement results for a constant quantity X is an important task. In the 'uncertainty approach', the individual measurement results will be different probability distributions that have been attributed to X, each distribution representing the information available to the contributing party. The laws of probability will be used to combine the contributing distributions to obtain the distribution that accurately represents all the information about X.

Consider the simple case in which the contributing distributions represent mutually exclusive sets of information. In this section, we show that if $p_i(\cdot)$ is the probability density function (pdf) representing the ith set of information about X then the unique pdf that represents the combined set of n unrelated pdfs is[5]

$$p(x) = \frac{\prod_{i=1}^{n} p_i(x)}{\int_{-\infty}^{\infty} \prod_{i=1}^{n} p_i(z)dz} \qquad -\infty < x < \infty. \tag{1}$$

Thus, (1) is applicable for the combination of n independently formed 'state of knowledge' distributions – if the idea of such distributions is valid.

2.1. *Example*

Suppose two scientists independently measure a quantity X using a device that incurs normal random error with known mean 0 and known standard deviation σ. Let the results be x_1 and x_2. The first scientist decides that X may be thought of as a normal random variable with mean x_1 and standard deviation σ, which has pdf

$$p_1(x) \propto \exp\left(\frac{-(x-x_1)^2}{2\sigma^2}\right) \qquad -\infty < x < +\infty. \tag{2}$$

Likewise, the second scientist attributes X the normal distribution with mean x_2 and standard deviation σ, with pdf

$$p_2(x) \propto \exp\left(\frac{-(x-x_2)^2}{2\sigma^2}\right) \qquad -\infty < x < +\infty. \tag{3}$$

Imagine that the scientists combine their information. Together their state of knowledge is the same as that of a single scientist who has a sample of two, $\{x_1, x_2\}$. Accordingly, they attribute to X the normal distribution with mean $\bar{x} = (x_1 + x_2)/2$ and standard deviation $\sigma/\sqrt{2}$, which has pdf

$$p(x) \propto \exp\left(\frac{-(x-\bar{x})^2}{\sigma^2}\right) \qquad -\infty < x < +\infty. \tag{4}$$

The right hand side of (4) is easily shown to be proportional to the product of the right hand sides of (2) and (3). So we see that, in this case, (1) is appropriate. (For another example in which pdfs are multiplied and the result is normalized, see Elster.[6])

2.2. *Proof that (1) is correct if the premise is correct*

The derivation of (1) *for this context* is given in stages. The corresponding equation for $n = 2$ discrete distributions is obtained first, and this equation is then used to obtain the result for $n = 2$ continuous distributions. Subsequently, this result is generalized to apply for all n.

Let $\acute{X}_1$ and $\acute{X}_2$ be the unknown true values of two quantities on a discrete scale with some unit of resolution δ. Imagine that two scientists, working independently from disjoint sets of information, assign the probability mass functions $p_1[x]$ and $p_2[x]$ to their respective unknowns. (We write $p[\cdot]$ for a probability mass function (pmf) and $p(\cdot)$ for a pdf.) Suppose the scientists combine their information. The two true values are seen as being different and the sets of information are disjoint. Therefore, $\acute{X}_1$ and $\acute{X}_2$ are regarded as independent random variables and the joint pmf for $\acute{X}_1$ and $\acute{X}_2$ is found by multiplication. Therefore, for these scientists,

$$\Pr(\acute{X}_1 = x_1 \cap \acute{X}_2 = x_2) = p_1[x_1]\, p_2[x_2].$$

Then, by the law of conditional probability $\Pr(\mathrm{A}|\mathrm{B}) = \Pr(\mathrm{A} \cap \mathrm{B})/\Pr(\mathrm{B})$, the probability that $\acute{X}_1 = a$ and $\acute{X}_2 = a$ on the condition that $\acute{X}_1 = \acute{X}_2$ is

$$\Pr\left(\acute{X}_1 = \acute{X}_2 = a \mid \acute{X}_1 = \acute{X}_2\right) = \frac{p_1[a]\, p_2[a]}{\sum_z p_1[z]\, p_2[z]},$$

which uses the diagonal elements of the matrix of joint probabilities.

Imagine that the scientists then receive the information that the quantities they have been studying are the same, i.e. that $\acute{X}_1 = \acute{X}_2$, and suppose that they now call this quantity $\acute{X}$. The pmf they attribute to $\acute{X}$ must then be

$$p[x] \equiv \Pr\left(\acute{X} = x\right) = \frac{p_1[x]p_2[x]}{\sum_z p_1[z]p_2[z]} \qquad x = \pm 0, \delta, 2\delta, \ldots. \tag{5}$$

However, the information about $\acute{X}$ that they now possess is exactly the same as the information they would have had if they had known that $\acute{X}_1 = \acute{X}_2$ at the outset. It follows that (5) is the pmf that would be obtained when combining two unrelated pmfs for the single quantity $\acute{X}$. This expression is (1) for the case of $n = 2$ discrete contributions.

To reach the same conclusion in an equivalent way, consider the variables $\acute{W}_1 = (\acute{X}_1 + \acute{X}_2)/2$ and $\acute{W}_2 = (\acute{X}_1 - \acute{X}_2)/2$. The event that $\acute{W}_1 = w$ and $\acute{W}_2 = 0$ is the event that both $\acute{X}_1 = w$ and $\acute{X}_2 = w$. So the joint pmf of $\acute{W}_1$ and $\acute{W}_2$ at the values w and 0 is given by $p[w, 0] = p_1[w]p_2[w]$, and the pmf of $\acute{W}_1$ on the condition that $\acute{W}_2 = 0$ is

$$\Pr\left(\acute{W}_1 = w \mid \acute{W}_2 = 0\right) = \frac{p_1[w]p_2[w]}{\sum_z p_1[z]p_2[z]}.$$

The information that the quantities studied are the same is the information that $\acute{W}_2 = 0$, in which case $\acute{W}_1 = \acute{X}$ and (5) is obtained.

To develop the result for two continuous distributions we write (5) as

$$\frac{p[x]}{\delta} = \frac{p_1[x]/\delta \times p_2[x]/\delta}{\sum_z p_1[z]/\delta \times p_2[z]/\delta \times \delta}. \tag{6}$$

Let X_1 and X_2 be measurands existing on a continuous scale and let them be represented by unrelated pdfs $p_1(x)$ and $p_2(x)$. Suppose these pdfs are discretized by integrating over contiguous units of extent δ, as in a rounding process. The resulting pmfs can be regarded as the unrelated pmfs of two discrete variables $\acute{X}_1$ and $\acute{X}_2$, for which (5) and (6) must hold. If the unit of discretization δ is reduced towards zero then $p_1[x]/\delta$ approaches $p_1(x)$ by definition of the concept of a pdf, $p_2[x]/\delta$ approaches $p_2(x)$ similarly, and (6) approaches the result that must be applicable in the continuous case for the random variables X_1 and X_2. This is the result

$$p(x) = \frac{p_1(x)\, p_2(x)}{\int_{-\infty}^{\infty} p_1(x)\, p_2(x)\, dx}, \tag{7}$$

which is a re-expression of (1) for $n = 2$. Therefore, (7) is the unique pdf to be obtained when combining two unrelated, i.e. independently formed, state-of-knowledge distributions for a single quantity X.

Equation (7) is easily extended to higher dimensions. We envisage a group of n experts being formed sequentially, with each expert bringing their pdf, $p_i(x)$, for their unknown, X_i. The first two experts meet and, when informed that $X_1 = X_2$, obtain the pdf (7) for the quantity $X \equiv X_1 = X_2$. Let us denote this pdf by $p_{1,2}(x)$. The third expert brings the pdf $p_3(x)$ for X_3 and, when the trio are informed that X_3 is equal to X, the combination of $p_3(x)$ with $p_{1,2}(x)$ using (7) gives

$$p_{1,2,3}(x) \quad \propto \quad p_{1,2}(x)\, p_3(x) \quad \propto \quad p_1(x)\, p_2(x)\, p_3(x).$$

Then $p_4(x)$ is combined with $p_{1,2,3}(x)$, and so on. The pdf finally obtained, $p_{1,\ldots,n}(x)$, is given by (1). All the available information has been used, each piece of information has been used only once, and nothing more has been assumed. So the resulting probability distribution must fully, accurately and uniquely represent the total set of information held about X. Straightforward analysis has shown that, *if the contributing pdfs are formed from disjoint sets of information and the premise of the analysis is accepted then there is no alternative.*

2.3. *Some background*

An earlier derivation[5] of (1) did not involve discrete variables. Instead, the reduction of dimension to a univariate distribution was carried out with continuous variables. As implied by Grientschnig and Lira,[7] this rendered the result uncertain because of the existence of Borel's paradox, which relates to calculating the probability of an event conditional on an event with probability zero. In our context, this can be understood as follows.

In the earlier derivation of (1), the argument was made that receiving the information that X_1 was the same as X_2 collapsed the bivariate continuous distribution into a univariate continuous distribution along the line $X_1 = X_2$, which is a line representing an event with probability zero. This event can be considered to be the limit as $\epsilon \to 0$ of the event $|X_1 - X_2| \leq \epsilon$ or the limit as $\epsilon \to 0$ of the event $|X_1/X_2 - 1| \leq \epsilon$. In general, the univariate distribution conditional on this event will depend on which construction is used, which means that there is ambiguity in the result.

The new derivation does not suffer from this problem because the event on which the final probabilities are conditioned, which is the event $\acute{X}_1 = \acute{X}_2$, (or $\acute{W}_2 = 0$), always has non-zero probability. The appropriate limiting operation does not involve reducing the dimensionality of a continuous distribution. Borel's paradox is no longer relevant.

3. A contradiction

So, under the premise of the analysis, we must regard (1) as the unique correct formula for the combination of pdfs that are said to represent disjoint sets of information about an unknown constant. This leads to a more significant conclusion.

Consider the Gaussian example involving (2)–(4), but suppose now that X is being measured to estimate $Y = f(X)$, where f is a monotonic non-linear function. With such a function, the pdf of Y is given by

$$p_Y(y) = \left|\frac{dx}{dy}\right| p_X(x), \tag{8}$$

with $y = f(x)$. Therefore, from (4) and (8), the pdf of $Y = \exp(X)$ for the pair of scientists in possession of both data will be the log-normal pdf

$$p_Y(y) \propto \frac{1}{y} \exp\left(\frac{-(\log y - \bar{x})^2}{\sigma^2}\right) \qquad y > 0, \tag{9}$$

and this is the result obtained when using (1) to combine (2) and (3) before applying (8). In contrast, when (8) is applied to (2) and (3) separately, the individual pdfs attributed to Y by the scientists will be the log-normal pdfs

$$p_{Yi}(y) \propto \frac{1}{y} \exp\left(\frac{-(\log y - x_i)^2}{2\sigma^2}\right) \qquad y > 0, \quad i = 1, 2$$

and the subsequent combination of these using (1) gives

$$p_Y(y) \propto \frac{1}{y^2} \exp\left(\frac{-(\log y - \bar{x})^2}{\sigma^2}\right) \qquad y > 0. \tag{10}$$

Equations (9) and (10) differ. In (9) the term $|dx/dy|$ from (8), which is $1/y$ in our example, appears raised to the power of 1 because (8) is applied to one variable – but in (10) this term appears raised to the power of 2 because there are two contributing pdfs that have been combined by multiplication, each of which has been formed using (8). So these applications of probability theory – both of which are correct – lead to a contradiction.

The inconsistency has been demonstrated with a measurand Y that is a function of X. However, it is also observed when X itself is the measurand and f is just an arbitrary non-linear function: obviously, applying (8) to transform both (9) and (10) back to pdfs for X gives different results. Moreover, when there are many contributing pdfs the discrepancy will be large, because the exponent of the term $|dx/dy|$ in (10) is n. Also, the effect will be observed with any non-linear function: our stipulation that f be monotonic is only for simplicity in (8).

3.1. *An implication*

So the combined pdf obtained by applying (1) to the individual pdfs and then applying (8) is not the same as the pdf obtained by applying (8) to the individual pdfs and then applying (1). Yet, according to the premises of the analysis and the laws of probability, both methods are legitimate and meaningful ways of obtaining the unique pdf that accurately represents the combined state of knowledge about X. One potential explanation is that (1) is only applicable in certain situations; however, Section 2 gave an easy demonstration of the fact that, with unrelated contributions, (1) is the general and unique correct equation. The only other possible cause seems to be fault in the premise stated earlier, which is that 'all observers with the same information about an unknown constant quantity can and should assign the same pdf to that quantity.' If this premise were true, we would not encounter this contradiction.

We can, in fact, go further and conclude that a distribution said to represent a set of information does not obey the laws of probability, and therefore it is not a probability distribution: it is something else. There remains the weaker premise that such a distribution represents the information only approximately, but the next section shows this, too, is unsound.

4. Discussion

Sections 2 and 3 contained the mathematical material of this article. However, correct inference is a matter of reasoning, not mathematics. In this section, we give other arguments to support the conclusion that the new approach to uncertainty analysis is based on a misconception.

4.1. *Weakening the premise*

It might be argued that the methodology in question does not rest on the strict premise of completeness and that any pdf could only be expected to represent a set of information approximately. The idea that there is a unique pdf for any set of information would no longer apply, and we could use any member of an infinite set of similar pdfs to represent the information, presumably only incurring an error that was adequately small. If this argument was acceptable, we could tolerate a small difference between the distributions obtained by the methods that gave (9) and (10). But could this principle account for the fact that the size of the discrepancy increases with the number of pdfs being combined? Generally speaking, results get better, not worse, when more data are combined! Furthermore, this argument does not explain how a difference between the pdfs can arise at all, nor can it justify the difference in the *forms* of the pdfs, where a factor is raised to a different power in (9) and (10).

Even the idea that "a pdf can represent a set of information approximately" would need clarification. If the implication is that there are unknown pdfs that would actually represent the sets of information exactly, then the analysis can be applied to those unknown pdfs, and the logical contradiction remains. Therefore, for this rebuttal to carry weight, the idea that "a pdf can represent a set of information approximately" must mean something else. Presumably, it would mean that a pdf can represent a set of information *partially*, in which case the pdf describes only some aspects of the information. The full set of information becomes something that cannot be accurately represented by a pdf, and the methodology fails through only addressing part of the problem. Even so, if we could conceive of sets of information that contain only such aspects that can be represented in a pdf then we would encounter the contradiction with these sets.

Further, the claim that in practice people do not have independent sets of information also fails. This claim seems reasonable, but it is not relevant to our conclusion, which is a conclusion about logic, not practice. Anyway, if we can envisage sets of information at all then we can easily envisage two disjoint sets: we consider two overlapping sets and use the disjoint sets that would remain if we learned that the information in the intersection was false. And if the methodology exhibits errant behaviour in the simplest case of disjoint sets of information then it will also be in error in more complicated cases, without this necessarily being detectable!

4.2. *Probability and belief*

So if a probability distribution for a constant quantity X does not represent information or knowledge about X then what does it represent? A probability distribution for a constant represents someone's *belief*, in accordance with the Bayesian understanding of probability as 'degree of belief'. Belief and knowledge are different in nature. Belief is intrinsically personal. As a concept of probability, a degree of belief is a subjective and personal thing, not a rational thing. The subjective-Bayesian definition of a numerical probability does not refer to knowledge or information: rather, the subjective probability that a person attributes to the idea that H is true is defined to be *the largest value of p for which that person will pay p small units of money in order to receive back 1 unit of money if H were found to be true.* Thus, the very definition of a probability as a degree of belief refers not to anything objective but only to hypothetical human behaviour for long-term gain in a series of opportunities. (And it is well known that humans do not behave in the manner implied.) The contradiction here falsifies the claim that, given a shared state of knowledge, all analysts will or should have the same 'rational' strength of belief expressible in a pdf.

5. Conclusion

In this paper, we have considered the starting point for the theory and methodology of uncertainty analysis currently favoured in official publications, where the metrologist is required to represent an unknown constant quantity X by a probability distribution for X. There are a number of overlapping premises that might underlie any appeal to this methodology. In order of decreasing strength, these premises are as follows.

(1) To any set of information about X, there corresponds a unique pdf.
(2) To any set, there corresponds one or more *approximating* pdfs.
(3) To any set, there corresponds one or more *useful* pdfs.
(4) It is legitimate to attribute a probability distribution to X.

The first of these premises has been shown to be false, and the second has been shown to be untenable. The third premise has not been discussed directly. It alludes to the idea that any method is to be fit for purpose, which depends on the context and objective. If the objective is to present a methodology with *prima facie* popular appeal then perhaps the approach succeeds. But if the purpose is to describe a methodology that is scientifically sound then it fails – the contradiction demonstrates this. The fourth premise is at the heart of controversies about the use of probability in scientific studies. It should not be assumed that this premise is valid.

Appendix A. Other viewpoints: discussion of review reports

Reports on this paper were received from two reviewers, both of whom focused on the task of combining data, perhaps because the original title of the manuscript encouraged that. The problem of combining data is important, but the actual subject of the paper is the (illogical) nature of 'logical probability'. So some revisions, including a change to the title, were made to emphasize this. One reviewer felt that (5) did not follow "without some further assumptions". As now described, the reviewer's argument served to uncover a difference between the idea of independent sets of information and the idea of independent random variables.

The initial manuscript described the component pdfs as 'independent pdfs' (not 'unrelated pdfs', which is the term preferred here). This expression invites confusion between the idea of independently formed distributions, i.e. distributions proposed from disjoint sets of information, and the familiar idea of independently distributed random variables. If two pdfs for different quantities are proposed from disjoint sets of information then, obviously, the random variables would be seen as independent, (and the joint pdf would be obtained by multiplication, as in Section 2.2). However, the converse is not true: if two quantities are assigned distributions and are deemed to be independently distributed then, as now illustrated, that does not imply that the distributions were formed independently.

The reviewer imagined that experts 1 and 2 study the coefficients of thermal expansion, a_1 and a_2, of two gauge blocks G1 and G2, nominally of the same length. The experts are told that the blocks are made from carbon stainless steel, and on this basis, they assign their respective coefficients Gaussian distributions with a mean and a standard deviation obtained from the standard reference book. The reviewer then imagined the experts learning that the blocks were selected from a batch made from exactly the same grade of carbon steel, giving $a_1 \approx a_2$. Despite the fact that the random variables might be considered to be independent, multiplying the densities to take this new information into account, (which would obtain a narrower pdf), does not seem appropriate. In this way, the reviewer could not agree with the claim that (5) was the unique correct solution.

The error in this argument is that, while the two initial variables might be deemed to be independent, the contributing pdfs are not unrelated: they were both formed from the same set of information, which was the data in the reference book. In that situation, (5) does not apply. As emphasized, (5) has been presented for when the sets of information are disjoint. The reviewer's scenario might be more realistic than the idealized scenario of separate sets of information, but the question of whether disjoint sets of information arise in practice is irrelevant to the conclusion. And, as has been stated, if the logical error is present in the simplest case, i.e. disjoint sets of information, then it will surely exist when some overlap is introduced.

The result proved in this paper will be unpalatable to individuals and research groups who have made informal commitments to the 'uncertainty approach' and to one or other Bayesian point of view in the analysis of measurement quality. But those who dislike it are asked to take a step back from what they consider to be their starting point, their premise(s), and consider whether they can return there in good scientific conscience.

So what are the possible starting points for the use of probability theory in scientific inference? This author finds meaning in the idea that the reporting of scientific results should involve probability only as a representation of a relative frequency. The chief competing viewpoints to this 'frequentist' view are those of 'objective' and 'subjective' Bayesian statistics. This paper presents a compelling argument against the logic of the 'objective Bayesian' position that 'people with the same information about an unknown constant should attribute it the same probability distribution.' In contrast, in 'subjective Bayesian' analysis a probability is inherently personal, being an attribute of (the behaviour claimed by) the individual, not an attribute of the object of study. That use of probability is cherished by some academics but, obviously, the results can be criticized as lacking scientific meaning. Some advocates of Bayesian methods seem reluctant to take either position, but they write without an appeal to any clear interpretation of a probability. The two Bayesian positions are qualitatively different. They are isolated: they are not the extremes on a continuum of interpretation on which people may position themselves. There is no such continuum. Bayesian writers who hold neither of these points of view must somehow describe the meaning they give to a probability.

A Bayesian approach has been advocated for the revision of the *Guide to the Expression of Uncertainty in Measurement.* If the premise involved is the premise examined in this paper then the proposed approach is scientifically unsound. If the premise is different then it must be articulated so that its implications can be examined and fully understood.

References

1. W. Bich, M. Cox and C. Michotte, *Metrologia* **53**, S149 (2016).
2. A. Possolo, *Simple Guide for Evaluating and Expressing the Uncertainty of NIST Measurement Results (NIST Technical Note 1900)* (National Institute of Standards and Technology, Gaithersburg, 2015).
3. V. Barnett, *Comparative Statistical Inference* (Wiley, 1973).
4. C. Ehrlich, R. Dybkaer and W. Wöger, *Accred Qual Assur* **12**, 201 (2007).
5. R. Willink, *Metrologia* **43**, 12 (2006).
6. C. Elster, *Metrologia* **44**, 111 (2007).
7. D. Grientschnig and I. Lira, *Int Stat Rev* **82**, 392 (2014).

F Pavese, A B Forbes, N F Zhang, A G Chunovkina (Eds.):
Advanced Mathematical and Computational Tools in Metrology and Testing XII
Series on Advances in Mathematics for Applied Sciences, Vol. 90

https://doi.org/10.1142/9789811242380_0033

Polycal - MATLAB algorithm for comparative polynomial calibration and its applications*

Viktor Witkovský

Institute of Measurement Science, Slovak Academy of Sciences, Bratislava, Slovakia
E-mail: witkovsky@savba.sk

Gejza Wimmer

Mathematical Institute, Slovak Academy of Sciences, Bratislava, Slovakia
E-mail: wimmer@mat.savba.sk

Linear function and polynomial function are the most frequently used calibration functions in comparative calibration. Calibration is a standard tool in measurement science and applications to specify our knowledge about the parameters of the considered calibration function. This information is further used for specifying our knowledge about the unobservable stimulus from the independent future indication (measured response) received by using the calibrated measurement device. Here we present a description of the MATLAB algorithm `PolyCal`, which is based on the EIV (Errors-In-Variables) modelling approach and the characteristic function approach. The algorithm is available in `CharFunTool` – the *Characteristic Functions Toolbox* for MATLAB, https://github.com/witkovsky/CharFunTool. The applicability of the algorithm is illustrated by a simple example.

Keywords: Polynomial comparative calibration; Characteristic function approach, MATLAB.

1. Introduction

Measurement procedures often involve fitting a function that best describes the relationship of one variable to another. Here we consider linear function and polynomial function of given order p and assume a situation when both variables entering the calibration are subject to errors. As the name implies, the comparative calibration is performed by comparing the quantity

*The work was supported by the Slovak Research and Development Agency, projects APVV-15-0295 and APVV-18-0066, and by the Scientific Grant Agency of the Ministry of Education of the Slovak Republic and the Slovak Academy of Sciences, projects VEGA 2/0081/19 and VEGA 2/0096/21.

values measured by a lower quality instrument (usually a cheap working instrument) with the values measured by a higher quality instrument (usually an expensive reference instrument). In a first step, calibration establishes a relation between the scales (units) of the considered instruments, based on a set of direct measurements performed by both instruments, which are specified by the state-of-knowledge distributions about the influencing factors which are derived by using Type A and Type B methods of evaluation.

In metrology, the straight-line calibration and the polynomial calibration problem is addressed in the ISO Technical Specifications ISO/TS 28037:2010 and ISO/TS 28038:2018 [1, 2] which are based on the GUM uncertainty framework using the law of propagation of uncertainty [3].

The algorithm presented in this paper is suggested as an alternative approach for estimating the calibration parameters and their associated uncertainties. The proposed algorithm is (it should be) numerically equivalent to the algorithms specified in the ISO Technical Specifications. The main advantage of this approach is that it allows a straightforward derivation of the state-of-knowledge distributions of the calibration function parameters (i.e., the distribution of values which could be attributed to the function parameters, based on the observed data and the associated expert knowledge). The algorithm `PolyCal` is based on the errors-in-variables (EIV) modeling approach with the iterative application of the best linear unbiased estimation (BLUE) in a linearized model with type II constraints [6] and on the characteristic function approach [10]. For more details on the suggested concept, see [11]. Efficiency and numerical precision comparisons of the alternative approaches under different scenarios are the subjects of further research.

As pointed out by the reviewer of this article, there is a severe problem of a different interpretation of results in modeling based on the classical statistical (frequentists) methods of data analysis and techniques based on the Bayesian perspective.

As far as possible, the authors prefer a statistical approach based on the classical statistical approach to the calibration problem, see, e.g., [7–9]. However, in this article, we consider an approach often referred to in the metrological literature as the Bayesian perspective (or paradigm), which is, however, not necessarily based on the application of the Bayesian theorem. This approach proves necessary if it is required to combine information from different sources (typically the empirical evidence observed from the data and the expert knowledge).

Hence, here we assume that our current knowledge of the factors that affect the process of direct measurements in the first step of the calibration is given and fully known through their probability density functions which are determined by Type A and Type B evaluation methods. Based on this knowledge and on the basis of the observed results of direct measurements of the considered quantities in the first step of calibration, it is then possible to determine the state-of-knowledge distribution of unknown values of measurands measured by both calibrated instruments, say $\mathcal{X}$ and $\mathcal{Y}$. Further, in calibration we can improve the current knowledge about these values by applying the required restrictions so that the values of the measured quantities meet the given type of calibration function.

In this article, we first describe the knowledge about the process of direct measurements by evaluation methods of Type A and Type B. In this model of the process of direct measurements, all factors (inputs) are fully described by their probability distribution (in accordance with our current knowledge of the measurement process). The only quantities that are considered as unknown parameters are the actual values of the considered measurands in units of the used measuring instruments. After performing the measurement (obtaining the observed value), by combining the knowledge of the input factors in the measurement process we construct the associated random variable, the probability distribution of which represents our (current) state-of-knowledge about the value of the measured quantity.

In order to simplify the mathematical expressions, we shall use the simplified vector-matrix notation.

Here we consider the following model representing the process of direct measurements in the first step of calibration,

$$\boldsymbol{X} = \boldsymbol{\mu} + \boldsymbol{E}_{\mathcal{X}} = \boldsymbol{\mu} + \left(\boldsymbol{X}^A + \boldsymbol{X}^B + X_0^B \mathbf{1}\right), \tag{1a}$$

$$\boldsymbol{Y} = \boldsymbol{\nu} + \boldsymbol{E}_{\mathcal{Y}} = \boldsymbol{\nu} + \left(\boldsymbol{Y}^A + \boldsymbol{Y}^B + Y_0^B \mathbf{1}\right), \tag{1b}$$

where $\boldsymbol{X} = (X_1, \ldots, X_m)'$ and $\boldsymbol{Y} = (Y_1, \ldots, Y_m)'$ represent our knowledge about the process of direct measurements of m objects (measurands) measured by the devices $\mathcal{X}$ and $\mathcal{Y}$, respectively. Here, $\boldsymbol{\mu} = (\mu_1, \ldots, \mu_m)'$ and $\boldsymbol{\nu} = (\nu_1, \ldots, \nu_m)'$ represent the true unknown values of the measurands, expressed in units of the measuring devices $\mathcal{X}$ and $\mathcal{Y}$, respectively, which are further assumed to be related by functional (polynomial) relation, symbolically expressed by

$$\boldsymbol{\nu} = \beta_0 \mathbf{1} + \beta_1 \boldsymbol{\mu} + \beta_2 \boldsymbol{\mu}^2 + \cdots + \beta_p \boldsymbol{\mu}^p, \tag{2}$$

where $\boldsymbol{\beta} = (\beta_0, \ldots, \beta_p)'$ is $(p+1)$-dimensional vector of parameters of the pth-order polynomial calibration function. Here, $\mathbf{1} = (1, \ldots, 1)'$ and $\boldsymbol{\mu}^k$

denotes the vector $(\mu_1^k, \ldots, \mu_m^k)'$. Note that $\boldsymbol{\mu}$, $\boldsymbol{\nu}$ and $\boldsymbol{\beta}$ are the unknown parameters of the calibration experiment and the functional relation (2) is nonlinear.

Here we assume that $\boldsymbol{E}_{\mathcal{X}}$ consists of $\boldsymbol{X}^A = (X_1^A, \ldots, X_m^A)'$, $\boldsymbol{X}^B = (X_1^B, \ldots, X_m^B)'$, and X_0^B, and $\boldsymbol{E}_{\mathcal{Y}}$ consists of $\boldsymbol{Y}^A = (Y_1^A, \ldots, Y_m^A)'$, $\boldsymbol{Y}^B = (Y_1^B, \ldots, Y_m^B)'$, and Y_0^B. All the components are assumed to be mutually independent zero-mean random variables. Their probability distributions represent our knowledge about possible measurement errors and other factors which possibly influence the result of direct measurements in the first step of the calibration (it is assumed that these probability distributions are fully specified and derived by Type A and Type B methods of evaluation). In particular, the components of $\boldsymbol{X}^A$ and $\boldsymbol{Y}^A$ are random variables representing our knowledge about components of measurement errors derived by Type A methods of evaluation. The components of $\boldsymbol{X}^B$ and $\boldsymbol{Y}^B$ are random variables representing our knowledge about components of measurement errors due to the 'individual' systematic effects of the measuring instruments, specific for each particular measurement derived by Type B methods of evaluation. The components X_0^B and Y_0^B are random variables representing our knowledge about components of measurement errors due to the 'common' systematic effects of the measuring instruments $\mathcal{X}$ and $\mathcal{Y}$ (common for all measurements performed by the instrument) derived by Type B methods of evaluation.

Let us denote by $\boldsymbol{x}$ the observed value of $\boldsymbol{X}$, and by $\boldsymbol{y}$, the observed value of $\boldsymbol{Y}$ (the best estimates of $\boldsymbol{X}$ and $\boldsymbol{Y}$). Given the observed values $\boldsymbol{x}$ and $\boldsymbol{y}$ and the full information about the factors influencing the direct measurements we define the associated random variables (vectors) $\tilde{\boldsymbol{\mu}}$ and $\tilde{\boldsymbol{\nu}}$ which represent our current state-of-knowledge about the values of $\boldsymbol{\mu}$ and $\boldsymbol{\nu}$ based on the observed results of direct measurements and based on our knowledge about the influencing factors (input variables). Thus,

$$\tilde{\boldsymbol{\mu}} = \boldsymbol{x} - \boldsymbol{E}_{\mathcal{X}} = \boldsymbol{x} - \left(\boldsymbol{X}^A + \boldsymbol{X}^B + X_0^B \mathbf{1}\right), \tag{3a}$$

$$\tilde{\boldsymbol{\nu}} = \boldsymbol{y} - \boldsymbol{E}_{\mathcal{Y}} = \boldsymbol{y} - \left(\boldsymbol{Y}^A + \boldsymbol{Y}^B + Y_0^B \mathbf{1}\right). \tag{3b}$$

The components of the random vectors $\tilde{\boldsymbol{\mu}}$ and $\tilde{\boldsymbol{\nu}}$ could be interpreted as the CD-random variables associated with the confidence distributions (CDs) of the respective parameters μ and ν. For a more detailed discussion about CDs and their interpretation see [13].

In order to estimate the calibration function parameters and their uncertainties (in fact, the full covariance matrix), we shall assume that the calibration model is specified by the linear regression model where expectation (mean value) and the covariance matrix of the variables representing the measurement process are given by

$$E\begin{pmatrix}\boldsymbol{X}\\ \boldsymbol{Y}\end{pmatrix}=\begin{pmatrix}\boldsymbol{\mu}\\ \boldsymbol{\nu}\end{pmatrix},\quad \mathrm{Cov}\begin{pmatrix}\boldsymbol{X}\\ \boldsymbol{Y}\end{pmatrix}=\boldsymbol{\Sigma}=\begin{pmatrix}\boldsymbol{\Sigma}_{\mathcal{X}} & \boldsymbol{\Sigma}_{\mathcal{X},\mathcal{Y}}\\ \boldsymbol{\Sigma}_{\mathcal{Y},\mathcal{X}} & \boldsymbol{\Sigma}_{\mathcal{Y}}\end{pmatrix} \tag{4}$$

with nonlinear constraints on the parameters, $\boldsymbol{\nu}=\beta_0\mathbf{1}+\cdots+\beta_p\boldsymbol{\mu}^p$, where $\boldsymbol{\Sigma}_{\mathcal{X}}$, $\boldsymbol{\Sigma}_{\mathcal{Y}}$, and $\boldsymbol{\Sigma}_{\mathcal{X},\mathcal{Y}}=\boldsymbol{\Sigma}_{\mathcal{Y},\mathcal{X}}$ are known covariance matrices derived from the model representing the process of direct measurements (1a)-(1b) and the specified uncertainties of the influencing factors $u(X_i^A)$, $u(X_i^B)$, $u(X_0^B)$, $u(Y_i^A)$, $u(Y_i^B)$, and $u(Y_0^B)$, with $i=1,\ldots,m$. Here we assume $\boldsymbol{\Sigma}_{\mathcal{X},\mathcal{Y}}=\boldsymbol{\Sigma}_{\mathcal{Y},\mathcal{X}}=\mathbf{0}$. For full description of the considered calibration model see [11].

2. Estimation of the parameters in the calibration model

Estimation of the model parameters $\boldsymbol{\mu}$, $\boldsymbol{\nu}$, and $\boldsymbol{\beta}$ is based on iterative method which requires repeated application the best linear unbiased estimation (BLUE) used iteratively in properly linearized regression model with type II constraints, as suggested in [6]. This approach allows to give explicit formulas of the estimator of the calibration function parameters, expressed as linear functions of the measurement vectors $\boldsymbol{X}$ and $\boldsymbol{Y}$.

Here we summarize the basic steps of the algorithm for evaluating the estimates of the calibration model parameters $\boldsymbol{\mu}$, $\boldsymbol{\nu}$, $\boldsymbol{\beta}$, and the estimate of the covariance matrix.

Algorithm 2.1.

(1) Initialization:

- *Set the initial values of the model parameters $\boldsymbol{\mu}_0$, $\boldsymbol{\nu}_0$, $\boldsymbol{\beta}_0$ used for the linearization of the model. In general, any starting values are possible. If there is no other prior information, we suggest to use the observed values as the starting points for the linearization of the model. In particular, $\boldsymbol{\mu}_0=\boldsymbol{x}$, $\boldsymbol{\nu}_0=\boldsymbol{y}$, and the least squares estimate of $\boldsymbol{\beta}$ based on the observed values, $\boldsymbol{\beta}_0=(\boldsymbol{B}_0'\boldsymbol{B}_0)^{-1}\boldsymbol{B}_0'\boldsymbol{\nu}_0$ with $\boldsymbol{B}_0=\left[\mathbf{1},\boldsymbol{\mu}_0,\boldsymbol{\mu}_0^2,\ldots,\boldsymbol{\mu}_0^p\right]$.*
- *Set the indicator of the iteration step, $k=1$.*
- *Set the initial value of the criterion function, $\mathrm{crit}=1$.*
- *Set the limit for the number of iteration steps, e.g. $k_{max}=100$.*
- *Set the preferred (small) tolerance value: $0<\varepsilon\ll 1$.*
- *Set the starting values of the parameter estimates: $\bar{\boldsymbol{\mu}}^{(0)}=\boldsymbol{\mu}_0$, $\bar{\boldsymbol{\nu}}^{(0)}=\boldsymbol{\nu}_0$, $\bar{\boldsymbol{\beta}}^{(0)}=\boldsymbol{\beta}_0$.*

(2) Convergence check and the results. Check, if the following criterion is true: crit $< \varepsilon$ *or* $k \geq k_{max}$.

- *If true, set and report the final results:*
  - *Save the best estimates of the vector parameters* $\boldsymbol{\mu}$, $\boldsymbol{\nu}$, *and* $\boldsymbol{\beta}$, *given by*

$$\bar{\boldsymbol{\mu}} = \boldsymbol{x} + \boldsymbol{\Sigma}_{\mathcal{X}} \boldsymbol{D}_*' \boldsymbol{Q}_* \left(\boldsymbol{y} - \boldsymbol{D}_*(\boldsymbol{x} - \boldsymbol{\mu}_*)\right), \tag{5a}$$

$$\bar{\boldsymbol{\nu}} = \boldsymbol{y} - \boldsymbol{\Sigma}_{\mathcal{Y}} \boldsymbol{Q}_* \left(\boldsymbol{y} - \boldsymbol{D}_*(\boldsymbol{x} - \boldsymbol{\mu}_*)\right), \tag{5b}$$

$$\bar{\boldsymbol{\beta}} = \left(\boldsymbol{B}_*' \boldsymbol{W}_*^{-1} \boldsymbol{B}_*\right)^{-1} \boldsymbol{B}_*' \boldsymbol{W}_*^{-1} \left(\boldsymbol{y} - \boldsymbol{D}_*(\boldsymbol{x} - \boldsymbol{\mu}_*)\right), \tag{5c}$$

  where $\boldsymbol{B}_* = [\mathbf{1}, \boldsymbol{\mu}_*, \ldots, \boldsymbol{\mu}_*^p]$, $\boldsymbol{W}_* = \boldsymbol{D}_* \boldsymbol{\Sigma}_{\mathcal{X}} \boldsymbol{D}_*' + \boldsymbol{\Sigma}_{\mathcal{Y}}$, $\boldsymbol{D}_* = \mathrm{Diag}(\beta_{*,1}\mathbf{1} + 2\beta_{*,2}\boldsymbol{\mu}_* + \cdots + p\beta_{*,p}\boldsymbol{\mu}_*^{p-1})$, $\boldsymbol{Q}_* = \boldsymbol{W}_*^{-1} - \boldsymbol{W}_*^{-1}\boldsymbol{B}_* \left(\boldsymbol{B}_*' \boldsymbol{W}_*^{-1} \boldsymbol{B}_*\right)^{-1} \boldsymbol{B}_*' \boldsymbol{W}_*^{-1}$, *and* $\boldsymbol{\mu}_*$, $\boldsymbol{\beta}_*$ *are the values used for linearization in the last iteration step.*
  - *Save the coefficient matrices* $\boldsymbol{K} = \left(\boldsymbol{B}_*' \boldsymbol{W}_*^{-1} \boldsymbol{B}_*\right)^{-1} \boldsymbol{B}_*' \boldsymbol{W}_*^{-1}$ *and* $\boldsymbol{L} = -\boldsymbol{K}\boldsymbol{D}_*$.
  - *Save the covariance matrix* $\boldsymbol{\Sigma}_{\widehat{\boldsymbol{\beta}}}$ *of the estimator (7), where*

$$\boldsymbol{\Sigma}_{\widehat{\boldsymbol{\beta}}} = \left(\boldsymbol{B}_*' \boldsymbol{W}_*^{-1} \boldsymbol{B}_*\right)^{-1} = \boldsymbol{L} \boldsymbol{\Sigma}_{\mathcal{X}} \boldsymbol{L}' + \boldsymbol{K} \boldsymbol{\Sigma}_{\mathcal{Y}} \boldsymbol{K}'. \tag{6}$$

  - *Stop the algorithm.*
- *If false, continue and go to the next step (3).*

(3) Linearization and estimation: For $k \geq 1$

- *Set* $\boldsymbol{\mu}_0 = \bar{\boldsymbol{\mu}}^{(k-1)}$ *and* $\boldsymbol{\beta}_0 = \bar{\boldsymbol{\beta}}^{(k-1)}$ *and evaluate the matrices* $\boldsymbol{B}_0 = [\mathbf{1}, \boldsymbol{\mu}_0, \ldots, \boldsymbol{\mu}_0^p]$, $\boldsymbol{W}_0 = \boldsymbol{D}_0 \boldsymbol{\Sigma}_{\mathcal{X}} \boldsymbol{D}_0' + \boldsymbol{\Sigma}_{\mathcal{Y}}$ *with* $\boldsymbol{D}_0 = \mathrm{Diag}(\beta_{0,1}\mathbf{1} + 2\beta_{0,2}\boldsymbol{\mu}_0 + \cdots + p\beta_{0,p}\boldsymbol{\mu}_0^{p-1})$, *and* $\boldsymbol{Q}_0 = \boldsymbol{W}_0^{-1} - \boldsymbol{W}_0^{-1}\boldsymbol{B}_0 \left(\boldsymbol{B}_0' \boldsymbol{W}_0^{-1} \boldsymbol{B}_0\right)^{-1} \boldsymbol{B}_0' \boldsymbol{W}_0^{-1}$.
- *Evaluate the locally best linear unbiased estimates of the model parameters* $\boldsymbol{\mu}$, $\boldsymbol{\nu}$, *and* $\boldsymbol{\beta}$, *given by*

$$\bar{\boldsymbol{\mu}} = \boldsymbol{x} + \boldsymbol{\Sigma}_{\mathcal{X}} \boldsymbol{D}_0' \boldsymbol{Q}_0 \left(\boldsymbol{y} - \boldsymbol{D}_0(\boldsymbol{x} - \boldsymbol{\mu}_0)\right),$$

$$\bar{\boldsymbol{\nu}} = \boldsymbol{y} - \boldsymbol{\Sigma}_{\mathcal{Y}} \boldsymbol{Q}_0 \left(\boldsymbol{y} - \boldsymbol{D}_0(\boldsymbol{x} - \boldsymbol{\mu}_0)\right),$$

$$\bar{\boldsymbol{\beta}} = \left(\boldsymbol{B}_0' \boldsymbol{W}_0^{-1} \boldsymbol{B}_0\right)^{-1} \boldsymbol{B}_0' \boldsymbol{W}_0^{-1} \left(\boldsymbol{y} - \boldsymbol{D}_0(\boldsymbol{x} - \boldsymbol{\mu}_0)\right).$$

(4) Update:

- *Set* $\bar{\boldsymbol{\mu}}_0^{(k)} = \bar{\boldsymbol{\mu}}$, $\bar{\boldsymbol{\nu}}_0^{(k)} = \bar{\boldsymbol{\nu}}$, *and* $\bar{\boldsymbol{\beta}}_0^{(k)} = \bar{\boldsymbol{\beta}}$.
- *Evaluate the criterion value* crit,

$$\mathrm{crit} = \frac{1}{d}\left(\|\bar{\boldsymbol{\mu}}^{(k)} - \bar{\boldsymbol{\mu}}^{(k-1)}\|^2 + \|\bar{\boldsymbol{\nu}}^{(k)} - \bar{\boldsymbol{\nu}}^{(k-1)}\|^2 + \|\bar{\boldsymbol{\beta}}^{(k)} - \bar{\boldsymbol{\beta}}^{(k-1)}\|^2\right),$$

where $\|\boldsymbol{z}\|^2$ *denotes the Euclidean norm of a vector* $\boldsymbol{z}$ *and* $d = 2m + (p+1)$ *is total dimension of the parameter vectors.*

- *Increase the iteration step, i.e. set* $k = k + 1$*, and go to (2) to check the convergence.*

The MATLAB implementation of the algorithm is available from the authors. The implementation of the algorithm `PolyCal` is also available as a part of the MATLAB toolbox `CharFunTool` [12].

For illustration purposes, Figure 1 presents simple MATLAB code for comparative calibration based on artificial data. Here we assume that the calibration experiment is based on simultaneous measurements by two instruments, say $\mathcal{X}$ and $\mathcal{Y}$, of 5 objects (measurands). From the uncertainty budgets, the best estimates $\boldsymbol{x}$ and $\boldsymbol{y}$ and the state-of-knowledge distributions about the components of the influencing measurement errors and factors are fully specified and given by their characteristic functions (here, the selection of the distributions is rather artificial to demonstrate the flexibility of the software). The results are presented graphically in Figure 2. For more details see the documentation of `PolyCal` and the toolbox `CharFunTool`.

3. State-of-knowledge distributions

The algorithm defines the $(\boldsymbol{\mu}_*, \boldsymbol{\nu}_*, \boldsymbol{\beta}_*)$-locally best linear unbiased estimator (LBLUE) of the calibration function parameters $\boldsymbol{\beta} = (\beta_0, \ldots, \beta_p)'$ given by

$$\widehat{\boldsymbol{\beta}} = \boldsymbol{L}(\boldsymbol{X} - \boldsymbol{\mu}_*) + \boldsymbol{K}\boldsymbol{Y}. \tag{7}$$

This estimator is a linear function of $\boldsymbol{X}$ and $\boldsymbol{Y}$ which represent our knowledge about the process of the direct measurements. The linearity of the estimator is approximate because it is based on the linearization of the model. The approximation is justified if the true parameters of the calibration model are *close* to the preselected values $(\boldsymbol{\mu}_*, \boldsymbol{\nu}_*, \boldsymbol{\beta}_*)$ where the model is linearized.

Based on that, the best estimate of the calibration function parameters is specified as the observed value of (7), i.e. by

$$\bar{\boldsymbol{\beta}} = \boldsymbol{L}(\boldsymbol{x} - \boldsymbol{\mu}_*) + \boldsymbol{K}\boldsymbol{y}, \tag{8}$$

and the current state-of-knowledge about the parameters can be specified by the probability distribution of the random variable $\widetilde{\boldsymbol{\beta}}$ given by

$$\widetilde{\boldsymbol{\beta}} = \bar{\boldsymbol{\beta}} - \boldsymbol{L}\boldsymbol{E}_{\mathcal{X}} - \boldsymbol{K}\boldsymbol{E}_{\mathcal{Y}}. \tag{9}$$

Note that $\bar{\boldsymbol{\beta}}$ is the non-stochastic part of $\widetilde{\boldsymbol{\beta}}$, representing our best estimate of the calibration parameter $\boldsymbol{\beta}$, and $\boldsymbol{L}\boldsymbol{E}_{\mathcal{X}} + \boldsymbol{K}\boldsymbol{E}_{\mathcal{Y}} = \boldsymbol{L}(\boldsymbol{X}^A + \boldsymbol{X}^B +$

```
% EXAMPLE - Comparative polynomial calibration by using the algorithm PolyCal
% Input data - the best estimates of the measured quantities X and Y
   x = [4.7, 5.5, 6.5, 7.5, 8.4]';
   y = [5.8, 7.9, 10.4, 12.7, 15.2]';
% Values where the calibration function is to be evaluated
   xFit = [5.0, 5.5, 6.0, 6.5, 7.0, 7.5, 8.0]';
% Further information is given via the structure options
% Order of the polynomial calibration function
   options.order = 1;

% Characteristic functions of the errors components XA, XB, XB0
   options.cfXA = { ...
       @(t)cf_Normal(0.01*t), ...
       @(t)cf_Normal(0.1*t), ...
       @(t)cf_Normal(0.1*t), ...
       @(t)cf_Student(0.1*t,5), ...
       @(t)cf_RectangularSymmetric(0.2*t)};
   options.cfXB = { ...
       @(t)cf_RectangularSymmetric(0.1*t), ...
       @(t)cf_RectangularSymmetric(0.1*t), ...
       @(t)cf_TriangularSymmetric(0.1*t), ...
       @(t)cf_TriangularSymmetric(0.1*t), ...
       @(t)cf_RectangularSymmetric(0.2*t)};
   options.cfXB0 = @(t)cf_Normal(0.02*t);
% Characteristic functions of the errors components YA, YB, YB0
   options.cfYA = { ...
       @(t)cf_Normal(0.15*t), ...
       @(t)cf_Normal(0.2*t), ...
       @(t)cf_Normal(0.2*t), ...
       @(t)cf_Normal(0.3*t), ...
       @(t)cf_Normal(0.3*t)};
   options.cfYB = { ...
       @(t)cf_RectangularSymmetric(0.2*t), ...
       @(t)cf_RectangularSymmetric(0.2*t), ...
       @(t)cf_TriangularSymmetric(0.3*t), ...
       @(t)cf_TriangularSymmetric(0.3*t), ...
       @(t)cf_RectangularSymmetric(0.3*t)};
   options.cfYB0 = @(t)cf_ArcsineSymmetric(0.1*t);
   options.tolDiff = 1e-2;

% Algorithm PolyCal and further application of the result
   result = PolyCal(x,y,xFit,options)
% Plot the PDF/CDF of the calibration function parameter beta0
% by inverting its characteristic function
   cfBeta0 = result.cfBeta{1};
       figure;
   resultBeta0 = cf2DistGP(cfBeta0,[],[],options);
% Plot the PDF/CDF of the calibration function parameter beta1
   cfBeta1 = result.cfBeta{2};
   figure;
   resultBeta1 = cf2DistGP(cfBeta1,[],[],options);
```

Fig. 1. MATLAB code used for comparative calibration based on artificial data, specified by the best estimates x and y and the state-of-knowledge distributions about the components of the measurement errors specified by their characteristic functions.

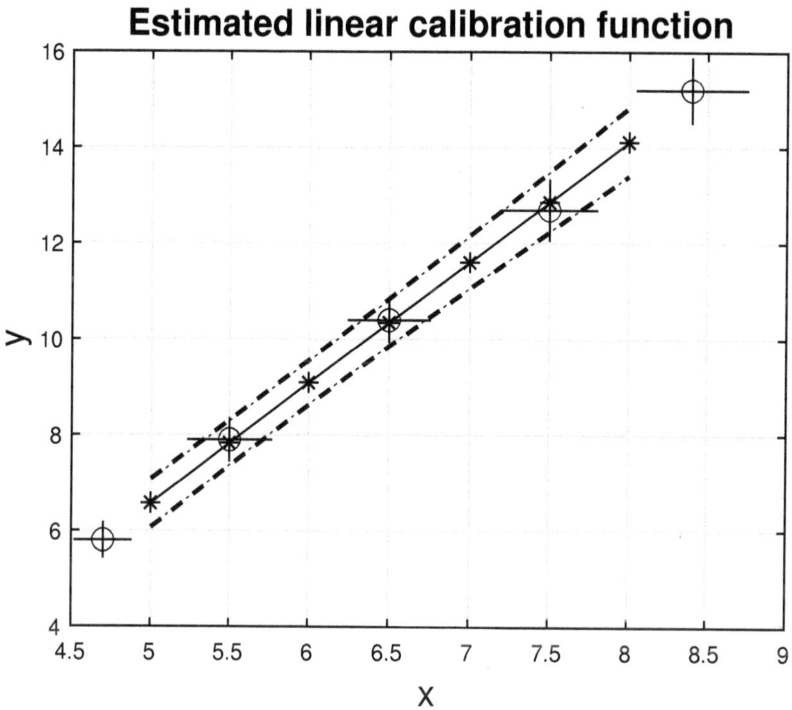

Fig. 2. Plot of the observed data (best estimates ∘) plotted with their expanded combined uncertainties (crossed solid lines), fitted linear calibration function (solid line), selected fitted values (∗), and coverage intervals of $\tilde{\nu} = \tilde{\beta}_0 + \tilde{\beta}_1 x$ calculated for different values of x from the state-of-knowledge distributions using (11) (dotted lines) as well as the coverage intervals of $\tilde{\nu}$ calculated for specified μ from the approximate Gaussian state-of-knowledge distribution using the corresponding mean and variance estimated by the algorithm (dashed lines). Here, the difference between the two approaches for computing the coverage intervals is visually indistinguishable.

$X_0^B\mathbf{1}) + \boldsymbol{K}(\boldsymbol{Y}^A + \boldsymbol{Y}^B + Y_0^B\mathbf{1})$ is the stochastic part of $\widetilde{\boldsymbol{\beta}}$ with zero-mean probability distribution, representing our knowledge about the associated uncertainty. The state-of-knowledge distribution about the linear combination of the calibration function parameters, say $\boldsymbol{w}'\boldsymbol{\beta} = \sum_{i=0}^{p} w_i\beta_i$, with $\boldsymbol{w} = (w_0, \ldots, w_p)'$, is given by the probability distribution of the random variable

$$\boldsymbol{w}'\widetilde{\boldsymbol{\beta}} = \boldsymbol{w}'\left(\bar{\boldsymbol{\beta}} - \boldsymbol{L}\boldsymbol{E}_{\mathcal{X}} - \boldsymbol{K}\boldsymbol{E}_{\mathcal{Y}}\right). \tag{10}$$

Frequently, we are interested in the value y_{Fit} of the calibration function for given value x_{Fit}. In such case $\boldsymbol{w} = (1, x_{Fit}, x_{Fit}^2, \ldots, x_{Fit}^p)'$. Let $\boldsymbol{w}_{\mathcal{X}} =$

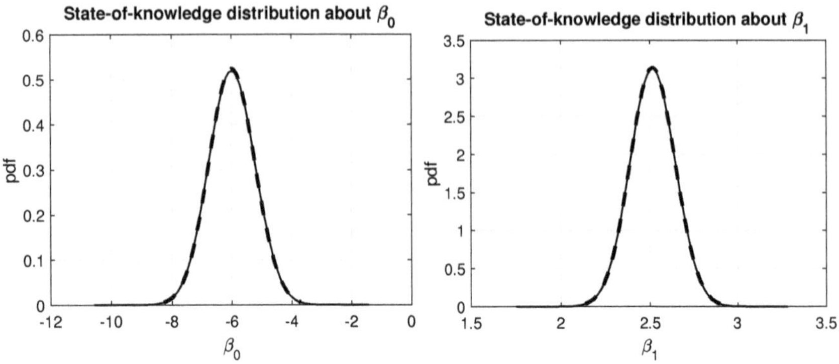

Fig. 3. Left panel: State-of-knowledge distribution about the parameter β_0 (i.e. the distribution of the random variable $\tilde{\beta}_0$ specified by (9) and depicted by solid line) together with the approximate distribution based on Gaussian approximation (dashed line). Right panel: State-of-knowledge distribution about the parameter β_1 (solid line) and the approximate distribution based on Gaussian approximation (dashed line). The best estimate of β_0 computed by using (8) is $\bar{\beta}_0 = -5.9973$ with $u(\bar{\beta}_0) \equiv \mathrm{std}(\tilde{\beta}_0) = 0.7613$ and best estimate of β_1 is $\bar{\beta}_1 = 2.5148$ with $u(\bar{\beta}_1) \equiv \mathrm{std}(\tilde{\beta}_1) = 0.1271$. Here, the difference between the *exact* and the approximate PDF is visually indistinguishable.

$\boldsymbol{w}'\boldsymbol{L} = (w_{\mathcal{X},1}, \ldots, w_{\mathcal{X},m})'$ and $\boldsymbol{w}_{\mathcal{Y}} = \boldsymbol{w}'\boldsymbol{K} = (w_{\mathcal{Y},1}, \ldots, w_{\mathcal{Y},m})'$, then we get

$$\sum_{i=0}^{p} w_i \widetilde{\beta}_i = \sum_{i=0}^{p} w_i \bar{\beta}_i - \sum_{i=1}^{m} w_{\mathcal{X},i} X_i^A - \sum_{i=1}^{m} w_{\mathcal{X},i} X_i^B - \left(\sum_{i=1}^{m} w_{\mathcal{X},i}\right) X_0^B - \sum_{i=1}^{m} w_{\mathcal{Y},i} Y_i^A - \sum_{i=1}^{m} w_{\mathcal{Y},i} Y_i^B - \left(\sum_{i=1}^{m} w_{\mathcal{Y},i}\right) Y_0^B. \tag{11}$$

The probability distribution (PDF and/or CDF) of $\widetilde{\beta}_0, \ldots, \widetilde{\beta}_p$ and $\sum_{i=0}^{p} w_i \widetilde{\beta}_i$ can be evaluated by propagating distributions of the input variables by using a Monte Carlo method, as specified in [4] and [5], or by using the characteristic functions approach, as suggested in [10]. The algorithm `PolyCal` evaluates the required state-of-knowledge distributions by the characteristic functions approach (if the characteristic functions of the input variables are fully specified).

As recommended by the first reviewer, it would be interesting to compare the state-of-knowledge distributions with a) those arising from an ordinary least squares fit, ignoring the errors-in-variables, and b) Gaussian distributions with the same mean and standard deviation. We plan to study the first issue in more detail in our future research. For comparison, in Figure 3 we present the state-of-knowledge distributions about the calibration function parameters β_0 and β_1 together with their approximations derived under the assumption of normality distribution with the mean value (best estimate) and variance (uncertainty) determined in the last iteration step of the algorithm.

4. Conclusion

The MATLAB algorithm `PolyCal` estimates the calibration function parameters $\beta_0, \ldots, \beta_p$ and the fitted values y_{Fit} of the calibration function evaluated at arbitrary specified values x_{Fit}, together with their uncertainties and the state-of-knowledge distributions represented by their characteristic functions. The calculated `PolyCal` result can be further used for evaluation of measurements with the calibrated device.

As mentioned before, the estimation part of the suggested algorithm is numerically equivalent with the algorithms specified in the ISO Technical Specifications [1, 2] which are based on iterative application of efficient numerical methods for the solution of the generalized Gauss-Markov regression. In typical applications, the algorithm converges quickly to the stationary solution (based on our experience the algorithm stops after 5-7 iterations), however a more detailed study about convergence properties is needed.

References

[1] ISO/TS 28037:2010. Determination and Use of Straight-Line Calibration Functions. International StandardsOrganization, Geneva, September 2010.

[2] ISO/TS 28038:2018. Determination and Use of Polynomial Calibration Functions. International Standards Organization, Geneva, December 2018.

[3] JCGM 100:2008. Evaluation of measurement data – Guide to the expression of uncertainty in measurement (GUM 1995 with minor corrections), ISO, BIPM, IEC, IFCC, ILAC, IUPAC, IUPAP and OIML, (2008).

[4] JCGM 101:2008. Evaluation of measurement data – Supplement 1 to the Guide to the expression of uncertainty in measurement – Propagation of distributions using a Monte Carlo method. ISO, BIPM, IEC, IFCC, ILAC, IUPAC, IUPAP and OIML, (2008).

[5] JCGM 102:2011. Evaluation of measurement data – Supplement 2 to the Guide to the expression of uncertainty in measurement – Extension to any number of output quantities. ISO, BIPM, IEC, IFCC, ILAC, IUPAC, IUPAP and OIML, (2011).

[6] Kubáček L. Foundations of Estimation Theory, Elsevier, Amsterdam (1988).

[7] Wimmer G. and Witkovský V. Scheff-type confidence region for the calibration line. *Austrian Journal of Statistics* **35** (2-3), (2006) 397–406.

[8] Wimmer G. and Witkovský V. Linear comparative calibration with correlated measurements. *Kybernetika* **43** (4), (2007) 443–452.

[9] Wimmer G. and Witkovský V. Univariate linear calibration via replicated errors-in-variables model. *Journal of Statistical Computation and Simulation* **77** (3), (2007) 213–227.

[10] Witkovský V. Numerical inversion of a characteristic function: An alternative tool to form the probability distribution of output quantity in linear measurement models. *ACTA IMEKO* **5** (3), (2016) 32–44.

[11] Witkovský V. and Wimmer G. Generalized polynomial comparative calibration: Parameter estimation and applications. In: *Advances in Measurements and Instrumentation: Reviews.* S.Y. Yurish (Ed.). Barcelona, Spain, IFSA (International Frequency Sensor Association Publishing) Publishing S.L., (2018) 15–52.

[12] Witkovský V. *CharFunTool: The Characteristic Functions Toolbox (MATLAB).* https://github.com/witkovsky/CharFunTool (2020).

[13] Xie M.G. and Singh K. Confidence distribution, the frequentist distribution estimator of a parameter: A review. *International Statistical Review*, **81** (1), (2013) 3–39.

F Pavese, A B Forbes, N F Zhang, A G Chunovkina (Eds):
Advanced Mathematical and Computational Tools in Metrology and Testing XII
Series on Advances in Mathematics for Applied Sciences, Vol. 90

https://doi.org/10.1142/9789811242380_0034

Statistical models for similarity score comparisons in firearm evidence identification

Nien Fan Zhang

Statistical Engineering Division, National Institute of Standards and Technology, Gaithersburg, MD 20899, USA
E-mail: nien-fan.zhang@nist.gov

In the branch of forensic science known as firearm evidence identification, various similarity scores have been proposed to compare firearm marks. Some similarity score comparisons, for example, congruent matching cells (CMC) method, are based on pass-or-fail tests. The CMC method deals with pairs of topography images of breech face impressions for which the similarity has been quantified. For an image pair, the CMC method determines certain number correlated cell pairs. In addition, each correlated pair is determined to be a congruent match cell (CMC) pair, or not based on several identification parameters. The number of CMC pairs as a threshold is crucial and required so that the two images of surface topographies can be identified as matching. The threshold would be determined after carefully designed measurements and error rate estimation.

To reliably estimate error rates and the associated uncertainties, the key is to find an appropriate probability distribution for the frequency distribution of the observed CMC results. This article discusses four statistical models for CMC measurements, which are binomial and three binomial-related probability distributions. In previous studies, for a sequence of binomial distributed or other binomial-related distributed random variables (r.v.), the number of Bernoulli trials N for each r.v. is assumed to be the same. However, in practice, N (the number of cell pairs in an image pair) varies from one r.v. to another. In that case, the term of frequency function of the CMC results is not appropriate. In this article, the general frequency function is introduced to depict the behavior of the CMC values and its limiting distribution is provided. Based on that, nonlinear regression models are used to estimate the model parameters. For illustration, the methodology is applied to a set of actual CMC results of fired cartridge cases.

Keywords: Binomial distribution; Congruent matching cells; Correlated binomial distribution; Maximum likelihood estimator; Nonlinear regression.

1. Introduction

In firearm evidence identification, the parts of the firearm make forcible contact with the bullets or cartridge cases when fired create characteristic tool marks or ballistic signatures on their surface. These signatures can be used for firearm evidence identifications [1]. In general, tool marks have so called class characteristics that are common to certain brands or models of firearms and individual characteristics arising from random variation in firearm manufacturing and wear. Figure 1 shows topography images of breech face impressions obtained from a pair of cartridge cases ejected from the same firearm. The image pair has several features in common.

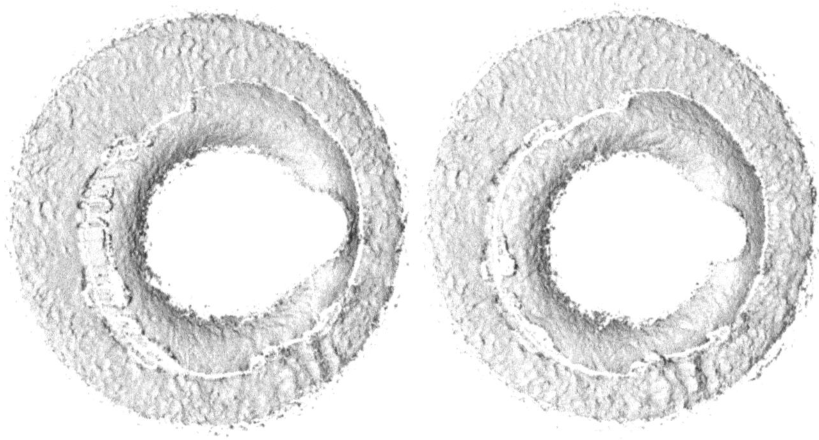

Figure 1. Topography images of breech face impressions obtained from a pair of cartridge cases fired from the same firearm.

In the area of firearm evidence identification, various similarity scores have been proposed to compare firearm marks for evidence identification, for example, the value of the normalized cross-correlation function for a pair of images [2], [3] and the congruent matching cells (CMC) method [4].

From a statistical point of view, both these similarity scores are based on pass-or-fail tests. In this article, we focus on the CMC method, which deals with pairs of topography images of breech face impressions for which the similarity or matching has been quantified. For a topography image pair, the CMC method divides one image, which is designated to be a reference image, into an array of rectangular cells as shown in Figure 2.

For each reference cell, a search for a matching cell is then conducted on the compared image to yield a correlated cell pair. Whether the cell pair is correlated or not is determined by the maximum cross-correlation function [4].

Next, each correlated pair is determined to be a congruent match cell (CMC) pair, or not based on several identification parameters. A certain number of

CMC pairs as a threshold is required so that the two images of surface topographies can be identified as matching. The details of the CMC method can be found in [4].

We discuss the error rates for firearm evidence identification. The CMC method is based on the pass-or-fail tests of individual cell pairs from an image pair of breech face impression.

When the CMC method is applied to a set of cartridge cases, the result, in general, includes certain known matching (KM) image pairs and certain known non-matching (KNM) image pairs.

The false positive error rate is the rate of pairs of images incorrectly judged as matches. It represents the expected frequency or probability of obtaining an erroneous result of identification (declared match) when the comparing samples are from different sources (KNM).

On the other hand, the false negative error rate is the rate of pairs of images incorrectly judged as non-matches. It represents the probability of obtaining an erroneous result of exclusion (declared non-match) when the comparing samples are from the same source (KM).

To reliably estimate error rates, the key is to find an appropriate probability distribution for the frequency distribution of the observed CMC results.

In Sec. 2, four statistical models for CMC measurements and their properties are discussed.

In Sec. 3, estimations of the parameters of the statistical models, in particular, the use of nonlinear regression models are discussed followed by discussions and conclusions in Sec. 4.

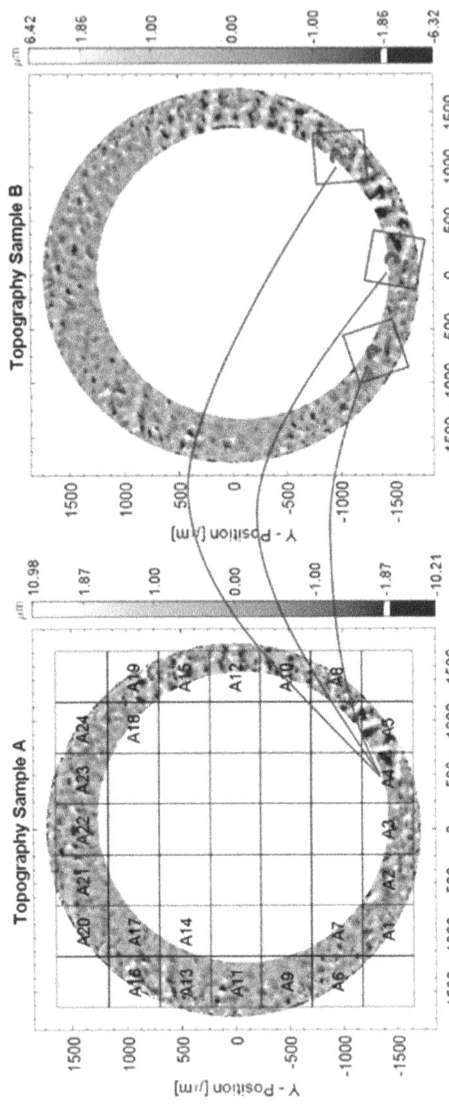

Figure. 2. Conceptual diagram of a topography image overlaid by a 7 × 7 grid, dividing the image into cells. Only cells with a sufficient fraction of measured pixels are used. Also shown is a schematic diagram of the automated search procedure to find an area in the compared image (top) that has a strong correlation with one of the cells in the reference image (bottom).

2. Statistical models for CMC measurements

2.1. *Binomial distribution*

A binomial probability distribution was proposed in [4] for the distribution of CMC measurements from an image pair comparison. Two assumptions are made: (1) The comparisons between cell pairs in an image pair are statistically independent from each other, (2) each cell pair comparison within an image pair has the same matching probability, p. Under these assumptions, for a given cell pair, a random variable (r.v.) X represents the outcome of the CMC method applied to the cell pair. When the CMC method determines that the cell pair is a matching cell pair, then $X=1$; otherwise, $X=0$. Denote the probability that $X=1$ by p. That is, $P(X=1)=p$, and $P(X=0)=1-p$. For an image pair with N cell pairs, we have a sequence of Bernoulli trials, $X_1,...,X_N$. The sum of $\{X_1,...,X_N\}$ denoted by Y is the number of the CMCs for the image pair. Under these two assumptions, $Y \sim Bin(N,p)$ is a binomially distributed r.v. with the probability mass function (pmf) given by

$$P_{[1]}(Y=k)=C_N^k p^k(1-p)^{N-k} \text{ for } k=0,1,...,N, \tag{1}$$

where $C_N^k = N!/[k!(N-k)!]$.

2.2 Correlated binomial distribution

For the binomial distribution, the assumption of independence among cell pair comparisons in an image pair, i.e., $\{X_1,...,X_N\}$, however, is most likely invalid in practice. For example, since the array of cells laid over the image is not precisely aligned with features in the image, neighboring cell pairs may share some individualizing features. Thus, we need to consider some type of dependent Bernoulli trials for cell pair comparisons. Reference [5] proposes a general model for dependent Bernoulli trials, which sometimes is called Bahadur-Lazarsfeld model [6]. The resultant distribution of Y, which is the sum of $\{X_1,...,X_N\}$ is called a correlated binomial distribution. Reference [7] provides a comprehensive discussion on the correlated binomial distribution and the corresponding parameter estimation.

For a sequence of dependent Bernoulli trials, $\{X_1,...,X_N\}$, where each X_i takes the value 0 or 1, with $P(X_i=1)=p$, and $P(X_i=0)=1-p$ for $i=1,...,N$, the second-order correlation between X_i and X_j, where $j \neq i$, is given by $r_{(2)}=\mathrm{Cov}[X_i,X_j]/\sigma^2$, where $\sigma=\sqrt{p(1-p)}$ is the standard deviation of X_i. Similarly, the third and higher-order correlations, up to the Nth order correlation, are defined. Y, the sum of $\{X_1,...,X_N\}$, has a correlated binomial distribution with pmf $P(Y)$. From [7], $P(Y)$ can be approximated by the second order approximation

$$\begin{aligned} P_{[2]}(Y) &= P_{[1]}(Y)\{1+r_{(2)}g_2(Y,p)\} \\ &= C_N^Y p^Y (1-p)^{N-Y}\{1+r_{(2)}g_2(Y,p)\}, \end{aligned} \tag{2}$$

where $P_{[1]}(Y)$ is the pmf of Y when $\{X_i\}$ are independent from each other as given in (1). Namely, $P_{[1]}(Y)$ is the pmf of the corresponding binomial distributed r.v., Y. $g_2(Y)$ is a second degree polynomial in Y and also a function of p. Denote $Y \sim \text{corr}.Bin(N,p,r_{(2)})$. Reference [7] discusses the properties of the second order approximation $P_{[2]}(Y)$. In particular, when $r_{(2)}=0$, $P_{[2]}(Y)$ equals $P_{[1]}(Y)$ Reference [7] demonstrates that the model of the second order approximation of a correlated binomial distribution fits a practical data set much better than that based on the binomial distribution. For simplicity, in this article, we call $P_{[2]}(Y)$ a correlated binomial distribution.

2.3 Beta-binomial distribution

In [4], it is proposed to relax the assumption of the binomial distribution when modeling the CMC measurements by using a beta-binomial distribution. In this case, we have a sequence of CMC results of image comparisons $\{Y_1,...,Y_M\}$, which are assumed to be independent from each other. Assume that within one image pair comparison, the probability p for all the independent Bernoulli trials is the same while for different image pairs, p varies and p is random with a beta distribution [8], i.e., $p \sim Beta(\alpha,\beta)$ with positive α and β. The pmf of the beta-binomial r.v. Y for given N, α and β is given by

$$P(Y=k \mid N,\alpha,\beta) = C_N^k\, B(k+\alpha, N-k+\beta)/B(\alpha,\beta), \tag{3}$$

where $B(\alpha,\beta)$ is a beta function [9] and $k=0,1,...,N$. In [4], beta-binomial distribution is applied to a sequence of CMC measurements. It shows that the beta-binomial distribution model fit a set of practical data better than the model based on the binomial distribution.

2.4 Beta-correlated binomial distribution

Although the use of the beta-binomial distribution relaxes the assumption of the same p for the binomial distribution for all image pair comparisons, it still assumes that within each image pair, all cell pair comparisons are independent from each other. To relax that, reference [7] proposes to use dependent Bernoulli trials for the cell pair comparisons within each image pair. The corresponding probabilities of the sum, Y are approximated by $P_{[2]}(Y)$ as given by (2) where only the second-order correlation with a constant $r_{(2)}$ is considered. As in the case of beta-binomial distribution for a sequence $\{Y_1,\ldots,Y_M\}$, assume that the parameter p in the correlated binomial distribution is random with a beta distribution. Namely, $Y_j \mid p_j, r_{(2)} \sim \text{corr.}Bin(N, p_j, r_{(2)})$, for $j = 1,\ldots,M$, where p has a beta distribution, *i.e.*, $p \sim Beta(\alpha,\beta)$. From [7], the probability mass function of Y for given N, α, β, and $r_{(2)}$ is given by

$$P(Y=k \mid N,\alpha,\beta,r_{(2)}) = \frac{C_N^k}{B(\alpha,\beta)} \int_0^1 p^{k+\alpha-1}(1-p)^{N-k+\beta-1}\{1+r_{(2)}g_2(k,p)\}dp, \quad (4)$$

where $g_2(k,p)$ is a second degree polynomial in Y and also a function of p. The marginal probability $P(Y=k \mid N,\alpha,\beta,r_{(2)})$ for $k=0,1,\ldots,N$ has no explicit expression. However, it can be calculated by numerical integration. In this case, the random variable Y has a compound probability distribution of beta-binomial and correlated binomial distributions called a beta-correlated binomial distribution.

3. Estimating the parameters of the statistical models

A data set of CMC measurements, which is a sequence of the CMC results from M image pair comparisons, usually consists of two columns. The first one is a vector of the number of cell pairs (N) for each image pair for a total of M image pairs. The second column is a vector of the corresponding CMC values for each image pair. From the practical image pair comparisons, it is not likely to have same cell pair number N for all M image pairs [4]. That is, the number of cell pairs for each image pair, i.e., N, varies from one image pair to another. Denote a sequence of M independent CMC values by $\{Y_{1,N_1},\ldots,Y_{j,N_j},\ldots,Y_{M,N_M}\}$, where Y_{j,N_j} indicates the CMC value for the jth image pair comparison which consists of N_j cell pairs.

In practice, given a data set of CMC measurements, statistical models such as binomial, correlated binomial, beta-binomial, and beta-correlated binomial distributions discussed in Sec. 2 can be applied. Under the assumption of binomial distribution, from [4], the maximum likelihood estimator (MLE) of p is given by

$$\hat{p} = \sum_{j=1}^{M} Y_{j,N_j} \Big/ \sum_{j=1}^{M} N_j . \tag{5}$$

In [7], when a correlated binomial distribution is assumed, the maximum likelihood estimator of p and $r_{(2)}$ is discussed for the case that N can be different for different image pairs. For the beta-binomial or beta-correlated binomial distributions, the maximum likelihood estimators of the corresponding parameters can be obtained, similarly.

In [7], nonlinear regression models are used to estimate p and $r_{(2)}$ under the assumptions: (1) a correlated binomial model, (2) the number of dependent Bernoulli trials N is the same for each Y. In this article, we discuss the case that N may vary for different image pairs.

For a sample from an independent sequence $\{Y_{1,N_1},...,Y_{j,N_j},...,Y_{M,N_M}\}$, we call $f_M(k)$ a generalized frequency function given by

$$f_M(k) = \frac{\text{number of elements in the sample} = \text{k}}{M} = \frac{\sum_{j=1}^{M} \mathbf{1}_{Y_{j,N_j}=k}}{M} \tag{6}$$

for $k = 0,...,\max\{N_j\}$ $(j = 1,...,M)$, where $\mathbf{1}_A$ is the indicator of event A. That is, $\mathbf{1}_{Y_{j,N_j}=k} = 1$ if $Y_{j,N_j} = k$ and $\mathbf{1}_{Y_{j,N_j}=k} = 0$ if $Y_{j,N_j} \neq k$ for $j = 1,...,M$. Assume that there are L distinct N's in $\{N_1,...,N_M\}$, denoted by $\{Nd_l\}$ $(l = 1,...,L)$ i.e., $Nd_1 \neq Nd_2 \neq ... \neq Nd_L$. Also assume that for each distinct Nd_l there are C_l indicators with the same Nd_l. Thus, $\sum_{l=1}^{L} C_l = M$. Here as an example, we assume Y has a correlated binomial distribution. Assume that (i) when $M \to \infty$, $C_l \to \infty$ for $l = 1,...,L$, (ii) when $M \to \infty$ and $C_l \to \infty$, $W_l = C_l / M$ remains fixed for each $l = 1,...,L$. By the law of large numbers [10], for every $Y = k$, it can be shown that when $M \to \infty$,

$$f_M(k) = \frac{\sum_{j=1}^{M} \mathbf{1}_{Y_{j,N_j}=k}}{M} \to \sum_{l=1}^{L} w_l \,\text{corr}.Bin(k, Nd_l, p) \text{ almost surely.} \tag{7}$$

Note that $\sum_{l=1}^{L} w_l \operatorname{corr}.Bin(k, Nd_l, p)$ is a proper probability mass function for $k = 0,1,\ldots,\max\{N_j\}$. Namely, the generalized frequency function will approach a mixture of correlated binomial pmf, which is a weighted mean of L correlated binomial pmf's. In addition, Eq. (7) also holds when Y either has a binomial distribution or a beta-binomial or a beta-correlated binomial distribution. Like Eq. 19 in [7], for a sample of $\{y_1,\ldots,y_M\}$ we have a nonlinear regression model

$$f_M(y) = \sum_{l=1}^{L} w_l P_{[1],Nd_l}(y)\{1 + r_{(2)} g_2(y, p, Nd_l)\} + \varepsilon, \tag{8}$$

where $f_M(y)$ is the generalized frequency function based on data from (6), $W_l = C_l / M$, and ε is a random error with zero mean. That is, $f_M(y)$ is approximated by a nonlinear function of p and $r_{(2)}$. This nonlinear regression model can be fitted to $f_M(y)$ to obtain optimal estimates of p and $r_{(2)}$ using Levenberg-Marquardt nonlinear least squares algorithm or other appropriate algorithms.

For illustration, an example is presented. The Weller data set of cartridge cases was obtained from a set of 11 firearm slides produced by the same manufacturer using the same process. The data set has 370 known matching (KM) image pairs and 4095 known non-matching (KNM) image pairs. The details of the data can be found in [4]. The KM data set consists of two vectors with lengths = 370, one for Y (CMC value) and another for N. Denote the second vector by $\{N\}$. For different image pairs, the number of cell pair N can be different. In fact, the M = 370 image pairs have L = 15 distinct N's, i.e., $Nd = (28,33,36,37,38,39,40,41,42,43,44,45,46,47,49)$. For each Nd_l $(l = 1,\ldots,L)$, the number of image pair with Nd_l cell pairs is C_l. That is, $C = (7,21,17,9,57,2,17,27,85,58,45,6,15,2,2)$. From (7), the generalized frequency function will approach a mixture or a weighted mean of 15 pmf's of the underlying distribution when $M \to \infty$.

First, we assume for each image pair, the distribution of CMC values is binomial, i.e., $Y_j \sim Bin(N_j, p)$ $(j = 1,\ldots,370)$. To estimate the parameter p, the maximum likelihood estimator can be applied from (5) with $\hat{p}$ = 0.908. Alternatively, assuming that the distribution of the CMC values Y is correlated binomial, the nonlinear regression estimates are $\tilde{p}$ = 0.9130 and $\tilde{r}_{(2)}$ = 0.0209. In addition, when the distribution of the CMC values is assumed to be beta-binomial, the corresponding estimates are $\tilde{\alpha} = 26.29$, $\tilde{\beta} = 2.62$ while for beta-correlated binomial distribution, the corresponding estimates are $\tilde{\alpha} = 64.12$, $\tilde{\beta} = 6.3$, and $\tilde{r}_{(2)} = 0.02$. The sums of squares of differences (SSD) between the generalized frequency function and each of the four mixture pmfs are calculated as given by 0.00496, 0.00204, 0.00185, and 0.00184, resepectively for the binomial, correlated binomial, beta-binomial, and beta-correlated binomial distributions. Using the smallest SSD as a performance criterion, we conclude that the mixtures of correlated binomial, beta-binomial, and beta-correlated binomial models fit the data better than the mixture binomial model. On the other hand, the pmfs of the three mixture distributions perform closely well.

The Weller data has 4095 known nonmatching (KNM) image pairs. For different image pairs, the number of cell pair N can be different. In fact, the $M = 4095$ image pairs have $L = 15$ distinct N's. From (7), the generalized frequency function will approach a pmf which is a mixture or a weighted mean of 15 pmf's of the underlying distribution when $M \to \infty$.

We consider using the binomial, correlated binomial, and beta-binomial distributions to fit the KNM set of the Weller data. Under the assumption of a binomial distribution, the MLE $\hat{p} = 0.0011209$. In addition, under the assumption of a correlated binomial distribution, the MLE $\hat{p} = 0.001121$, $\hat{r}_{(2)} = 0.000165$ while for the assumption of a beta-binomial distribution, $\hat{\alpha} = 7.2594$ and $\hat{\beta} = 6469.527$. The generalized frequency function and the mixtures of the three pmf's when $Y = 0, 1, 2, 3, 4$ are listed in Table 1. The SSD between the generalized function data and each of the three mixture pmf's are also listed in Table 1. It shows that the models of the mixtures of the three probability mass functions fit the generalized frequency function closely well.

Table 1. The generalized function for CMC values from 0 to 4 based on Weller's KNM data set and the corresponding mixture pmf's of the binomial, correlated binomial, and beta-binomial distributions.

Y (CMC)	Generalized frequency function	Binomial pmf	Correlated binomial pmf	Beta-binomial pmf
0	0.954572	0.954410	0.954556	0.954552
1	0.044192	0.044553	0.044264	0.044278
2	0.001212	0.001021	0.001158	0.001147
3	0	0.000015	0.000021	0.000022
4	0	0	0	0
SSD		$1.93 \cdot 10^{-7}$	$9.05 \cdot 10^{-9}$	$1.27 \cdot 10^{-8}$

4 Discussions and conclusions

This article discusses four statistical models for the results of similarity score comparisons in firearm evidence identification, specifically the congruent matching cells (CMC) method. Because the assumption of independence among the cell pair comparisons from the CMC method is most likely invalid in practice, the correlated binomial distribution is proposed. On the other hand, although using the beta-binomial distribution can relax the assumption of the same p for all image pair comparisons, it still assumes that within each image pair, all cell pair comparisons are independent from each other. To relax this assumption, the beta-correlated binomial as a compound probability distribution of the beta-binomial and correlated binomial distributions is proposed. Theoretically, the beta-correlated binomial, beta-binomial, and correlated binomial distributions are more complete than the binomial distribution because they contain fewer assumptions.

In previous studies, for simplicity, the number of cell pairs (N) for a sequence of image pairs is assumed to be the same. However, in practice, N varies from one image pair to another. In that case, the term of frequency function of the CMC results is not appropriate, and the nonlinear regression estimator proposed in [7] cannot be applied. In Sec. 3, the generalized frequency function is introduced to depict the behavior of the CMC values. By the law of large numbers, the generalized frequency function approaches a mixture probability mass function of the underlying probability distribution when the number of image pairs increases. Based on that, maximum likelihood and nonlinear regression estimators of the model parameters are proposed for the CMC data with variable N. Finally, the methodology is applied to an actual CMC data (Weller data) of fired cartridge cases with variable N to check the performance of the different statistical models based on the comparisons between the generalized frequency function and the corresponding limiting probability mass functions. For the known matching (KM) data set of the Weller data, the beta-correlated binomial, correlated binomial, and beta binomial distributions fit the data better than the binomial distributions. On the other hand, for the known non-matching (KNM) set, each of the binomial, correlated binomial, and beta-binomial distributions fits the data well. Since for the KNM data, the value of correlation $r_{(2)}$ for the correlated binomial distribution is often very small and the pmf's of the beta-binomial, correlated binomial, and binomial distributions are almost indistinguishable, the binomial distribution with only one single unknown parameter may be good enough for the fit to the KNM data.

References

[1] Gerules G, Bhatia SK, Jackson D (2015) A survey of image processing techniques and statistics for ballistic specimens. *Science and Justice* 53: 236-250.

[2] Song J, Whitenton E, Kelley D, Clary R, Ma L, Ballou S (2004) SRM 2460/2461 standard bullets and cartridge cases project, *Journal of Research of the National Institute of Standards and Technology*, 109 (6), 533-542.

[3] NIST SRM 2460/2461 Standard Bullet and Cartridge Cases, available at http://www.nist.gov/pml/div683/grp02/sbc.cfm (updated, May 2017).

[4] Song J, Vorburger TV, Chu W, Yen J, Soons JA, Ott DB, Zhang NF (2018) Estimating error rates for firearm evidence identification in forensic science. *Forensic Science International* 284: 15-32.

[5] Bahadur RR (1961) A representation of the joint distribution of the response to n dichotomous items. *Studies in Item Analysis and Prediction*, 158-168, ed. Solomon H, Stanford University Press, Stanford, CA.

[6] van der Geest PAG. (2005) The binomial distribution with dependent Bernoulli trials, *Journal of Statistical Computation and Simulation*, 75(2), 141-154.

[7] Zhang NF (2019) The use of correlated binomial distribution in estimating error rates for firearm evidence identification, *Journal of Research of the National Institute of Standards and Technology*, 124, Article No 124026.

[8] Johnson NL, Kotz S, Balakrishna N (1995) *Continuous Univariate Distributions*, Vol. 2 (2nd ed.), Wiley.

[9] Askey RA, Roy R (2010) Beta function, *NIST Handbook of Mathematical Functions*, eds. Lozier DM, Boisvert, Clark CW, Cambridge University Press

[10] van der Vaart AW (1998) *Asymptotic Statistics*, Cambridge University Press, Cambridge, UK, 265-266.

F Pavese, A B Forbes, N F Zhang, A G Chunovkina (Eds):
Advanced Mathematical and Computational Tools in Metrology and Testing XII
Series on Advances in Mathematics for Applied Sciences, Vol. 90

https://doi.org/10.1142/9789811242380_bmatter

Keyword Index

F Pavese, A B Forbes, N F Zhang, A G Chunovkina (Eds):
Advanced Mathematical and Computational Tools in Metrology and Testing XII
Series on Advances in Mathematics for Applied Sciences, Vol. 90

https://doi.org/10.1142/9789811242380_bmatter

Author Index

Series on Advances in Mathematics for Applied Sciences

Aims and Scope

This Series reports on new developments in mathematical research relating to methods, qualitative and numerical analysis, mathematical modeling in the applied and the technological sciences. Contributions rlated to constitutive theories, fluid dynamics, kinetic and transport theories, solid mechanics, system theory and mathematical methods for the applications are welcomed.

This Series includes books, lecture notes, proceedings, collections of research papers. Monograph collections on specialized topics of current interest are particularly encouraged. Both the proceedings and monograph collections will generally be edited by a Guest editor.

High quality, novelty of the content and potential for the applications to modern problems in applied science will be the guidelines for the selection of the content of this series.

Instructions for Authors

Submission of proposals should be addressed to the editors-in-charge or to any member of the editorial board. In the latter, the authors should also notify the proposal to one of the editors-in-charge. Acceptance of books and lecture notes will generally be based on the description of the general content and scope of the book or lecture notes as well as on sample of the parts judged to be more significantly by the authors.

Acceptance of proceedings will be based on relevance of the topics and of the lecturers contributing to the volume.

Acceptance of monograph collections will be based on relevance of the subject and of the authors contributing to the volume.

Authors are urged, in order to avoid re-typing, not to begin the final preparation of the text until they received the publisher's guidelines. They will receive from World Scientific the instructions for preparing camera-ready manuscript.

Series on Advances in Mathematics for Applied Sciences

Published:*

Vol. 75 Applied and Industrial Mathematics in Italy II
eds. V. Cutello et al.

Vol. 76 Geometric Control and Nonsmooth Analysis
eds. F. Ancona et al.

Vol. 77 Continuum Thermodynamics
by K. Wilmanski

Vol. 78 Advanced Mathematical and Computational Tools in Metrology and Testing
eds. F. Pavese et al.

Vol. 79 From Genetics to Mathematics
eds. M. Lachowicz and J. Miękisz

Vol. 80 Inelasticity of Materials: An Engineering Approach and a Practical Guide
by A. R. Srinivasa and S. M. Srinivasan

Vol. 81 Stability Criteria for Fluid Flows
by A. Georgescu and L. Palese

Vol. 82 Applied and Industrial Mathematics in Italy III
eds. E. De Bernardis, R. Spigler and V. Valente

Vol. 83 Linear Inverse Problems: The Maximum Entropy Connection
by H. Gzyl and Y. Velásquez

Vol. 84 Advanced Mathematical and Computational Tools in Metrology and Texting IX
eds. F. Pavese et al.

Vol. 85 Continuum Thermodynamics
Part II: Applications and Examples
by B. Albers and K. Wilmanski

Vol. 86 Advanced Mathematical and Computational Tools in Metrology and Testing X
eds. F. Pavese et al.

Vol. 87 Mathematical Methods for the Natural and Engineering Sciences (2nd Edition)
by R. E. Mickens

Vol. 88 Road Vehicle Dynamics: Fundamentals of Modeling and Simulation
by G. Genta and A. Genta

Vol. 89 Advanced Mathematical and Computational Tools in Metrology and Testing XI
eds. A. B. Forbes et al.

Vol. 90 Advanced Mathematical and Computational Tools in Metrology and Testing XII
eds. F. Pavese et al.

*To view the complete list of the published volumes in the series, please visit:
https://www.worldscibooks.com/series/samas_series.shtml

Lightning Source UK Ltd.
Milton Keynes UK
UKHW021242190122
397359UK00003B/251

9 789811 242373